# STATA BASE REFERENCE MANUAL

## VOLUME 3

## R–Z

## RELEASE 9

A Stata Press Publication
StataCorp LP
College Station, Texas

Stata Press, 4905 Lakeway Drive, College Station, Texas 77845


Version 9
Typeset in TeX
Printed in the United States of America

10 9 8 7 6 5 4 3 2 1

ISBN 1-881228-94-0 (volumes 1–3)
ISBN 1-881228-91-6 (volume 1)
ISBN 1-881228-92-4 (volume 2)
ISBN 1-881228-93-2 (volume 3)



The suggested citation for this software is

StataCorp. 2005. *Stata Statistical Software: Release 9*. College Station, TX: StataCorp LP.

# Title

**ranksum** — Equality tests on unmatched data

# Syntax

*Wilcoxon rank-sum test*

ranksum *varname* [*if*] [*in*] , by(*groupvar*) [porder]

*Nonparametric equality-of-medians test*

median *varname* [*if*] [*in*] [*weight*] , by(*groupvar*) [*median_options*]

| ranksum options | description |
|---|---|
| Main | |
| * by(*groupvar*) | grouping variable |
| porder | probability that variable for first group is larger than variable for second group |

| *median_options* | description |
|---|---|
| Main | |
| * by(*groupvar*) | grouping variable |
| exact | perform Fisher's exact test |
| medianties(below) | assign values equal to the median to below group |
| medianties(above) | assign values equal to the median to above group |
| medianties(drop) | drop values equal to the median from the analysis |
| medianties(split) | split values equal to the median equally between the two groups |

*by(*groupvar*) is required.
by may be used with ranksum and median; see [D] **by**.
fweights are allowed with median; see [U] **11.1.6 weight**.

# Description

ranksum tests the hypothesis that two independent samples (i.e., *unmatched* data) are from populations with the same distribution using the Wilcoxon rank-sum test, which is also known as the Mann–Whitney two-sample statistic (Wilcoxon 1945; Mann and Whitney 1947).

median performs a nonparametric $k$-sample test on the equality of medians. It tests the null hypothesis that the $k$ samples were drawn from populations with the same median. In the case of two samples, the test chi-squared statistic is computed both with and without a continuity correction.

ranksum and median are for use with *unmatched* data. For equality tests on matched data, see [R] **signrank**.

# Options for ranksum

Main

`by(`*groupvar*`)` is required. It specifies the name of the grouping variable.

`porder` displays an estimate of the probability that the variable for the first group is larger than the variable for the second group.

# Options for median

Main

`by(`*groupvar*`)` is required. It specifies the name of the grouping variable.

`exact` displays the significance calculated by Fisher's exact test. In the case of two samples, both one- and two-sided probabilities are displayed.

`medianties(below|above|drop|split)` specifies how values equal to the overall median are to be handled. The median test computes the median for *varname* using all observations and then divides the observations into those falling above the median and those falling below the median. When the value for an observation is equal to the sample median, they can be dropped from the analysis by specifying `medianties(drop)`; added to the group above or below the median by specifying `medianties(above)` or `medianties(below)`, respectively; or if there is more than one observation with values equal to the median, they can be equally divided into the two groups by specifying `medianties(split)`. If this option is not specified, `medianties(below)` is assumed.

# Remarks

## ▷ Example 1

We are testing the effectiveness of a new fuel additive. We run an experiment with 24 cars: 12 cars with the fuel treatment and 12 cars without. We input these data by creating a dataset with 24 observations. `mpg` records the mileage rating, and `treat` records 0 if the mileage corresponds to untreated fuel and 1 if it corresponds to treated fuel.

```
. use http://www.stata-press.com/data/r9/fuel2
. ranksum mpg, by(treat)
Two-sample Wilcoxon rank-sum (Mann-Whitney) test
       treat |      obs    rank sum    expected
-------------+---------------------------------
           0 |       12         128         150
           1 |       12         172         150
-------------+---------------------------------
    combined |       24         300         300
unadjusted variance      300.00
adjustment for ties       -4.04
                     ----------
adjusted variance        295.96
Ho: mpg(treat==0) = mpg(treat==1)
             z =  -1.279
    Prob > |z| =   0.2010
```

These results indicate that the medians are not statistically different at any level smaller than 20.1%. Similarly, the median test

```
. median mpg, by(treat) exact
Median test

   Greater |
  than the |         treat
    median |         0          1 |     Total
-----------+----------------------+----------
        no |         7          5 |        12
       yes |         5          7 |        12
-----------+----------------------+----------
     Total |        12         12 |        24
          Pearson chi2(1) =   0.6667   Pr = 0.414
           Fisher's exact =                 0.684
   1-sided Fisher's exact =                 0.342

   Continuity corrected:
          Pearson chi2(1) =   0.1667   Pr = 0.683
```

fails to reject the null hypothesis that there is no difference between the two fuel additives.

Compare these results from these two tests with those obtained from the `signrank` and `signtest` where we found significant differences; see [R] **signrank**. An experiment run on 24 different cars is not as powerful as a before-and-after comparison using the same 12 cars.

◁

## Saved Results

`ranksum` saves in `r()`:

Scalars

| | | | |
|---|---|---|---|
| `r(N_1)` | sample size $n_1$ | `r(group1)` | value of variable for first group |
| `r(N_2)` | sample size $n_2$ | `r(sum_obs)` | actual sum of ranks for first group |
| `r(z)` | $z$ statistic | `r(sum_exp)` | expected sum of ranks for first group |
| `r(Var_a)` | adjusted variance | | |

`median` saves in `r()`:

Scalars

| | | | |
|---|---|---|---|
| `r(N)` | sample size | `r(groups)` | number of groups compared |
| `r(chi2)` | Pearson's $\chi^2$ | `r(chi2_cc)` | continuity-corrected Pearson's $\chi^2$ |
| `r(p)` | significance of Pearson's $\chi^2$ | `r(p_cc)` | continuity-corrected significance |
| `r(p_exact)` | Fisher's exact $p$ | `r(p1_exact)` | one-sided Fisher's exact $p$ |

## Methods and Formulas

`ranksum` and `median` are implemented as ado-files.

For a practical introduction to these techniques with an emphasis on examples rather than theory, see Bland (2000) or Sprent (1993). For a summary of these tests, see Snedecor and Cochran (1989).

## ranksum

For the Wilcoxon rank-sum test, there are two independent random variables $X_1$ and $X_2$, and we test the null hypothesis that $X_1 \sim X_2$. We have a sample of size $n_1$ from $X_1$ and another of size $n_2$ from $X_2$.

The data are then ranked without regard to the sample to which they belong. If the data are tied, averaged ranks are used. Wilcoxon's test statistic (1945) is the sum of the ranks for the observations in the first sample:

$$T = \sum_{i=1}^{n_1} R_{1i}$$

Mann and Whitney's $U$ statistic (1947) is the number of pairs $(X_{1i}, X_{2j})$ such that $X_{1i} > X_{2j}$. These statistics differ only by a constant:

$$U = T - \frac{n_1(n_1+1)}{2}$$

Again Fisher's Principle of Randomization provides a method for calculating the distribution of the test statistic, ties or not. The randomization distribution consists of the $\binom{n}{n_1}$ ways to choose $n_1$ ranks from the set of all $n = n_1 + n_2$ ranks and assign them to the first sample.

It is a straightforward exercise to verify that

$$E(T) = \frac{n_1(n+1)}{2} \qquad \text{and} \qquad \text{Var}(T) = \frac{n_1 n_2 s^2}{n}$$

where $s$ is the standard deviation of the combined ranks $r_i$ for both groups:

$$s^2 = \frac{1}{n-1}\sum_{i=1}^{n}(r_i - \overline{r})^2$$

This formula for the variance is exact and holds both when there are no ties and when there are ties and we use averaged ranks. (Indeed, the variance formula holds for the randomization distribution of choosing $n_1$ numbers from any set of $n$ numbers.)

Using a normal approximation, we calculate

$$z = \frac{T - E(T)}{\sqrt{\text{Var}(T)}}$$

## median

The median test examines whether it is likely that two or more samples came from populations with the same median. The null hypothesis is that the samples were drawn from populations with the same median. The alternative hypothesis is that at least one sample was drawn from a population with a different median. The test should be used only with ordinal or interval data.

Assume that there are score values for *k* independent samples to be compared. The median test is performed by first computing the median score for all observations combined, regardless of the sample group. Each score is compared to this computed grand median and is classified as being above the grand median, below the grand median, or equal to the grand median. Observations with scores equal to the grand median can be dropped, added to the "above" group, added to the "below" group, or split between the two groups.

Once all observations are classified, the data are cast into a $2 \times k$ contingency table, and a Pearson's chi-squared test or Fisher's exact test is performed.

Henry Berthold Mann (1905–2000) was born in Vienna, where he completed a doctorate in algebraic number theory. He moved to the United States in 1938 and for several years earned a living by tutoring in New York. During this time he proved a celebrated conjecture in number theory and studied statistics at Columbia with Abraham Wald, with whom he wrote three papers. After the war he taught at Ohio State and the Universities of Wisconsin and Arizona. In addition to his work in number theory and statistics, he made major contributions to algebra and combinatorics.

Donald Ransom Whitney (1915– ) studied at Princeton and Ohio State Universities and worked at the latter throughout his career. His Ph.D. thesis under Henry Mann was on nonparametric statistics and produced the test that bears his name.

## References


Bland, M. 2000. *An Introduction to Medical Statistics*. 3rd ed. Oxford: Oxford University Press.

Feiveson, A. H. 2002. Power by simulation. *Stata Journal* 2: 107–124.

Fisher, R. A. 1935. *Design of Experiments*. Edinburgh: Oliver & Boyd.

Goldstein, R. 1997. sg69: Immediate Mann–Whitney and binomial effect-size display. *Stata Technical Bulletin* 36: 29–31. Reprinted in *Stata Technical Bulletin Reprints*, vol. 6, pp. 187–189.

Kruskal, W. H. 1957. Historical notes on the Wilcoxon unpaired two-sample test. *Journal of the American Statistical Association* 52: 356–360.

Mann, H. B. and D. R. Whitney. 1947. On a test of whether one of two random variables is stochastically larger than the other. *Annals of Mathematical Statistics* 18: 50–60.

Newson, R. 2000a. snp15: somersd—Confidence intervals for nonparametric statistics and their differences. *Stata Technical Bulletin* 55: 47–55. Reprinted in *Stata Technical Bulletin Reprints*, vol. 10, pp. 312–322.

——. 2000b. snp15.1: Update to somersd. *Stata Technical Bulletin* 57: 35. Reprinted in *Stata Technical Bulletin Reprints*, vol. 10, pp. 322–323.

——. 2000c. snp15.2: Update to somersd. *Stata Technical Bulletin* 58: 30. Reprinted in *Stata Technical Bulletin Reprints*, vol. 10, p. 323.

——. 2001. snp15.3: Update to somersd. *Stata Technical Bulletin* 61: 22. Reprinted in *Stata Technical Bulletin Reprints*, vol. 10, p. 324.

Perkins, A. M. 1998. snp14: A two-sample multivariate nonparametric test. *Stata Technical Bulletin* 42: 47–49. Reprinted in *Stata Technical Bulletin Reprints*, vol. 7, pp. 243–245.

Snedecor, G. W. and W. G. Cochran. 1989. *Statistical Methods*. 8th ed. Ames, IA: Iowa State University Press.

Sprent, P. and N. C. Smeeton. 2001. *Applied Nonparametric Statistical Methods*. 3rd ed. Boca Raton, FL: Chapman & Hall/CRC.

Sribney, W. M. 1995. crc40: Correcting for ties and zeros in sign and rank tests. *Stata Technical Bulletin* 26: 2–4. Reprinted in *Stata Technical Bulletin Reprints*, vol. 5, pp. 5–8.

Wilcoxon, F. 1945. Individual comparisons by ranking methods. *Biometrics* 1: 80–83.


## Also See

**Related:** [R] **kwallis**, [R] **nptrend**, [R] **runtest**, [R] **signrank**, [R] **ttest**

# Title

**ratio** — Estimate ratios

# Syntax

*Basic syntax*

`ratio` [*name*`:`] *varname* [`/`] *varname*

*Full syntax*

`ratio` `(`[*name*`:`] *varname* [`/`] *varname*`)`
[ `(`[*name*`:`] *varname* [`/`] *varname*`)` ... ] [*if*] [*in*] [*weight*] [`,` *options*]

| *options* | description |
|---|---|
| Model | |
| `stdize(`*varname*`)` | variable identifying strata for standardization |
| `stdweight(`*varname*`)` | weight variable for standardization |
| `nostdrescale` | do not rescale the standard weight variable |
| if/in/over | |
| `over(`*varlist*[`,` `nolabel`]`)` | group over subpopulations defined by *varlist*; optionally, suppress group labels |
| SE/Cluster | |
| `vce(`*vcetype*`)` | *vcetype* may be `bootstrap` or `jackknife` |
| `cluster(`*varname*`)` | adjust standard errors for intragroup correlation |
| Reporting | |
| `level(`*#*`)` | set confidence level; default is `level(95)` |
| `noheader` | suppress the table header |
| `nolegend` | suppress the table legend |

`svy` may be used with `ratio`; see [SVY] **svy: ratio**.
`fweight`s, `iweight`s, and `pweight`s are allowed; see [U] **11.1.6 weight**.
See [U] **20 Estimation and postestimation commands** for additional capabilities of estimation commands.

# Description

`ratio` produces estimates of ratios, along with standard errors.

# Options

Model

`stdize(`*varname*`)` specifies that the point estimates be adjusted by direct standardization across the strata identified by *varname*. This option requires the `stdweight()` option.

`stdweight(`*varname*`)` specifies the weight variable associated with the strata identified in the `stdize()` option. The standardization weights must be constant within the strata identified in the `stdize()` option.

`nostdrescale` prevents the standardization weights from being rescaled within the `over()` groups. This option requires `stdize()` but is ignored if the `over()` option is not specified.

if/in/over

`over(`*varlist* [ , `nolabel` ]`)` specifies that estimates be computed for multiple subpopulations, which are identified by the different values of the variables in *varlist*.

When this option is supplied with a single variable name, such as `over(`*varname*`)`, the value labels of *varname* are used to identify the subpopulations. If *varname* does not have labeled values (or there are unlabeled values), the values themselves are used, provided that they are non-negative integers. Noninteger values, negative values, and labels that are not valid Stata names are substituted with a default identifier.

When `over()` is supplied with multiple variable names, each subpopulation is assigned a unique default identifier.

`nolabel` requests that value labels attached to the variables identifying the subpopulations be ignored.

SE/Cluster

`vce(`*vcetype*`)`; see [R] ***vce_option***.

`cluster(`*varname*`)`; see [R] **estimation options**.

Reporting

`level(`*#*`)`; see [R] **estimation options**.

`noheader` prevents the table header from being displayed. This option implies `nolegend`.

`nolegend` prevents the table legend identifying the subpopulations from being displayed.

# Remarks

## ▷ Example 1

Using the fuel data from [R] **ttest**, we estimate the ratio of mileage for the cars without the fuel treatment (`mpg1`) to those with the fuel treatment (`mpg2`).

```
. use http://www.stata-press.com/data/r9/fuel
. ratio myratio: mpg1/mpg2
Ratio estimation                    Number of obs    =      12
      myratio: mpg1/mpg2
```

| | Ratio | Linearized Std. Err. | [95% Conf. | Interval] |
|---|---|---|---|---|
| myratio | .9230769 | .032493 | .8515603 | .9945936 |

Using these results, we can test to see if this ratio is significantly different from one.

```
. test _b[myratio] = 1
 ( 1)  myratio = 1
       F(  1,    11) =    5.60
            Prob > F =    0.0373
```

We find that the ratio is different from one at the 5% significance level but not at the 1% significance level.

◁

## ▷ Example 2

Using state-level census data, we want to test whether the marriage rate is equal to the death rate.

```
. use http://www.stata-press.com/data/r9/census2
(1980 Census data by state)
. ratio (deathrate: death/pop) (marrate: marriage/pop)
Ratio estimation                    Number of obs    =      50
    deathrate: death/pop
      marrate: marriage/pop
```

| | Ratio | Linearized Std. Err. | [95% Conf. | Interval] |
|---|---|---|---|---|
| deathrate | .0087368 | .0002052 | .0083244 | .0091492 |
| marrate | .0105577 | .0006184 | .009315 | .0118005 |

```
. test _b[deathrate]  =  _b[marrate]
 ( 1)  deathrate - marrate = 0
       F(  1,    49) =    6.93
            Prob > F =    0.0113
```

◁

See [SVY] **direct standardization** for information about standardized ratios.

## Saved Results

ratio saves in e():

Scalars

| | | | |
|---|---|---|---|
| e(N) | number of observations | e(df_r) | sample degrees of freedom |
| e(N_over) | number of subpopulations | e(N_clust) | number of clusters |
| e(N_stdize) | number of standard strata | | |

Macros

| | | | |
|---|---|---|---|
| e(cmd) | ratio | e(over) | *varlist* from over() |
| e(varlist) | *varlist* | e(over_labels) | labels from over() variables |
| e(stdize) | *varname* from stdize() | e(over_namelist) | names from e(over_labels) |
| e(stdweight) | *varname* from stdweight() | e(vce) | *vcetype* specified in vce() |
| e(wtype) | weight type | e(vcetype) | title used to label Std. Err. |
| e(wexp) | weight expression | e(estat_cmd) | program used to implement estat |
| e(title) | title in estimation output | e(namelist) | ratio identifiers |
| e(cluster) | name of cluster variable | e(properties) | b V |

Matrices

| | |
|---|---|
| e(b) | vector of mean estimates |
| e(V) | (co)variance estimates |
| e(_N) | vector of numbers of nonmissing observations |
| e(_N_stdsum) | number of nonmissing observations within the standard strata |
| e(_p_stdize) | standardizing proportions |

Functions

| | |
|---|---|
| e(sample) | marks estimation sample |

## Methods and Formulas

ratio is implemented as an ado-file.

Let $R = Y/X$ be the ratio to be estimated, where $Y$ and $X$ are totals; see [R] **total**. The estimate for $R$ is $\widehat{R} = \widehat{Y}/\widehat{X}$ (the ratio of the sample totals). Using the delta method (i.e., a first-order Taylor expansion), the approximate variance of the sampling distribution of the linearized $\widehat{R}$ is

$$V(\widehat{R}) \approx \frac{1}{X^2}\left\{V(\widehat{Y}) - 2R\mathrm{Cov}(\widehat{Y},\widehat{X}) + R^2V(\widehat{X})\right\}$$

Direct substitution of $\widehat{X}$, $\widehat{R}$, and the estimated variances and covariance of $\widehat{X}$ and $\widehat{Y}$ leads to the following variance estimator:

$$\widehat{V}(\widehat{R}) = \frac{1}{\widehat{X}^2}\left\{\widehat{V}(\widehat{Y}) - 2\widehat{R}\widehat{\mathrm{Cov}}(\widehat{Y},\widehat{X}) + \widehat{R}^2\widehat{V}(\widehat{X})\right\} \qquad (1)$$

See [SVY] **direct standardization** for information about standardized ratios.

## References

Cochran, W. G. 1977. *Sampling Techniques*. 3rd ed. New York: Wiley.

Stuart, A. and J. K. Ord. 1994. *Kendall's Advanced Theory of Statistics, Vol. I*. 6th ed. London: Arnold.

## Also See

**Complementary:** [R] **ratio postestimation**

**Related:** [SVY] **svy: ratio**;
[R] **mean**, [R] **proportion**, [R] **total**

**Background:** [U] **20 Estimation and postestimation commands**,
[R] **estimation options**, [R] ***vce_option***

## Title

**ratio postestimation** — Postestimation tools for ratio

## Description

The following postestimation commands are available for `ratio`:

| command | description |
|---|---|
| `estat` | VCE |
| `estimates` | cataloging estimation results |
| `lincom` | point estimates, standard errors, testing, and inference for linear combinations of coefficients |
| `nlcom` | point estimates, standard errors, testing, and inference for nonlinear combinations of coefficients |
| `test` | Wald tests for simple and composite linear hypotheses |
| `testnl` | Wald tests of nonlinear hypotheses |

See the corresponding entries in the *Stata Base Reference Manual* for details.

## Methods and Formulas

All postestimation commands listed above are implemented as ado-files.

## Also See

**Complementary:** [R] **ratio**; [R] **estimates**, [R] **lincom**, [R] **nlcom**, [R] **test**, [R] **testnl**

**Related:** [SVY] **svy: ratio postestimation**

**Background:** [U] **13.5 Accessing coefficients and standard errors**,
[U] **13.6 Accessing results from Stata commands**,
[R] **estat**

## Title

**reg3** — Three-stage estimation for systems of simultaneous equations

## Syntax

*Basic syntax*

reg3 (*depvar*$_1$ *varlist*$_1$) (*depvar*$_2$ *varlist*$_2$) ...(*depvar*$_N$ *varlist*$_N$) [*if*] [*in*] [*weight*]

*Full syntax*

reg3 ([*eqname*$_1$:]*depvar*$_{1a}$ [*depvar*$_{1b}$ ... =]*varlist*$_1$ [, noconstant])

([*eqname*$_2$:]*depvar*$_{2a}$ [*depvar*$_{2b}$ ... =]*varlist*$_2$ [, noconstant])

...

([*eqname*$_N$:]*depvar*$_{Na}$ [*depvar*$_{Nb}$ ... =]*varlist*$_N$ [, noconstant])

[*if*] [*in*] [*weight*] [, *options*]

(*Continued on next page*)

| *options* | description |
|---|---|
| Model | |
| ireg3 | iterate until estimates converge |
| constraints(*constraints*) | apply specified linear constraints |
| Model 2 | |
| exog(*varlist*) | exogenous variables not specified in system equations |
| endog(*varlist*) | additional RHS endogenous variables |
| inst(*varlist*) | full list of exogenous variables |
| allexog | all right-hand-side variables are exogenous |
| noconstant | suppress constant term from instrument list |
| Est. method | |
| 3sls | three-stage least squares; the default |
| 2sls | two-stage least squares |
| ols | ordinary least squares |
| sure | seemingly unrelated regression |
| mvreg | sure with OLS degree-of-freedom adjustment |
| corr(*correlation*) | unstructured or independent correlation structure; default is unstructured |
| df adj. | |
| small | report small-sample statistics |
| dfk | adjust for number of covariates when computing disturbance covariance |
| dfk2 | use mean residual degrees of freedom when computing disturbance covariance |
| Reporting | |
| level(*#*) | set confidence level; default is level(95) |
| first | report first-stage regression |
| Opt options | |
| *optimization_options* | control the optimization process; seldom used |
| †noheader | suppress display of header |
| †notable | suppress display of coefficient table |
| †nofooter | suppress display of footer |

†noheader, notable, and nofooter do not appear in the dialog box.

*depvar* and *varlist* may contain time-series operators; see [U] **11.4.3 Time-series varlists**.

bootstrap, by, jackknife, rolling, statsby, and xi are allowed; see [U] **11.1.10 Prefix commands**.

aweights and fweights are allowed; see [U] **11.1.6 weight**.

See [U] **20 Estimation and postestimation commands** for additional capabilities of estimation commands.

Explicit equation naming (*eqname*:) cannot be combined with multiple dependent variables in an equation specification.

# Description

reg3 estimates a system of structural equations, where some equations contain endogenous variables among the explanatory variables. Estimation is via three-stage least squares (3SLS); see Zellner and Theil (1962). Typically, the endogenous explanatory variables are dependent variables from other equations in the system. reg3 supports iterated GLS estimation and linear constraints.

reg3 can also estimate systems of equations by seemingly unrelated regression (SURE), multivariate regression (MVREG), and equation-by-equation ordinary least squares (OLS) or two-stage least squares (2SLS).

## Nomenclature

Under 3SLS or 2SLS estimation, a *structural equation* is defined as one of the equations specified in the system. *Dependent variable* will have its usual interpretation as the left-hand-side variable in an equation with an associated disturbance term. All dependent variables are explicitly taken to be *endogenous* to the system and are treated as correlated with the disturbances in the system's equations. Unless specified in an endog() option, all other variables in the system are treated as *exogenous* to the system and uncorrelated with the disturbances. The exogenous variables are taken to be *instruments* for the endogenous variables.

# Options

Model

ireg3 causes reg3 to iterate over the estimated disturbance covariance matrix and parameter estimates until the parameter estimates converge. Although the iteration is usually successful, there is no guarantee that it will converge to a stable point. Under seemingly unrelated regression, this iteration converges to the maximum likelihood estimates.

constraints(*constraints*); see [R] **estimation options**.

Model 2

exog(*varlist*) specifies additional exogenous variables that are not included in any of the system equations. This can occur when the system contains identities that are not estimated. If implicitly exogenous variables from the equations are listed here, reg3 will just ignore the additional information. Specified variables will be added to the exogenous variables in the system and used in the "first stage" as instruments for the endogenous variables. By specifying dependent variables from the structural equations, you can use exog() to override their endogeneity.

endog(*varlist*) identifies variables in the system that are not dependent variables, but are endogenous to the system. These variables must appear in the variable list of at least one equation in the system. Again the need for this identification often occurs when the system contains identities. For example, a variable that is the sum of an exogenous variable and a dependent variable may appear as an explanatory variable in some equations.

inst(*varlist*) specifies a full list of all exogenous variables and may not be used with the endog() or exog() options. It must contain a full list of variables to be used as instruments for the endogenous regressors. Like exog(), the list may contain variables not specified in the system of equations. This option can be used to achieve the same results as the endog() and exog() options, and the choice is a matter of convenience. Any variable not specified in the *varlist* of the inst() option is assumed to be endogenous to the system. As with exog(), including the dependent variables from the structural equations will override their endogeneity.

`allexog` indicates that all right-hand-side variables are to be treated as exogenous—even if they appear as the dependent variable of another equation in the system. This option can be used to enforce a seemingly unrelated regression or multivariate regression estimation even when some dependent variables appear as regressors.

`noconstant`; see [R] **estimation options**.

Est. method

`3sls` specifies the full three-stage least-squares estimation of the system and is the default for `reg3`.

`2sls` causes `reg3` to perform equation-by-equation two-stage least squares on the full system of equations. This option implies `dfk`, `small`, and `corr(independent)`.

Note that cross-equation testing should not be performed after estimation with this option. With `2sls`, no covariance is estimated between the parameters of the equations. For cross-equation testing, use full `3sls`.

`ols` causes `reg3` to perform equation-by-equation OLS on the system—even if dependent variables appear as regressors or the regressors differ for each equation; see [R] **mvreg**. `ols` implies `allexog`, `dfk`, `small`, and `corr(independent)`; `nodfk` and `nosmall` may be specified to override `dfk` and `small`.

Note that the covariance of the coefficients between equations is not estimated under this option and that cross-equation tests should not be performed after estimation with `ols`. For cross-equation testing, use `sureg` or `3sls` (the default).

`sure` causes `reg3` to perform a seemingly unrelated regression estimation of the system—even if dependent variables from some equations appear as regressors in other equations; see [R] **sureg**. `sure` is a synonym for `allexog`.

`mvreg` is identical to `sure`, except that the disturbance covariance matrix is estimated with an OLS degrees-of-freedom adjustment—the `dfk` option. If the regressors are identical for all equations, the parameter point estimates will be the standard multivariate regression results. If any of the regressors differ, the point estimates are those for seemingly unrelated regression with an OLS degrees-of-freedom adjustment in computing the covariance matrix. `nodfk` and `nosmall` may be specified to override `dfk` and `small`.

`corr(`*correlation*`)` specifies the assumed form of the correlation structure of the equation disturbances and is rarely requested explicitly. For the family of models fitted by `reg3`, the only two allowable correlation structures are `independent` and `unstructured`. The default is `unstructured`.

This option is used almost exclusively to estimate a system of equations by two-stage least squares or to perform OLS regression with `reg3` on multiple equations. In these cases, the correlation is set to `independent`, forcing `reg3` to treat the covariance matrix of equation disturbances as diagonal in estimating model parameters. Thus a set of two-stage coefficient estimates can be obtained if the system contains endogenous right-hand-side variables, or OLS regression can be imposed, even if the regressors differ across equations. Without imposing independent disturbances, `reg3` would estimate the former by three-stage least squares and the latter by seemingly unrelated regression.

Note that any tests performed after estimation with the `independent` option will treat coefficients in different equations as having no covariance; cross-equation tests should not be used after specifying `corr(independent)`.

df adj.

`small` specifies that small-sample statistics be computed. It shifts the test statistics from $\chi^2$ and $Z$ statistics to $F$ statistics and $t$ statistics. This option is primarily intended to support multivariate regression. While the standard errors from each equation are computed using the degrees of freedom for the equation, the degrees of freedom for the $t$ statistics are all taken to be those for the first equation. This poses no problem under multivariate regression because the regressors are the same across equations.

`dfk` specifies the use of an alternative divisor in computing the covariance matrix for the equation residuals. As an asymptotically justified estimator, `reg3` by default uses the number of sample observations $n$ as a divisor. When the `dfk` option is set, a small-sample adjustment is made, and the divisor is taken to be $\sqrt{(n-k_i)(n-k_j)}$, where $k_i$ and $k_j$ are the numbers of parameters in equations $i$ and $j$, respectively.

`dfk2` specifies the use of an alternative divisor in computing the covariance matrix for the equation errors. When the `dfk2` option is set, the divisor is taken to be the mean of the residual degrees of freedom from the individual equations.

Reporting

`level(#)`; see [R] **estimation options**.

`first` requests that the first-stage regression results be displayed during estimation.

Opt options

*optimization_options* control the iterative process that minimizes the sum of squared errors when `ireg3` is specified. These options are seldom used.

`iterate(#)` specifies the maximum number of iterations. When the number of iterations equals #, the optimizer stops and presents the current results, even if the convergence tolerance has not been reached. The default value of `iterate()` is the current value of `set maxiter`, which is `iterate(16000)` if `maxiter` has not been changed.

`trace` adds to the iteration log a display of the current parameter vector

`nolog` suppresses the display of the iteration log.

`tolerance(#)` specifies the tolerance for the coefficient vector. When the relative change in the coefficient vector from one iteration to the next is less than or equal to #, the optimization process is stopped. `tolerance(1e-6)` is the default.

The following options are available with `reg3` but are not shown in the dialog box:

`noheader` suppresses display of the header reporting the estimation method and the table of equation summary statistics.

`notable` suppresses display of the coefficient table.

`nofooter` suppresses display of the footer reporting the list of endogenous variables in the model.

## Remarks

`reg3` estimates systems of structural equations where some equations contain endogenous variables among the explanatory variables. Generally, these endogenous variables are the dependent variables of other equations in the system, though not always. The disturbance is correlated with the endogenous variables—violating the assumptions of ordinary least squares. Further, since some of the explanatory variables are the dependent variables of other equations in the system, the error terms among the equations are expected to be correlated. `reg3` uses an instrumental variable approach to produce consistent estimates and generalized least squares (GLS) to account for the correlation structure in the disturbances across the equations. Good general references on three-stage estimation include Kmenta (1997) and Greene (2003, 405–407).

Three-stage least squares can be thought of as producing estimates from a three-step process.

**Stage 1.** Develop instrumented values for all endogenous variables. These instrumented values can simply be considered as the predicted values resulting from a regression of each endogenous variable on all exogenous variables in the system. This stage is identical to the first step in two-stage least squares and is critical for the consistency of the parameter estimates.

**Stage 2.** Obtain a consistent estimate for the covariance matrix of the equation disturbances. These estimates are based on the residuals from a two-stage least-squares estimation of each structural equation.

**Stage 3.** Perform a GLS-type estimation using the covariance matrix estimated in the second stage and with the instrumented values in place of the right-hand-side endogenous variables.

### ❑ Technical Note

The estimation and use of the covariance matrix of disturbances in three-stage estimation is almost identical to the seemingly unrelated regression (SURE) method—`sureg`. As with SURE, the use of this covariance matrix improves the efficiency of the three-stage estimator. Even without the use of the covariance matrix, the estimates would be consistent. (They would be two-stage least-squares estimates.) This improvement in efficiency comes with a caveat. All the parameter estimates now depend on the consistency of the covariance matrix estimates. If a single equation in the system is misspecified, the disturbance covariance estimates will be inconsistent, and the resulting coefficients will be biased and inconsistent. Alternatively, if each equation is estimated separately by two-stage least squares ([R] **regress**), only the coefficients in the misspecified equation are affected.

❑

### ❑ Technical Note

If an equation is just identified, the three-stage least-squares point estimates for that equation are identical to the two-stage least-squares estimates. However, as with `sureg`, even if all equations are just identified, fitting the model via `reg3` has at least one advantage over fitting each equation separately via `ivreg`; by using `reg3`, tests involving coefficients in different equations can be performed easily using `test` or `testnl`.

❑

### ▷ Example 1

A very simple macroeconomic model relates consumption (`consump`) to private and government wages paid (`wagepriv` and `wagegovt`). Simultaneously, private wages depend on consumption, total government expenditures (`govt`), and the lagged stock of capital in the economy (`capital1`). While this is not a very plausible model, it does meet the criterion of being simple. This model could be written as

$$\texttt{consump} = \beta_0 + \beta_1\,\texttt{wagepriv} + \beta_2\,\texttt{wagegovt} + \epsilon_1$$
$$\texttt{wagepriv} = \beta_3 + \beta_4\,\texttt{consump} + \beta_5\,\texttt{govt} + \beta_6\,\texttt{capital1} + \epsilon_2$$

Assuming that this is the full system, `consump` and `wagepriv` will be endogenous variables, with `wagegovt`, `govt`, and `capital1` exogenous. Data for the US economy on these variables are taken from Klein (1950). This model can be fitted with `reg3` by typing

```
. use http://www.stata-press.com/data/r9/klein
. reg3 (consump wagepriv wagegovt) (wagepriv consump govt capital1)
Three-stage least-squares regression
```

| Equation | Obs | Parms | RMSE | "R-sq" | chi2 | P |
|---|---|---|---|---|---|---|
| consump | 22 | 2 | 1.776297 | 0.9388 | 208.02 | 0.0000 |
| wagepriv | 22 | 3 | 2.372443 | 0.8542 | 80.04 | 0.0000 |

| | Coef. | Std. Err. | z | P>\|z\| | [95% Conf. | Interval] |
|---|---|---|---|---|---|---|
| consump | | | | | | |
| wagepriv | .8012754 | .1279329 | 6.26 | 0.000 | .5505314 | 1.052019 |
| wagegovt | 1.029531 | .3048424 | 3.38 | 0.001 | .432051 | 1.627011 |
| _cons | 19.3559 | 3.583772 | 5.40 | 0.000 | 12.33184 | 26.37996 |
| wagepriv | | | | | | |
| consump | .4026076 | .2567312 | 1.57 | 0.117 | -.1005764 | .9057916 |
| govt | 1.177792 | .5421253 | 2.17 | 0.030 | .1152461 | 2.240338 |
| capital1 | -.0281145 | .0572111 | -0.49 | 0.623 | -.1402462 | .0840173 |
| _cons | 14.63026 | 10.26693 | 1.42 | 0.154 | -5.492552 | 34.75306 |

```
Endogenous variables:  consump wagepriv
Exogenous variables:   wagegovt govt capital1
```

Without showing the two-stage least-squares results, we note that the consumption function in this system falls under the conditions noted earlier. That is, the two-stage and three-stage least-squares coefficients for the equation are identical.

◁

## ▷ Example 2

Some of the most common simultaneous systems encountered are supply-and-demand models. A very simple system could be specified as

$$\texttt{qDemand} = \beta_0 + \beta_1\,\texttt{price} + \beta_2\,\texttt{pcompete} + \beta_3\,\texttt{income} + \epsilon_1$$
$$\texttt{qSupply} = \beta_4 + \beta_5\,\texttt{price} + \beta_6\,\texttt{praw} + \epsilon_2$$

$$\texttt{Equilibrium condition: quantity} = \texttt{qDemand} = \texttt{qSupply}$$

where

`quantity` is the quantity of a product produced and sold
`price` is the price of the product
`pcompete` is the price of a competing product
`income` is the average income level of consumers
`praw` is the price of raw materials used to produce the product

In this system, price is assumed to be determined simultaneously with demand. The important statistical implications are that price is not a predetermined variable and that it is correlated with the disturbances of both equations. The system is somewhat unusual: quantity is associated with two disturbances. This really poses no problem because the disturbances are specified on the behavioral demand and supply equations—two separate entities. Often one of the two equations is rewritten to place price on the left-hand side making this endogeneity explicit in the specification.

To provide a concrete illustration of the effects of simultaneous equations, we can simulate data for the above system using known coefficients and disturbance properties. Specifically, we will simulate the data as

$$\texttt{qDemand} = 40 - 1.0\,\texttt{price} + 0.25\,\texttt{pcompete} + 0.5\,\texttt{income} + \epsilon_1$$

$$\texttt{qSupply} = 0.5\,\texttt{price} - 0.75\,\texttt{praw} + \epsilon_2$$

where

$$\epsilon_1 \sim N(0, 2.4)$$

$$\epsilon_2 \sim N(0, 3.8)$$

For comparison, we can estimate the supply and demand equations separately by OLS. The estimates for the demand equation are

```
. use http://www.stata-press.com/data/r9/supDem
. regress quantity price pcompete income
      Source |       SS       df       MS              Number of obs =      49
-------------+------------------------------           F(  3,    45) =    1.00
       Model |  23.1579302     3  7.71931008           Prob > F      =  0.4004
    Residual |  346.459313    45  7.69909584           R-squared     =  0.0627
-------------+------------------------------           Adj R-squared =  0.0002
       Total |  369.617243    48  7.70035923           Root MSE      =  2.7747

------------------------------------------------------------------------------
    quantity |      Coef.   Std. Err.      t    P>|t|     [95% Conf. Interval]
-------------+----------------------------------------------------------------
       price |   .1186265   .1716014     0.69   0.493    -.2269965    .4642496
    pcompete |   .0946416   .1200815     0.79   0.435    -.1472149    .3364981
      income |   .0785339   .1159867     0.68   0.502    -.1550754    .3121432
       _cons |   7.563261   5.019479     1.51   0.139     -2.54649    17.67301
------------------------------------------------------------------------------
```

The OLS estimates for the supply equation are

```
. regress quantity price praw
      Source |       SS       df       MS              Number of obs =      49
-------------+------------------------------           F(  2,    46) =   35.71
       Model |  224.819549     2  112.409774           Prob > F      =  0.0000
    Residual |  144.797694    46  3.14777596           R-squared     =  0.6082
-------------+------------------------------           Adj R-squared =  0.5912
       Total |  369.617243    48  7.70035923           Root MSE      =  1.7742

------------------------------------------------------------------------------
    quantity |      Coef.   Std. Err.      t    P>|t|     [95% Conf. Interval]
-------------+----------------------------------------------------------------
       price |    .724675   .1095657     6.61   0.000     .5041307    .9452192
        praw |  -.8674796   .1066114    -8.14   0.000    -1.082077    -.652882
       _cons |   -6.97291   3.323105    -2.10   0.041    -13.66197    -.283847
------------------------------------------------------------------------------
```

Examining the coefficients from these regressions, we note that they are not very close to the known parameters used to generate the simulated data. In particular, the positive coefficient on `price` in the demand equation stands out. We constructed our simulated data to be consistent with economic theory—people demand less of a product if its price rises and more if their personal income rises. Although the `price` coefficient is statistically insignificant, the positive value contrasts starkly with what is predicted from economic price theory and the $-1.0$ value that we used in the simulation. Likewise, we are disappointed with the insignificance and level of the coefficient on average `income`. The supply equation has correct signs on the two main parameters, but their levels are quite different from the known values. In fact, the coefficient on `price` (.724675) is different from the simulated parameter (0.5) at the 5% level of significance.

All these problems are to be expected. We explicitly constructed a simultaneous system of equations that violated one of the assumptions of least squares. Specifically, the disturbances were correlated with one of the regressors—`price`.

Two-stage least squares can be used to address the correlation between regressors and disturbances. Using instruments for the endogenous variable, `price`, two-stage least squares will produce consistent estimates of the parameters in the system. Let's use `ivreg` to see how our simulated system behaves when fitted using two-stage least squares.

```
. ivreg quantity (price = praw) pcompete income

Instrumental variables (2SLS) regression

      Source |       SS       df       MS              Number of obs =      49
-------------+------------------------------           F(  3,    45) =    2.68
       Model | -313.325605     3 -104.441868           Prob > F      =  0.0579
    Residual |  682.942847    45  15.1765077           R-squared     =       .
-------------+------------------------------           Adj R-squared =       .
       Total |  369.617243    48  7.70035923           Root MSE      =  3.8957

------------------------------------------------------------------------------
    quantity |      Coef.   Std. Err.      t    P>|t|     [95% Conf. Interval]
-------------+----------------------------------------------------------------
       price |  -1.015817   .3904865    -2.60   0.013    -1.802297    -.229337
    pcompete |   .3319504   .1804334     1.84   0.072    -.031461     .6953619
      income |   .5090607   .2002977     2.54   0.015     .1056405    .9124809
       _cons |   39.89988   11.24242     3.55   0.001     17.25648    62.54329
------------------------------------------------------------------------------
Instrumented:  price
Instruments:   pcompete income praw
------------------------------------------------------------------------------
```

```
. ivreg quantity (price = pcompete income) praw

Instrumental variables (2SLS) regression

      Source |       SS       df       MS              Number of obs =      49
-------------+------------------------------           F(  2,    46) =   18.42
       Model |  219.125463     2  109.562732           Prob > F      =  0.0000
    Residual |  150.491779    46  3.27156042           R-squared     =  0.5928
-------------+------------------------------           Adj R-squared =  0.5751
       Total |  369.617243    48  7.70035923           Root MSE      =  1.8087

------------------------------------------------------------------------------
    quantity |      Coef.   Std. Err.      t    P>|t|     [95% Conf. Interval]
-------------+----------------------------------------------------------------
       price |   .5773133   .1806137     3.20   0.003     .2137567    .9408698
        praw |  -.7835496   .1354534    -5.78   0.000    -1.056203   -.5108961
       _cons |  -2.550694   5.442299    -0.47   0.642    -13.50547    8.404086
------------------------------------------------------------------------------
Instrumented:  price
Instruments:   praw pcompete income
------------------------------------------------------------------------------
```

We are now much happier with the estimation results. All the coefficients from both equations are quite close to the true parameter values for the system. In particular, the coefficients are all well within 95% confidence intervals for the parameters. We do note that the missing $R$-squared in the demand equation seems unusual; we will discuss that more later.

Finally, this system could be estimated using three-stage least squares. To demonstrate how large systems might be handled and to avoid multiline commands, we will use global macros (see [P] **macro**) to hold the specifications for our equations.

```
. global demand "(qDemand: quantity price pcompete income)"
. global supply "(qSupply: quantity price praw)"
. reg3 $demand $supply, endog(price)
```

Note that we must specify `price` as endogenous since it does not appear as a dependent variable in either equation. Without this option, `reg3` would assume that there are no endogenous variables in the system and produce seemingly unrelated regression (`sureg`) estimates. The `reg3` output from our series of commands is

```
Three-stage least-squares regression
----------------------------------------------------------------------
Equation          Obs  Parms        RMSE    "R-sq"       chi2        P
----------------------------------------------------------------------
qDemand            49      3    3.739686   -0.8540       8.68   0.0338
qSupply            49      2    1.752501    0.5928      39.25   0.0000
----------------------------------------------------------------------

------------------------------------------------------------------------------
             |      Coef.   Std. Err.      z    P>|z|     [95% Conf. Interval]
-------------+----------------------------------------------------------------
qDemand      |
       price |  -1.014345   .3742036    -2.71   0.007     -1.74777   -.2809194
    pcompete |   .2647206   .1464194     1.81   0.071    -.0222561    .5516973
      income |   .5299146   .1898161     2.79   0.005     .1578819    .9019472
       _cons |   40.08749   10.77072     3.72   0.000     18.97726    61.19772
-------------+----------------------------------------------------------------
qSupply      |
       price |   .5773133   .1749974     3.30   0.001     .2343247    .9203019
        praw |  -.7835496   .1312414    -5.97   0.000    -1.040778   -.5263213
       _cons |  -2.550694   5.273067    -0.48   0.629    -12.88571    7.784327
------------------------------------------------------------------------------
Endogenous variables:  quantity price
Exogenous variables:   pcompete income praw
------------------------------------------------------------------------------
```

As noted earlier, the use of three-stage least squares over two-stage least squares is essentially an efficiency issue. The coefficients of the demand equation from three-stage least squares are very close to the coefficients from two-stage least squares, and those of the supply equation are identical. The latter case was mentioned earlier for systems with some exactly identified equations. However, even for the demand equation, we do not expect the coefficients to change systematically. What we do expect from three-stage least squares are more precise estimates of the parameters given the validity of our specification and `reg3`'s use of the covariances among the disturbances. This increased precision is exactly what is observed in the three-stage results. The standard errors of the three-stage estimates are 3 to 20% smaller than those for the two-stage estimates.

Let's summarize the results. With OLS, we got obviously biased estimates of the parameters. No amount of data would have improved the OLS estimates—they are inconsistent in the face of the violated OLS assumptions. With two-stage least squares, we obtained consistent estimates of the parameters, and these would have improved with more data. With three-stage least squares, we obtained consistent estimates of the parameters that are more efficient than those obtained by two-stage least squares. ◁

❑ Technical Note

We noted earlier that the $R$-squared was missing from the two-stage estimates of the demand equation. Now we see that the $R$-squared is negative for the three-stage estimates of the same equation. How can we have a negative $R$-squared?

In most estimators, other than least squares, the $R$-squared is no more than a summary measure of the overall in-sample predictive power of the estimator. The computational formula for $R$-squared is $R\text{-squared} = 1 - RSS/TSS$, where $RSS$ is the residual sum of squares (sum of squared residuals) and $TSS$ is the total sum of squared deviations about the mean of the dependent variable. In a standard linear model with a constant, the model from which the $TSS$ is computed is nested within the full model from which $RSS$ is computed—they both have a constant term based on the same data. Thus it must be that $TSS \geq RSS$ and $R$-squared is constrained between 0 and 1.

For two- and three-stage least squares, some of the regressors enter the model as instruments when the parameters are estimated. However, since our goal is to fit the structural model, the actual values, not the instruments for the endogenous right-hand-side variables, are used to determine $R$-squared. The model residuals are computed over a different set of regressors from those used to fit the model. The two- and/or three-stage estimates are no longer nested within a constant-only model of the dependent variable, and the residual sum of squares is no longer constrained to be smaller than the total sum of squares.

A negative $R$-squared in three-stage least squares should be taken for exactly what it is—an indication that the structural model predicts the dependent variable worse than a constant-only model. Is this a problem? It depends on the application. Note that three-stage least squares applied to our contrived supply-and-demand example produced very good estimates of the known true parameters. Still, the demand equation produced an $R$-squared of $-0.854$. How do we feel about our parameter estimates? This should be determined by the estimates themselves, their associated standard errors, and the overall model significance. On this basis, negative $R$-squared and all, we feel pretty good about all the parameter estimates for both the supply and demand equations. Would we want to make predictions about equilibrium quantity using the demand equation alone? Probably not. Would we want to make these quantity predictions using the supply equation? Possibly, because based on in-sample predictions, they seem better than those from the demand equations. However, both the supply and demand estimates are based on limited information. If we are interested in predicting quantity, a reduced-form equation containing all our independent variables would usually be preferred.

❑

❑ Technical Note

As a matter of syntax, we could have specified the supply-and-demand model on a single line without using global macros.

(*Continued on next page*)

```
. reg3 (quantity price pcompete income) (quantity price praw), endog(price)
Three-stage least-squares regression
----------------------------------------------------------------------
Equation          Obs  Parms        RMSE    "R-sq"       chi2        P
----------------------------------------------------------------------
quantity           49      3    3.739686   -0.8540       8.68   0.0338
2quantity          49      2    1.752501    0.5928      39.25   0.0000
----------------------------------------------------------------------

------------------------------------------------------------------------------
             |      Coef.   Std. Err.      z    P>|z|     [95% Conf. Interval]
-------------+----------------------------------------------------------------
quantity     |
       price |  -1.014345   .3742036    -2.71   0.007     -1.74777   -.2809194
    pcompete |   .2647206   .1464194     1.81   0.071    -.0222561    .5516973
      income |   .5299146   .1898161     2.79   0.005     .1578819    .9019472
       _cons |   40.08749   10.77072     3.72   0.000     18.97726    61.19772
-------------+----------------------------------------------------------------
2quantity    |
       price |   .5773133   .1749974     3.30   0.001     .2343247    .9203019
        praw |  -.7835496   .1312414    -5.97   0.000    -1.040778   -.5263213
       _cons |  -2.550694   5.273067    -0.48   0.629    -12.88571    7.784327
------------------------------------------------------------------------------
Endogenous variables:  quantity price
Exogenous variables:   pcompete income praw
------------------------------------------------------------------------------
```

However, in this case, **reg3** has been forced to create a unique equation name for the supply equation—**2quantity**. Both the supply and demand equations could not be designated as **quantity**, so a number was prefixed to the name for the supply equation.

We could have specified

```
. reg3 (qDemand: quantity price pcompete income) (qSupply: quantity price praw),
> endog(price)
```

and obtained exactly the same results and equation labeling as when we used global macros to hold the equation specifications.

In the absence of explicit equation names, **reg3** always assumes that the dependent variable should be used to name equations. When each equation has a different dependent variable, this rule causes no problems and produces easily interpreted result tables. If the same dependent variable appears in more than one equation, however, **reg3** will create a unique equation name based on the dependent variable name. Since equation names must be used for cross-equation tests, you have more control in this situation if explicit names are placed on the equations.

❑

## ▷ Example 3: Using the full syntax of reg3

Klein's (1950) model of the US economy is often used to demonstrate system estimators. It contains several common features that will serve to demonstrate the full syntax of **reg3**. The Klein model is defined by the following seven relationships.

*(Continued on next page)*

$$\texttt{c} = \beta_0 + \beta_1\texttt{p} + \beta_2\texttt{p1} + \beta_3\texttt{w} + \epsilon_1 \tag{1}$$
$$\texttt{i} = \beta_4 + \beta_5\texttt{p} + \beta_6\texttt{p1} + \beta_7\texttt{k1} + \epsilon_2 \tag{2}$$
$$\texttt{wp} = \beta_8 + \beta_9\texttt{y} + \beta_{10}\texttt{y1} + \beta_{11}\texttt{yr} + \epsilon_3 \tag{3}$$
$$\texttt{y} = \texttt{c} + \texttt{i} + \texttt{g} \tag{4}$$
$$\texttt{p} = \texttt{y} - \texttt{t} - \texttt{wp} \tag{5}$$
$$\texttt{k} = \texttt{k1} + \texttt{i} \tag{6}$$
$$\texttt{w} = \texttt{wg} + \texttt{wp} \tag{7}$$

The variables in the model are listed below. Two sets of variable names are shown. The concise first name uses traditional economics mnemonics, while the second name provides more guidance for everyone else. The concise names serve to keep the specification of the model small (and quite understandable to economists).

| Short Name | Long Name | Variable Definition | Type |
|---|---|---|---|
| c | consump | Consumption | endogenous |
| p | profits | Private industry profits | endogenous |
| p1 | profits1 | Last year's private industry profits | exogenous |
| wp | wagepriv | Private wage bill | endogenous |
| wg | wagegovt | Government wage bill | exogenous |
| w | wagetot | Total wage bill | endogenous |
| i | invest | Investment | endogenous |
| k1 | capital1 | Last year's level of capital stock | exogenous |
| y | totinc | Total income/demand | endogenous |
| y1 | totinc1 | Last year's total income | exogenous |
| g | govt | Government spending | exogenous |
| t | taxnetx | Indirect bus. taxes + net exports | exogenous |
| yr | year | Year—1931 | exogenous |

Equations (1)–(3) are behavioral and contain explicit disturbances ($\epsilon_1$, $\epsilon_2$, and $\epsilon_3$). The remaining equations are identities that specify additional variables in the system and their "accounting" relationships with the variables in the behavioral equations. Some variables are explicitly endogenous by appearing as dependent variables in (1)–(3). Others are implicitly endogenous as linear combinations that contain other endogenous variables (e.g., w and p). Still other variables are implicitly exogenous by appearing in the identities but not in the behavioral equations (e.g., wg and g).

Using the concise names, Klein's model may be fitted with the following command:

```
. use http://www.stata-press.com/data/r9/kleinAbr
. reg3 (c p p1 w) (i p p1 k1) (wp y y1 yr), endog(w p y) exog(t wg g)
Three-stage least-squares regression
------------------------------------------------------------------------
Equation          Obs  Parms        RMSE    "R-sq"       chi2        P
------------------------------------------------------------------------
c                  21      3    .9443305    0.9801     864.59   0.0000
i                  21      3    1.446736    0.8258     162.98   0.0000
wp                 21      3    .7211282    0.9863    1594.75   0.0000
------------------------------------------------------------------------

------------------------------------------------------------------------------
             |      Coef.   Std. Err.      z    P>|z|     [95% Conf. Interval]
-------------+----------------------------------------------------------------
c            |
           p |   .1248904   .1081291     1.16   0.248    -.0870387    .3368194
          p1 |   .1631439   .1004382     1.62   0.104    -.0337113    .3599992
           w |    .790081   .0379379    20.83   0.000     .7157244    .8644379
       _cons |   16.44079   1.304549    12.60   0.000     13.88392    18.99766
-------------+----------------------------------------------------------------
i            |
           p |  -.0130791   .1618962    -0.08   0.936    -.3303898    .3042316
          p1 |   .7557238   .1529331     4.94   0.000     .4559805    1.055467
          k1 |  -.1948482   .0325307    -5.99   0.000    -.2586072   -.1310893
       _cons |   28.17785   6.793768     4.15   0.000     14.86231    41.49339
-------------+----------------------------------------------------------------
wp           |
           y |   .4004919   .0318134    12.59   0.000     .3381388     .462845
          y1 |    .181291   .0341588     5.31   0.000     .1143411    .2482409
          yr |    .149674   .0279352     5.36   0.000      .094922    .2044261
       _cons |   1.797216   1.115854     1.61   0.107    -.3898181    3.984251
------------------------------------------------------------------------------
Endogenous variables:  c i wp w p y
Exogenous variables:   p1 k1 y1 yr t wg g
------------------------------------------------------------------------------
```

We used the `exog()` option to identify `t`, `wg`, and `g` as exogenous variables in the system. These variables must be identified because they are part of the system but do not appear directly in any of the behavioral equations. Without this option, `reg3` would not know they were part of the system. The `endog()` option specifying `w`, `p`, and `y` is also required. Without this information, `reg3` would be unaware that these variables are linear combinations that include endogenous variables.

## ❑ Technical Note

Rather than listing additional endogenous and exogenous variables, we could specify the full list of exogenous variables in an `inst()` option,

```
. reg3 (c p p1 w) (i p p1 k1) (wp y y1 yr), inst(g t wg yr p1 k1 y1)
```

or, equivalently,

```
. global conseqn "(c p p1 w)"
. global inveqn  "(i p p1 k1)"
. global wageqn  "(wp y y1 yr)"
. global inlist  "g t wg yr p1 k1 y1"
. reg3 $conseqn $inveqn $wageqn, inst($inlist)
```

Macros and explicit equations can also be mixed in the specification

```
. reg3 $conseqn (i p p1 k1) $wageqn, endog(w p y) exog(t wg g)
```

or

```
. reg3 (c p p1 w) $inveqn (wp y y1 yr), endog(w p y) exog(t wg g)
```

Placing the equation-binding parentheses in the global macros was also arbitrary. We could have used

```
. global consump  "c p p1 w"
. global invest   "i p p1 k1"
. global wagepriv "wp y y1 yr"
. reg3 ($consump) ($invest) ($wagepriv), endog(w p y) exog(t wg g)
```

`reg3` is tolerant of all combinations, and these commands will produce identical output.

❑

Switching to the full variable names, we can fit Klein's model with the commands below. We will use global macros to store the lists of endogenous and exogenous variables. Again this is not necessary: these lists could have been typed directly on the command line. However, assigning the lists to local macros makes additional processing easier if alternative models are to be fitted. We will also use the `ireg3` option to produce the iterated estimates.

```
. use http://www.stata-press.com/data/r9/klein
. global conseqn "(consump profits profits1 wagetot)"
. global inveqn  "(invest profits profits1 capital1)"
. global wageqn "(wagepriv totinc totinc1 year)"

. global enlist "wagetot profits totinc"
. global exlist "taxnetx wagegovt govt"
```

(*Continued on next page*)

```
. reg3 $conseqn $inveqn $wageqn, endog($enlist) exog($exlist) ireg3
Iteration 1:   tolerance =  .3712549
Iteration 2:   tolerance =  .1894712
Iteration 3:   tolerance =  .1076401
 (output omitted)
Iteration 24:   tolerance =  7.049e-07

Three-stage least-squares regression, iterated
```

| Equation | Obs | Parms | RMSE | "R-sq" | chi2 | P |
|---|---|---|---|---|---|---|
| consump | 21 | 3 | .9565088 | 0.9796 | 970.31 | 0.0000 |
| invest | 21 | 3 | 2.134327 | 0.6209 | 56.78 | 0.0000 |
| wagepriv | 21 | 3 | .7782334 | 0.9840 | 1312.19 | 0.0000 |

| | Coef. | Std. Err. | z | P>\|z\| | [95% Conf. | Interval] |
|---|---|---|---|---|---|---|
| consump | | | | | | |
| profits | .1645096 | .0961979 | 1.71 | 0.087 | -.0240348 | .3530539 |
| profits1 | .1765639 | .0901001 | 1.96 | 0.050 | -.0000291 | .3531569 |
| wagetot | .7658011 | .0347599 | 22.03 | 0.000 | .6976729 | .8339294 |
| _cons | 16.55899 | 1.224401 | 13.52 | 0.000 | 14.15921 | 18.95877 |
| invest | | | | | | |
| profits | -.3565316 | .2601568 | -1.37 | 0.171 | -.8664296 | .1533664 |
| profits1 | 1.011299 | .2487745 | 4.07 | 0.000 | .5237098 | 1.498888 |
| capital1 | -.2602 | .0508694 | -5.12 | 0.000 | -.3599022 | -.1604978 |
| _cons | 42.89629 | 10.59386 | 4.05 | 0.000 | 22.13271 | 63.65987 |
| wagepriv | | | | | | |
| totinc | .3747792 | .0311027 | 12.05 | 0.000 | .3138191 | .4357394 |
| totinc1 | .1936506 | .0324018 | 5.98 | 0.000 | .1301443 | .257157 |
| year | .1679262 | .0289291 | 5.80 | 0.000 | .1112263 | .2246261 |
| _cons | 2.624766 | 1.195559 | 2.20 | 0.028 | .2815124 | 4.968019 |

```
Endogenous variables:  consump invest wagepriv wagetot profits totinc
Exogenous variables:   profits1 capital1 totinc1 year taxnetx wagegovt govt
```

◁

## ▷ Example 4: Constraints with reg3

As a simple example of constraints, (1) above may be rewritten with both wages explicitly appearing (rather than as a variable containing the sum). Using the longer variable names, we have

$$\texttt{consump} = \beta_0 + \beta_1\,\texttt{profits} + \beta_2\,\texttt{profits1} + \beta_3\,\texttt{wagepriv} + \beta_{12}\,\texttt{wagegovt} + \epsilon_1$$

To retain the effect of the identity in (7), we need $\beta_3 = \beta_{12}$ as a constraint on the system. We obtain this result by defining the constraint in the usual way and then specifying its use in `reg3`. Since `reg3` is a system estimator, we will need to use the full equation syntax of `constraint`. Note the assumption that the following commands are entered after the model above has been estimated. We are simply changing the definition of the consumption equation (`consump`) and adding a constraint on two of its parameters. The remainder of the model definition is carried forward.

```
. global conseqn "(consump profits profits1 wagepriv wagegovt)"
. constraint define 1 [consump]wagepriv = [consump]wagegovt
```

```
. reg3 $conseqn $inveqn $wageqn, endog($enlist) exog($exlist) constr(1) ireg3
note:  additional endogenous variables not in the system have no effect
       and are ignored:  wagetot
Iteration 1:   tolerance =   .3712547
Iteration 2:   tolerance =    .189471
Iteration 3:   tolerance =     .10764
  (output omitted)
Iteration 24:   tolerance =  7.049e-07

Three-stage least-squares regression, iterated

Constraints:
 ( 1)  [consump]wagepriv - [consump]wagegovt = 0
----------------------------------------------------------------------
Equation          Obs  Parms        RMSE    "R-sq"       chi2        P
----------------------------------------------------------------------
consump            21      3    .9565086    0.9796     970.31   0.0000
invest             21      3    2.134326    0.6209      56.78   0.0000
wagepriv           21      3    .7782334    0.9840    1312.19   0.0000
----------------------------------------------------------------------

------------------------------------------------------------------------------
             |      Coef.   Std. Err.      z    P>|z|     [95% Conf. Interval]
-------------+----------------------------------------------------------------
consump      |
     profits |   .1645097   .0961978     1.71   0.087    -.0240346     .353054
    profits1 |   .1765639   .0901001     1.96   0.050    -.0000291    .3531568
    wagepriv |   .7658012   .0347599    22.03   0.000     .6976729    .8339294
    wagegovt |   .7658012   .0347599    22.03   0.000     .6976729    .8339294
       _cons |   16.55899   1.224401    13.52   0.000      14.1592    18.95877
-------------+----------------------------------------------------------------
invest       |
     profits |  -.3565311   .2601567    -1.37   0.171    -.8664288    .1533666
    profits1 |   1.011298   .2487744     4.07   0.000     .5237096    1.498887
    capital1 |  -.2601999   .0508694    -5.12   0.000     -.359902   -.1604977
       _cons |   42.89626   10.59386     4.05   0.000     22.13269    63.65984
-------------+----------------------------------------------------------------
wagepriv     |
      totinc |   .3747792   .0311027    12.05   0.000      .313819    .4357394
     totinc1 |   .1936506   .0324018     5.98   0.000     .1301443     .257157
        year |   .1679262   .0289291     5.80   0.000     .1112263    .2246261
       _cons |   2.624766   1.195559     2.20   0.028      .281512    4.968019
------------------------------------------------------------------------------
Endogenous variables:  consump invest wagepriv wagetot profits totinc
Exogenous variables:   profits1 wagegovt capital1 totinc1 year taxnetx govt
------------------------------------------------------------------------------
```

As expected, none of the parameter or standard error estimates has changed from the previous estimates (before the seventh significant digit). We have simply decomposed the total wage variable into its two parts and constrained the coefficients on these parts. The warning about additional endogenous variables was just `reg3`'s way of letting us know that we had specified some information that was irrelevant to the estimation of the system. We had left the variable `wagetot` in our `endog` macro. It does not mean anything to the system to specify `wagetot` as endogenous since it is no longer in the system. That's fine with `reg3` and fine for our current purposes.

We can also impose constraints across the equations. For example, the admittedly meaningless constraint of requiring `profits` to have the same effect in both the consumption and investment equations could be imposed. Retaining the constraint on the wage coefficients, we would estimate this constrained system.

```
. constraint define 2 [consump]profits = [invest]profits
```

```
. reg3 $conseqn $inveqn $wageqn, endog($enlist) exog($exlist) constr(1 2) ireg3
note:  additional endogenous variables not in the system have no effect
       and are ignored:  wagetot
Iteration 1:   tolerance =   .1427927
Iteration 2:   tolerance =    .032539
Iteration 3:   tolerance =  .00307811
Iteration 4:   tolerance =  .00016903
Iteration 5:   tolerance =  .00003409
Iteration 6:   tolerance =  7.763e-06
Iteration 7:   tolerance =  9.240e-07

Three-stage least-squares regression, iterated

Constraints:
 ( 1)  [consump]wagepriv - [consump]wagegovt = 0
 ( 2)  [consump]profits - [invest]profits = 0

------------------------------------------------------------------
Equation          Obs  Parms        RMSE    "R-sq"       chi2        P
------------------------------------------------------------------
consump            21      3    .9504669    0.9798    1019.54   0.0000
invest             21      3    1.247066    0.8706     144.57   0.0000
wagepriv           21      3    .7225276    0.9862    1537.45   0.0000
------------------------------------------------------------------

------------------------------------------------------------------------------
             |      Coef.   Std. Err.      z    P>|z|     [95% Conf. Interval]
-------------+----------------------------------------------------------------
consump      |
     profits |   .1075413   .0957767     1.12   0.262    -.0801777    .2952602
    profits1 |   .1712756   .0912613     1.88   0.061    -.0075932    .3501444
    wagepriv |    .798484   .0340876    23.42   0.000     .7316734    .8652946
    wagegovt |    .798484   .0340876    23.42   0.000     .7316734    .8652946
       _cons |    16.2521   1.212157    13.41   0.000     13.87631    18.62788
-------------+----------------------------------------------------------------
invest       |
     profits |   .1075413   .0957767     1.12   0.262    -.0801777    .2952602
    profits1 |   .6443378   .1058682     6.09   0.000       .43684    .8518356
    capital1 |  -.1766669   .0261889    -6.75   0.000    -.2279962   -.1253375
       _cons |   24.31931   5.284325     4.60   0.000     13.96222     34.6764
-------------+----------------------------------------------------------------
wagepriv     |
      totinc |   .4014106   .0300552    13.36   0.000     .3425035    .4603177
     totinc1 |   .1775359   .0321583     5.52   0.000     .1145068     .240565
        year |   .1549211   .0282291     5.49   0.000      .099593    .2102492
       _cons |   1.959788    1.14467     1.71   0.087    -.2837242    4.203299
------------------------------------------------------------------------------
Endogenous variables:  consump invest wagepriv wagetot profits totinc
Exogenous variables:   profits1 wagegovt capital1 totinc1 year taxnetx govt
------------------------------------------------------------------------------
```

◁

## ❑ Technical Note

Identification in a system of simultaneous equations involves the notion that there is sufficient information to estimate the parameters of the model given the specified functional form. Under-identification usually manifests itself as a singular matrix in the three-stage least-squares computations. The most commonly violated order condition for two- or three-stage least squares involves the number of endogenous and exogenous variables. There must be at least as many noncollinear exogenous variables in the remaining system as there are endogenous right-hand-side variables in an equation. This condition must hold for each structural equation in the system.

Put as a set of rules the following:

1. Count the number of right-hand-side endogenous variables in an equation and call this $m_i$.
2. Count the number of exogenous variables in the same equation and call this $k_i$.
3. Count the total number of exogenous variables in all the structural equations plus any additional variables specified in an `exog()` or `inst()` option and call this $K$.
4. If $m_i > (K - k_i)$ for any structural equation ($i$), then the system is underidentified and cannot be estimated by three-stage least squares.

We are also possibly in trouble if any of the exogenous variables are linearly dependent. We must have $m_i$ linearly independent variables among the exogenous variables represented by $(K - k_i)$.

The complete conditions for identification involve rank-order conditions on several matrices. For a full treatment, see Theil (1971) or Greene (2003, 405–407).

❑

## Saved Results

`reg3` saves in `e()`:

Scalars

| | | | |
|---|---|---|---|
| `e(N)` | number of observations | `e(F_#)` | $F$ statistic for eqn. # (`small`) |
| `e(k_eq)` | number of equations | `e(rmse_#)` | root mean squared error for eqn. # |
| `e(mss_#)` | model sum of squares for eqn. # | `e(ll)` | log likelihood |
| `e(df_m#)` | model degrees of freedom for eqn. # | `e(chi2_#)` | $\chi^2$ for equation # |
| `e(rss_#)` | residual sum of squares for eqn. # | `e(p_#)` | significance for equation # |
| `e(df_r)` | residual degrees of freedom (`small`) | `e(ic)` | number of iterations |
| `e(r2_#)` | $R$-squared for equation # | `e(cons_#)` | 1 when equation # has a constant; 0 otherwise |

Macros

| | | | |
|---|---|---|---|
| `e(cmd)` | `reg3` | `e(wtype)` | weight type |
| `e(depvar)` | names of dependent variables | `e(wexp)` | weight expression |
| `e(exog)` | names of exogenous variables | `e(method)` | `3sls`, `2sls`, `ols`, `sure`, or `mvreg` |
| `e(endog)` | names of endogenous variables | `e(small)` | `small` |
| `e(eqnames)` | names of equations | `e(properties)` | `b V` |
| `e(corr)` | correlation structure | `e(predict)` | program used to implement `predict` |

Matrices

| | | | |
|---|---|---|---|
| `e(b)` | coefficient vector | `e(V)` | variance–covariance matrix of the estimators |
| `e(Sigma)` | $\widehat{\Sigma}$ matrix | | |

Functions

| | |
|---|---|
| `e(sample)` | marks estimation sample |

## Methods and Formulas

`reg3` is implemented as an ado-file.

The most concise way to represent a system of equations for three-stage least squares requires thinking of the individual equations and their associated data as being stacked. `reg3` does not expect the data in this format, but it is a convenient shorthand. The system could then be formulated as

$$\begin{bmatrix} \mathbf{y}_1 \\ \mathbf{y}_2 \\ \vdots \\ \mathbf{y}_M \end{bmatrix} = \begin{bmatrix} \mathbf{Z}_1 & 0 & \ldots & 0 \\ 0 & \mathbf{Z}_2 & \ldots & 0 \\ \vdots & \vdots & \ddots & \vdots \\ 0 & 0 & \ldots & \mathbf{Z}_M \end{bmatrix} \begin{bmatrix} \beta_1 \\ \beta_2 \\ \vdots \\ \beta_M \end{bmatrix} + \begin{bmatrix} \epsilon_1 \\ \epsilon_2 \\ \vdots \\ \epsilon_M \end{bmatrix}$$

In full matrix notation, this is just

$$\mathbf{y} = \mathbf{Z}\,\mathbf{B} + \epsilon$$

The $\mathbf{Z}$ elements in these matrices represent both the endogenous and the exogenous right-hand-side variables in the equations.

Also assume that there will be correlation between the disturbances of the equations so that

$$E(\epsilon\epsilon') = \mathbf{\Sigma}$$

where the disturbances are further assumed to have an expected value of 0; $E(\epsilon) = 0$.

The "first stage" of three-stage least-squares regression requires developing instrumented values for the endogenous variables in the system. These can be derived as the predictions from a linear regression of each endogenous regressor on all exogenous variables in the system, or, more succinctly, as the projection of each regressor through the projection matrix of all exogenous variables onto the regressors. Designating the set of all exogenous variables as $\mathbf{X}$ results in

$$\widehat{\mathbf{z}}_i = \mathbf{X}(\mathbf{X}'\mathbf{X})^{-1}\mathbf{X}'\mathbf{z}_i \quad \textit{for each } i$$

Taken collectively, these $\widehat{\mathbf{Z}}$ contain the instrumented values for all the regressors. They take on the actual values for the exogenous variables and first-stage predictions for the endogenous variables. Given these instrumented variables, a generalized least squares (GLS) or Aitken (1935) estimator can be formed for the parameters of the system

$$\widehat{\mathbf{B}} = \left\{\widehat{\mathbf{Z}}'(\mathbf{\Sigma}^{-1} \otimes \mathbf{I})\widehat{\mathbf{Z}}\right\}^{-1} \widehat{\mathbf{Z}}'(\mathbf{\Sigma}^{-1} \otimes \mathbf{I})\mathbf{y}$$

All that remains is to obtain a consistent estimator for $\mathbf{\Sigma}$. This estimate can be formed from the residuals of two-stage least squares estimates of each equation in the system. Alternately, and identically, the residuals can be computed from the estimates formed by taking $\mathbf{\Sigma}$ to be an identity matrix. This maintains the full system of coefficients and allows constraints to be applied when the residuals are computed.

Taking $\mathbf{E}$ to be the matrix of residuals from these estimates, a consistent estimate of $\mathbf{\Sigma}$ is

$$\widehat{\mathbf{\Sigma}} = \frac{\mathbf{E}'\mathbf{E}}{n}$$

where $n$ is the number of observations in the sample. An alternative divisor for this estimate can be obtained with the `dfk` option as outlined under options.

With the estimate of $\widehat{\mathbf{\Sigma}}$ placed into the GLS estimating equation

$$\widehat{\mathbf{B}} = \left\{\widehat{\mathbf{Z}}'(\widehat{\boldsymbol{\Sigma}}^{-1} \otimes \mathbf{I})\widehat{\mathbf{Z}}\right\}^{-1} \widehat{\mathbf{Z}}'(\widehat{\boldsymbol{\Sigma}}^{-1} \otimes \mathbf{I})\mathbf{y}$$

is the three-stage least-squares estimates of the system parameters.

The asymptotic variance–covariance matrix of the estimator is just the standard formulation for a GLS estimator

$$\mathbf{V}_{\widehat{\mathbf{B}}} = \left\{\widehat{\mathbf{Z}}'(\widehat{\boldsymbol{\Sigma}}^{-1} \otimes \mathbf{I})\widehat{\mathbf{Z}}\right\}^{-1}$$

Iterated three-stage least-squares estimates can be obtained by computing the residuals from the three-stage parameter estimates, using these to formulate a new $\widehat{\boldsymbol{\Sigma}}$, and recomputing the parameter estimates. This process is repeated until the estimates $\widehat{\mathbf{B}}$ converge—if they converge. Convergence is not guaranteed. When estimating a system by SURE, these iterated estimates will be the maximum likelihood estimates for the system. The iterated solution can also be used to produce estimates that are invariant to choice of system and restriction parameterization for many linear systems under full three-stage least squares.

The exposition above follows the parallel developments in Greene (2003) and Kmenta (1997).

## References

Aitken, A. C. 1935. On least squares and linear combination of observations. *Proceedings, Royal Society of Edinburgh* 55: 42–48.

Greene, W. H. 2003. *Econometric Analysis*. 5th ed. Upper Saddle River, NJ: Prentice Hall.

Klein, L. 1950. *Economic Fluctuations in the United States 1921–1941*. New York: Wiley.

Kmenta, J. 1997. *Elements of Econometrics*. 2nd ed. Ann Arbor: University of Michigan Press.

Theil, H. 1971. *Principles of Econometrics*. New York: Wiley.

Weesie, J. 1999. sg121: Seemingly unrelated estimation and the cluster-adjusted sandwich estimator. *Stata Technical Bulletin* 52: 34–47. Reprinted in *Stata Technical Bulletin Reprints*, vol. 9, pp. 231–248.

Zellner, A. and H. Theil. 1962. Three stage least squares: Simultaneous estimate of simultaneous equations. *Econometrica* 29: 63–68.

## Also See

**Complementary:** [R] **reg3 postestimation**; [R] **constraint**

**Related:** [R] **biprobit**, [R] **cnsreg**, [R] **ivreg**, [R] **mvreg**, [R] **regress**, [R] **sureg**

**Background:** [U] **11.1.10 Prefix commands**,
[U] **20 Estimation and postestimation commands**,
[R] **estimation options**

# Title

**reg3 postestimation** — Postestimation tools for reg3

# Description

The following postestimation commands are available for `reg3`:

| command | description |
|---|---|
| * `estat` | AIC, BIC, VCE, and estimation sample summary |
| `estimates` | cataloging estimation results |
| `hausman` | Hausman's specification test |
| `lincom` | point estimates, standard errors, testing, and inference for linear combinations of coefficients |
| `mfx` | marginal effects or elasticities |
| `nlcom` | point estimates, standard errors, testing, and inference for nonlinear combinations of coefficients |
| `predict` | predictions, residuals, influence statistics, and other diagnostic measures |
| `predictnl` | point estimates, standard errors, testing, and inference for generalized predictions |
| `test` | Wald tests for simple and composite linear hypotheses |
| `testnl` | Wald tests of nonlinear hypotheses |

* `estat ic` may not be used after `reg3, 2sls`.

See the corresponding entries in the *Stata Base Reference Manual* for details.

# Syntax for predict

`predict` [*type*] *newvar* [*if*] [*in*] [, `equation(`*eqno*[`,`*eqno*]`)` *statistic*]

| *statistic* | description |
|---|---|
| `xb` | $\mathbf{x}_j\mathbf{b}$, fitted values; the default |
| `stdp` | standard error of the prediction |
| `residuals` | residuals |
| `difference` | difference between the linear predictions of two equations |
| `stddp` | standard error of the difference in linear predictions |

These statistics are available both in and out of sample; type `predict ... if e(sample) ...` if wanted only for the estimation sample.

(*Continued on next page*)

## Options for predict

equation(*eqno*[ ,*eqno* ]) specifies to which equation you are referring.

equation() is filled in with one *eqno* for options xb, stdp, and residuals. equation(#1) would mean the calculation is to be made for the first equation, equation(#2) would mean the second, and so on. Alternatively, you could refer to the equations by their names. equation(income) would refer to the equation named income and equation(hours) to the equation named hours.

If you do not specify equation(), results are the same as if you specified equation(#1).

difference and stddp refer to between-equation concepts. To use these options, you must specify two equations, e.g., equation(#1,#2) or equation(income,hours). When two equations must be specified, equation() is required.

xb, the default, calculates the fitted values—the prediction of $\mathbf{x}_j\mathbf{b}$ for the specified equation.

stdp calculates the standard error of the prediction for the specified equation. It can be thought of as the standard error of the predicted expected value or mean for the observation's covariate pattern. This is also referred to as the standard error of the fitted value.

residuals calculates the residuals.

difference calculates the difference between the linear predictions of two equations in the system. With equation(#1,#2), difference computes the prediction of equation(#1) minus the prediction of equation(#2).

stddp is allowed only after you have previously fitted a multiple-equation model. The standard error of the difference in linear predictions $(\mathbf{x}_{1j}\mathbf{b} - \mathbf{x}_{2j}\mathbf{b})$ between equations 1 and 2 is calculated.

For more information on using predict after multiple-equation estimation commands, see [R] **predict**.

## Methods and Formulas

All postestimation commands listed above are implemented as ado-files.

The computational formulas for the statistics produced by predict can be found in [R] **predict** and [R] **regress postestimation**.

## Also See

**Complementary:** [R] **reg3**; [R] **estimates**, [R] **hausman**, [R] **lincom**, [R] **mfx**, [R] **nlcom**, [R] **predictnl**, [R] **test**, [R] **testnl**

**Background:** [U] **11.1.10 Prefix commands**, [U] **13.5 Accessing coefficients and standard errors**, [U] **20 Estimation and postestimation commands**, [R] **estat**, [R] **predict**

# Title

**regress** — Linear regression

# Syntax

`regress` *depvar* [*indepvars*] [*if*] [*in*] [*weight*] [, *options*]

| *options* | description |
|---|---|
| Model | |
| `noconstant` | suppress constant term |
| `hascons` | has user-supplied constant |
| `tsscons` | compute total sum of squares with constant; seldom used |
| SE/Robust | |
| `vce(`*vcetype*`)` | *vcetype* may be `robust`, `bootstrap` or `jackknife` |
| `robust` | synonym for `vce(robust)` |
| `cluster(`*varname*`)` | adjust standard errors for intragroup correlation |
| `mse1` | force mean squared error to 1 |
| `hc2` | use $u_j^2/(1 - h_{jj})$ as observation's variance |
| `hc3` | use $u_j^2/(1 - h_{jj})^2$ as observation's variance |
| Reporting | |
| `level(`*#*`)` | set confidence level; default is `level(95)` |
| `beta` | report standardized beta coefficients |
| `eform(`*string*`)` | report exponentiated coefficients and label as *string* |
| `noheader` | suppress the table header |
| `plus` | make table extendable |
| `depname(`*varname*`)` | substitute dependent variable name; programmer's option |

*depvar* and the *varlist* following *depvar* may contain time-series operators; see [U] **11.4.3 Time-series varlists**.

`bootstrap`, `by`, `jackknife`, `rolling`, `statsby`, `stepwise`, `svy`, and `xi` are allowed; see [U] **11.1.10 Prefix commands**.

`aweight`s, `fweight`s, `iweight`s, and `pweight`s are allowed; see [U] **11.1.6 weight**.

See [U] **20 Estimation and postestimation commands** for additional capabilities of estimation commands.

# Description

`regress` fits a model of *depvar* on *varlist* using linear regression.

Here is a short list of other regression commands that may be of interest. See [I] **estimation commands** for a complete list.

| | | |
|---|---|---|
| areg | [R] **areg** | an easier way to fit regressions with many dummy variables |
| arch | [TS] **arch** | regression models with ARCH errors |
| arima | [TS] **arima** | ARIMA models |
| boxcox | [R] **boxcox** | Box–Cox regression models |
| cnreg | [R] **cnreg** | censored-normal regression |
| cnsreg | [R] **cnsreg** | constrained linear regression |
| eivreg | [R] **eivreg** | errors-in-variables regression |
| frontier | [R] **frontier** | stochastic frontier models |
| heckman | [R] **heckman** | Heckman selection model |
| intreg | [R] **intreg** | interval regression |
| ivreg | [R] **ivreg** | instrumental variables (2SLS) regression |
| ivtobit | [R] **ivtobit** | tobit regression with endogenous variables |
| newey | [TS] **newey** | regression with Newey–West standard errors |
| qreg | [R] **qreg** | quantile (including median) regression |
| reg3 | [R] **reg3** | three-stage least-squares (3SLS) regression |
| rreg | [R] **rreg** | a type of robust regression |
| sureg | [R] **sureg** | seemingly unrelated regression |
| svy: * | [SVY] **survey** | regression models for survey data |
| tobit | [R] **tobit** | tobit regression |
| treatreg | [R] **treatreg** | treatment-effects model |
| truncreg | [R] **truncreg** | truncated regression |
| xtabond | [XT] **xtabond** | Arellano–Bond linear, dynamic panel-data estimator |
| xtfrontier | [XT] **xtfrontier** | panel-data stochastic frontier models |
| xtgls | [XT] **xtgls** | panel-data GLS models |
| xthtaylor | [XT] **xthtaylor** | Hausman–Taylor estimator for error-components models |
| xtintreg | [XT] **xtintreg** | panel-data interval regression models |
| xtivreg | [XT] **xtivreg** | panel-data instrumental variables (2SLS) regression |
| xtpcse | [XT] **xtpcse** | OLS or Prais–Winsten models with panel-corrected standard errors |
| xtreg | [XT] **xtreg** | fixed- and random-effects linear models |
| xtregar | [XT] **xtregar** | fixed- and random-effects linear models with an AR(1) disturbance |
| xttobit | [XT] **xttobit** | panel-data tobit models |

## Options

Model

noconstant; see [R] **estimation options**.

hascons indicates that a user-defined constant or its equivalent is specified among the independent variables in *varlist*. Some caution is recommended when specifying this option, as resulting estimates may not be as accurate as they otherwise would be. Use of this option requires "sweeping" the constant last, so the moment matrix must be accumulated in absolute rather than deviation form.

This option may be safely specified when the means of the dependent and independent variables are all “reasonable” and there are not large amounts of collinearity between the independent variables. The best procedure is to view `hascons` as a reporting option—estimate with and without `hascons` and verify that the coefficients and standard errors of the variables not affected by the identity of the constant are unchanged.

`tsscons` forces the total sum of squares to be computed as though the model has a constant, that is, as deviations from the mean of the dependent variable. This is a rarely used option that has an effect only when specified with `noconstant`. It affects the total sum of squares and all results derived from the total sum of squares.

SE/Robust

`vce(`*vcetype*`)`; see [R] ***vce_option***.

`robust`, `cluster(`*varname*`)`; see [R] **estimation options**. `cluster()` can be used with `pweights` to produce estimates for unstratified cluster-sampled data, but see [SVY] **svy: regress** for a command especially designed for survey data.

`mse1` is used only in programs and ado-files that employ `regress` to fit models other than linear regression. `mse1` sets the mean squared error to 1, forcing the variance–covariance matrix of the estimators to be $(\mathbf{X}'\mathbf{D}\mathbf{X})^{-1}$ (see *Methods and Formulas* below) and affecting calculated standard errors. Degrees of freedom for $t$ statistics are calculated as $n$ rather than $n-k$.

`hc2` and `hc3` specify an alternate bias correction for the `robust` variance calculation. `hc2` and `hc3` may not be specified with `cluster()`. In the unclustered case, `robust` uses $\widehat{\sigma}_j^2 = \{n/(n-k)\}u_j^2$ as an estimate of the variance of the $j$th observation, where $u_j$ is the calculated residual and $n/(n-k)$ is included to improve the overall estimate's small-sample properties.

`hc2` instead uses $u_j^2/(1-h_{jj})$ as the observation's variance estimate, where $h_{jj}$ is the diagonal element of the hat (projection) matrix. This is unbiased if the model really is homoskedastic. `hc2` tends to produce slightly more conservative confidence intervals.

`hc3` uses $u_j^2/(1-h_{jj})^2$ as suggested by Davidson and MacKinnon (1993), who report that this tends to produce better results when the model really is heteroskedastic. `hc3` produces confidence intervals that tend to be even more conservative.

See Davidson and MacKinnon (1993, 554–556) for more discussion on these two bias corrections. The names `hc2` and `hc3` are based on their notation, page 554; `robust` without `hc2` or `hc3` corresponds to HC1 in their notation.

`hc2` and `hc3` imply `robust`; typing `robust hc2` (or `robust hc3`) is equivalent to typing `hc2` (or `hc3`) by itself.

Reporting

`level(`*#*`)`; see [R] **estimation options**.

`beta` asks that standardized beta coefficients be reported instead of confidence intervals. The beta coefficients are the regression coefficients obtained by first standardizing all variables to have a mean of 0 and a standard deviation of 1. `beta` may not be specified with `cluster()`.

`eform(`*string*`)` is used only in programs and ado-files that employ `regress` to fit models other than linear regression. `eform()` specifies that the coefficient table be displayed in “exponentiated form” as defined in [R] **maximize** and that *string* be used to label the exponentiated coefficients in the table.

`noheader` suppresses the display of the ANOVA table and summary statistics at the top of the output; only the coefficient table is displayed. This option is often used in programs and ado-files.

`plus` specifies that the output table be made extendable. This option is often used in programs and ado-files.

`depname(`*varname*`)` is used only in programs and ado-files that employ `regress` to fit models other than linear regression. `depname()` may be specified only at estimation time. *varname* is recorded as the identity of the dependent variable, even though the estimates are calculated using *depvar*. This affects the labeling of the output—not the results calculated—but could affect subsequent calculations made by `predict`, where the residual would be calculated as deviations from *varname* rather than *depvar*. `depname()` is most typically used when *depvar* is a temporary variable (see [P] **macro**) used as a proxy for *varname*.

# Remarks

Remarks are presented under the headings



`regress` performs linear regression, including ordinary least squares and weighted least squares. For a general discussion of linear regression, see Draper and Smith (1998), Johnston and DiNardo (1997), or Kmenta (1997).

See Wooldridge (2002a) for an excellent treatment of estimation, inference, interpretation, and specification testing in linear regression models. This presentation stands out for its clarification of the statistical issues, as opposed to the algebraic issues. See Wooldridge (2002b, chapter 4) for a more advanced discussion along the same lines.

See Hamilton (2004, chapter 6) for an introduction to linear regression using Stata. Dohoo, Martin, and Stryhn (2003) discuss linear regression using examples from epidemiology, and Stata datasets and do-files used in the text are available.

Chatterjee, Hadi, and Price (2000) explain regression analysis using examples containing typical problems that you might encounter when performing exploratory data analysis. We also recommend Weisberg (1985), who emphasizes the importance of the assumptions of linear regression and problems resulting from these assumptions. For a discussion of model-selection techniques and exploratory data analysis, see Mosteller and Tukey (1977). For a mathematically rigorous treatment, see Peracchi (2001, chapter 6). Finally, see Plackett (1972) if you are interested in the history of regression. Least squares, which dates back to the 1790s, was discovered independently by Legendre and Gauss.

## Ordinary least squares

### ▷ Example 1

Suppose that we have data on the mileage rating and weight of 74 automobiles. The variables in our data are `mpg`, `weight`, and `foreign`. The last variable assumes the value 1 for foreign and 0 for domestic automobiles. We wish to fit the model

$$\texttt{mpg} = \beta_0 + \beta_1\texttt{weight} + \beta_2\texttt{weight}^2 + \beta_3\texttt{foreign} + \epsilon$$

We do this by creating a new variable called `weightsq` and then typing `regress mpg weight weightsq foreign`:

```
. use http://www.stata-press.com/data/r9/auto
(1978 Automobile Data)

. generate weightsq=weight^2

. regress mpg weight weightsq foreign

      Source |       SS       df       MS              Number of obs =      74
-------------+------------------------------           F(  3,    70) =   52.25
       Model |  1689.15372     3  563.05124            Prob > F      =  0.0000
    Residual |   754.30574    70  10.7757963           R-squared     =  0.6913
-------------+------------------------------           Adj R-squared =  0.6781
       Total |  2443.45946    73  33.4720474           Root MSE      =  3.2827

------------------------------------------------------------------------------
         mpg |      Coef.   Std. Err.      t    P>|t|     [95% Conf. Interval]
-------------+----------------------------------------------------------------
      weight |  -.0165729   .0039692    -4.18   0.000    -.0244892   -.0086567
    weightsq |   1.59e-06   6.25e-07     2.55   0.013     3.45e-07    2.84e-06
     foreign |    -2.2035   1.059246    -2.08   0.041      -4.3161   -.0909002
       _cons |   56.53884   6.197383     9.12   0.000     44.17855    68.89913
------------------------------------------------------------------------------
```

◁

`regress` produces a variety of summary statistics along with the table of regression coefficients. At the upper left, `regress` reports an analysis-of-variance (ANOVA) table. The column headings `SS`, `df`, and `MS` stand for "Sum of Squares", "degrees of freedom", and "Mean Square", respectively. In the previous example, the total sum of squares is 2,443.5: 1,689.2 accounted for by the model and 754.3 left unexplained. Since the regression included a constant, the total sum reflects the sum after removal of means, as does the sum of squares due to the model. The table also reveals that there are 73 total degrees of freedom (counted as 74 observations less 1 for the mean removal), of which 3 are consumed by the model, leaving 70 for the residual.

To the right of the ANOVA table are presented other summary statistics. The $F$ statistic associated with the ANOVA table is 52.25. The statistic has 3 numerator and 70 denominator degrees of freedom. The $F$ statistic tests the hypothesis that all coefficients *excluding the constant* are zero. The chance of observing an $F$ statistic that large or larger is reported as 0.0000, which is Stata's way of indicating a number smaller than 0.00005. The $R$-squared ($R^2$) for the regression is 0.6913, and the $R$-squared adjusted for degrees of freedom ($R_a^2$) is 0.6781. The root mean squared error, labeled `Root MSE`, is 3.2827. It is the square root of the mean squared error reported for the residual in the ANOVA table.

Finally, Stata produces a table of the estimated coefficients. The first line of the table indicates that the left-hand-side variable is `mpg`. Thereafter follow the four estimated coefficients. Our fitted model is

$$\texttt{mpg\_hat} = 56.54 - 0.0166\,\texttt{weight} + 1.59 \cdot 10^{-6}\,\texttt{weightsq} - 2.20\,\texttt{foreign}$$

Reported to the right of the coefficients in the output are the standard errors. For instance, the standard error for the coefficient on `weight` is 0.0039692. The corresponding $t$ statistic is $-4.175$, which has a two-sided significance level of 0.000. This number indicates that the significance is less than 0.0005. The 95% confidence interval for the coefficient is $[-.024, -.009]$.

### ▷ Example 2

`regress` shares the features of all estimation commands. Among other things, this means that after running a regression, we can use `test` to test hypotheses about the coefficients, `estat vce` to examine the covariance matrix of the estimators, and `predict` to obtain predicted values, residuals, and influence statistics. See [U] **20 Estimation and postestimation commands**. Options that affect how estimates are displayed, such as `beta` or `level`, can be used when replaying results.

Suppose that we meant to specify the `beta` option to obtain beta coefficients (regression coefficients normalized by the ratio of the standard deviation of the regressor to the standard deviation of the dependent variable). Even though we forgot, we can specify the option now:

```
. regress, beta

      Source |       SS       df       MS              Number of obs =      74
-------------+------------------------------           F(  3,    70) =   52.25
       Model |  1689.15372     3  563.05124           Prob > F      =  0.0000
    Residual |   754.30574    70  10.7757963           R-squared     =  0.6913
-------------+------------------------------           Adj R-squared =  0.6781
       Total |  2443.45946    73  33.4720474           Root MSE      =  3.2827

------------------------------------------------------------------------------
         mpg |      Coef.   Std. Err.      t    P>|t|                     Beta
-------------+----------------------------------------------------------------
      weight |  -.0165729   .0039692    -4.18   0.000                -2.226321
    weightsq |   1.59e-06   6.25e-07     2.55   0.013                  1.32654
     foreign |    -2.2035   1.059246    -2.08   0.041                  -.17527
       _cons |   56.53884   6.197383     9.12   0.000                        .
------------------------------------------------------------------------------
```

◁

## Treatment of the constant

By default, `regress` includes an intercept (constant) term in the model. The `noconstant` option suppresses it, and the `hascons` option tells `regress` that the model already has one.

### ▷ Example 3

We wish to fit a regression of the `weight` of an automobile against its `length`, and we wish to impose the constraint that the weight is zero when the length is zero.

If we simply type `regress weight length`, we are fitting the model

$$\texttt{weight} = \beta_0 + \beta_1\,\texttt{length} + \epsilon$$

In this case, a `length` of zero corresponds to a `weight` of $\beta_0$. We want to force $\beta_0$ to be zero or, equivalently, estimate an equation that does not include an intercept:

$$\texttt{weight} = \beta_1\,\texttt{length} + \epsilon$$

We do this by specifying the `noconstant` option:

```
. regress weight length, noconstant

      Source |       SS       df       MS              Number of obs =      74
-------------+------------------------------           F(  1,    73) = 3450.13
       Model |   703869302     1   703869302           Prob > F      =  0.0000
    Residual |  14892897.8    73  204012.299           R-squared     =  0.9793
-------------+------------------------------           Adj R-squared =  0.9790
       Total |   718762200    74   9713002.7           Root MSE      =  451.68

------------------------------------------------------------------------------
      weight |      Coef.   Std. Err.      t    P>|t|     [95% Conf. Interval]
-------------+----------------------------------------------------------------
      length |   16.29829   .2774752    58.74   0.000     15.74528     16.8513
------------------------------------------------------------------------------
```

In our data, **length** is measured in inches and **weight** in pounds. We discover that each inch of length adds 16 pounds to the weight.

◁

In some cases, there is no need for Stata to include a constant term in the model. Most commonly, this occurs when the model contains a set of mutually exclusive indicator variables. **hascons** is a variation of the **noconstant** option—it tells Stata not to add a constant to the regression because the regression specification already has one, either directly or indirectly.

For instance, we now refit our model of **weight** as a function of **length** and include separate constants for foreign and domestic cars. Our dataset already has an indicator variable called **foreign** marking foreign-made automobiles, so we create a corresponding **domestic** variable and fit the regression:

```
. generate domestic=!foreign

. regress weight length domestic foreign, hascons

      Source |       SS       df       MS              Number of obs =      74
-------------+------------------------------           F(  2,    71) =  316.54
       Model |  39647744.7     2  19823872.3           Prob > F      =  0.0000
    Residual |   4446433.7    71  62625.8268           R-squared     =  0.8992
-------------+------------------------------           Adj R-squared =  0.8963
       Total |  44094178.4    73  604029.841           Root MSE      =  250.25

------------------------------------------------------------------------------
      weight |      Coef.   Std. Err.      t    P>|t|     [95% Conf. Interval]
-------------+----------------------------------------------------------------
      length |   31.44455   1.601234    19.64   0.000     28.25178    34.63732
    domestic |   -2850.25   315.9691    -9.02   0.000    -3480.274   -2220.225
     foreign |  -2983.927   275.1041   -10.85   0.000    -3532.469   -2435.385
------------------------------------------------------------------------------
```

## ❑ Technical Note

There is a subtle distinction between the **hascons** and **noconstant** options. We can most easily reveal it by refitting the last regression, specifying **noconstant** rather than **hascons**:

(*Continued on next page*)

```
. regress weight length domestic foreign, noconstant

      Source |       SS       df       MS              Number of obs =      74
-------------+------------------------------           F(  3,    71) = 3802.03
       Model |  714315766      3  238105255            Prob > F      =  0.0000
    Residual |  4446433.7     71  62625.8268           R-squared     =  0.9938
-------------+------------------------------           Adj R-squared =  0.9936
       Total |  718762200     74  9713002.7            Root MSE      =  250.25

------------------------------------------------------------------------------
      weight |      Coef.   Std. Err.      t    P>|t|     [95% Conf. Interval]
-------------+----------------------------------------------------------------
      length |   31.44455   1.601234    19.64   0.000     28.25178    34.63732
    domestic |   -2850.25   315.9691    -9.02   0.000    -3480.274   -2220.225
     foreign |  -2983.927   275.1041   -10.85   0.000    -3532.469   -2435.385
------------------------------------------------------------------------------
```

Comparing this output with that produced by the previous **regress** command, we see that they are almost, but not quite, identical. The parameter estimates and their associated statistics—the second half of the output—are identical. The overall summary statistics and the ANOVA table—the first half of the output—are different, however.

In the first case, the $R^2$ is shown as 0.8992; in this case, it is shown as 0.9938. In the first case, the $F$ statistic is 316.54; now it is 3802.03. Notice that the numerator degrees of freedom are different as well. In the first case, the numerator degrees of freedom are 2; now they are 3. Which is correct?

Both are. Specifying the **hascons** option causes **regress** to adjust the ANOVA table and its associated statistics for the explanatory power of the constant. The regression in effect has a constant; it is just written in such a way that a separate constant is unnecessary. No such adjustment is made with the **noconstant** option.

❑

## ❑ Technical Note

When the **hascons** option is specified, **regress** checks to make sure the model does in fact have a constant term. If **regress** cannot find a constant term, it automatically adds one. Fitting a model of **weight** on **length** and specifying the **hascons** option, we obtain

```
. regress weight length, hascons
(note: hascons false)

      Source |       SS       df       MS              Number of obs =      74
-------------+------------------------------           F(  1,    72) =  613.27
       Model |  39461306.8     1  39461306.8           Prob > F      =  0.0000
    Residual |  4632871.55    72  64345.4382           R-squared     =  0.8949
-------------+------------------------------           Adj R-squared =  0.8935
       Total |  44094178.4    73  604029.841           Root MSE      =  253.66

------------------------------------------------------------------------------
      weight |      Coef.   Std. Err.      t    P>|t|     [95% Conf. Interval]
-------------+----------------------------------------------------------------
      length |   33.01988   1.333364    24.76   0.000     30.36187    35.67789
       _cons |  -3186.047   252.3113   -12.63   0.000     -3689.02   -2683.073
------------------------------------------------------------------------------
```

Even though we specified **hascons**, **regress** included a constant, anyway. It also added a note to our output: "note: hascons false".

❑

### ❑ Technical Note

Even if the model specification effectively includes a constant term, we need not specify the `hascons` option. `regress` is always on the lookout for collinear variables and drops them as necessary. For instance,

```
. regress weight length domestic foreign
      Source |       SS       df       MS              Number of obs =      74
-------------+------------------------------           F(  2,    71) =  316.54
       Model |  39647744.7     2  19823872.3           Prob > F      =  0.0000
    Residual |   4446433.7    71  62625.8268           R-squared     =  0.8992
-------------+------------------------------           Adj R-squared =  0.8963
       Total |  44094178.4    73  604029.841           Root MSE      =  250.25

------------------------------------------------------------------------------
      weight |      Coef.   Std. Err.      t    P>|t|     [95% Conf. Interval]
-------------+----------------------------------------------------------------
      length |   31.44455   1.601234    19.64   0.000     28.25178    34.63732
    domestic |  (dropped)
     foreign |  -133.6775   77.47615    -1.73   0.089    -288.1605    20.80555
       _cons |   -2850.25   315.9691    -9.02   0.000    -3480.274   -2220.225
------------------------------------------------------------------------------
```

❑

## Robust standard errors

`regress` with the `robust` option substitutes a robust variance matrix calculation for the conventional calculation and, if `cluster()` is specified, allows relaxing the assumption of independence within groups. How this works is explained in [U] **20.14 Obtaining robust variance estimates**. Below we show how well this works.

### ▷ Example 4

Specifying the `robust` option (without `cluster()`) is equivalent to requesting White-corrected standard errors in the presence of heteroskedasticity. We use the automobile data and, in the process of looking at the energy efficiency of cars, analyze a variable with considerable heteroskedasticity.

We will examine the amount of energy—measured in gallons of gasoline—that the cars in the data need to move 1,000 pounds of their weight 100 miles. We are going to examine the relative efficiency of foreign and domestic cars.

```
. gen gpmw = ((1/mpg)/weight)*100*1000

. summarize gpmw

    Variable |       Obs        Mean    Std. Dev.       Min        Max
-------------+--------------------------------------------------------
        gpmw |        74    1.682184    .2426311    1.09553    2.30521
```

In these data, the engines consume between 1.10 and 2.31 gallons of gas to move 1,000 pounds of the car's weight 100 miles. If we ran a regression with conventional standard errors of `gpmw` on `foreign`, we would obtain

(*Continued on next page*)

```
. regress gpmw foreign

      Source |       SS       df       MS              Number of obs =      74
-------------+------------------------------           F(  1,    72) =   20.07
       Model |  .936705572     1  .936705572           Prob > F      =  0.0000
    Residual |  3.36079459    72  .046677703           R-squared     =  0.2180
-------------+------------------------------           Adj R-squared =  0.2071
       Total |  4.29750017    73  .058869865           Root MSE      =  .21605

------------------------------------------------------------------------------
        gpmw |      Coef.   Std. Err.      t    P>|t|     [95% Conf. Interval]
-------------+----------------------------------------------------------------
     foreign |   .2461526   .0549487     4.48   0.000     .1366143    .3556909
       _cons |   1.609004   .0299608    53.70   0.000     1.549278     1.66873
------------------------------------------------------------------------------
```

**regress** with the **robust** option, on the other hand, reports

```
. regress gpmw foreign, robust

Linear regression                                      Number of obs =      74
                                                       F(  1,    72) =   13.13
                                                       Prob > F      =  0.0005
                                                       R-squared     =  0.2180
                                                       Root MSE      =  .21605

------------------------------------------------------------------------------
             |               Robust
        gpmw |      Coef.   Std. Err.      t    P>|t|     [95% Conf. Interval]
-------------+----------------------------------------------------------------
     foreign |   .2461526   .0679238     3.62   0.001     .1107489    .3815563
       _cons |   1.609004   .0234535    68.60   0.000      1.56225    1.655758
------------------------------------------------------------------------------
```

The point estimates are the same (foreign cars need one-quarter gallon more gas), but the standard errors differ by roughly 20 percent. Conventional regression reports the 95% confidence interval as $[.14, .36]$, whereas the robust standard errors make the interval $[.11, .38]$.

Which is right? Notice that **gpmw** is a variable with considerable heteroskedasticity:

```
. tabulate foreign, summarize(gpmw)

            |        Summary of gpmw
   Car type |        Mean   Std. Dev.       Freq.
------------+------------------------------------
   Domestic |   1.6090039   .16845182          52
    Foreign |   1.8551565   .30186861          22
------------+------------------------------------
      Total |   1.6821844   .24263113          74
```

Thus, in this case we favor the robust standard errors. In [U] **20.14 Obtaining robust variance estimates**, we show another example using linear regression where it makes little difference whether we specify **robust**. The linear-regression assumptions were true, and we obtained nearly linear-regression results. The advantage of the robust estimate is that in neither case did we have to check assumptions.

◁

## ❑ Technical Note

**robust**, **regress** purposefully suppresses displaying the ANOVA table when **robust** is specified, as it is no longer appropriate in a statistical sense, even though, mechanically, the numbers would be unchanged. That is, sums of squares remain unchanged, but the meaning of those sums is no longer relevant. The $F$ statistic, for instance, is no longer based on sums of squares; it becomes a Wald test based on the robustly estimated variance matrix. Nevertheless, `regress` continues to report the $R^2$ and the root MSE even though both numbers are based on sums of squares and are, strictly speaking, irrelevant. In this, the root MSE is more in violation of the spirit of the robust estimator than is $R^2$. As a goodness-of-fit statistic, $R^2$ is still fine; just do not use it in formulas to obtain $F$ statistics because those formulas no longer apply. The root MSE is valid in a literal sense—it is the square root of the mean squared error, but it is no longer an estimate of $\sigma$ because there is no single $\sigma$; the variance of the residual varies observation by observation.

❑

## ▷ Example 5

Options `hc2` and `hc3` modify the robust variance calculation. In the context of linear regression without clustering, the idea behind the robust calculation is somehow to measure $\sigma_j^2$, the variance of the residual associated with the $j$th observation, and then to use that estimate to improve the estimated variance of $\widehat{\beta}$. Since residuals have (theoretically and practically) mean 0, one estimate of $\sigma_j^2$ is the observation's squared residual itself—$u_j^2$. A finite-sample correction could improve that by multiplying $u_j^2$ by $n/(n-k)$, and, as a matter of fact, `robust` uses $\{n/(n-k)\}u_j^2$ as its estimate of the residual's variance.

`hc2` and `hc3` use alternative estimators of the observation-specific variances. For instance, if the residuals are homoskedastic, we can show that the expected value of $u_j^2$ is $\sigma^2(1-h_{jj})$, where $h_{jj}$ is the $j$th diagonal element of the projection (hat) matrix. $h_{jj}$ has average value $k/n$, so $1-h_{jj}$ has average value $1-k/n = (n-k)/n$. Thus the default robust estimator $\widehat{\sigma}_j = \{n/(n-k)\}u_j^2$ amounts to dividing $u_j^2$ by the average of the expectation.

`hc2` divides $u_j^2$ by $1-h_{jj}$ itself, so it should yield better estimates if the residuals really are homoskedastic. `hc3` divides $u_j^2$ by $(1-h_{jj})^2$ and has no such clean interpretation. Davidson and MacKinnon (1993) show that $u_j^2/(1-h_{jj})^2$ approximates a more complicated estimator that they obtain by jackknifing (MacKinnon and White 1985).

Here are the results of refitting our efficiency model using `hc2` and `hc3`:

```
. regress gpmw foreign, hc2
Linear regression                                      Number of obs =      74
                                                       F(  1,     72) =   12.93
                                                       Prob > F      =  0.0006
                                                       R-squared     =  0.2180
                                                       Root MSE      =  .21605

------------------------------------------------------------------------------
             |             Robust HC2
        gpmw |      Coef.   Std. Err.      t    P>|t|     [95% Conf. Interval]
-------------+----------------------------------------------------------------
     foreign |   .2461526   .0684669     3.60   0.001     .1096662    .3826389
       _cons |   1.609004   .0233601    68.88   0.000     1.562437    1.655571
------------------------------------------------------------------------------
```

(Continued on next page)

```
. regress gpmw foreign, hc3

Linear regression                                      Number of obs =      74
                                                       F(  1,     72) =   12.38
                                                       Prob > F      =  0.0008
                                                       R-squared     =  0.2180
                                                       Root MSE      =  .21605

------------------------------------------------------------------------------
             |             Robust HC3
        gpmw |      Coef.   Std. Err.      t    P>|t|     [95% Conf. Interval]
-------------+----------------------------------------------------------------
     foreign |   .2461526    .069969     3.52   0.001     .1066719    .3856332
       _cons |   1.609004    .023588    68.21   0.000     1.561982    1.656026
------------------------------------------------------------------------------
```

◁

## ▷ Example 6

With `cluster()`, `robust` is able to relax the assumption of independence. Below we have 28,101 observations on 4,711 women aged 16 to 46. Data were collected on these women between 1970 and 1988. We are going to fit a classic earnings model, and we begin by ignoring the fact that each woman appears an average of 5.96 times in the data:

```
. use http://www.stata-press.com/data/r9/regsmpl
(NLS Women 14-26 in 1968)

. regress ln_wage age age2 tenure

      Source |       SS       df       MS              Number of obs =   28101
-------------+------------------------------           F(  3, 28097) = 1842.45
       Model |  1054.52501     3  351.508335           Prob > F      =  0.0000
    Residual |  5360.43962 28097  .190783344           R-squared     =  0.1644
-------------+------------------------------           Adj R-squared =  0.1643
       Total |  6414.96462 28100  .228290556           Root MSE      =  .43679

------------------------------------------------------------------------------
     ln_wage |      Coef.   Std. Err.      t    P>|t|     [95% Conf. Interval]
-------------+----------------------------------------------------------------
         age |   .0752172   .0034736    21.65   0.000     .0684088    .0820257
        age2 |  -.0010851   .0000575   -18.86   0.000    -.0011979   -.0009724
      tenure |   .0390877   .0007743    50.48   0.000     .0375699    .0406054
       _cons |   .3339821   .0504413     6.62   0.000     .2351148    .4328495
------------------------------------------------------------------------------
```

We can be reasonably certain that the standard errors reported above are meaningless. Without a doubt, a woman with higher-than-average wages in one year typically has higher-than-average wages in other years, and so the residuals are not independent. One way to deal with this would be to fit a random-effects model—and we are going to do that—but first we fit the model using `regress` specifying `robust` and `cluster(id)`, which treats only observations with differing person `id`s as truly independent:

```
. regress ln_wage age age2 tenure, robust cluster(id)
Linear regression                                      Number of obs =   28101
                                                       F(  3,  4698) =  748.82
                                                       Prob > F      =  0.0000
                                                       R-squared     =  0.1644
Number of clusters (idcode) = 4699                     Root MSE      =  .43679

------------------------------------------------------------------------------
             |               Robust
     ln_wage |      Coef.   Std. Err.      t    P>|t|     [95% Conf. Interval]
-------------+----------------------------------------------------------------
         age |   .0752172   .0045711    16.45   0.000     .0662557    .0841788
        age2 |  -.0010851   .0000778   -13.94   0.000    -.0012377   -.0009325
      tenure |   .0390877   .0014425    27.10   0.000     .0362596    .0419157
       _cons |   .3339821   .0641918     5.20   0.000      .208136    .4598282
------------------------------------------------------------------------------
```

For comparison, we focus on the tenure coefficient, which, in economics jargon, can be interpreted as the rate of return for keeping your job. The 95% confidence interval we previously estimated—an interval we do not believe—is [.038, .041]. The robust interval is twice as wide, being [.036, .042].

As we said, one "correct" way to fit this model is by random-effects regression. Here is the random-effects result:

```
. xtreg ln_wage age age2 tenure, re
Random-effects GLS regression                   Number of obs      =     28101
Group variable (i): idcode                      Number of groups   =      4699

R-sq:  within  = 0.1370                         Obs per group: min =         1
       between = 0.2154                                        avg =       6.0
       overall = 0.1608                                        max =        15

Random effects u_i ~ Gaussian                   Wald chi2(3)       =   4717.05
corr(u_i, X)       = 0 (assumed)                Prob > chi2        =    0.0000

------------------------------------------------------------------------------
     ln_wage |      Coef.   Std. Err.      z    P>|z|     [95% Conf. Interval]
-------------+----------------------------------------------------------------
         age |   .0568296   .0026958    21.08   0.000     .0515459    .0621132
        age2 |  -.0007566   .0000447   -16.93   0.000    -.0008441    -.000669
      tenure |   .0260135   .0007477    34.79   0.000     .0245481    .0274789
       _cons |   .6136792   .0394611    15.55   0.000     .5363368    .6910216
-------------+----------------------------------------------------------------
     sigma_u |  .33542449
     sigma_e |  .29674679
         rho |  .56095413   (fraction of variance due to u_i)
------------------------------------------------------------------------------
```

Robust regression estimated the 95% interval [.036, .042], and `xtreg` estimates [.024, .027]. Which is better? The random-effects regression estimator assumes a lot. We can check some of these assumptions by performing a Hausman test. Using `estimates`, we save the random-effects estimation results, and then we run the required fixed-effects regression in order to perform the test.

*(Continued on next page)*

```
. estimates store random
. xtreg ln_wage age age2 tenure, fe
Fixed-effects (within) regression               Number of obs      =     28101
Group variable (i): idcode                      Number of groups   =      4699
R-sq:  within  = 0.1375                         Obs per group: min =         1
       between = 0.2066                                        avg =       6.0
       overall = 0.1568                                        max =        15
                                                F(3,23399)         =   1243.00
corr(u_i, Xb)  = 0.1380                         Prob > F           =    0.0000

     ln_wage       Coef.   Std. Err.      t    P>|t|     [95% Conf. Interval]

         age    .0522751    .002783    18.78   0.000     .0468202      .05773
        age2   -.0006717   .0000461   -14.56   0.000    -.0007621   -.0005813
      tenure     .021738    .000799    27.21   0.000      .020172     .023304
       _cons     .687178   .0405944    16.93   0.000     .6076103    .7667456

     sigma_u   .38743138
     sigma_e   .29674679
         rho   .6302569   (fraction of variance due to u_i)

F test that all u_i=0:     F(4698, 23399) =     7.98         Prob > F = 0.0000
. hausman . random
                 ---- Coefficients ----
                 (b)          (B)            (b-B)     sqrt(diag(V_b-V_B))
                  .          random       Difference          S.E.

         age    .0522751     .0568296       -.0045545        .0006913
        age2   -.0006717    -.0007566        .0000849        .0000115
      tenure     .021738     .0260135       -.0042756        .0002816

                        b = consistent under Ho and Ha; obtained from xtreg
         B = inconsistent under Ha, efficient under Ho; obtained from xtreg
    Test:  Ho:  difference in coefficients not systematic
                  chi2(3) = (b-B)'[(V_b-V_B)^(-1)](b-B)
                          =      336.62
                Prob>chi2 =      0.0000
```

The Hausman test casts grave suspicions on the random-effects model we just fitted, so we should be very careful in interpreting those results.

Meanwhile, our robust regression results still stand, as long as we are careful about the interpretation. The correct interpretation is that, if the data collection were repeated (on women sampled the same way as in the original sample), and if we were to refit the model, 95% of the time we would expect the estimated coefficient on tenure to be in the range [.036, .042].

Even with robust regression, we must be careful about going beyond that statement. In this case, the Hausman test is probably picking up something that differs within and between person, which would cast doubt on our robust regression model in terms of interpreting [.036, .042] to contain the rate of return for keeping a job, economywide, for all women, without exception.

◁

## Weighted regression

`regress` can perform weighted and unweighted regression. We indicate the weight by specifying the [*weight*] qualifier. By default, `regress` assumes analytic weights; see the technical note below.

### ▷ Example 7

We have census data recording the death rate (`drate`) and median age (`medage`) for each state. The data also record the region of the country in which each state is located and the overall population of the state:

```
. use http://www.stata-press.com/data/r9/census9
(1980 Census data by state)

. describe

Contains data from http://www.stata-press.com/data/r9/census9.dta
  obs:            50                          1980 Census data by state
 vars:             5                          13 Jan 2005 22:03
 size:         1,550 (99.9% of memory free)
-------------------------------------------------------------------------------
              storage  display     value
variable name   type   format      label      variable label
-------------------------------------------------------------------------------
state           str14  %-14s                  State
drate           float  %9.0g                  Death Rate
pop             long   %12.0gc                Population
medage          float  %9.2f                  Median age
region          byte   %-8.0g      cenreg     Census region
-------------------------------------------------------------------------------
Sorted by:
```

We can use the `xi` command to automatically create and include dummy variables for region. Since the variables in the regression reflect means rather than individual observations, the appropriate method of estimation is analytically weighted least squares (Johnston and DiNardo 1997), where the weight is total population:

```
. xi: regress drate medage I.region [w=pop]
I.region          _Iregion_1-4        (naturally coded; _Iregion_1 omitted)
(analytic weights assumed)
(sum of wgt is   2.2591e+08)

      Source |       SS       df       MS              Number of obs =      50
-------------+------------------------------           F(  4,    45) =   37.21
       Model |  4096.6093     4  1024.15232            Prob > F      =  0.0000
    Residual |  1238.40987    45  27.5202192           R-squared     =  0.7679
-------------+------------------------------           Adj R-squared =  0.7472
       Total |  5335.01916    49  108.877942           Root MSE      =   5.246

------------------------------------------------------------------------------
       drate |      Coef.   Std. Err.      t    P>|t|     [95% Conf. Interval]
-------------+----------------------------------------------------------------
      medage |   4.283183   .5393329     7.94   0.000     3.196911    5.369455
  _Iregion_2 |   .3138738   2.456431     0.13   0.899    -4.633632     5.26138
  _Iregion_3 |  -1.438452   2.320244    -0.62   0.538    -6.111663    3.234758
  _Iregion_4 |  -10.90629   2.681349    -4.07   0.000    -16.30681   -5.505777
       _cons |  -39.14727   17.23613    -2.27   0.028    -73.86262   -4.431915
------------------------------------------------------------------------------
```

To weight the regression by population, we added the qualifier `[w=pop]` to the end of the `regress` command. Our qualifier was vague (we did not say `[aweight=pop]`), but unless told otherwise, Stata assumes analytic weights in the case of `regress`. Stata informed us that the sum of the weight is $2.2591 \cdot 10^8$; there were approximately 226 million people residing in the U.S. according to our 1980 data.

`xi` provides one way to include dummy variables and can be used with any estimation command. In the special case of linear regression, another alternative would be to use `anova` with the `regress` option. This would probably be better, but only because `anova` has special logic to fit models with many dummy variables, which uses less memory and computer time.

◁

### ❑ Technical Note

Once we fit a weighted regression, we can obtain the appropriately weighted variance–covariance matrix of the estimators using `estat vce` and perform appropriately weighted hypothesis tests using `test`.

In the weighted regression in the previous example, we see that `_Iregion_4` is statistically significant but that `_Iregion_2` and `_Iregion_3` are not. We use `test` to test the joint significance of the region variables:

```
. test _Iregion_2 _Iregion_3 _Iregion_4
 ( 1)  _Iregion_2 = 0
 ( 2)  _Iregion_3 = 0
 ( 3)  _Iregion_4 = 0
       F(  3,    45) =    9.84
            Prob > F =    0.0000
```

The results indicate that the region variables are jointly significant.

❑

`regress` also accepts frequency weights (`fweights`). Frequency weights are appropriate when the data do not reflect cell means, but instead represent replicated observations. Specifying `aweights` or `fweights` will not change the parameter estimates, but it will change the corresponding significance levels.

For instance, if we specified `[fweight=pop]` in the weighted regression example above—which would be statistically incorrect—Stata would treat the data as if the data represented 226 million independent observations on death rates and median age. The data most certainly do not represent that—they represent 50 observations on state averages.

With `aweights`, Stata treats the number of observations on the process as the number of observations in the data. When we specify `fweights`, Stata treats the number of observations as if it were equal to the sum of the weights; see *Methods and Formulas* below.

### ❑ Technical Note

A popular request on the help line is to describe the effect of specifying `[aweight=`*exp*`]` with `regress` in terms of transformation of the dependent and independent variables. The mechanical answer is that typing

```
. regress y x1 x2 [aweight=n]
```

is equivalent to fitting the model

$$y_j\sqrt{n_j} = \beta_0\sqrt{n_j} + \beta_1 x_{1j}\sqrt{n_j} + \beta_2 x_{2j}\sqrt{n_j} + u_j\sqrt{n_j}$$

This regression will reproduce the coefficients and covariance matrix produced by the aweighted regression. The mean squared errors (estimates of the variance of the residuals) will, however, be different. The transformed regression reports $s_t^2$, an estimate of $\mathrm{Var}(u_j\sqrt{n_j})$. The aweighted regression reports $s_a^2$, an estimate of $\mathrm{Var}(u_j\sqrt{n_j}\sqrt{N/\sum_k n_k})$, where $N$ is the number of observations. Thus

$$s_a^2 = \frac{N}{\sum_k n_k} s_t^2 = \frac{s_t^2}{\overline{n}} \tag{1}$$

The logic for this adjustment is as follows: Consider the model

$$y = \beta_0 + \beta_1 x_1 + \beta_2 x_2 + u$$

Assume that, were this model fitted on individuals, $\mathrm{Var}(u) = \sigma_u^2$, a constant. Assume that individual data are not available; what is available are averages $(\overline{y}_j, \overline{x}_{1j}, \overline{x}_{2j})$ for $j = 1, \dots, N$, and each average is calculated over $n_j$ observations. Then it is still true that

$$\overline{y}_j = \beta_0 + \beta_1 \overline{x}_{1j} + \beta_2 \overline{x}_{2j} + \overline{u}_j$$

where $\overline{u}_j$ is the average of $n_j$ mean 0, variance $\sigma_u^2$ deviates and has variance $\sigma_{\overline{u}}^2 = \sigma_u^2/n_j$. Thus multiplying through by $\sqrt{n_j}$ produces

$$\overline{y}_j\sqrt{n_j} = \beta_0\sqrt{n_j} + \beta_1\overline{x}_{1j}\sqrt{n_j} + \beta_2\overline{x}_{2j}\sqrt{n_j} + \overline{u}_j\sqrt{n_j}$$

and $\mathrm{Var}(\overline{u}_j\sqrt{n_j}) = \sigma_u^2$. The mean squared error, $s_t^2$, reported by fitting this transformed regression is an estimate of $\sigma_u^2$. Alternatively, the coefficients and covariance matrix could be obtained by aweighted regress. The only difference would be in the reported mean squared error, which from (1) is $\sigma_u^2/\overline{n}$. On average, each observation in the data reflects the averages calculated over $\overline{n} = \sum_k n_k/N$ individuals, and thus this reported mean squared error is the average variance of an observation in the dataset. We can retrieve the estimate of $\sigma_u^2$ by multiplying the reported mean squared error by $\overline{n}$.

More generally, aweights are used to solve general heteroskedasticity problems. In these cases, we have the model

$$y_j = \beta_0 + \beta_1 x_{1j} + \beta_2 x_{2j} + u_j$$

and the variance of $u_j$ is thought to be proportional to $a_j$. If the variance is proportional to $a_j$, it is also proportional to $\alpha a_j$, where $\alpha$ is any positive constant. Not quite arbitrarily, but with no loss of generality, we could choose $\alpha = \sum_k (1/a_k)/N$, the average value of the inverse of $a_j$. We can then write $\mathrm{Var}(u_j) = k\alpha a_j\sigma^2$, where $k$ is the constant of proportionality that is no longer a function of the scale of the weights.

Dividing this regression through by the $\sqrt{a_j}$,

$$y_j/\sqrt{a_j} = \beta_0/\sqrt{a_j} + \beta_1 x_{1j}/\sqrt{a_j} + \beta_2 x_{2j}/\sqrt{a_j} + u_j/\sqrt{a_j}$$

produces a model with $\mathrm{Var}(u_j/\sqrt{a_j}) = k\alpha\sigma^2$, which is the constant part of $\mathrm{Var}(u_j)$. Notice in particular that this variance is a function of $\alpha$, the average of the reciprocal weights; if the weights are scaled arbitrarily, then so is this variance.

We can also fit this model by typing

```
. regress y x1 x2 [aweight=1/a]
```

This will produce the same estimates of the coefficients and covariance matrix; the reported mean squared error is, from (1), $\{N/\sum_k(1/a_k)\}k\alpha\sigma^2 = k\sigma^2$. Note that this variance is independent of the scale of $a_j$.

❑

### Instrumental variables and two-stage least-squares regression

An alternate syntax for `regress` can be used to produce instrumental variables (two-stage least squares) estimates.

`regress` *depvar* [ *varlist*$_1$ [ (*varlist*$_2$) ] ] [ *if* ] [ *in* ] [ *weight* ] [ , *regress_options* ]

This syntax is mainly used by programmers developing estimators using the instrumental variables estimates as intermediate results. `ivreg` is normally used to directly fit these models; see [R] **ivreg**.

With this syntax, `regress` fits a structural equation of *depvar* on *varlist*$_1$ using instrumental variables regression; (*varlist*$_2$) indicates the list of instrumental variables. With the exception of `hc2` and `hc3`, all standard `regress` options are allowed.

## Saved Results

`regress` saves in `e()`:

| | | | |
|---|---|---|---|
| Scalars | | | |
| `e(N)` | number of observations | `e(r2_a)` | adjusted $R$-squared |
| `e(mss)` | model sum of squares | `e(F)` | $F$ statistic |
| `e(df_m)` | model degrees of freedom | `e(rmse)` | root mean squared error |
| `e(rss)` | residual sum of squares | `e(ll)` | log likelihood |
| `e(df_r)` | residual degrees of freedom | `e(ll_0)` | log likelihood, constant-only model |
| `e(r2)` | $R$-squared | `e(N_clust)` | number of clusters |
| Macros | | | |
| `e(cmd)` | `regress` | `e(vce)` | *vcetype* specified in `vce()` |
| `e(depvar)` | name of dependent variable | `e(vcetype)` | title used to label Std. Err. |
| `e(model)` | `ols` or `iv` | `e(properties)` | `b V` |
| `e(wtype)` | weight type | `e(estat_cmd)` | program used to implement `estat` |
| `e(wexp)` | weight expression | `e(predict)` | program used to implement `predict` |
| `e(clustvar)` | name of cluster variable | | |
| Matrices | | | |
| `e(b)` | coefficient vector | `e(V)` | variance–covariance matrix of the estimators |
| Functions | | | |
| `e(sample)` | marks estimation sample | | |

## Methods and Formulas

Variables printed in lowercase and not boldfaced (e.g., $x$) are scalars. Variables printed in lowercase and boldfaced (e.g., $\mathbf{x}$) are column vectors. Variables printed in uppercase and boldfaced (e.g., $\mathbf{X}$) are matrices.

Let $\mathbf{v}$ be a column vector of weights specified by the user. If no weights are specified, $\mathbf{v} = \mathbf{1}$. Let $\mathbf{w}$ be a column vector of normalized weights. If no weights are specified or if the user specified `fweights` or `iweights`, $\mathbf{w} = \mathbf{v}$. Otherwise, $\mathbf{w} = \{\mathbf{v}/(\mathbf{1}'\mathbf{v})\}(\mathbf{1}'\mathbf{1})$.

The *number of observations*, $n$, is defined as $\mathbf{1}'\mathbf{w}$. In the case of `iweights`, this is truncated to an integer. The *sum of the weights* is $\mathbf{1}'\mathbf{v}$. Define $c = 1$ if there is a constant in the regression and zero otherwise. Define $k$ as the number of right-hand-side (rhs) variables (including the constant).

Let $\mathbf{X}$ denote the matrix of observations on the rhs variables, $\mathbf{y}$ the vector of observations on the left-hand-side (lhs) variable, and $\mathbf{Z}$ the matrix of observations on the instruments. If the user specifies no instruments, then $\mathbf{Z} = \mathbf{X}$. In the following formulas, if the user specifies weights, then $\mathbf{X}'\mathbf{X}$, $\mathbf{X}'\mathbf{y}$, $\mathbf{y}'\mathbf{y}$, $\mathbf{Z}'\mathbf{Z}$, $\mathbf{Z}'\mathbf{X}$, and $\mathbf{Z}'\mathbf{y}$ are replaced by $\mathbf{X}'\mathbf{D}\mathbf{X}$, $\mathbf{X}'\mathbf{D}\mathbf{y}$, $\mathbf{y}'\mathbf{D}\mathbf{y}$, $\mathbf{Z}'\mathbf{D}\mathbf{Z}$, $\mathbf{Z}'\mathbf{D}\mathbf{X}$, and $\mathbf{Z}'\mathbf{D}\mathbf{y}$, respectively, where $\mathbf{D}$ is a diagonal matrix whose diagonal elements are the elements of $\mathbf{w}$. We suppress the $\mathbf{D}$ below to simplify the notation.

If no instruments are specified, define $\mathbf{A}$ as $\mathbf{X}'\mathbf{X}$ and $\mathbf{a}$ as $\mathbf{X}'\mathbf{y}$. Otherwise, define $\mathbf{A}$ as $\mathbf{X}'\mathbf{Z}(\mathbf{Z}'\mathbf{Z})^{-1}(\mathbf{X}'\mathbf{Z})'$ and $\mathbf{a}$ as $\mathbf{X}'\mathbf{Z}(\mathbf{Z}'\mathbf{Z})^{-1}\mathbf{Z}'\mathbf{y}$.

The coefficient vector $\mathbf{b}$ is defined as $\mathbf{A}^{-1}\mathbf{a}$. Although not shown in the notation, unless `hascons` is specified, $\mathbf{A}$ and $\mathbf{a}$ are accumulated in deviation form and the constant is calculated separately. This comment applies to all statistics listed below.

The *total sum of squares*, TSS, equals $\mathbf{y}'\mathbf{y}$ if there is no intercept and $\mathbf{y}'\mathbf{y} - \left\{(\mathbf{1}'\mathbf{y})^2/n\right\}$ otherwise. The *degrees of freedom* are $n - c$.

The *error sum of squares*, ESS, is defined as $\mathbf{y}'\mathbf{y} - 2\mathbf{b}\mathbf{X}'\mathbf{y} + \mathbf{b}'\mathbf{X}'\mathbf{X}\mathbf{b}$ if there are instruments and as $\mathbf{y}'\mathbf{y} - \mathbf{b}'\mathbf{X}'\mathbf{y}$ otherwise. The *degrees of freedom* are $n - k$.

The *model sum of squares*, MSS, equals TSS − ESS. The *degrees of freedom* are $k - c$.

The *mean squared error*, $s^2$, is defined as $\text{ESS}/(n-k)$. The *root mean squared error* is $s$, its square root.

The $F$ statistic with $k - c$ and $n - k$ degrees of freedom is defined as

$$F = \frac{\text{MSS}}{(k-c)s^2}$$

if no instruments are specified. If instruments are specified and $c = 1$, then $F$ is defined as

$$F = \frac{(\mathbf{b}-\mathbf{c})'\mathbf{A}(\mathbf{b}-\mathbf{c})}{(k-1)s^2}$$

where $\mathbf{c}$ is a vector of $k - 1$ zeros and $k$th element $\mathbf{1}'\mathbf{y}/n$. Otherwise, $F$ is defined as *missing*. (In this case, you may use the `test` command to construct any $F$ test you wish.)

The *R-squared*, $R^2$, is defined as $R^2 = 1 - \text{ESS}/\text{TSS}$.

The *adjusted R-squared*, $R^2_a$, is $1 - (1 - R^2)(n-c)/(n-k)$.

If `robust` is not specified, the conventional estimate of variance is $s^2\mathbf{A}^{-1}$. The handling of `robust` is described below.

## A general notation for the robust variance calculation

Put aside all context of linear regression and the notation that goes with it—we will return to it. First, we are going to establish a notation for describing robust variance calculations.

The calculation formula for the robust variance calculation is

$$\widehat{\mathcal{V}} = q_c \widehat{\mathbf{V}} \Big( \sum_{k=1}^{M} \mathbf{u}_k^{(G)\prime} \mathbf{u}_k^{(G)} \Big) \widehat{\mathbf{V}}$$

where

$$\mathbf{u}_k^{(G)} = \sum_{j \in G_k} w_j \mathbf{u}_j$$

$G_1, G_2, \ldots, G_M$ are the clusters specified by `cluster()`, and $w_j$ are the user-specified weights, normalized if `fweights` are specified and equal to 1 if no weights are specified. (In the case of `fweights`, the formula for $\widehat{\mathcal{V}}$ is modified to produce the same results as if the dataset were expanded and the calculation made on unweighted data, meaning that $w_k^{(G)} = \sum_{j \in G_k} w_j$ is introduced into the denominator.)

If `cluster()` is not specified, $M = N$, and each cluster contains one observation. The inputs into this calculation are

1. $\widehat{\mathbf{V}}$, which is typically a conventionally calculated variance matrix;
2. $\mathbf{u}_j$, $j = 1, \ldots, N$, a row vector of scores; and
3. $q_\mathrm{c}$, a constant finite-sample adjustment.

Thus we can now describe how estimators apply the robust calculation formula by defining $\widehat{\mathbf{V}}$, $\mathbf{u}_j$, and $q_\mathrm{c}$.

Two definitions are popular enough for $q_\mathrm{c}$ to deserve a name. The regression-like formula for $q_\mathrm{c}$ (Fuller et al. 1986) is

$$q_\mathrm{c} = \frac{N-1}{N-k}\,\frac{M}{M-1}$$

where $M$ is the number of clusters and $N$ is the number of observations. In the case of weights, $N$ refers to the sum of the weights if weights are frequency weights and the number of observations in the dataset (ignoring weights) in all other cases. Also note that, weighted or not, $M = N$ when `cluster()` is not specified, and, in that case, $q_\mathrm{c} = N/(N-k)$.

The asymptotic-like formula for $q_\mathrm{c}$ is

$$q_\mathrm{c} = \frac{M}{M-1}$$

where $M = N$ if `cluster()` is not specified.

See [U] **20.14 Obtaining robust variance estimates** and [P] **_robust** for a discussion of the robust variance estimator and a development of these formulas.

### Robust calculation for regress

In the case of `regress`, $\widehat{\mathbf{V}} = \mathbf{A}^{-1}$. The other terms are

No instruments, `robust`, but not `hc2` or `hc3`,

$$\mathbf{u}_j = (y_j - \mathbf{x}_j\mathbf{b})\mathbf{x}_j$$

and $q_\mathrm{c}$ is given by its regression-like definition.

No instruments, `hc2`,

$$\mathbf{u}_j = \frac{1}{\sqrt{1-h_{jj}}}(y_j - \mathbf{x}_j\mathbf{b})\mathbf{x}_j$$

where $q_\mathrm{c} = 1$ and $h_{jj} = \mathbf{x}_j(\mathbf{X}'\mathbf{X})^{-1}\mathbf{x}_j{}'$.

No instruments, `hc3`,

$$\mathbf{u}_j = \frac{1}{1-h_{jj}}(y_j - \mathbf{x}_j\mathbf{b})\mathbf{x}_j$$

where $q_c = 1$ and $h_{jj} = \mathbf{x}_j(\mathbf{X}'\mathbf{X})^{-1}\mathbf{x}_j{}'$.

Instrumental variables,

$$\mathbf{u}_j = (y_j - \mathbf{x}_j\mathbf{b})\widehat{\mathbf{x}}_j$$

where $q_c$ is given by its regression-like definition, and

$$\widehat{\mathbf{x}}_j' = \mathbf{P}\mathbf{z}_j{}'$$

where $\mathbf{P} = (\mathbf{X}'\mathbf{Z})(\mathbf{Z}'\mathbf{Z})^{-1}$.

## Acknowledgments

The robust estimate of variance was first implemented in Stata by Mead Over, Dean Jolliffe, and Andrew Foster (1996).

## References

Alexandersson, A. 1998. gr32: Confidence ellipses. *Stata Technical Bulletin* 46: 10–13. Reprinted in *Stata Technical Bulletin Reprints*, vol. 8, pp. 54–57.

Chatterjee, S., A. S. Hadi, and B. Price. 2000. *Regression Analysis by Example*. 3rd ed. New York: Wiley.

Davidson, R. and J. G. MacKinnon. 1993. *Estimation and Inference in Econometrics*. New York: Oxford University Press.

Dohoo, I., W. Martin, and H. Stryhn. 2003. *Veterinary Epidemiologic Research*. Charlottetown, Prince Edward Island: AVC.

Draper, N. and H. Smith. 1998. *Applied Regression Analysis*. 3rd ed. New York: Wiley.

Dunnington, G. W. 2004. *Gauss: Titan of Science*. Washington, DC: Mathematical Association of America.

Fuller, W. A., W. J. Kennedy, D. Schnell, G. Sullivan, H. J. Park. 1986. *PC Carp*. Ames, IA: Statistical Laboratory, Iowa State University.

Gillham, N. W. 2001. *A Life of Sir Francis Galton: From African Exploration to the Birth of Eugenics*. New York: Oxford University Press.

Gillispie, C. C. 1997. *Pierre-Simon Laplace, 1749–1827: A Life in Exact Science*. Princeton: Princeton University Press.

Hald, A. 1998. *A History of Mathematical Statistics from 1750 to 1930*. New York: Wiley.

Hamilton, L. C. 2004. *Statistics with Stata*. Belmont, CA: Brooks/Cole.

Johnston, J. and J. DiNardo. 1997. *Econometric Methods*. 4th ed. New York: McGraw–Hill.

Kmenta, J. 1997. *Elements of Econometrics*. 2nd ed. Ann Arbor: University of Michigan Press.

Long, J. S. and J. Freese. 2000. sg152: Listing and interpreting transformed coefficients from certain regression models. *Stata Technical Bulletin* 57: 27–34. Reprinted in *Stata Technical Bulletin Reprints*, vol. 10, pp. 231–240.

MacKinnon, J. G. and H. White. 1985. Some heteroskedasticity consistent covariance matrix estimators with improved finite sample properties. *Journal of Econometrics* 29: 305–325.

Mosteller, F. and J. W. Tukey. 1977. *Data Analysis and Regression*. Reading, MA: Addison–Wesley.

Over, M., D. Jolliffe, and A. Foster. 1996. sg46: Huber correction for two-stage least-squares estimates. *Stata Technical Bulletin* 29: 24–25. Reprinted in *Stata Technical Bulletin Reprints*, vol. 5, pp. 140–142.

Peracchi, F. 2001. *Econometrics*. Chichester, UK: Wiley.

Plackett, R. L. 1972. The discovery of the method of least squares. *Biometrika* 59: 239–251.

Rogers, W. H. 1991. smv2: Analyzing repeated measurements—some practical alternatives. *Stata Technical Bulletin* 4: 10–16. Reprinted in *Stata Technical Bulletin Reprints*, vol. 1, pp. 123–131.

Royston, P. and G. Ambler. 1998. sg79: Generalized additive models. *Stata Technical Bulletin* 42: 38–43. Reprinted in *Stata Technical Bulletin Reprints*, vol. 7, pp. 217–224.

Stigler, S. M. 1986. *The History of Statistics: The Measurement of Uncertainty before 1900*. Cambridge, MA: Harvard University Press.

Tyler, J. H. 1997. sg73: Table making program. *Stata Technical Bulletin* 40: 18–23. Reprinted in *Stata Technical Bulletin Reprints*, vol. 7, pp. 186–192.

Weesie, J. 1998. sg77: Regression analysis with multiplicative heteroskedasticity. *Stata Technical Bulletin* 42: 28–32. Reprinted in *Stata Technical Bulletin Reprints*, vol. 7, pp. 204–210.

Weisberg, S. 1985. *Applied Linear Regression*. 2nd ed. New York: Wiley.

Wooldridge, J. M. 2002a. *Introductory Econometrics: A Modern Approach*. 2nd ed. Cincinnati, OH: South-Western.

——. 2002b. *Econometric Analysis of Cross Section and Panel Data*. Cambridge, MA: MIT Press.

Zimmerman, F. 1998. sg93: Switching regressions. *Stata Technical Bulletin* 45: 30–33. Reprinted in *Stata Technical Bulletin Reprints*, vol. 8, pp. 183–186.

## Also See

**Complementary:** [R] **regress postestimation**, [R] **regress postestimation time series**

**Related:** [R] **anova**, [R] **areg**, [R] **cnsreg**, [R] **heckman**, [R] **ivreg**, [R] **mvreg**, [R] **qreg**, [R] **reg3**, [R] **rreg**, [R] **sureg**, [R] **tobit**, [R] **truncreg**, [P] **_robust**, [SVY] **svy: regress**, [TS] **arch**, [TS] **arima**, [TS] **newey**, [TS] **prais**, [XT] **xtgee**, [XT] **xtgls**, [XT] **xtintreg**, [XT] **xtivreg**, [XT] **xtpcse**, [XT] **xtreg**, [XT] **xtregar**, [XT] **xttobit**

**Background:** [U] **11.1.10 Prefix commands**, [U] **20 Estimation and postestimation commands**, [R] **estimation options**, [R] ***vce_option***

*(Continued on next page)*

The history of regression is long and complicated: the books by Stigler (1986) and Hald (1998) are largely devoted to the story. Legendre published first on least squares in 1805. Gauss published later in 1809, but he had the idea earlier. Gauss, and especially Laplace, tied least squares to a normal errors assumption. The idea of the normal distribution can itself be traced back to De Moivre in 1733. Laplace discussed a variety of other estimation methods and error assumptions over his long career, while linear models long predate either innovation. Most of this work was linked to problems in astronomy and geodesy.

A second wave of ideas started when Galton used graphical and descriptive methods on data bearing on heredity to develop what he called regression. His term reflects the common phenomenon that characteristics of offspring are positively correlated with those of parents, but with regression slope such that offspring "regress towards the mean". Galton's work was rather intuitive: contributions from Pearson, Edgeworth, Yule, and others introduced more formal machinery, developed related ideas on correlation, and extended application into the biological and social sciences. So most of the elements of regression as we know it were in place by 1900.

Pierre-Simon Laplace (1749–1827) was born in Normandy and was early recognized as a remarkable mathematician. He weathered a changing political climate well enough to rise to Minister of the Interior under Napoleon in 1799 (although only for 6 weeks) and to be made a Marquis by Louis XVIII in 1817. He made many contributions to mathematics and physics, his two main interests being theoretical astronomy and probability theory (including statistics). Laplace transforms are named for him.

Adrien-Marie Legendre (1752–1833) was born in Paris (or possibly in Toulouse) and educated in mathematics and physics. He worked in number theory, geometry, differential equations, calculus, function theory, applied mathematics, and geodesy. The Legendre polynomials are named for him. His main contribution to statistics is as one of the discoverers of least squares. He died in poverty, having refused to bow to political pressures.

Johann Carl Friedrich Gauss (1777–1855) was born in Braunschweig (Brunswick), now in Germany. He studied there and at Göttingen. His doctoral dissertation at the University of Helmstedt was a discussion of the fundamental theorem of algebra. He made many fundamental contributions to geometry, number theory, algebra, real analysis, differential equations, numerical analysis, statistics, astronomy, optics, geodesy, mechanics, and magnetism. An outstanding genius, Gauss worked mostly in isolation in Göttingen.

Francis Galton (1822–1911) was born in Birmingham, England, into a well-to-do family with many connections: he and Charles Darwin were first cousins. After an unsuccessful foray into medicine, he became independently wealthy at the death of his father. Galton traveled widely in Europe, the Middle East, and Africa, and became celebrated as an explorer and geographer. His pioneering work on weather maps helped in the identification of anticyclones, which he named. From about 1865, most of his work was centered on quantitative problems in biology, anthropology, and psychology. In a sense, Galton (re-)invented regression, and he certainly named it. Galton also promoted the normal distribution, correlation approaches, and the use of median and selected quantiles as descriptive statistics. He was knighted in 1909.

# Title

**regress postestimation** — Postestimation tools for regress

# Description

The following postestimation commands are of special interest after `regress`:

| command | description |
|---|---|
| `dfbeta` | DFBETA influence statistics |
| `estat hettest` | tests for heteroskedasticity |
| `estat imtest` | information matrix test |
| `estat ovtest` | Ramsey regression specification-error test for omitted variables |
| `estat szroeter` | Szroeter's rank test for heteroskedasticity |
| `estat vif` | variance inflation factors for the independent variables |
| `acprplot` | augmented component-plus-residual plot |
| `avplot` | added-variable plot |
| `avplots` | all added-variables plots in a single image |
| `cprplot` | component-plus-residual plot |
| `lvr2plot` | leverage-versus-squared-residual plot |
| `rvfplot` | residual-versus-fitted plot |
| `rvpplot` | residual-versus-predictor plot |

For information about these commands, see below.

In addition, the following standard postestimation commands are available:

(*Continued on next page*)

| command | description |
|---|---|
| `adjust` | adjusted predictions of $\mathbf{x}\boldsymbol{\beta}$ |
| `estat` | AIC, BIC, VCE, and estimation sample summary |
| `estimates` | cataloging estimation results |
| `hausman` | Hausman's specification test |
| `lincom` | point estimates, standard errors, testing, and inference for linear combinations of coefficients |
| `linktest` | link test for model specification |
| `lrtest` | likelihood-ratio test |
| `mfx` | marginal effects or elasticities |
| `nlcom` | point estimates, standard errors, testing, and inference for nonlinear combinations of coefficients |
| `predict` | predictions, residuals, influence statistics, and other diagnostic measures |
| `predictnl` | point estimates, standard errors, testing, and inference for generalized predictions |
| `suest` | seemingly unrelated estimation |
| `test` | Wald tests for simple and composite linear hypotheses |
| `testnl` | Wald tests of nonlinear hypotheses |

See the corresponding entries in the *Stata Base Reference Manual* for details.

For postestimation tests specific to time series, see [R] **regress postestimation time series**.

## Special-interest postestimation commands

These commands provide tools for diagnosing sensitivity to individual observations, analyzing residuals, and assessing specification.

`dfbeta` will calculate one, more than one, or all the DFBETAs after `regress`. Although `predict` will also calculate DFBETAs, `predict` can do this for only one variable at a time. `dfbeta` is a convenience tool for those who want to calculate DFBETAs for multiple variables. The names for the new variables created are chosen automatically and begin with the letters `DF`.

`estat hettest` performs two flavors of the Breusch–Pagan (1979) and Cook and Weisberg (1983) test for heteroskedasticity. This test amounts to testing $\mathbf{t} = \mathbf{0}$ in $\text{Var}(e) = \sigma^2\exp(\mathbf{zt})$. If *varlist* is not specified, the fitted values are used for $\mathbf{z}$. If *varlist* or the option `rhs` is specified, the variables specified are used for $\mathbf{z}$.

`estat imtest` performs an information matrix test for the regression model and an orthogonal decomposition into tests for heteroskedasticity, skewness, and kurtosis due to Cameron and Trivedi (1990); White's test for homoskedasticity against unrestricted forms of heteroskedasticity (1980) is available as an option. White's test is usually very similar to the first term of the Cameron–Trivedi decomposition.

`estat ovtest` performs two flavors of the Ramsey (1969) regression specification error test (RESET) for omitted variables. This test amounts to fitting $y = \mathbf{xb} + \mathbf{zt} + u$ and then testing $\mathbf{t} = \mathbf{0}$. If option `rhs` is not specified, powers of the fitted values are used for $\mathbf{z}$. If `rhs` is specified, powers of the individual elements of $\mathbf{x}$ are used.

`estat szroeter` performs Szroeter's rank test for heteroskedasticity for each of the variables in *varlist* or for the explanatory variables of the regression if `rhs` is specified.

`estat vif` calculates the variance inflation factors (VIFs) for the independent variables specified in a linear regression model.

`acprplot` graphs an augmented component-plus-residual plot (a.k.a. augmented partial residual plot) as described by Mallows (1986). This seems to work better than the component-plus-residual plot for identifying nonlinearities in the data.

`avplot` graphs an added-variable plot (a.k.a. partial-regression leverage plot, partial regression plot, or adjusted partial residual plot) after `regress`. *indepvar* may be an independent variable (a.k.a. predictor, carrier, or covariate) that is currently in the model or not.

`avplots` graphs all the added-variable plots in a single image.

`cprplot` graphs a component-plus-residual plot (a.k.a. partial residual plot) after `regress`. *indepvar* must be an independent variable that is currently in the model.

`lvr2plot` graphs a leverage-versus-squared-residual plot (a.k.a. L-R plot).

`rvfplot` graphs a residual-versus-fitted plot, a graph of the residuals against the fitted values.

`rvpplot` graphs a residual-versus-predictor plot (a.k.a. independent variable plot or carrier plot), a graph of the residuals against the specified predictor.

(*Continued on next page*)

## Syntax for predict

`predict` [*type*] *newvar* [*if*] [*in*] [, *statistic*]

| *statistic* | description |
|---|---|
| `xb` | $\mathbf{x}_j\mathbf{b}$, fitted values; the default |
| `residuals` | residuals |
| `score` | score; equivalent to `residuals` |
| `rstandard` | standardized residuals |
| `rstudent` | studentized (jackknifed) residuals |
| `cooksd` | Cook's distance |
| `leverage` \| `hat` | leverage (diagonal elements of hat matrix) |
| `pr(`*a*`,`*b*`)` | $\Pr(y_j \mid a < y_j < b)$ |
| `e(`*a*`,`*b*`)` | $E(y_j \mid a < y_j < b)$ |
| `ystar(`*a*`,`*b*`)` | $E(y_j^*)$, $y_j^* = \max\{a, \min(y_j, b)\}$ |
| * `dfbeta(`*varname*`)` | DFBETA for *varname* |
| `stdp` | standard error of the prediction |
| `stdf` | standard error of the forecast |
| `stdr` | standard error of the residual |
| * `covratio` | COVRATIO |
| * `dfits` | DFITS |
| * `welsch` | Welsch distance |

Unstarred statistics are available both in and out of sample; type `predict ... if e(sample) ...` if wanted only for the estimation sample. Starred statistics are calculated only for the estimation sample, even when `if e(sample)` is not specified.

where *a* and *b* may be numbers or variables; *a* missing ($a \geq .$) means $-\infty$, and *b* missing ($b \geq .$) means $+\infty$; see [U] **12.2.1 Missing values**.

`cooksd`, `leverage`, `rstandard`, `rstudent`, `stdf`, `stdr`, `covratio`, `dfbeta()`, `dfits`, and `welsch` are not available if `robust`, `cluster()`, `hc2`, or `hc3` were specified with `regress`.

## Options for predict

`xb`, the default, calculates the linear prediction.

`residuals` calculates the residuals.

`score` is equivalent to `residuals` in linear regression.

`rstandard` calculates the standardized residuals.

`rstudent` calculates the studentized (jackknifed) residuals.

`cooksd` calculates the Cook's $D$ influence statistic (Cook 1977).

`leverage` or `hat` calculates the diagonal elements of the projection hat matrix.

(*Continued on next page*)

`pr(`*a*`,`*b*`)` calculates $\Pr(a < \mathbf{x}_j\mathbf{b} + u_j < b)$, the probability that $y_j|\mathbf{x}_j$ would be observed in the interval $(a, b)$.

*a* and *b* may be specified as numbers or variable names; *lb* and *ub* are variable names;
`pr(20,30)` calculates $\Pr(20 < \mathbf{x}_j\mathbf{b} + u_j < 30)$;
`pr(`*lb*`,`*ub*`)` calculates $\Pr(lb < \mathbf{x}_j\mathbf{b} + u_j < ub)$;
and `pr(20,`*ub*`)` calculates $\Pr(20 < \mathbf{x}_j\mathbf{b} + u_j < ub)$.

*a* missing ($a \geq .$) means $-\infty$; `pr(.,30)` calculates $\Pr(-\infty < \mathbf{x}_j\mathbf{b} + u_j < 30)$;
`pr(`*lb*`,30)` calculates $\Pr(-\infty < \mathbf{x}_j\mathbf{b} + u_j < 30)$ in observations for which $lb \geq .$
and calculates $\Pr(lb < \mathbf{x}_j\mathbf{b} + u_j < 30)$ elsewhere.

*b* missing ($b \geq .$) means $+\infty$; `pr(20,.)` calculates $\Pr(+\infty > \mathbf{x}_j\mathbf{b} + u_j > 20)$;
`pr(20,`*ub*`)` calculates $\Pr(+\infty > \mathbf{x}_j\mathbf{b} + u_j > 20)$ in observations for which $ub \geq .$
and calculates $\Pr(20 < \mathbf{x}_j\mathbf{b} + u_j < ub)$ elsewhere.

`e(`*a*`,`*b*`)` calculates $E(\mathbf{x}_j\mathbf{b} + u_j \mid a < \mathbf{x}_j\mathbf{b} + u_j < b)$, the expected value of $y_j|\mathbf{x}_j$ conditional on $y_j|\mathbf{x}_j$ being in the interval $(a, b)$, meaning that $y_j|\mathbf{x}_j$ is censored.
*a* and *b* are specified as they are for `pr()`.

`ystar(`*a*`,`*b*`)` calculates $E(y_j^*)$, where $y_j^* = a$ if $\mathbf{x}_j\mathbf{b} + u_j \leq a$, $y_j^* = b$ if $\mathbf{x}_j\mathbf{b} + u_j \geq b$, and $y_j^* = \mathbf{x}_j\mathbf{b} + u_j$ otherwise, meaning that $y_j^*$ is truncated. *a* and *b* are specified as they are for `pr()`.

`dfbeta(`*varname*`)` calculates the DFBETA for *varname*, the difference between the regression coefficient when the $j$th observation is included and excluded, said difference being scaled by the estimated standard error of the coefficient. *varname* must have been included among the regressors in the previously fitted model. The calculation is automatically restricted to the estimation subsample.

`stdp` calculates the standard error of the prediction, which can be thought of as the standard error of the predicted expected value or mean for the observation's covariate pattern. This is also referred to as the standard error of the fitted value.

`stdf` calculates the standard error of the forecast, which is the standard error of the point prediction for a single observation. It is commonly referred to as the standard error of the future or forecast value. By construction, the standard errors produced by `stdf` are always larger than those produced by `stdp`; see *Methods and Formulas.*

`stdr` calculates the standard error of the residuals.

`covratio` calculates COVRATIO (Belsley, Kuh, and Welsch 1980), a measure of the influence of the $j$th observation based on considering the effect on the variance–covariance matrix of the estimates. The calculation is automatically restricted to the estimation subsample.

`dfits` calculates DFITS (Welsch and Kuh 1977) and attempts to summarize the information in the leverage versus residual-squared plot into a single statistic. The calculation is automatically restricted to the estimation subsample.

`welsch` calculates Welsch distance (Welsch 1982) and is a variation on `dfits`. The calculation is automatically restricted to the estimation subsample.

## Syntax for dfbeta

`dfbeta` [*indepvar* [*indepvar* [...]]]

## Syntax for estat hettest

estat hettest [*varlist*] [, rhs mtest[(*spec*)]]

## Options for estat hettest

rhs specifies that tests for heteroskedasticity be performed for the right-hand-side (explanatory) variables of the fitted regression model. Option rhs may be combined with a *varlist*.

mtest[(*spec*)] specifies that multiple testing be performed. The argument specifies how $p$-values are adjusted. The following specifications *spec* are supported:

| | |
|---|---|
| bonferroni | Bonferroni's multiple testing adjustment |
| holm | Holm's multiple testing adjustment |
| sidak | Šidák's multiple testing adjustment |
| noadjust | no adjustment is made for multiple testing |

mtest may be specified without an argument. This is equivalent to specifying mtest(noadjust), that is, tests for the individual variables should be performed with unadjusted $p$-values. By default, estat hettest does not perform multiple testing.

## Syntax for estat imtest

estat imtest [, preserve white]

## Options for estat imtest

preserve specifies that the data in memory be preserved, all variables and cases that are not needed in the calculations be dropped, and at the conclusion the original data be restored. This is costly for large datasets. However, as estat imtest has to perform an auxiliary regression on $k(k+1)/2$ temporary variables, where $k$ is the number of regressors, it may not be able to perform the test otherwise.

white specifies that White's original heteroskedasticity test also be performed.

## Syntax for estat ovtest

estat ovtest [, rhs]

## Option for estat ovtest

rhs specifies that powers of the right-hand-side (explanatory) variables be used in the test rather than powers of the fitted values.

## Syntax for estat szroeter

`estat szroeter [varlist] [, rhs mtest(spec)]`

## Options for estat szroeter

`rhs` specifies that tests for heteroskedasticity be performed for the right-hand-side (explanatory) variables of the fitted regression model. Option `rhs` may be combined with a *varlist*.

`mtest[(spec)]` specifies that multiple testing be performed. The argument specifies how *p*-values are adjusted. The following specifications *spec* are supported:

| | |
|---|---|
| `bonferroni` | Bonferroni's multiple testing adjustment |
| `holm` | Holm's multiple testing adjustment |
| `sidak` | Šidák's multiple testing adjustment |
| `noadjust` | no adjustment is made for multiple testing |

`estat szroeter` always performs multiple testing. By default, it does not adjust the *p*-values.

## Syntax for estat vif

`estat vif`

## Syntax for acprplot

`acprplot indepvar [, acprplot_options]`

| *acprplot_options* | description |
|---|---|
| Plot | |
| *marker_options* | change look of markers (color, size, etc.) |
| *marker_label_options* | add marker labels; change look or position |
| Reference line | |
| `rlopts(cline_options)` | affect rendition of the reference line |
| Lowess | |
| `lowess` | add a lowess smooth of the plotted points |
| `lsopts(lowess_options)` | affect rendition of the lowess smooth |
| Spline | |
| `mspline` | add median spline of the plotted points |
| `msopts(mspline_options)` | affect rendition of the spline |
| Add plot | |
| `addplot(plot)` | add other plots to the generated graph |
| Y-Axis, X-Axis, Title, Caption, Legend, Overall | |
| *twoway_options* | any options other than `by()` documented in [G] ***twoway_options*** |

## Options for acprplot

Plot

*marker_options* affect the rendition of markers drawn at the plotted points, including their shape, size, color, and outline; see [G] ***marker_options***.

*marker_label_options* specify if and how the markers are to be labeled; see [G] ***marker_label_options***.

Reference line

`rlopts(`*cline_options*`)` affect the rendition of the reference line. See [G] ***cline_options***.

Lowess

`lowess` adds a lowess smooth of the plotted points in order to assist in detecting nonlinearities.

`lsopts(`*lowess_options*`)` affect the rendition of the lowess smooth. For an explanation of these options, especially the `bwidth()` option, see [R] **lowess**. Specifying `lsopts()` implies the `lowess` option.

Spline

`mspline` adds a median spline of the plotted points in order to assist in detecting nonlinearities.

`msopts(`*mspline_options*`)` affect the rendition of the spline. For an explanation of these options, especially the `bands()` option, see [G] **graph twoway mspline**. Specifying `msopts()` implies the `mspline` option.

Add plot

`addplot(`*plot*`)` provides a way to add other plots to the generated graph. See [G] ***addplot_option***.

Y-Axis, X-Axis, Title, Caption, Legend, Overall

*twoway_options* are any of the options documented in [G] ***twoway_options***, excluding `by()`. These include options for titling the graph (see [G] ***title_options***) and options for saving the graph to disk (see [G] ***saving_option***).

## Syntax for avplot

`avplot` *indepvar* [ , *avplot_options* ]

| *avplot_options* | description |
|---|---|
| Plot | |
| *marker_options* | change look of markers (color, size, etc.) |
| *marker_label_options* | add marker labels; change look or position |
| Reference line | |
| `rlopts(`*cline_options*`)` | affect rendition of the reference line |
| Add plot | |
| `addplot(`*plot*`)` | add other plots to the generated graph |
| Y-Axis, X-Axis, Title, Caption, Legend, Overall | |
| *twoway_options* | any options other than `by()` documented in [G] ***twoway_options*** |

## Options for avplot

Plot

*marker_options* affect the rendition of markers drawn at the plotted points, including their shape, size, color, and outline; see [G] ***marker_options***.

*marker_label_options* specify if and how the markers are to be labeled; see [G] ***marker_label_options***.

Reference line

`rlopts(`*cline_options*`)` affect the rendition of the reference line. See [G] ***cline_options***.

Add plot

`addplot(`*plot*`)` provides a way to add other plots to the generated graph. See [G] ***addplot_option***.

Y-Axis, X-Axis, Title, Caption, Legend, Overall

*twoway_options* are any of the options documented in [G] ***twoway_options***, excluding `by()`. These include options for titling the graph (see [G] ***title_options***) and options for saving the graph to disk (see [G] ***saving_option***).

## Syntax for avplots

`avplots` [ , *avplots_options* ]

| *avplots_options* | description |
|---|---|
| Plot | |
| *marker_options* | change look of markers (color, size, etc.) |
| *marker_label_options* | add marker labels; change look or position |
| *combine_options* | any of the options documented in [G] **graph combine** |
| Reference line | |
| `rlopts(`*cline_options*`)` | affect rendition of the reference line |

## Options for avplots

Plot

*marker_options* affect the rendition of markers drawn at the plotted points, including their shape, size, color, and outline; see [G] ***marker_options***.

*marker_label_options* specify if and how the markers are to be labeled; see [G] ***marker_label_options***.

*combine_options* are any of the options documented in [G] **graph combine**. These include options for titling the graph (see [G] ***title_options***) and options for saving the graph to disk (see [G] ***saving_option***).

Reference line

rlopts(*cline_options*) affect the rendition of the reference line. See [G] ***cline_options***.

## Syntax for cprplot

cprplot *indepvar* [ , *cprplot_options* ]

| *cprplot_options* | description |
|---|---|
| Plot | |
| *marker_options* | change look of markers (color, size, etc.) |
| *marker_label_options* | add marker labels; change look or position |
| Reference line | |
| rlopts(*cline_options*) | affect rendition of the reference line |
| Lowess | |
| lowess | add a lowess smooth of the plotted points |
| lsopts(*lowess_options*) | affect rendition of the lowess smooth |
| Spline | |
| mspline | add median spline of the plotted points |
| msopts(*mspline_options*) | affect rendition of the spline |
| Add plot | |
| addplot(*plot*) | add other plots to the generated graph |
| Y-Axis, X-Axis, Title, Caption, Legend, Overall | |
| *twoway_options* | any options other than by() documented in [G] ***twoway_options*** |

## Options for cprplot

Plot

*marker_options* affect the rendition of markers drawn at the plotted points, including their shape, size, color, and outline; see [G] ***marker_options***.

*marker_label_options* specify if and how the markers are to be labeled; see [G] ***marker_label_options***.

Reference line

rlopts(*cline_options*) affect the rendition of the reference line. See [G] ***cline_options***.

Lowess

lowess adds a lowess smooth of the plotted points in order to assist in detecting nonlinearities.

lsopts(*lowess_options*) affect the rendition of the lowess smooth. For an explanation of these options, especially the bwidth() option, see [R] **lowess**. Specifying lsopts() implies the lowess option.

Spline

`mspline` adds a median spline of the plotted points in order to assist in detecting nonlinearities.

`msopts(`*mspline_options*`)` affect the rendition of the spline. For an explanation of these options, especially the `bands()` option, see [G] **graph twoway mspline**. Specifying `msopts()` implies the `mspline` option.

Add plot

`addplot(`*plot*`)` provides a way to add other plots to the generated graph. See [G] ***addplot_option***.

Y-Axis, X-Axis, Title, Caption, Legend, Overall

*twoway_options* are any of the options documented in [G] ***twoway_options***, excluding `by()`. These include options for titling the graph (see [G] ***title_options***) and options for saving the graph to disk (see [G] ***saving_option***).

# Syntax for lvr2plot

`lvr2plot` [ , *lvr2plot_options* ]

| *lvr2plot_options* | description |
|---|---|
| Plot | |
| *marker_options* | change look of markers (color, size, etc.) |
| *marker_label_options* | add marker labels; change look or position |
| Add plot | |
| `addplot(`*plot*`)` | add other plots to the generated graph |
| Y-Axis, X-Axis, Title, Caption, Legend, Overall | |
| *twoway_options* | any options other than `by()` documented in [G] ***twoway_options*** |

# Options for lvr2plot

Plot

*marker_options* affect the rendition of markers drawn at the plotted points, including their shape, size, color, and outline; see [G] ***marker_options***.

*marker_label_options* specify if and how the markers are to be labeled; see [G] ***marker_label_options***.

Add plot

`addplot(`*plot*`)` provides a way to add other plots to the generated graph; see [G] ***addplot_option***.

Y-Axis, X-Axis, Title, Caption, Legend, Overall

*twoway_options* are any of the options documented in [G] ***twoway_options***, excluding `by()`. These include options for titling the graph (see [G] ***title_options***) and options for saving the graph to disk (see [G] ***saving_option***).

## Syntax for rvfplot

`rvfplot` [ , *rvfplot_options* ]

| *rvfplot_options* | description |
|---|---|
| Plot | |
| *marker_options* | change look of markers (color, size, etc.) |
| *marker_label_options* | add marker labels; change look or position |
| Add plot | |
| `addplot(`*plot*`)` | add plots to the generated graph |
| Y-Axis, X-Axis, Title, Caption, Legend, Overall | |
| *twoway_options* | any options other than `by()` documented in [G] ***twoway_options*** |

## Options for rvfplot

Plot

*marker_options* affect the rendition of markers drawn at the plotted points, including their shape, size, color, and outline; see [G] ***marker_options***.

*marker_label_options* specify if and how the markers are to be labeled; see [G] ***marker_label_options***.

Add plot

`addplot(`*plot*`)` provides a way to add plots to the generated graph. See [G] ***addplot_option***.

Y-Axis, X-Axis, Title, Caption, Legend, Overall

*twoway_options* are any of the options documented in [G] ***twoway_options***, excluding `by()`. These include options for titling the graph (see [G] ***title_options***) and options for saving the graph to disk (see [G] ***saving_option***).

## Syntax for rvpplot

`rvpplot` *indepvar* [ , *rvpplot_options* ]

| *rvpplot_options* | description |
|---|---|
| Plot | |
| *marker_options* | change look of markers (color, size, etc.) |
| *marker_label_options* | add marker labels; change look or position |
| Add plot | |
| `addplot(`*plot*`)` | add other plots to the generated graph |
| Y-Axis, X-Axis, Title, Caption, Legend, Overall | |
| *twoway_options* | any options other than `by()` documented in [G] ***twoway_options*** |

## Options for rvpplot

Plot

*marker_options* affect the rendition of markers drawn at the plotted points, including their shape, size, color, and outline; see [G] ***marker_options***.

*marker_label_options* specify if and how the markers are to be labeled; see [G] ***marker_label_options***.

Add plot

`addplot(`*plot*`)` provides a way to add other plots to the generated graph; see [G] ***addplot_option***.

Y-Axis, X-Axis, Title, Caption, Legend, Overall

*twoway_options* are any of the options documented in [G] ***twoway_options***, excluding `by()`. These include options for titling the graph (see [G] ***title_options***) and options for saving the graph to disk (see [G] ***saving_option***).

## Remarks

Remarks are presented under the headings

*Fitted values and residuals*
*Prediction standard errors*
*Prediction with weighted data*
*Residual-versus-fitted plots*
*Added-variable plots*
*Component-plus-residual plots*
*Residual-versus-predictor plots*
*Leverage statistics*
*L-R plots*
*Standardized and studentized residuals*
*DFITS, Cook's Distance, and Welsch Distance*
*COVRATIO*
*DFBETAs*
*Formal tests for violations of assumptions*
*Variance inflation factors*

Many of these commands concern identifying influential data in linear regression. This is, unfortunately, a field that is dominated by jargon, codified and partially begun by Belsley, Kuh, and Welsch (1980). In the words of Chatterjee and Hadi (1986, 416), "Belsley, Kuh, and Welsch's book, *Regression Diagnostics*, was a very valuable contribution to the statistical literature, but it unleashed on an unsuspecting statistical community a computer speak (à la Orwell), the likes of which we have never seen". Things have only gotten worse since then. Chatterjee and Hadi's (1986, 1988) own attempts to clean up the jargon did not improve matters (see Hoaglin and Kempthorne 1986, Velleman 1986, and Welsch 1986). We apologize for the jargon, and, for our contribution to the jargon in the form of inelegant command names, we apologize most of all.

Model *sensitivity* refers to how estimates are affected by subsets of our data. Imagine data on $y$ and $x$, and assume that the data are to be fit by the regression $y_i = \alpha + \beta x_i + \epsilon_i$. The regression estimates of $\alpha$ and $\beta$ are $a$ and $b$, respectively. Now imagine that the estimated $a$ and $b$ would be very different if a small portion of the dataset, perhaps even a single observation, were deleted. As a data analyst, you would like to think that you are summarizing tendencies that apply to all the data, but you have just been told that the model you fitted is unduly influenced by a single point or just a few points and that, as a matter of fact, there is another model that applies to the rest of the data—a model you have ignored. The search for subsets of the data that, if deleted, would change the results markedly is a predominant theme of this entry.

There are three key issues in identifying model sensitivity to individual observations, which go by the names *residuals*, *leverage*, and *influence*. In our $y_i = a + bx_i + e_i$ regression, the residuals are, of course, $e_i$—they reveal how much our fitted value $\widehat{y}_i = a + bx_i$ differs from the observed $y_i$. A point $(x_i, y_i)$ with a corresponding large residual is called an outlier. Say that you are interested in outliers because you somehow feel that such points will exert undue influence on your estimates. Your feelings are generally right, but there are exceptions. A point might have a huge residual and yet not affect the estimated $b$ at all. Nevertheless, studying observations with large residuals almost always pays off.

$(x_i, y_i)$ can be an outlier in another way—just as $y_i$ can be far from $\widehat{y}_i$, $x_i$ can be far from the center of mass of the other $x$s. Such an "outlier" should interest you just as much as the more traditional outliers. Picture a scatterplot of $y$ against $x$ with thousands of points in some sort of mass at the lower left of the graph and a single point at the upper right of the graph. Now run a regression line through the points—the regression line will come very close to the point at the upper right of the graph, and may, in fact, go through it. That is, this isolated point will not appear as an outlier as measured by residuals because its residual will be small. Yet this point might have a dramatic effect on our resulting estimates in the sense that, were you to delete the point, the estimates would change markedly. Such a point is said to have high leverage. Just as with traditional outliers, a high leverage point does not necessarily have an undue effect on regression estimates, but if it does not, it is more the exception than the rule.

Now all of this is a most unsatisfactory state of affairs. Points with large residuals may, but need not, have a large effect on our results, and points with small residuals may still have a large effect. Points with high leverage may, but need not, have a large effect on our results, and points with low leverage may still have a large effect. Can't you identify the influential points and simply have the computer list them for you? You can, but you will have to define what you mean by "influential".

"Influential" is defined with respect to some statistic. For instance, you might ask which points in your data have a large effect on your estimated $a$, which points have a large effect on your estimated $b$, which points have a large effect on your estimated standard error of $b$, and so on, but do not be surprised when the answers to these questions are different. In any case, obtaining such measures is not difficult—all you have to do is fit the regression excluding each observation one at a time and record the statistic of interest which, in the day of the modern computer, is not too onerous. Moreover, you can save considerable computer time by doing algebra ahead of time and working out formulas that will calculate the same answers as if you ran each of the regressions. (Ignore the question of pairs of observations that, together, exert undue influence, and triples, and so on, which remains largely unsolved and for which the brute force fit-every-possible-regression procedure is not a viable alternative.)

## Fitted values and residuals

Typing `predict` *newvar* with no options creates *newvar* containing the fitted values. Typing `predict` *newvar*`, resid` creates *newvar* containing the residuals.

### ▷ Example 1

Using the example from [R] **predict**, we have data on automobiles, including the mileage rating (`mpg`), the car's weight (`weight`), and whether the car is foreign (`foreign`). We wish to fit the following model:

$$\texttt{mpg} = \beta_1\texttt{weight} + \beta_2\texttt{weight}^2 + \beta_3\texttt{foreign} + \beta_4$$

We first create the `weight`$^2$ variable and then type the `regress` command:

```
. use http://www.stata-press.com/data/r9/auto
(1978 Automobile Data)

. generate weight2 = weight^2

. regress mpg weight weight2 foreign

      Source |       SS       df       MS              Number of obs =      74
-------------+------------------------------           F(  3,    70) =   52.25
       Model |  1689.15372     3  563.05124            Prob > F      =  0.0000
    Residual |   754.30574    70  10.7757963           R-squared     =  0.6913
-------------+------------------------------           Adj R-squared =  0.6781
       Total |  2443.45946    73  33.4720474           Root MSE      =  3.2827

------------------------------------------------------------------------------
         mpg |      Coef.   Std. Err.      t    P>|t|     [95% Conf. Interval]
-------------+----------------------------------------------------------------
      weight |  -.0165729   .0039692    -4.18   0.000    -.0244892   -.0086567
     weight2 |   1.59e-06   6.25e-07     2.55   0.013     3.45e-07    2.84e-06
     foreign |    -2.2035   1.059246    -2.08   0.041      -4.3161   -.0909002
       _cons |   56.53884   6.197383     9.12   0.000     44.17855    68.89913
------------------------------------------------------------------------------
```

That done, we can now obtain the predicted values from the regression. We will store them in a new variable called `pmpg` by typing `predict pmpg`. Since `predict` produces no output, we will follow that by summarizing our predicted and observed values.

```
. predict pmpg
(option xb assumed; fitted values)

. summarize pmpg mpg

    Variable |       Obs        Mean    Std. Dev.       Min        Max
-------------+--------------------------------------------------------
        pmpg |        74     21.2973    4.810311   13.59953   31.86288
         mpg |        74     21.2973    5.785503         12         41
```

◁

## ▷ Example 2: Out-of-sample predictions

We can just as easily obtain predicted values from the model using a wholly different dataset from the one on which the model was fitted. The only requirement is that the data have the necessary variables, which in this case are `weight`, `weight2`, and `foreign`.

Using the data on two new cars (the Pontiac Sunbird and the Volvo 260) from the `newautos.dta` dataset, we can obtain out-of-sample predictions (or forecasts) by typing

```
. use http://www.stata-press.com/data/r9/newautos, clear
(New Automobile Models)

. generate weight2=weight^2

. predict mpg
(option xb assumed; fitted values)

. list, divider
```

|  | make | weight | foreign | weight2 | mpg |
|---|---|---|---|---|---|
| 1. | Pont. Sunbird | 2690 | Domestic | 7236100 | 23.47137 |
| 2. | Volvo 260 | 3170 | Foreign | 1.00e+07 | 17.78846 |

The Pontiac Sunbird has a predicted mileage rating of 23.5 mpg, whereas the Volvo 260 has a predicted rating of 17.8 mpg. In comparison, the actual mileage ratings are 24 for the Pontiac and 17 for the Volvo.

◁

## Prediction standard errors

`predict` can calculate the standard error of the forecast (`stdf` option), the standard error of the prediction (`stdp` option), and the standard error of the residual (`stdr` option). It is easy to confuse `stdf` and `stdp` because both are often called the prediction error. Consider the prediction $\widehat{y}_j = \mathbf{x}_j\mathbf{b}$, where $\mathbf{b}$ is the estimated coefficient (column) vector and $\mathbf{x}_j$ is a (row) vector of independent variables for which you want the prediction. First, $\widehat{y}_j$ has a variance due to the variance of the estimated coefficient vector $\mathbf{b}$,

$$\mathrm{Var}(\widehat{y}_j) = \mathrm{Var}(\mathbf{x}_j\mathbf{b}) = s^2 h_j$$

where $h_j = \mathbf{x}_j(\mathbf{X}'\mathbf{X})^{-1}\mathbf{x}_j'$ and $s^2$ is the mean squared error of the regression. Do not panic over the algebra—just remember that $\mathrm{Var}(\widehat{y}_j) = s^2 h_j$, whatever $s^2$ and $h_j$ are. `stdp` calculates this quantity. This is the error in the prediction due to the uncertainty about $\mathbf{b}$.

If you are about to hand this number out as your forecast, however, there is another error. According to your model, the true value of $y_j$ is given by

$$y_j = \mathbf{x}_j\mathbf{b} + \epsilon_j = \widehat{y}_j + \epsilon_j$$

and thus, the $\mathrm{Var}(y_j) = \mathrm{Var}(\widehat{y}_j) + \mathrm{Var}(\epsilon_j) = s^2 h_j + s^2$, which is the square of `stdf`. `stdf`, then, is the sum of the error in the prediction plus the residual error.

`stdr` has to do with an analysis-of-variance decomposition of $s^2$, the estimated variance of $y$. The standard error of the prediction is $s^2 h_j$, and therefore $s^2 h_j + s^2(1 - h_j) = s^2$ decomposes $s^2$ into the prediction and residual variances.

### ▷ Example 3: standard error of the forecast

Returning to our model of `mpg` on `weight`, `weight`$^2$, and `foreign`, we previously predicted the mileage rating for the Pontiac Sunbird and Volvo 260 as 23.5 and 17.8 mpg, respectively. We now want to put a standard error around our forecast. Remember, the data for these two cars were in `newautos.dta`:

```
. use http://www.stata-press.com/data/r9/newautos, clear
(New Automobile Models)
. gen weight2=weight*weight
. predict mpg
(option xb assumed; fitted values)
. predict se_mpg, stdf
. list, divider
```

| | make | weight | foreign | weight2 | mpg | se_mpg |
|---|---|---|---|---|---|---|
| 1. | Pont. Sunbird | 2690 | Domestic | 7236100 | 23.47137 | 3.341823 |
| 2. | Volvo 260 | 3170 | Foreign | 1.00e+07 | 17.78846 | 3.438714 |

Thus an approximate 95% confidence interval for the mileage rating of the Volvo 260 is $17.8 \pm 2 \cdot 3.44 = [\,10.92, 24.68\,]$.

◁

## Prediction with weighted data

predict can be used after frequency-weighted (fweight) estimation, just as it is used after unweighted estimation. The technical note below concerns the use of predict after analytically weighted (aweight) estimation.

### ❑ Technical Note

After analytically weighted estimation, predict is only willing to calculate the prediction (no options), residual (residual option), standard error of the prediction (stdp option), and diagonal elements of the projection matrix (hat option). Moreover, the results produced by hat need to be adjusted, as will be described. For analytically weighted estimation, the standard error of the forecast and residuals, the standardized and studentized residuals, and Cook's $D$ are not statistically well-defined concepts.

To obtain the correct values of the diagonal elements of the hat matrix, you can use predict with the hat option to make a first, partially adjusted calculation, and then follow that by completing the adjustment. Assume that you are fitting a linear regression model weighting the data with the variable w ([aweight=w]). Begin by creating a new variable w0:

```
. predict resid if e(sample), resid
. summarize w if resid < . & e(sample)
. gen w0=w/r(mean)
```

Some caution is necessary at this step—the summarize w must be performed on the same sample that was used to fit the model, which means that you must include if e(sample) to restrict the prediction to the estimation sample. You created the residual and then included the modifier 'if resid < .' so that if the dependent variable or any of the independent variables is missing, the corresponding observations will be excluded from the calculation of the average value of the original weight.

To correct predict's hat calculation, multiply the result by w0:

```
. predict myhat, hat
. replace myhat = w0 * myhat
```

❑

## Residual-versus-fitted plots

### ▷ Example 4: rvfplot

Using the automobile dataset described in [U] **1.2.1 Sample datasets**, we will use regress to fit a model of price on weight, mpg, foreign, and the interaction of foreign with mpg.

```
. use http://www.stata-press.com/data/r9/auto, clear
(1978 Automobile Data)
. generate forXmpg=foreign*mpg
```

```
. regress price weight mpg forXmpg foreign
      Source |       SS       df       MS              Number of obs =      74
-------------+------------------------------           F(  4,    69) =   21.22
       Model |   350319665     4  87579916.3           Prob > F      =  0.0000
    Residual |   284745731    69  4126749.72           R-squared     =  0.5516
-------------+------------------------------           Adj R-squared =  0.5256
       Total |   635065396    73  8699525.97           Root MSE      =  2031.4

------------------------------------------------------------------------------
       price |      Coef.   Std. Err.      t    P>|t|     [95% Conf. Interval]
-------------+----------------------------------------------------------------
      weight |   4.613589   .7254961     6.36   0.000     3.166263    6.060914
         mpg |   263.1875   110.7961     2.38   0.020     42.15527    484.2197
     forXmpg |  -307.2166   108.5307    -2.83   0.006    -523.7294   -90.70368
     foreign |   11240.33   2751.681     4.08   0.000     5750.878    16729.78
       _cons |  -14449.58    4425.72    -3.26   0.002    -23278.65    -5620.51
------------------------------------------------------------------------------
```

Once we have fitted a model, we may use any of the `regression diagnostics` commands. `rvfplot` (read residual-versus-fitted plot) graphs the residuals against the fitted values:

```
. rvfplot, yline(0)
```

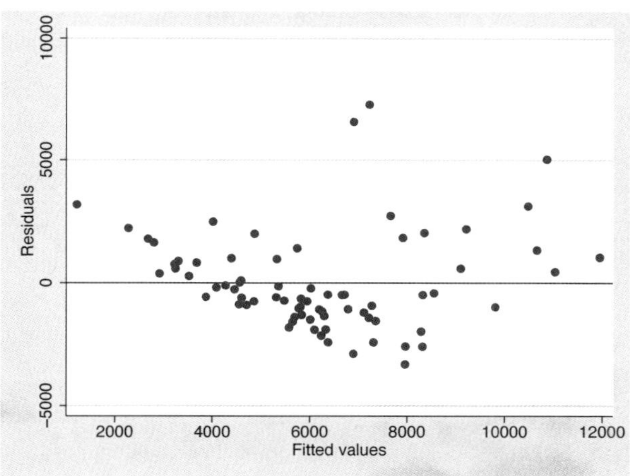


All the diagnostic plot commands allow the options of `graph twoway` and `graph twoway scatter`; we specified a `yline(0)` to draw a line across the graph at $y = 0$; see [G] **graph twoway scatter**.

In a well-fitted model, there should be no pattern to the residuals plotted against the fitted values—something not true of our model. Ignoring the two outliers at the top center of the graph, we see curvature in the pattern of the residuals, suggesting a violation of the assumption that `price` is linear in our independent variables. Alternatively, we might have seen increasing or decreasing variation in the residuals—heteroskedasticity. Any pattern whatsoever indicates a violation of the least-squares assumptions.

◁

## Added-variable plots

### ▷ Example 5: avplot

Continuing with our price model, another diagnostic graph is provided by `avplot` (read added-variable plot, also known as the partial-regression leverage plot).

One of the wonderful features of one-regressor regressions (regressions of $y$ on a single $x$) is that we can graph the data and the regression line. There is no easier way to understand the regression than to examine such a graph. Unfortunately, we cannot do this when we have more than one regressor. With two regressors, it is still theoretically possible—the graph must be drawn in three dimensions, but with three or more regressors no graph is possible.

The added-variable plot is an attempt to project multidimensional data back to the two-dimensional world for each of the original regressors. This is, of course, impossible without making some concessions. Call the coordinates on an added-variable plot $y$ and $x$. The added-variable plot has the following properties:

1. There is a one-to-one correspondence between $(x_i, y_i)$ and the $i$th observation used in the original regression.
2. A regression of $y$ on $x$ has the same coefficient and standard error (up to a degree-of-freedom adjustment) as the estimated coefficient and standard error for the regressor in the original regression.
3. The "outlierness" of each observation in determining the slope is in some sense preserved.

It is equally important to note the properties that are not listed. The $y$ and $x$ coordinates of the added-variable plot cannot be used to identify functional form, or, at least, not very well (see Mallows 1986). In the construction of the added-variable plot, the relationship between $y$ and $x$ is forced to be linear.

Let us examine the added-variable plot for `mpg` in our regression of `price` on `weight`, `mpg`, `forXmpg`, and `foreign`:

```
. avplot mpg
```

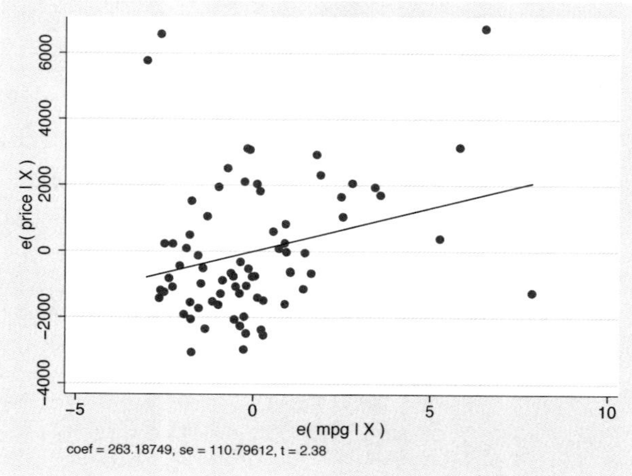


This graph suggests a problem in the determination of the coefficient on `mpg`. Were this a one-regressor regression, the two points at the top-left corner and the one at the top right would cause us concern, and so it does in our more complicated multiple-regressor case. To identify the problem points, we retyped our command, modifying it to read `avplot mpg, mlabel(make)`, and discovered that the two cars at the top left are the Cadillac Eldorado and the Lincoln Versailles; the point at the top right is the Cadillac Seville. These three cars account for 100% of the luxury cars in our data, suggesting that our model is misspecified. By the way, the point at the lower right of the graph, also cause for concern, is the Plymouth Arrow, our data-entry error.

◁

## ❑ Technical Note

Stata's `avplot` command can be used with regressors already in the model, as we just did, or with potential regressors not yet in the model. In either case, `avplot` will produce the correct graph. The name "added-variable plot" is unfortunate in the case when the variable is already among the list of regressors but is, we feel, still preferable to the name "partial-regression leverage plot" assigned by Belsley, Kuh, and Welsch (1980, 30) and more in the spirit of the original use of such plots by Mosteller and Tukey (1977, 271–279). Welsch (1986, 403), however, disagrees: "I am sorry to see that Chatterjee and Hadi [1986] endorse the term 'added-variable plot' when $X_j$ is part of the original model" and goes on to suggest the name "adjusted partial residual plot".

❑

## ▷ Example 6: avplots

Added-variable plots are so useful that we should look at them for every regressor in the data. `avplots` makes this easy:

```
. avplots
```

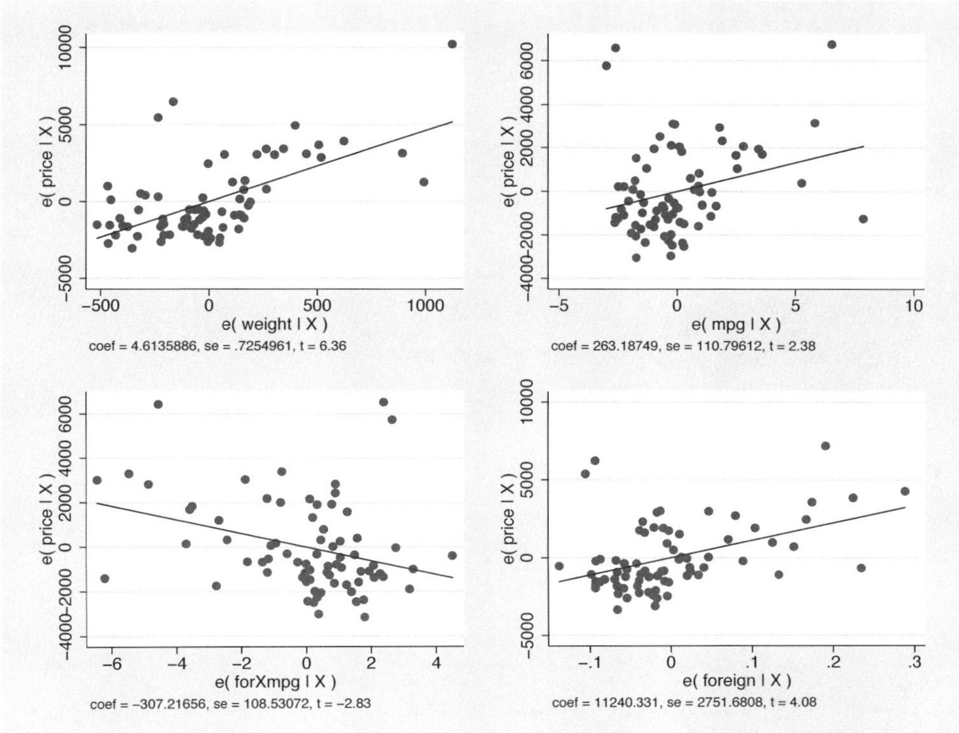


◁

## Component-plus-residual plots

Added-variable plots are quite successful at identifying outliers, but they cannot be used to identify functional form. The component-plus-residual plot (Ezekiel 1924; Larsen and McCleary 1972) is another attempt at projecting multidimensional data into a two-dimensional form, but with different properties. While the added-variable plot can identify outliers, the component-plus-residual plot cannot. It can, however, be used to examine the functional-form assumptions of the model. Both plots have the property that a regression line through the coordinates has a slope equal to the estimated coefficient in the regression model.

### ▷ Example 7: cprplot and acprplot

To illustrate this, we begin with a different model:

```
. use http://www.stata-press.com/data/r9/auto1, clear
(Automobile Models)
. regress price mpg weight
      Source |       SS       df       MS              Number of obs =      74
-------------+------------------------------           F(  2,    71) =   14.90
       Model |  187716578     2  93858289              Prob > F      =  0.0000
    Residual |  447348818    71  6300687.58            R-squared     =  0.2956
-------------+------------------------------           Adj R-squared =  0.2757
       Total |  635065396    73  8699525.97            Root MSE      =  2510.1

------------------------------------------------------------------------------
       price |      Coef.   Std. Err.      t    P>|t|     [95% Conf. Interval]
-------------+----------------------------------------------------------------
         mpg |   -55.9393   75.24136    -0.74   0.460    -205.9663    94.08771
      weight |   1.710992   .5861682     2.92   0.005     .5422063    2.879779
       _cons |     2197.9   3190.768     0.69   0.493    -4164.311     8560.11
------------------------------------------------------------------------------
```

In fact, we know that the effects of `mpg` in this model are nonlinear—if we added `mpg` squared to the model, its coefficient would have a $t$ statistic of 2.38, the $t$ statistic on `mpg` would become $-2.48$, and `weight`'s effect would become about one-third of its current value and become statistically insignificant. Pretend that we do not know this.

The component-plus-residual plot for `mpg` is

```
. cprplot mpg, mspline msopts(bands(13))
```

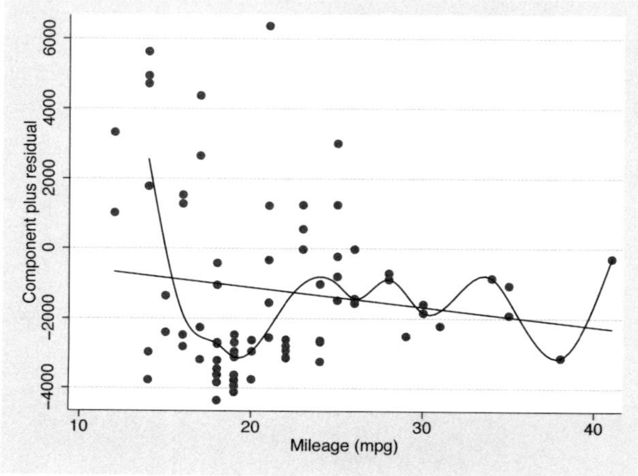

We are supposed to examine the above graph for nonlinearities, or, equivalently, ask if the regression line, which has slope equal to the estimated effect of `mpg` in the original model, fits the data adequately. To assist our eyes, we added a median spline. Perhaps some people may detect nonlinearity based on this graph, but we assert that if we had not previously revealed the nonlinearity of `mpg` and if we had not added the median spline, we would not be overly bothered by the graph.

Mallows (1986) proposed an augmented component-plus-residual plot that is often more sensitive to detecting nonlinearity:

```
. acprplot mpg, mspline msopts(bands(13))
```

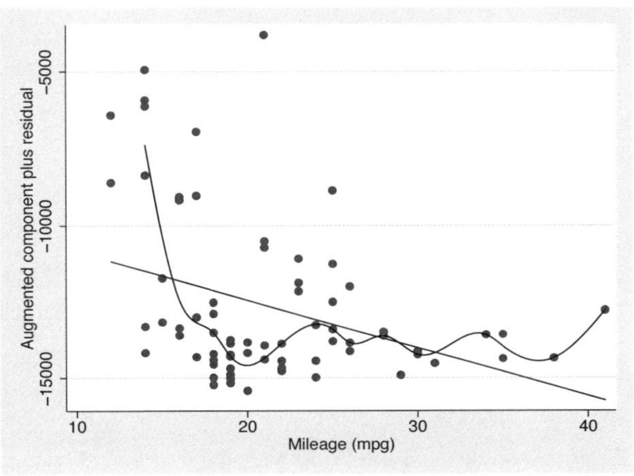


It does do somewhat better.

◁

## Residual-versus-predictor plots

### ▷ Example 8: rvpplot

The residual-versus-predictor plot is a simple way to look for violations of the regression assumptions. If the assumptions are correct, there should be no pattern in the graph. Using our `price` on `mpg` and `weight` model, we type

*(Continued on next page)*

```
. rvpplot mpg, yline(0)
```

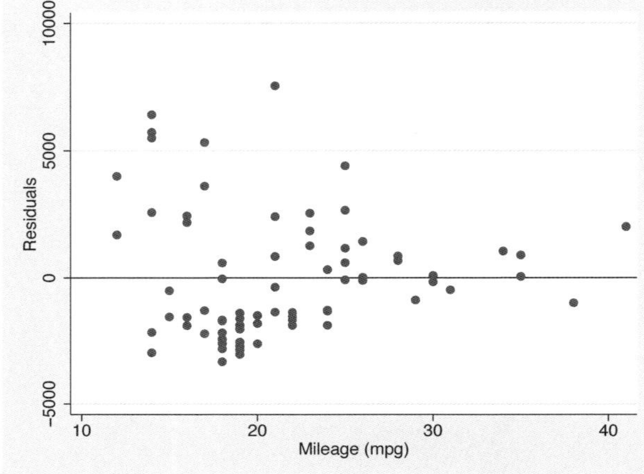


Remember, any pattern counts as a problem, and in this graph, we see that the variation in the residuals decreases as `mpg` increases.

◁

## Leverage statistics

In addition to providing fitted values and the associated standard errors, the `predict` command can also be used to generate various statistics used to detect the influence of individual observations. This section provides a brief introduction to leverage (hat) statistics, and some of the following subsections discuss other influence statistics produced by `predict`.

### ▷ Example 9: diagonal elements of projection matrix

The diagonal elements of the projection matrix, obtained by the `hat` option, are a measure of distance in explanatory variable space. `leverage` is a synonym for `hat`.

```
. use http://www.stata-press.com/data/r9/auto
(1978 Automobile Data)
. generate weight2 = weight^2
. regress mpg weight weight2 foreign
  (output omitted)
. predict xdist, hat
. summarize xdist, detail
                            Leverage
-------------------------------------------------------------
      Percentiles      Smallest
 1%     .0251334       .0251334
 5%     .0255623       .0251334
10%     .0259213       .0253883       Obs                  74
25%     .0278442       .0255623       Sum of Wgt.          74

50%       .04103                      Mean           .0540541
                        Largest       Std. Dev.      .0459218
75%     .0631279       .1593606
90%     .0854584       .1593606       Variance       .0021088
95%     .1593606       .2326124       Skewness       3.440809
99%     .3075759       .3075759       Kurtosis       16.95135
```

Some 5% of our sample has an `xdist` measure in excess of 0.15. Let's force them to reveal their identities:

```
. list foreign make mpg if xdist>.15, divider
```

| | foreign | make | mpg |
|---|---|---|---|
| 24. | Domestic | Ford Fiesta | 28 |
| 26. | Domestic | Linc. Continental | 12 |
| 27. | Domestic | Linc. Mark V | 12 |
| 43. | Domestic | Plym. Champ | 34 |

To understand why these cars are on this list, we must remember that the explanatory variables in our model are `weight` and `foreign`, and that `xdist` measures distance in this metric. The Ford Fiesta and the Plymouth Champ are the two lightest domestic cars in our data. The Lincolns are the two heaviest domestic cars.

◁

## L-R plots

### ▷ Example 10: lvr2plot

One of the most useful diagnostic graphs is provided by `lvr2plot` (read leverage-versus-residual-squared plot), a graph of leverage against the (normalized) residuals squared.

```
. use http://www.stata-press.com/data/r9/auto, clear
(1978 Automobile Data)
. generate forXmpg=foreign*mpg
. regress price weight mpg forXmpg foreign
  (output omitted)
. lvr2plot
```

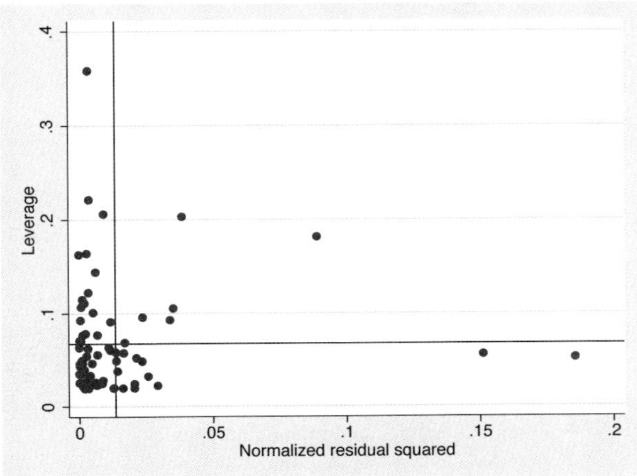


The lines on the chart show the average values of leverage and the (normalized) residuals squared. Points above the horizontal line have higher-than-average leverage; points to the right of the vertical line have larger-than-average residuals.

One point immediately catches our eye, and four more make us pause. The point at the top of the graph has high leverage and a smaller-than-average residual. The other points that bother us all have higher-than-average leverage, two with smaller-than-average residuals and two with larger-than-average residuals.

A less pretty but more useful version of the above graph specifies that `make` be used as the symbol (see [G] ***marker_label_options***):

```
. lvr2plot, mlabel(make) mlabp(0) m(none) mlabsize(small)
```

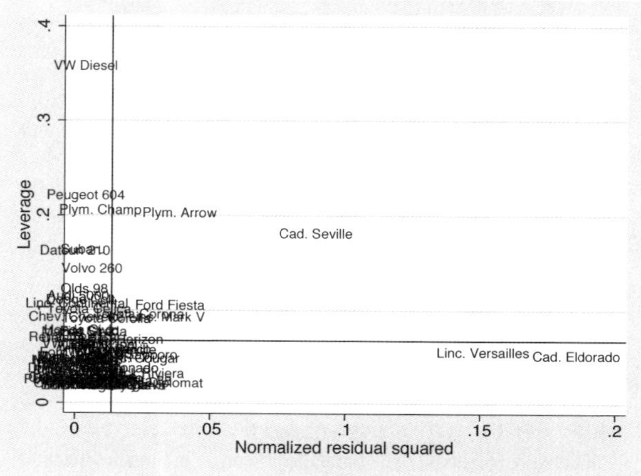


The VW Diesel, Plymouth Champ, Plymouth Arrow, and Peugeot 604 are the points that cause us the most concern. When we further examine our data, we discover that the VW Diesel is the only diesel in our data, and that the data for the Plymouth Arrow were entered incorrectly into the computer. No such simple explanations were found for the Plymouth Champ and Peugeot 604.

◁

## Standardized and studentized residuals

The terms standardized and studentized residuals have meant different things to different authors. In Stata, `predict` defines the standardized residual as $\widehat{e}_i = e_i/(s\sqrt{1-h_i})$ and the studentized residual as $r_i = e_i/(s_{(i)}\sqrt{1-h_i})$, where $s_{(i)}$ is the root mean squared error of a regression with the $i$th observation removed. Stata's definition of the studentized residual is the same as the one given in Bollen and Jackman (1990, 264) and is what Chatterjee and Hadi (1988, 74) call the "externally studentized" residual. Stata's "standardized" residual is the same as what Chatterjee and Hadi (1988, 74) call the "internally studentized" residual.

Standardized and studentized residuals are attempts to adjust residuals for their standard errors. Although the $\epsilon_i$ theoretical residuals are homoskedastic by assumption (i.e., they all have the same variance), the calculated $e_i$ are not. In fact,

$$\mathrm{Var}(e_i) = \sigma^2(1-h_i)$$

where $h_i$ are the leverage measures obtained from the diagonal elements of hat matrix. Thus observations with the greatest leverage have corresponding residuals with the smallest variance.

Standardized residuals use the root mean squared error of the regression for $\sigma$. Studentized residuals use the root mean squared error of a regression omitting the observation in question for $\sigma$. In general, studentized residuals are preferable to standardized residuals for purposes of outlier identification. Studentized residuals can be interpreted as the $t$ statistic for testing the significance of a dummy variable equal to 1 in the observation in question and 0 elsewhere (Belsley, Kuh, and Welsch 1980). Such a dummy variable would effectively absorb the observation and so remove its influence in determining the other coefficients in the model. Caution must be exercised here, however, because of the simultaneous testing problem. You cannot simply list the residuals that would be individually significant at the 5% level—their joint significance would be far less (their joint significance *level* would be far greater).

### ▷ Example 11: standardized and studentized residuals

In the opening remarks for this entry, we distinguished residuals from leverage and speculated on the impact of an observation with a small residual but large leverage. If we had adjusted the residuals for their standard errors, however, the adjusted residual would have been (relatively) larger and perhaps large enough so that we could simply examine the adjusted residuals. Taking our `price` on `weight`, `mpg`, `forXmpg`, and `foreign` model, we can obtain the in-sample standardized and studentized residuals by typing

```
. predict esta if e(sample), rstandard
. predict estu if e(sample), rstudent
```

Under the subheading *L-R plots*, we discovered that the VW Diesel has the highest leverage in our data, but a corresponding small residual. The standardized and studentized residuals for the VW Diesel are

```
. list make price esta estu if make=="VW Diesel"
```

| | make | price | esta | estu |
|---|---|---|---|---|
| 74. | VW Diesel | 5,397 | .6142691 | .6114758 |

The studentized residual of 0.611 can be interpreted as the $t$ statistic for including a dummy variable for VW Diesel in our regression. Such a variable would not be significant.

◁

## DFITS, Cook's Distance, and Welsch Distance

DFITS (Welsch and Kuh 1977), Cook's Distance (Cook 1977), and Welsch Distance (Welsch 1982) are three attempts to summarize the information in the leverage versus residual-squared plot into a single statistic. That is, the goal is to create an index that is affected by the size of the residuals—outliers—and the size of $h_i$—leverage. Viewed mechanically, one way to write DFITS (Bollen and Jackman 1990, 265) is

$$\text{DFITS}_i = r_i \sqrt{\frac{h_i}{1-h_i}}$$

where $r_i$ are the studentized residuals. Thus large residuals increase the value of DFITS, as do large values of $h_i$. Viewed more traditionally, DFITS is a scaled difference between predicted values for the $i$th case when the regression is fitted with and without the $i$th observation, hence the name.

The mechanical relationship between DFITS and Cook's Distance $D_i$ (Bollen and Jackman 1990, 266) is

$$D_i = \frac{1}{k} \frac{s^2_{(i)}}{s^2} \text{DFITS}_i^2$$

where $k$ is the number of variables (including the constant) in the regression, $s$ is the root mean squared error of the regression, and $s_{(i)}$ is the root mean squared error when the $i$th observation is omitted. Viewed more traditionally, $D_i$ is a scaled measure of the distance between the coefficient vectors when the $i$th observation is omitted.

The mechanical relationship between DFITS and Welsch's Distance $W_i$ (Chatterjee and Hadi 1988, 123) is

$$W_i = \text{DFITS}_i \sqrt{\frac{n-1}{1-h_i}}$$

The interpretation of $W_i$ is more difficult, as it is based on the empirical influence curve. Note that whereas DFITS and Cook's distance are quite similar, the Welsch distance measure includes another normalization by leverage.

Belsley, Kuh, and Welsch (1980, 28) suggest that DFITS values greater than $2\sqrt{k/n}$ deserve further investigation, and so values of Cook's distance greater than $4/n$ should also be examined (Bollen and Jackman 1990, 265–266). Following similar logic, the cutoff for Welsch distance is approximately $3\sqrt{k}$ (Chatterjee and Hadi 1988, 124).

## ▷ Example 12: DFITS influence measure

Using our model of `price` on `weight`, `mpg`, `forXmpg`, and `foreign`, we can obtain the DFITS influence measure:

```
. use http://www.stata-press.com/data/r9/auto, clear
(1978 Automobile Data)
. generate forXmpg = foreign*mpg
. regress price weight mpg forXmpg foreign
  (output omitted)
. predict e if e(sample), resid
. predict dfits, dfits
```

Note that we did not specify `if e(sample)` in computing the DFITS statistic. DFITS is only available over the estimation sample, so specifying `if e(sample)` would have been redundant. It would have done no harm, but it would not have changed the results.

Our model has $k = 5$ independent variables ($k$ includes the constant) and $n = 74$ observations; following the $2\sqrt{k/n}$ cutoff advice, we type

```
. list make price e dfits if abs(dfits) > 2*sqrt(5/74), divider
```

| | make | price | e | dfits |
|---|---|---|---|---|
| 12. | Cad. Eldorado | 14,500 | 7271.96 | .9564455 |
| 13. | Cad. Seville | 15,906 | 5036.348 | 1.356619 |
| 24. | Ford Fiesta | 4,389 | 3164.872 | .5724172 |
| 27. | Linc. Mark V | 13,594 | 3109.193 | .5200413 |
| 28. | Linc. Versailles | 13,466 | 6560.912 | .8760136 |
| 42. | Plym. Arrow | 4,647 | -3312.968 | -.9384231 |

We calculate Cook's distance and list the observations greater than the suggested $4/n$ cutoff:

```
. predict cooksd if e(sample), cooksd
. list make price e cooksd if cooksd > 4/74, divider
```

| | make | price | e | cooksd |
|---|---|---|---|---|
| 40. | Cad. Eldorado | 14,500 | 7271.96 | .1492676 |
| 43. | Linc. Versailles | 13,466 | 6560.912 | .1308004 |
| 62. | Ford Fiesta | 4,389 | 3164.872 | .0638815 |
| 70. | Cad. Seville | 15,906 | 5036.348 | .3328515 |
| 71. | Plym. Arrow | 4,647 | -3312.968 | .1700736 |

Here we used `if e(sample)` because Cook's distance is not restricted to the estimation sample by default. It is worth comparing this list with the preceding one.

Finally, we use Welsch distance and the suggested $3\sqrt{k}$ cutoff:

```
. predict wd, welsch
. list make price e wd if abs(wd)>3*sqrt(5), divider
```

| | make | price | e | wd |
|---|---|---|---|---|
| 12. | Cad. Eldorado | 14,500 | 7271.96 | 8.394372 |
| 13. | Cad. Seville | 15,906 | 5036.348 | 12.81125 |
| 28. | Linc. Versailles | 13,466 | 6560.912 | 7.703005 |
| 42. | Plym. Arrow | 4,647 | -3312.968 | -8.981481 |

Note that here we did not need to specify `if e(sample)` since `welsch` automatically restricts the prediction to the estimation sample.

◁

## COVRATIO

COVRATIO (Belsley, Kuh, and Welsch 1980) measures the influence of the $i$th observation by considering the effect on the variance–covariance matrix of the estimates. The measure is the ratio of the determinants of the covariances matrix, with and without the $i$th observation. The resulting formula is

$$\text{COVRATIO}_i = \frac{1}{1-h_i}\left(\frac{n-k-\widehat{e}_i^{\,2}}{n-k-1}\right)^k$$

where $\widehat{e}_i$ is the standardized residual.

For noninfluential observations, the value of COVRATIO is approximately 1. Large values of the residuals or large values of leverage will cause deviations from 1, although if both are large, COVRATIO may tend back toward 1 and therefore not identify such observations (Chatterjee and Hadi 1988, 139).

Belsley, Kuh, and Welsch (1980) suggest that observations for which

$$|\text{COVRATIO}_i - 1| \geq \frac{3k}{n}$$

are worthy of further examination.

### ▷ Example 13: COVRATIO influence measure

Using our model of `price` on `weight`, `mpg`, `forXmpg`, and `foreign`, we can obtain the COVRATIO measure and list the observations outside the suggested cutoff by typing

```
. predict covr, covratio
. list make price e covr if abs(covr-1) >= 3*5/74, divider
```

| | make | price | e | covr |
|---|---|---|---|---|
| 40. | Cad. Eldorado | 14,500 | 7271.96 | .3814242 |
| 43. | Linc. Versailles | 13,466 | 6560.912 | .4761695 |
| 65. | Audi 5000 | 9,690 | 591.2883 | 1.206842 |
| 67. | Volvo 260 | 11,995 | 1327.668 | 1.211888 |
| 68. | Datsun 210 | 4,589 | 19.81829 | 1.284801 |
| 69. | Subaru | 3,798 | -909.5894 | 1.264677 |
| 70. | Cad. Seville | 15,906 | 5036.348 | .7386969 |
| 72. | Plym. Champ | 4,425 | 1621.747 | 1.27782 |
| 73. | Peugeot 604 | 12,990 | 1037.184 | 1.348219 |
| 74. | VW Diesel | 5,397 | 999.7209 | 1.630653 |

The `covratio` option automatically restricts the prediction to the estimation sample.

◁

## DFBETAs

DFBETAs are perhaps the most direct influence measure of interest to model builders. DFBETAs focus on one coefficient and measure the difference between the regression coefficient when the $i$th observation is included and excluded, the difference being scaled by the estimated standard error of the coefficient. Belsley, Kuh, and Welsch (1980, 28) suggest observations with $|\text{DFBETA}_i| > 2/\sqrt{n}$ as deserving special attention, but it is also common practice to use 1 (Bollen and Jackman 1990, 267), meaning that the observation shifted the estimate at least one standard error.

### ▷ Example 14: DFBETAs influence measure; the dfbeta() option

Using our model of `price` on `weight`, `mpg`, `forXmpg`, and `foreign`, let us first ask which observations have the greatest impact on the determination of the coefficient on `foreign`. We will use the suggested $2/\sqrt{n}$ cutoff:

```
. sort foreign make
. predict dfor, dfbeta(foreign)
. list make price foreign dfor if abs(dfor) > 2/sqrt(74), divider
```

| | make | price | foreign | dfor |
|---|---|---|---|---|
| 12. | Cad. Eldorado | 14,500 | Domestic | -.5290519 |
| 13. | Cad. Seville | 15,906 | Domestic | .8243419 |
| 28. | Linc. Versailles | 13,466 | Domestic | -.5283729 |
| 42. | Plym. Arrow | 4,647 | Domestic | -.6622424 |
| 43. | Plym. Champ | 4,425 | Domestic | .2371104 |
| 64. | Peugeot 604 | 12,990 | Foreign | .2552032 |
| 69. | Toyota Corona | 5,719 | Foreign | -.256431 |

Note that the Cadillac Seville shifted the **foreign** coefficient .82 standard deviations!

Now let us ask which observations have the greatest effect on the **mpg** coefficient:

```
. predict dmpg, dfbeta(mpg)
. list make price mpg dmpg if abs(dmpg) > 2/sqrt(74), divider
```

| | make | price | mpg | dmpg |
|---|---|---|---|---|
| 12. | Cad. Eldorado | 14,500 | 14 | -.5970351 |
| 13. | Cad. Seville | 15,906 | 21 | 1.134269 |
| 28. | Linc. Versailles | 13,466 | 14 | -.6069287 |
| 42. | Plym. Arrow | 4,647 | 28 | -.8925859 |
| 43. | Plym. Champ | 4,425 | 34 | .3186909 |

Once again we see the Cadillac Seville heading the list, suggesting that our regression results may be dominated by this one car.

◁

## ▷ Example 15: DFBETAs influence measure; the dfbeta command

We can use **predict, dfbeta()** or the **dfbeta** command to generate the DFBETAs. **dfbeta** makes up names for the new variables automatically and, without arguments, generates the DFBETAs for all the variables in the regression:

```
. dfbeta
                               DFweight:  DFbeta(weight)
                                  DFmpg:  DFbeta(mpg)
                              DFforXmpg:  DFbeta(forXmpg)
                              DFforeign:  DFbeta(foreign)
```

**dfbeta** created four new variables in our dataset: **DFweight**, containing the DFBETAs for **weight**; **DFmpg**, containing the DFBETAs for **mpg**; and so on. Alternatively, if we had only wanted the DFBETAs for **mpg** and **weight**, we might have typed

```
. dfbeta mpg weight
                                  DFmpg:  DFbeta(mpg)
                               DFweight:  DFbeta(weight)
```

In the example above, we typed **dfbeta mpg weight** instead of **dfbeta**—if we had typed **dfbeta** followed by **dfbeta mpg weight**, here is what would have happened:

```
. dfbeta
                               DFweight:  DFbeta(weight)
                                  DFmpg:  DFbeta(mpg)
                              DFforXmpg:  DFbeta(forXmpg)
                              DFforeign:  DFbeta(foreign)
. dfbeta mpg weight
                                    DF1:  DFbeta(mpg)
                                    DF2:  DFbeta(weight)
```

**dfbeta** would have made up different names for the new variables. **dfbeta** never replaces existing variables—it instead makes up a different name, so we need to pay attention to **dfbeta**'s output.

◁

## Formal tests for violations of assumptions

This section introduces some regression diagnostic commands that are designed to test for certain violations that `rvfplot` less formally attempts to detect. `estat ovtest` provides Ramsey's test for omitted variables—a pattern in the residuals. `estat hettest` provides a test for heteroskedasticity—the increasing or decreasing variation in the residuals with fitted values, with respect to the explanatory variables, or with respect to yet other variables. The score test implemented in `estat hettest` (Breusch and Pagan 1979; Cook and Weisberg 1983) performs a score test for the null hypothesis that $b = 0$ against the alternative hypothesis of multiplicative heteroskedasticity. `estat szroeter` provides a rank test for heteroskedasticity, which is an alternative to the score test computed by `estat hettest`. Finally, `estat imtest` computes an information matrix test, including an orthogonal decomposition into tests for heteroskedasticity, skewness, and kurtosis (Cameron and Trivedi 1990). The heteroskedasticity test computed by `estat imtest` is very similar to the general test for heteroskedasticity that was proposed by White (1980).

### ▷ Example 16: estat ovtest, estat hettest, estat szroeter, and estat imtest

We run these commands just mentioned on our model:

```
. estat ovtest

Ramsey RESET test using powers of the fitted values of price
       Ho:  model has no omitted variables
                 F(3, 66) =      7.77
                 Prob > F =      0.0002

. estat hettest

Breusch-Pagan / Cook-Weisberg tests for heteroskedasticity
         Ho: Constant variance
         variables: fitted values of price

         chi2(1)      =     6.50
         Prob > chi2  =   0.0108
```

Testing for heteroskedasticity in the right-hand-side variables is requested by specifying the option `rhs`. By specifying the option `mtest(bonferroni)`, we request that tests be conducted for each of the variables, with a Bonferroni-adjustment for the $p$-values to accommodate the fact that we are testing multiple hypotheses.

```
. estat hettest, rhs mtest(bonf)

Breusch-Pagan / Cook-Weisberg tests for heteroskedasticity
         Ho: Constant variance

---------------------------------------
    Variable |      chi2   df      p
-------------+-------------------------
      weight |     15.24    1   0.0004 #
         mpg |      9.04    1   0.0106 #
     forXmpg |      6.02    1   0.0566 #
     foreign |      6.15    1   0.0525 #
-------------+-------------------------
simultaneous |     15.60    4   0.0036
---------------------------------------
         # Bonferroni adjusted p-values
```

```
. estat szroeter, rhs mtest(holm)
Szroeter's test for homoskedasticity

    Ho: variance constant
    Ha: variance monotonic in variable
```

| Variable | chi2 | df | p |
|---|---|---|---|
| weight | 17.07 | 1 | 0.0001 # |
| mpg | 11.45 | 1 | 0.0021 # |
| forXmpg | 6.17 | 1 | 0.0260 # |
| foreign | 6.15 | 1 | 0.0131 # |

*# Holm adjusted p-values*

Finally, we request the information-matrix test, which is a conditional moments test with second-, third-, and fourth-order moment conditions.

```
. estat imtest
Cameron & Trivedi's decomposition of IM-test
```

| Source | chi2 | df | p |
|---|---|---|---|
| Heteroskedasticity | 18.86 | 10 | 0.0420 |
| Skewness | 11.69 | 4 | 0.0198 |
| Kurtosis | 2.33 | 1 | 0.1273 |
| Total | 32.87 | 15 | 0.0049 |

We find evidence for omitted variables, heteroskedasticity, and non-normal skewness.

So, why bother with the various graphical commands when the tests seem so much easier to interpret? In part, it is a matter of taste: both are designed to uncover the same problem, and both are, in fact, going about it in similar ways. One is based on a formal calculation, while the other is based on personal judgment in evaluating a graph. On the other hand, the tests are seeking evidence of quite specific problems while judgment is more general. The careful analyst will use both.

Note that we performed the omitted-variable test first. Omitted variables are a more serious problem than heteroskedasticity or the violations of higher moment conditions tested by `estat imtest`. If this were not a manual, having found evidence of omitted variables, we would never have run the `estat hettest`, `estat szroeter`, and `estat imtest` commands, at least not until we solved the omitted-variable problem.

◁

## ❑ Technical Note

`estat ovtest` and `estat hettest` both perform two flavors of their respective tests. By default, `estat ovtest` looks for evidence of omitted variables by fitting the original model augmented by $\widehat{y}^2$, $\widehat{y}^3$, and $\widehat{y}^4$, which are the fitted values from the original model. Under the assumption of no misspecification, the coefficients on the powers of the fitted values will be zero. With the `rhs` option, `estat ovtest` instead augments the original model with powers (second through fourth) of the explanatory variables (except for dummy variables).

`estat hettest`, by default, looks for heteroskedasticity by modeling the variance as a function of the fitted values. If, however, we specify a variable or variables, the variance will be modeled as a function of the specified variables. In our example, if we had, *a priori*, some reason to suspect heteroskedasticity and that the heteroskedasticity is a function of a car's weight, then using a test that focuses on weight would be more powerful than the more general tests such as White's test or the first term in the Cameron–Trivedi decomposition test.

❑

## Variance inflation factors

Problems arise in regression when the predictors are highly correlated. In this situation, there may be a significant change in the regression coefficients if you add or delete an independent variable. The estimated standard errors of the fitted coefficients are inflated, or the estimated coefficients may not be statistically significant even though a statistical relation exists between the dependent and independent variables.

Data analysts rely on these facts to check informally for the presence of multicollinearity. `estat vif`, another command for use after `regress`, calculates the variance inflation factors and tolerances for each of the independent variables.

The output shows the variance inflation factors together with their reciprocals. Some analysts compare the reciprocals with a predetermined tolerance. In the comparison, if the reciprocal of the VIF is smaller than the tolerance, the associated predictor variable is removed from the regression model. However, most analysts rely on informal rules of thumb applied to the VIF; see Chatterjee, Hadi, and Price (2000). According to these rules, there is evidence of multicollinearity if

1. The largest VIF is greater than 10 (some choose a more conservative threshold value of 30).
2. The mean of all the VIFs is considerably larger than 1.

### ▷ Example 17: estat vif

We examine a regression model fitted using the ubiquitous automobile dataset:

```
. regress price mpg rep78 trunk headroom length turn displ gear_ratio
      Source |       SS       df       MS              Number of obs =      69
-------------+------------------------------           F(  8,    60) =    6.33
       Model |  264102049     8  33012756.2           Prob > F      =  0.0000
    Residual |  312694909    60  5211581.82           R-squared     =  0.4579
-------------+------------------------------           Adj R-squared =  0.3856
       Total |  576796959    68  8482308.22           Root MSE      =  2282.9

------------------------------------------------------------------------------
       price |      Coef.   Std. Err.      t    P>|t|     [95% Conf. Interval]
-------------+----------------------------------------------------------------
         mpg |    -144.84   82.12751    -1.76   0.083    -309.1195    19.43948
       rep78 |   727.5783   337.6107     2.16   0.035     52.25638      1402.9
       trunk |   44.02061    108.141     0.41   0.685    -172.2935    260.3347
    headroom |  -807.0996   435.5802    -1.85   0.069      -1678.39    64.19061
      length |  -8.688914   34.89848    -0.25   0.804    -78.49626    61.11843
        turn |  -177.9064   137.3455    -1.30   0.200    -452.6383    96.82551
displacement |   30.73146   7.576952     4.06   0.000      15.5753    45.88762
  gear_ratio |   1500.119   1110.959     1.35   0.182    -722.1303    3722.368
       _cons |   6691.976   7457.906     0.90   0.373    -8226.057    21610.01
------------------------------------------------------------------------------
```

```
. estat vif

    Variable |       VIF       1/VIF
-------------+----------------------
      length |      8.22    0.121614
displacement |      6.50    0.153860
        turn |      4.85    0.205997
  gear_ratio |      3.45    0.290068
         mpg |      3.03    0.330171
       trunk |      2.88    0.347444
    headroom |      1.80    0.554917
       rep78 |      1.46    0.686147
-------------+----------------------
    Mean VIF |      4.02
```

The results are mixed. Although we do not have any VIFs greater than 10, the mean VIF is greater than 1, though not considerably so. We could continue the investigation of collinearity, but given that other authors advise that collinearity is only a problem when VIFs exist that are greater than 30 (contradicting our rule above), we will not do so here.

◁

## ▷ Example 18: estat vif, with strong evidence of multicollinearity

This example comes from a dataset described in Neter, Wasserman, and Kutner (1989) that examines body fat as modeled by caliper measurements on the triceps, midarm, and thigh.

```
. use http://www.stata-press.com/data/r9/bodyfat, clear
(Body Fat)

. regress bodyfat tricep thigh midarm

      Source |       SS       df       MS              Number of obs =      20
-------------+------------------------------           F(  3,    16) =   21.52
       Model |  396.984607     3  132.328202           Prob > F      =  0.0000
    Residual |  98.4049068    16  6.15030667           R-squared     =  0.8014
-------------+------------------------------           Adj R-squared =  0.7641
       Total |  495.389513    19  26.0731323           Root MSE      =    2.48

------------------------------------------------------------------------------
     bodyfat |      Coef.   Std. Err.      t    P>|t|     [95% Conf. Interval]
-------------+----------------------------------------------------------------
     triceps |   4.334085   3.015511     1.44   0.170    -2.058512    10.72668
       thigh |  -2.856842   2.582015    -1.11   0.285    -8.330468    2.616785
      midarm |  -2.186056   1.595499    -1.37   0.190    -5.568362     1.19625
       _cons |   117.0844   99.78238     1.17   0.258    -94.44474    328.6136
------------------------------------------------------------------------------

. estat vif

    Variable |       VIF       1/VIF
-------------+----------------------
     triceps |    708.84    0.001411
       thigh |    564.34    0.001772
      midarm |    104.61    0.009560
-------------+----------------------
    Mean VIF |    459.26
```

In this example, we see very strong evidence of multicollinearity in our model. Further investigation reveals that the measurements on the thigh and the triceps are highly correlated:

```
. corr triceps thigh midarm
(obs=20)

             |  triceps    thigh   midarm
-------------+---------------------------
     triceps |   1.0000
       thigh |   0.9238   1.0000
      midarm |   0.4578   0.0847   1.0000
```

If we remove the predictor `tricep` from the model (since it had the highest VIF), we get

```
. regress bodyfat thigh midarm

      Source |       SS       df       MS              Number of obs =      20
-------------+------------------------------           F(  2,    17) =   29.40
       Model |  384.279748     2  192.139874           Prob > F      =  0.0000
    Residual |  111.109765    17  6.53586854           R-squared     =  0.7757
-------------+------------------------------           Adj R-squared =  0.7493
       Total |  495.389513    19  26.0731323           Root MSE      =  2.5565

------------------------------------------------------------------------------
     bodyfat |      Coef.   Std. Err.      t    P>|t|     [95% Conf. Interval]
-------------+----------------------------------------------------------------
       thigh |   .8508818   .1124482     7.57   0.000     .6136367    1.088127
      midarm |   .0960295   .1613927     0.60   0.560    -.2444792    .4365383
       _cons |  -25.99696    6.99732    -3.72   0.002    -40.76001   -11.2339
------------------------------------------------------------------------------

. estat vif

    Variable |       VIF       1/VIF
-------------+----------------------
      midarm |      1.01    0.992831
       thigh |      1.01    0.992831
-------------+----------------------
    Mean VIF |      1.01
```

Note how the coefficients change and how the estimated standard errors for each of the regression coefficients become much smaller. The calculated value of $R^2$ for the overall regression for the subset model does not appreciably decline when we remove the correlated predictor. Removing an independent variable from the model is one way to deal with multicollinearity. Other methods include ridge regression, weighted least squares, and restricting the use of the fitted model to data that follow the same pattern of multicollinearity. In economic studies, it is sometimes possible to estimate the regression coefficients from different subsets of the data using cross-section and time series.

◁

# Saved Results

`estat hettest` saves results for the (multivariate) score test in `r()`:

Scalars

| | |
|---|---|
| `r(chi2)` | test statistic |
| `r(df)` | #df for the asymptotic $\chi^2$ distribution under $H_0$ |
| `r(p)` | $p$-value |

`estat hettest` (if `mtest` is specified) and `estat szroeter` save in `r()`:

Matrix

| | |
|---|---|
| `r(mtest)` | a matrix of test results, with rows corresponding to the univariate tests |
| `mtest[.,1]` | chi2 test statistic |
| `mtest[.,2]` | #df |
| `mtest[.,3]` | unadjusted $p$-value |
| `mtest[.,4]` | adjusted $p$-value (if an `mtest()` adjustment method is specified) |

Macros

| | |
|---|---|
| `r(mtmethod)` | adjustment method for $p$-values |

`estat imtest` saves in `r()`:

Scalars

| | |
|---|---|
| `r(chi2_t)` | IM-test statistic (= `r(chi2_h)` + `r(chi2_s)` + `r(chi2_k)`) |
| `r(df_t)` | df for limiting $\chi^2$ distribution under $H_0$ (= `r(df_h)` + `r(df_s)` + `r(df_k)`) |
| `r(chi2_h)` | heteroskedasticity test statistic |
| `r(df_h)` | df for limiting $\chi^2$ distribution under $H_0$ |
| `r(chi2_s)` | skewness test statistic |
| `r(df_s)` | df for limiting $\chi^2$ distribution under $H_0$ |
| `r(chi2_k)` | kurtosis test statistic |
| `r(df_k)` | df for limiting $\chi^2$ distribution under $H_0$ |
| `r(chi2_w)` | White's heteroskedasticity test (if `white` specified) |
| `r(df_w)` | df for limiting $\chi^2$ distribution under $H_0$ |

`estat ovtest` saves in `r()`:

Scalars

| | | | |
|---|---|---|---|
| `r(p)` | two-sided $p$-value | `r(df)` | degrees of freedom |
| `r(F)` | $F$ statistic | `r(df_r)` | residual degrees of freedom |

# Methods and Formulas

All regression fit and diagnostic commands are implemented as ado-files.

See Hamilton (2004, chapter 7) for an overview of regression diagnostics using Stata. See Peracchi (2001, chapter 8) for a mathematically rigorous discussion of diagnostics.

(*Continued on next page*)

## Methods and formulas for predict

Assume that you have already fitted the regression model

$$\mathbf{y} = \mathbf{X}\mathbf{b} + \mathbf{e}$$

where $\mathbf{X}$ is $n \times k$.

Denote the previously estimated coefficient vector by $\mathbf{b}$ and its estimated variance matrix by $\mathbf{V}$. `predict` works by recalling various aspects of the model, such as $\mathbf{b}$, and combining that information with the data currently in memory. Let $\mathbf{x}_j$ be the $j$th observation currently in memory, and let $s^2$ be the mean squared error of the regression.

Let $\mathbf{V} = s^2(\mathbf{X}'\mathbf{X})^{-1}$. Let $k$ be the number of independent variables including the intercept, if any, and let $y_j$ be the observed value of the dependent variable.

The *predicted value* (`xb` option) is defined $\widehat{y}_j = \mathbf{x}_j\mathbf{b}$.

Let $\ell_j$ represent a lower bound for an observation $j$ and $u_j$ represent an upper bound. The probability that $y_j|\mathbf{x}_j$ would be observed in the interval $(\ell_j, u_j)$—option `pr(`$\ell, u$`)`—is

$$P(\ell_j, u_j) = \Pr(\ell_j < \mathbf{x}_j\mathbf{b} + e_j < u_j) = \Phi\left(\frac{u_j - \widehat{y}_j}{s}\right) - \Phi\left(\frac{\ell_j - \widehat{y}_j}{s}\right)$$

where for the options `pr(`$\ell, u$`)`, `e(`$\ell, u$`)`, and `ystar(`$\ell, u$`)`, $\ell_j$ and $u_j$ can be anywhere in the range $(-\infty, +\infty)$.

The option `e(`$\ell, u$`)` computes the expected value of $y_j|\mathbf{x}_j$ conditional on $y_j|\mathbf{x}_j$ being in the interval $(\ell_j, u_j)$, that is, when $y_j|\mathbf{x}_j$ is censored. It can be expressed as

$$E(\ell_j, u_j) = E(\mathbf{x}_j\mathbf{b} + e_j \mid \ell_j < \mathbf{x}_j\mathbf{b} + e_j < u_j) = \widehat{y}_j - s\frac{\phi\left(\frac{u_j - \widehat{y}_j}{s}\right) - \phi\left(\frac{\ell_j - \widehat{y}_j}{s}\right)}{\Phi\left(\frac{u_j - \widehat{y}_j}{s}\right) - \Phi\left(\frac{\ell_j - \widehat{y}_j}{s}\right)}$$

where $\phi$ is the normal density and $\Phi$ is the cumulative normal.

You can also compute `ystar(`$\ell, u$`)`—the expected value of $y_j|\mathbf{x}_j$, where $y_j$ is assumed truncated at $\ell_j$ and $u_j$:

$$y_j^* = \begin{cases} \ell_j & \text{if } \mathbf{x}_j\mathbf{b} + e_j \le \ell_j \\ \mathbf{x}_j\mathbf{b} + u & \text{if } \ell_j < \mathbf{x}_j\mathbf{b} + e_j < u_j \\ u_j & \text{if } \mathbf{x}_j\mathbf{b} + e_j \ge u_j \end{cases}$$

This computation can be expressed in several ways, but the most intuitive formulation involves a combination of the two statistics just defined:

$$y_j^* = P(-\infty, \ell_j)\ell_j + P(\ell_j, u_j)E(\ell_j, u_j) + P(u_j, +\infty)u_j$$

A diagonal element of the projection matrix (`hat`) or (`leverage`) is given by

$$h_j = \mathbf{x}_j(\mathbf{X}'\mathbf{X})^{-1}\mathbf{x}_j'$$

The *standard error of the prediction* (`stdp`) is defined as $s_{p_j} = \sqrt{\mathbf{x}_j\mathbf{V}\mathbf{x}_j'}$

and can also be written $s_{p_j} = s\sqrt{h_j}$.

The *standard error of the forecast* (`stdf`) is defined as $s_{f_j} = s\sqrt{1+h_j}$.

The *standard error of the residual* (`stdr`) is defined as $s_{r_j} = s\sqrt{1-h_j}$.

The *residuals* (`residuals`) are defined as $\widehat{e}_j = y_j - \widehat{y}_j$.

The *standardized residuals* (`rstandard`) are defined as $\widehat{e}_{s_j} = \widehat{e}_j / s_{r_j}$.

The *studentized residuals* (`rstudent`) are defined as

$$r_j = \frac{\widehat{e}_j}{s_{(j)}\sqrt{1-h_j}}$$

where $s_{(j)}$ represents the root mean squared error with the $j$th observation removed, which is given by

$$s^2_{(j)} = \frac{s^2(T-k)}{T-k-1} - \frac{\widehat{e}^2_j}{(T-k-1)(1-h_j)}$$

Cook's $D$ (`cooksd`) is given by

$$D_j = \frac{\widehat{e}^2_{s_j}(s_{p_j}/s_{r_j})^2}{k} = \frac{h_j \widehat{e}^2_j}{ks^2(1-h_j)^2}$$

DFITS (`dfits`) is given by

$$\text{DFITS}_j = r_j\sqrt{\frac{h_j}{1-h_j}}$$

Welsch distance (`welsch`) is given by

$$W_j = \frac{r_j\sqrt{h_j(n-1)}}{1-h_j}$$

COVRATIO (`covratio`) is given by

$$\text{COVRATIO}_j = \frac{1}{1-h_j}\left(\frac{n-k-\widehat{e}^2_j}{n-k-1}\right)^k$$

The DFBETAs (`dfbeta`) for a particular regressor $x_i$ are given by

$$\text{DFBETA}_j = \frac{r_j u_j}{\sqrt{U^2(1-h_j)}}$$

where $u_j$ are the residuals obtained from a regression of $x_j$ on the remaining $x$s and $U^2 = \sum_j u^2_j$.

The omitted-variable test (Ramsey 1969) reported by `ovtest` fits the regression $y_i = \mathbf{x}_i\mathbf{b} + \mathbf{z}_i\mathbf{t} + u_i$ and then performs a standard $F$ test of $\mathbf{t} = \mathbf{0}$. The default test uses $\mathbf{z}_i = (\widehat{y}^2_i, \widehat{y}^3_i, \widehat{y}^4_i)$. If `rhs` is specified, $\mathbf{z}_i = (x^2_{1i}, x^3_{1i}, x^4_{1i}, x^2_{2i}, \ldots, x^4_{mi})$. In either case, the variables are normalized to have minimum 0 and maximum 1 before powers are calculated.

The test for heteroskedasticity (Breusch and Pagan 1979; Cook and Weisberg 1983) models $\mathrm{Var}(e_i) = \sigma^2 \exp(\mathbf{zt})$, where $\mathbf{z}$ is either a variable list specified by the user, the list of right-hand-side variables, or the fitted values $\mathbf{x}\widehat{\beta}$. The test is of $\mathbf{t} = \mathbf{0}$. Mechanically, `estat hettest` fits the model $\widehat{e}_i^2/\widehat{\sigma}^2 = a + \mathbf{z}_i\mathbf{t} + v_i$ and then forms the score test statistic $S$ equal to the model sum of squares divided by 2. Under the null hypothesis, $S$ has the $\chi^2$ distribution with $m$ degrees of freedom, where $m$ is the number of columns of $\mathbf{z}$.

Szroeter's class of tests for homoskedasticity against the alternative that the residual variance increases in some variable $x$ is defined in terms of

$$H = \frac{\sum_{i=1}^{n} h(x_i) e_i^2}{\sum_{i=1}^{n} e_i^2}$$

where $h(x)$ is some weight function that increases in $x$ (Szroeter 1978). $H$ is a weighted average of the $h(x)$, with the squared residuals serving as weights. Under homoskedasticity, $H$ should be approximately equal to the unweighted average of $h(x)$. Large values of $H$ suggest that $e_i^2$ tends to be large where $h(x)$ is large, i.e., the variance indeed increases in $x$, while small values of $H$ suggest that the variance actually decreases in $x$. `estat szroeter` uses $h(x_i) = \mathrm{rank}(x_i \text{ in } x_1 \ldots x_n)$; see Judge 1985, 452 for details. `estat szroeter` displays a normalized version of $H$,

$$Q = \sqrt{\frac{6n}{n^2 - 1}} H$$

which is approximately $N(0, 1)$ distributed under the null (homoskedasticity).

`estat hettest` and `estat szroeter` provide adjustments of $p$-values for multiple testing. The supported methods are described in [R] **test**.

`estat imtest` performs the information matrix test for the regression model, as well as an orthogonal decomposition into tests for heteroskedasticity $\delta_1$, non-normal skewness $\delta_2$, and non-normal kurtosis $\delta_3$ (Cameron and Trivedi 1990; Long and Trivedi 1992). The decomposition is obtained via three auxiliary regressions. Let $e$ be the regression residuals, $\widehat{\sigma}^2$ be the maximum likelihood estimate of $\sigma^2$ in the regression, $n$ be the number of observations, $X$ be the set of $k$ variables specified with `estat imtest`, and $R^2_{\mathrm{un}}$ be the uncentered $R^2$ from a regression. $\delta_1$ is obtained as $nR^2_{\mathrm{un}}$ from a regression of $e^2 - \widehat{\sigma}^2$ on the cross-products of the variables in $X$. $\delta_2$ is computed as $nR^2_{\mathrm{un}}$ from a regression of $e^3 - 3\widehat{\sigma}^2 e$ on $X$. Finally, $\delta_3$ is obtained as $nR^2_{\mathrm{un}}$ from a regression of $e^4 - 6\widehat{\sigma}^2 e^2 - 3\widehat{\sigma}^4$ on $X$. $\delta_1$, $\delta_2$, and $\delta_3$ are asymptotically $\chi^2$-distributed with $\frac{1}{2}k(k+1)$, $K$, and 1 degrees of freedom. The information test statistic $\delta = \delta_1 + \delta_2 + \delta_3$ is asymptotically $\chi^2$-distributed with $\frac{1}{2}k(k+3)$ degrees of freedom. White's test for heteroskedasticity is computed as $nR^2$ from a regression of $\widehat{u}^2$ on $X$ and the cross-products of the variables in $X$. This test statistic is usually very close to $\delta_1$.

The variance inflation factor (VIF) (Chatterjee, Hadi, and Price 2000, 240–242) for $x_j$ is given by

$$\mathrm{VIF}(x_j) = \frac{1}{1 - R_j^2}$$

where $R_j^2$ is the square of the multiple correlation coefficient that results when $x_j$ is regressed against all the other explanatory variables.

## Acknowledgments

estat ovtest and estat hettest are based on programs originally written by Richard Goldstein (1991, 1992). estat imtest, estat szroeter, and the current version of estat hettest were written by Jeroen Weesie; estat imtest is based in part on code written by J. Scott Long.

## References

Baum, C. F., N. J. Cox, and V. Wiggins. 2000. sg137: Tests for heteroskedasticity in regression error distribution. *Stata Technical Bulletin* 55: 15–17. Reprinted in *Stata Technical Bulletin Reprints*, vol. 10, pp. 147–149.

Baum, C. F. and V. Wiggins. 2000a. sg135: Test for autoregressive conditional heteroskedasticity in regression error distribution. *Stata Technical Bulletin* 55: 13–14. Reprinted in *Stata Technical Bulletin Reprints*, vol. 10, pp. 143–144.

——. 2000b. sg136: Tests for serial correlation in regression error distribution. *Stata Technical Bulletin* 55: 14–15. Reprinted in *Stata Technical Bulletin Reprints*, vol. 10, pp. 145–147.

Belsley, D. A., E. Kuh, and R. E. Welsch. 1980. *Regression Diagnostics*. New York: Wiley.

Bollen, K. A. and R. W. Jackman. 1990. Regression diagnostics: An expository treatment of outliers and influential cases. In *Modern Methods of Data Analysis*, ed. J. Fox and J. S. Long, 257–291. Newbury Park, CA: Sage.

Breusch, T. and A. Pagan. 1979. A simple test for heteroscedasticity and random coefficient variation. *Econometrica* 47: 1287–1294.

Cameron, A. C. and P. K. Trivedi. 1990. *The Information Matrix Test and Its Applied Alternative Hypotheses*. Working Paper.

Chatterjee, S. and A. S. Hadi. 1986. Influential observations, high leverage points, and outliers in linear regression. *Statistical Science* 1: 379–416.

——. 1988. *Sensitivity Analysis in Linear Regression*. New York: Wiley.

Chatterjee, S., A. S. Hadi, and B. Price. 2000. *Regression Analysis by Example*. 3rd ed. New York: Wiley.

Cook, R. D. 1977. Detection of influential observations in linear regression. *Technometrics* 19: 15–18.

Cook, R. D. and S. Weisberg. 1982. *Residuals and Influence in Regression*. New York: Chapman & Hall.

——. 1983. Diagnostics for heteroscedasticity in regression. *Biometrika* 70: 1–10.

Cox, N. J. 2004 Speaking Stata: Graphing model diagnostics. *Stata Journal* 4: 449–475.

Durbin, J. and G. S. Watson. 1950. Testing for serial correlation in least-squares regression. *Biometrika* 37: 409–428.

——. 1951. Testing for serial correlation in least-squares regression. *Biometrika* 38: 159–178.

Ezekiel, M. 1924. A method of handling curvilinear correlation for any number of variables. *Journal of the American Statistical Association* 19: 431–453.

Garrett, J. M. 2000. sg157: Predicted values calculated from linear or logistic regression models. *Stata Technical Bulletin* 58: 27–30. Reprinted in *Stata Technical Bulletin Reprints*, vol. 10, pp. 258–261.

Goldstein, R. 1991. srd5: Ramsey test for heteroskedasticity and omitted variables. *Stata Technical Bulletin* 2: 27. Reprinted in *Stata Technical Bulletin Reprints*, vol. 1, p. 177.

——. 1992. srd14: Cook–Weisberg test of heteroskedasticity. *Stata Technical Bulletin* 10: 27–28. Reprinted in *Stata Technical Bulletin Reprints*, vol. 2, pp. 183–184.

Hamilton, L. C. 1992. *Regression with Graphics*. Pacific Grove, CA: Brooks/Cole.

——. 2004. *Statistics with Stata*. Belmont, CA: Brooks/Cole.

Hardin, J. W. 1995. sg32: Variance inflation factors and variance-decomposition proportions. *Stata Technical Bulletin* 24: 17–22. Reprinted in *Stata Technical Bulletin Reprints*, vol. 4, pp. 154–160.

Hoaglin, D. C. and P. J. Kempthorne. 1986. Comment [on Chatterjee and Hadi 1986]. *Statistical Science* 1: 408–412.

Hoaglin, D. C. and R. E. Welsch. 1978. The hat matrix in regression and ANOVA. *The American Statistician* 32: 17–22.

Judge, G. G., W. E. Griffiths, R. C. Hill, H. Lütkepohl, and T. C. Lee. 1985. *The Theory and Practice of Econometrics*. 2nd ed. New York: Wiley.

Larsen, W. A. and S. J. McCleary. 1972. The use of partial residual plots in regression analysis. *Technometrics* 14: 781–790.

Long, J. S. and J. Freese. 2000. sg145: Scalar measures of fit for regression models. *Stata Technical Bulletin* 56: 34–40. Reprinted in *Stata Technical Bulletin Reprints*, vol. 10, pp. 197–205.

Long, J. S. and P. K. Trivedi. 1992. Some specification tests for the linear regression model. *Sociological Methods and Research* 21 (2): 161–204. Reprinted in *Testing Structural Equation Models*, ed. K. A. Bollen and J. S. Long, 66–110. Newbury Park, CA: Sage.

Mallows, C. L. 1986. Augmented partial residuals. *Technometrics* 28: 313–319.

Mosteller, F. and J. W. Tukey. 1977. *Data Analysis and Regression*. Reading, MA: Addison–Wesley.

Neter, J., W. Wasserman, and M. H. Kutner. 1989. *Applied Linear Regression Models*. Homewood, IL: Irwin.

Peracchi, F. 2001. *Econometrics*. Chichester, UK: Wiley.

Ramsey, J. B. 1969. Tests for specification errors in classical linear least-squares regression analysis. *Journal of the Royal Statistical Society, Series B* 31: 350–371.

Ramsey, J. B. and P. Schmidt. 1976. Some further results on the use of OLS and BLUS residuals in specification error tests. *Journal of the American Statistical Association* 71: 389–390.

Rousseeuw, P. J. and A. M. Leroy. 1987. *Robust Regression and Outlier Detection*. New York: Wiley.

Ryan, T. P. 1997. *Modern Regression Methods*. New York: Wiley.

Szroeter, J. 1978. A class of parametric tests for heteroscedasticity in linear econometric models. *Econometrica* 46, 1311–28.

Velleman, P. F. 1986. Comment [on Chatterjee and Hadi 1986]. *Statistical Science* 1: 412–413.

Velleman, P. F. and R. E. Welsch. 1981. Efficient computing of regression diagnostics. *The American Statistician* 35: 234–242.

Weesie, J. 2001. sg161: Analysis of the turning point of a quadratic specification. *Stata Technical Bulletin* 60: 18–20. Reprinted in *Stata Technical Bulletin Reprints*, vol. 10, pp. 273–277.

Weisberg, S. 1985. *Applied Linear Regression*. 2nd ed. New York: Wiley.

Welsch, R. E. 1982. Influence functions and regression diagnostics. In *Modern Data Analysis*, ed. R. L. Launer and A. F. Siegel, 149–169. New York: Academic Press.

——. 1986. Comment [on Chatterjee and Hadi 1986]. *Statistical Science* 1: 403–405.

Welsch, R. E. and E. Kuh. 1977. *Technical Report 923-77: Linear Regression Diagnostics*. Cambridge, MA: Sloan School of Management, Massachusetts Institute of Technology.

White, H. 1980. A heteroskedasticity-consistent covariance matrix estimator and a direct test for heteroskedasticity. *Econometrica*, 48: 817–838.

## Also See

**Complementary:** [R] **regress**; [R] **adjust**, [R] **estimates**, [R] **hausman**, [R] **lincom**, [R] **linktest**, [R] **lrtest**, [R] **mfx**, [R] **nlcom**, [R] **predictnl**, [R] **suest**, [R] **test**, [R] **testnl**

**Related:** [R] **regress postestimation time series**

**Background:** [U] **20 Estimation and postestimation commands**, [R] **estat**, [R] **predict**, *Stata Graphics Reference Manual*

# Title

**regress postestimation time series** — Postestimation tools for regress with time series

# Description

The following postestimation commands for time series are available for `regress`:

| command | description |
|---|---|
| `estat archlm` | test for ARCH effects in the residuals |
| `estat bgodfrey` | Breusch–Godfrey test for higher-order serial correlation |
| `estat durbinalt` | Durbin's alternative test for serial correlation |
| `estat dwatson` | Durbin–Watson *d* statistic to test for first-order serial correlation |

These commands provide regression diagnostic tools specific to time series. You must `tsset` your data before using these commands.

`estat archlm` tests for time-dependent volatility. `estat bgodfrey`, `estat durbinalt`, and `estat dwatson` test for serial correlation in the residuals of a linear regression. For non-time-series regression diagnostic tools, see [R] **regress postestimation**.

`estat archlm` performs Engle's LM test for the presence of autoregressive conditional heteroskedasticity.

`estat bgodfrey` performs the Breusch–Godfrey test for higher-order serial correlation in the disturbance. This test does not require that all the regressors be strictly exogenous.

`estat durbinalt` performs Durbin's alternative test for serial correlation in the disturbance. This test does not require that all the regressors be strictly exogenous.

`estat dwatson` computes the Durbin–Watson *d* statistic (Durbin and Watson 1950) to test for first-order serial correlation in the disturbance when all the regressors are strictly exogenous.

# Syntax for estat archlm

`estat archlm` [ , *archlm_options* ]

| *archlm_options* | description |
|---|---|
| `lags(`*numlist*`)` | test *numlist* lag orders |
| `force` | allow test after `regress, robust` |

# Options for estat archlm

`lags(`*numlist*`)` specifies a list of numbers, indicating the lag orders to be tested. The test will be performed separately for each order. The default is order one.

`force` allows the test to be run after `regress, robust`. The command will not work if the `cluster` option is specified with [R] **regress**.

## Syntax for estat bgodfrey

`estat bgodfrey [ , bgodfrey_options ]`

| *bgodfrey_options* | description |
|---|---|
| `lags(`*numlist*`)` | test *numlist* lag orders |
| `nomiss0` | do not use Davidson and MacKinnon's approach |
| `small` | obtain $p$-values using the $F$ or $t$ distribution |

## Options for estat bgodfrey

`lags(`*numlist*`)` specifies a list of numbers, indicating the lag orders to be tested. The test will be performed separately for each order. The default is order one.

`nomiss0` specifies that Davidson and MacKinnon's approach (1993, 358), which replaces the missing values in the initial observations on the lagged residuals in the auxiliary regression with zeros, not be used.

`small` specifies that the $p$-values of the test statistics be obtained using the $F$ or $t$ distribution instead of the default chi-squared or normal distribution. This option may not be specified with `robust`, which always uses an $F$ or $t$ distribution.

## Syntax for estat durbinalt

`estat durbinalt [ , durbinalt_options ]`

| *durbinalt_options* | description |
|---|---|
| `lags(`*numlist*`)` | test *numlist* lag orders |
| `nomiss0` | do not use Davidson and MacKinnon's approach |
| `robust` | compute standard errors using the robust/sandwich estimator |
| `small` | obtain $p$-values using the $F$ or $t$ distribution |
| `force` | allow test after `regress, robust` or after `newey` |

## Options for estat durbinalt

`lags(`*numlist*`)` specifies a list of numbers, indicating the lag orders to be tested. The test will be performed separately for each order. The default is order one.

`nomiss0` specifies that Davidson and MacKinnon's approach (1993, 358), which replaces the missing values in the initial observations on the lagged residuals in the auxiliary regression with zeros, not be used.

`robust` specifies that the Huber/White/sandwich robust estimator of the variance–covariance matrix be used in Durbin's alternative test.

`small` specifies that the $p$-values of the test statistics be obtained using the $F$ or $t$ distribution instead of the default chi-squared or normal distribution. This option may not be specified with `robust`, which always uses an $F$ or $t$ distribution.

`force` allows the test to be run after `regress, robust` and after `newey` (see [R] **regress** and [TS] **newey**). The command will not work if the `cluster` option is specified with [R] **regress**.

## Syntax for estat dwatson

```
estat dwatson
```

## Remarks

The Durbin–Watson test is used to determine whether the error term in a linear regression model follows an AR(1) process. For the linear model

$$y_t = \mathbf{x}_t\boldsymbol{\beta} + u_t$$

the AR(1) process can be written as

$$u_t = \rho u_{t-1} + \epsilon_t$$

In general, an AR(1) process requires only that $\epsilon_t$ be independently and identically distributed (i.i.d.). The Durbin–Watson test, however, requires $\epsilon_t$ to be distributed $\mathrm{N}(0, \sigma^2)$ for the statistic to have an exact distribution. Also, the Durbin–Watson test can only be applied when the regressors are strictly exogenous. A regressor $x$ is strictly exogenous if $\mathrm{Corr}(x_s, u_t) = 0$ for all $s$ and $t$, which precludes the use of the Durbin–Watson statistic with models where lagged values of the dependent variable are included as regressors.

The null hypothesis of the test is that there is no first-order autocorrelation. The Durbin–Watson $d$ statistic can take on values between 0 and 4 and under the null $d$ is equal to 2. Values of $d$ less than two suggest positive autocorrelation ($\rho > 0$), while values of $d$ greater than two suggest negative autocorrelation ($\rho < 0$). Calculating the exact distribution of the $d$ statistic is difficult, but empirical upper and lower bounds have been established based on the sample size and the number of regressors. Extended tables for the $d$ statistic have been published by Savin and White (1977). For example, suppose you have a model with 30 observations and 3 regressors (including the constant term). For a test of the null hypothesis of no autocorrelation versus the alternative of positive autocorrelation, the lower bound of the $d$ statistic is 1.284, and the upper bound is 1.567 at the 5% significance level. You would reject the null if $d < 1.284$, and you would fail to reject if $d > 1.567$. A value falling within the range (1.284, 1.567) leads to no conclusion about whether or not to reject the null hypothesis.

When lagged dependent variables are included among the regressors, the past values of the error term are correlated with those lagged variables at time $t$, implying that they are not strictly exogenous regressors. The inclusion of covariates that are not strictly exogenous causes the $d$ statistic to be biased toward the acceptance of the null hypothesis. Durbin (1970) suggested an alternative test for models with lagged dependent variables and extended that test to the more general AR($p$) serial correlation process

$$u_t = \rho_1 u_{t-1} + \cdots + \rho_p u_{t-p} + \epsilon_t$$

where $\epsilon_t$ is independently and identically distributed with variance $\sigma^2$ but is not assumed or required to be normal for the test.

The null hypothesis of Durbin's alternative test is

$$H_0 : \rho_1 = 0, \ldots, \rho_p = 0$$

and the alternative is that at least one of the $\rho$'s is nonzero. Although the null hypothesis was originally derived for an AR($p$) process, this test turns out to have power against MA($p$) processes as well. Hence, the actual null of this test is that there is no serial correlation up to order $p$. The reason for this serendipitous result is that the MA($p$) and the AR($p$) models are locally equivalent alternatives under the null. See Godfrey (1988, 113–115) for a discussion of this result.

Durbin's alternative test is in fact a Lagrange-multiplier (LM) test, but it is most easily computed with a Wald test on the coefficients of the lagged residuals in an auxiliary OLS regression of the residuals on their lags and all the covariates in the original regression. Consider the linear regression model

$$y_t = \beta_1 x_{1t} + \cdots + \beta_k x_{kt} + u_t \tag{1}$$

in which the covariates $x_1$ through $x_k$ are not assumed to be strictly exogenous and $u_t$ is assumed to be i.i.d and to have finite variance. The process is also assumed to be stationary. (See Wooldridge [2002] for a discussion of stationarity.) Estimating the parameters in (1) by OLS obtains the residuals $\widehat{u}_t$. Next another OLS regression is performed of $\widehat{u}_t$ on $\widehat{u}_{t-1}, \ldots, \widehat{u}_{t-p}$ and the other regressors,

$$\widehat{u}_t = \gamma_1 \widehat{u}_{t-1} + \cdots + \gamma_p \widehat{u}_{t-p} + \beta_1 x_{1t} + \cdots + \beta_k x_{kt} + \epsilon_t \tag{2}$$

where $\epsilon_t$ stands for the random-error term in this auxiliary OLS regression. Durbin's alternative test is then obtained by performing a Wald test that $\gamma_1, \ldots, \gamma_p$ are jointly zero. The test can be made robust to an unknown form of heteroskedasticity by using a robust VCE estimator when estimating the regression in (2). When there are only strictly exogenous regressors and $p = 1$, this test is asymptotically equivalent to the Durbin–Watson test.

The Breusch–Godfrey test is also an LM test of the null hypothesis of no autocorrelation versus the alternative that $u_t$ follows an AR($p$) or MA($p$) process. Like Durbin's alternative test, it is based on the auxiliary regression (2), and it is computed as $NR^2$, where $N$ is the number of observations and $R^2$ is the simple $R^2$ from the regression. This test and Durbin's alternative test are asymptotically equivalent. The test statistic $NR^2$ has an asymptotic $\chi^2$ distribution with $p$ degrees of freedom. It is valid with or without the strict exogeneity assumption but is not robust to conditional heteroskedasticity, even if a robust VCE is used when fitting (2).

In fitting (2), the values of the lagged residuals will be missing in the initial time periods. As noted by Davidson and MacKinnon (1993), the residuals will not be orthogonal to the other covariates in the model in this restricted sample. This implies that the $R^2$ from the auxiliary regression will not be zero when the lagged residuals are left out. Hence Breusch and Godfrey's $NR^2$ version of the test may over-reject in small samples. To correct this problem, Davidson and MacKinnon (1993) recommend setting the missing values of the lagged residuals to zero and running the auxiliary regression in (2) over the full sample used in (1). This small-sample correction has become conventional for both the Breusch–Godfrey and Durbin's alternative test, and it is the default for both commands. Specifying the `nomiss0` option overrides this default behavior and treats the initial missing values generated by regressing on the lagged residuals as missing. Hence, `miss0` causes these initial observations to be dropped from the sample of the auxiliary regression.

Durbin's alternative test and the Breusch–Godfrey test were originally derived for the case covered by `regress` without the `robust` option. However, after `regress, robust` and `newey`, Durbin's alternative test is still valid and can be invoked if the `robust` and `force` options are specified.

## ▷ Example 1: tests for serial correlation

Using data from Klein (1950), we first fit an OLS regression of consumption on the government wage bill:

```
. use http://www.stata-press.com/data/r9/klein
. tsset yr
        time variable:  yr, 1920 to 1941
. regress consump wagegovt

      Source |       SS       df       MS              Number of obs =      22
-------------+------------------------------           F(  1,    20) =   17.72
       Model |  532.567711     1  532.567711           Prob > F      =  0.0004
    Residual |  601.207167    20  30.0603584           R-squared     =  0.4697
-------------+------------------------------           Adj R-squared =  0.4432
       Total |  1133.77488    21  53.9892799           Root MSE      =  5.4827

------------------------------------------------------------------------------
     consump |      Coef.   Std. Err.      t    P>|t|     [95% Conf. Interval]
-------------+----------------------------------------------------------------
    wagegovt |    2.50744   .5957173     4.21   0.000     1.264796    3.750085
       _cons |   40.84699   3.192183    12.80   0.000     34.18821    47.50577
------------------------------------------------------------------------------
```

If we assume that wagegov is a strictly exogenous variable, we can use the Durbin–Watson test to check for first-order serial correlation in the errors.

```
. estat dwatson
Durbin-Watson d-statistic(  2,    22) =  .3217998
```

The Durbin–Watson $d$ statistic 0.32 is far from the center of its distribution ($d = 2.0$). Given 22 observations and 2 regressors (including the constant term) in the model, the lower 5% bound is about 0.997, much greater than the computed $d$ statistic. Assuming that wagegov is strictly exogenous, we can reject the null of no first-order serial correlation. Note that rejecting the null hypothesis does not necessarily mean an AR process; other forms of misspecification may also lead to a significant test statistic. If we are willing to assume that the errors follow an AR(1) process and that wagegov is strictly exogenous, we could refit the model using arima or prais and model the error process explicitly; see [TS] **arima** and [TS] **prais**.

If we are not willing to assume that wagegov is strictly exogenous, we could instead use Durbin's alternative test or the Breusch–Godfrey to test for first-order serial correlation. Because we have only 22 observations, we will use the small option.

```
. estat durbinalt, small
Durbin's alternative test for autocorrelation

---------------------------------------------------------------------------
    lags(p)  |          F                 df                  Prob > F
-------------+-------------------------------------------------------------
       1     |       35.035           (  1,    19 )            0.0000
---------------------------------------------------------------------------
                         H0: no serial correlation
. estat bgodfrey, small
Breusch-Godfrey LM test for autocorrelation

---------------------------------------------------------------------------
    lags(p)  |          F                 df                  Prob > F
-------------+-------------------------------------------------------------
       1     |       14.264           (  1,    19 )            0.0013
---------------------------------------------------------------------------
                         H0: no serial correlation
```

Both tests strongly reject the null of no first-order serial correlation, so we decide to refit the model with two lags of consump included as regressors and then rerun estat durbinalt and estat bgodfrey. Because the revised model includes lagged values of the dependent variable, the Durbin–Watson test is not applicable.

```
. regress consump wagegovt L.consump L2.consump

      Source |       SS       df       MS              Number of obs =      20
-------------+------------------------------           F(  3,    16) =   44.01
       Model |  702.660311     3  234.220104           Prob > F      =  0.0000
    Residual |  85.1596011    16  5.32247507           R-squared     =  0.8919
-------------+------------------------------           Adj R-squared =  0.8716
       Total |  787.819912    19  41.4642059           Root MSE      =   2.307

------------------------------------------------------------------------------
     consump |      Coef.   Std. Err.      t    P>|t|     [95% Conf. Interval]
-------------+----------------------------------------------------------------
    wagegovt |   .6904282   .3295485     2.10   0.052    -.0081835     1.38904
     consump |
         L1. |   1.420536    .197024     7.21   0.000     1.002864    1.838208
         L2. |   -.650888   .1933351    -3.37   0.004     -1.06074   -.241036
       _cons |   9.209073   5.006701     1.84   0.084    -1.404659    19.82281
------------------------------------------------------------------------------

. estat durbinalt, small lags(1/2)
Durbin's alternative test for autocorrelation
---------------------------------------------------------------------------
    lags(p)  |          F                 df                   Prob > F
-------------+-------------------------------------------------------------
       1     |        0.080          (  1,    15 )               0.7805
       2     |        0.260          (  2,    14 )               0.7750
---------------------------------------------------------------------------
                        H0: no serial correlation
. estat bgodfrey, small lags(1/2)
Breusch-Godfrey LM test for autocorrelation
---------------------------------------------------------------------------
    lags(p)  |          F                 df                   Prob > F
-------------+-------------------------------------------------------------
       1     |        0.107          (  1,    15 )               0.7484
       2     |        0.358          (  2,    14 )               0.7056
---------------------------------------------------------------------------
                        H0: no serial correlation
```

Although `wagegov` and the constant term are no longer statistically different from zero at the 5% level, the output from `estat durbinalt` and `estat bgodfrey` indicates that including the two lags of `consump` has removed any serial correlation from the errors.

◁

Engle (1982) suggests an LM test for checking for autoregressive conditional heteroskedasticity (ARCH) in the errors. The $p$th order ARCH model can be written as

$$\begin{aligned}\sigma_t^2 &= \mathrm{E}(u_t^2 | u_{t-1}, \dots, u_{t-p}) \\ &= \gamma_0 + \gamma_1 u_{t-1}^2 + \dots + \gamma_p u_{t-p}^2\end{aligned}$$

To test the null hypothesis of no autoregressive conditional heteroskedasticity (i.e., $\gamma_1 = \dots = \gamma_p = 0$), we first fit the OLS model (1), obtain the residuals $\widehat{u}_t$, and run another OLS regression on the lagged residuals:

$$\widehat{u}_t^2 = \gamma_0 + \gamma_1 \widehat{u}_{t-1}^2 + \dots + \gamma_p \widehat{u}_{t-p}^2 + \epsilon \qquad (3)$$

The test statistic is $NR^2$, where $N$ is the number of observations in the sample and $R^2$ is the $R^2$ from the regression in (3). Under the null hypothesis, the test statistic follows a $\chi_p^2$ distribution.

▷ Example 2: estat archlm

We refit the original model that does not include the two lags of `consump` and then use `estat archlm` to see if there is any evidence that the errors are autoregressive conditional heteroskedastic.

```
. regress consump wagegovt
      Source |       SS       df       MS              Number of obs =      22
-------------+------------------------------           F(  1,    20) =   17.72
       Model |  532.567711     1  532.567711           Prob > F      =  0.0004
    Residual |  601.207167    20  30.0603584           R-squared     =  0.4697
-------------+------------------------------           Adj R-squared =  0.4432
       Total |  1133.77488    21  53.9892799           Root MSE      =  5.4827

------------------------------------------------------------------------------
     consump |      Coef.   Std. Err.      t    P>|t|     [95% Conf. Interval]
-------------+----------------------------------------------------------------
    wagegovt |    2.50744   .5957173     4.21   0.000     1.264796    3.750085
       _cons |   40.84699   3.192183    12.80   0.000     34.18821    47.50577
------------------------------------------------------------------------------
. estat archlm, lags(1 2 3)
LM test for autoregressive conditional heteroskedasticity (ARCH)
---------------------------------------------------------------------------
    lags(p)  |          chi2               df                 Prob > chi2
-------------+-------------------------------------------------------------
       1     |          5.543               1                   0.0186
       2     |          9.431               2                   0.0090
       3     |          9.039               3                   0.0288
---------------------------------------------------------------------------
                H0: no ARCH effects        vs.  H1: ARCH(p) disturbance
```

`estat archlm` shows the results for tests of ARCH(1), ARCH(2), and ARCH(3) effects, respectively. At the 5% significance level, all three tests reject the null hypothesis that the errors are not autoregressive conditional heteroskedastic. See [TS] **arch** for information on fitting ARCH models.

◁

# Saved Results

`estat archlm` saves in `r()`:

Scalars

| | | | |
|---|---|---|---|
| `r(N)` | number of observations | `r(N_gaps)` | number of gaps |
| `r(k)` | number of regressors | | |

Macros

| | |
|---|---|
| `r(lags)` | orders of lags |

Matrices

| | | | |
|---|---|---|---|
| `r(arch)` | test statistic for each order of lags | `r(p)` | two-sided $p$-values |
| `r(df)` | degrees of freedom | | |

`estat bgodfrey` saves in `r()`:

Scalars

| | | | |
|---|---|---|---|
| `r(N)` | number of observations | `r(N_gaps)` | number of gaps |
| `r(k)` | number of regressors | | |

Macros

| | |
|---|---|
| `r(lags)` | orders of lag |

Matrices

| | | | |
|---|---|---|---|
| `r(chi2)` | $\chi^2$ statistic for each order of lags | `r(p)` | two-sided $p$-values |
| `r(F)` | $F$ statistic for each order of lags (`small` only) | `r(df)` | degrees of freedom |
| `r(df_r)` | residual degrees of freedom (`small` only) | | |

`estat durbinalt` saves in `r()`:

Scalars

| | | | |
|---|---|---|---|
| `r(N)` | number of observations | `r(N_gaps)` | number of gaps |
| `r(k)` | number of regressors | | |

Macros

| | |
|---|---|
| `r(lags)` | orders of lags |

Matrices

| | | | |
|---|---|---|---|
| `r(chi2)` | $\chi^2$ statistic for each order of lags | `r(p)` | two-sided $p$-values |
| `r(F)` | $F$ statistic for each order of lags (`small` only) | `r(df)` | degrees of freedom |
| `r(df_r)` | residual degrees of freedom (`small` only) | | |

`estat dwatson` saves in `r()`:

Scalars

| | | | |
|---|---|---|---|
| `r(N)` | number of observations | `r(N_gaps)` | number of gaps |
| `r(k)` | number of regressors | `r(dw)` | Durbin–Watson statistic |

## Methods and Formulas

`estat archlm`, `estat bgodfrey`, `estat durbinalt`, and `estat dwatson` are implemented as ado-files.

Consider the regression

$$y_t = \beta_1 x_{1t} + \cdots + \beta_k x_{kt} + u_t \tag{4}$$

in which some of the covariates are not strictly exogenous. In particular, some of the $x_{it}$ may be lags of the dependent variable. We are interested in whether the $u_t$ are serially correlated.

The Durbin–Watson $d$ statistic reported by `estat dwatson` is

$$d = \frac{\sum\limits_{t=1}^{n-1} (\widehat{u}_{t+1} - \widehat{u}_t)^2}{\sum\limits_{t=1}^{n} \widehat{u}_t^2}$$

where $\widehat{u}_t$ represents the residual of the $t$th observation.

To compute Durbin's alternative test and the Breusch–Godfrey test against the null hypothesis that there is no $p$th order serial correlation, we fit the regression in (4), compute the residuals, and then fit the following auxiliary regression of the residuals $\widehat{u}_t$ on $p$ lags of $\widehat{u}_t$ and on all the covariates in the original regression in (4):

$$\widehat{u}_t = \gamma_1 \widehat{u}_{t-1} + \cdots + \gamma_p \widehat{u}_{t-p} + \beta_1 x_{1t} + \cdots + \beta_k x_{kt} + \epsilon \tag{5}$$

Durbin's alternative test is computed by performing a Wald test to determine whether the coefficients of $\widehat{u}_{t-1}, \ldots, \widehat{u}_{t-p}$ are jointly different from zero. By default, the statistic is assumed to be distributed $\chi^2(\mathrm{p})$. When `small` is specified, the statistic is assumed to follow an $F(p, N-p-k)$ distribution. The reported $p$-value is a two-sided $p$-value. When `robust` is specified, the Wald test is performed using the Huber/White/sandwich estimator of the variance–covariance matrix, and the test is robust to an unspecified form of heteroskedasticity.

The Breusch–Godfrey test is computed as $NR^2$, where $N$ is the number of observations in the auxiliary regression (5) and $R^2$ is the $R^2$ from the same regression (5). Like Durbin's alternative test, the Breusch–Godfrey test is asymptotically distributed $\chi^2(p)$, but specifying `small` causes the $p$-value to be computed using an F($p$, $N-p-k$).

By default, the initial missing values of the lagged residuals are replaced with zeros, and the auxiliary regression is run over the full sample used in the original regression of (4). Specifying the `nomiss0` option causes these missing values to be treated as missing values, and the observations are dropped from the sample.

Engle's LM test for ARCH($p$) effects fits an OLS regression of $\widehat{u}_t^2$ on $\widehat{u}_{t-1}^2, \ldots, \widehat{u}_{t-p}^2$:

$$\widehat{u}_t^2 = \gamma_0 + \gamma_1 \widehat{u}_{t-1}^2 + \cdots + \gamma_p \widehat{u}_{t-p}^2 + \epsilon$$

The test statistic is $nR^2$ and is asymptotically distributed $\chi^2(p)$.

# Acknowledgment

The original versions of `estat archlm`, `estat bgodfrey`, and `estat durbinalt` were written by Christopher F. Baum, Boston College.

# References

Baum, C. F. and V. L. Wiggins. 2000a. sg135: Test for autoregressive conditional heteroskedasticity in regression error distribution. *Stata Technical Bulletin* 55: 13–14. Reprinted in *Stata Technical Bulletin Reprints*, vol. 10, 143–144.

——. 2000b. sg136: Tests for serial correlation in regression error distribution. *Stata Technical Bulletin* 55: 14–15. Reprinted in *Stata Technical Bulletin Reprints*, vol. 10, 145–147.

Beran, R. J. and N. I. Fisher. 1998. A conversation with Geoff Watson. *Statistical Science* 13: 75–93.

Davidson, R. and J. G. MacKinnon. 1993. *Estimation and Inference in Econometrics*. New York: Oxford University Press.

Durbin, J. 1970. Testing for serial correlation in least-squares regressions when some of the regressors are lagged dependent variables. *Econometrica* 38: 410–421.

Engle, R. F. 1982. Autoregressive conditional heteroskedasticity with estimates of the variance of United Kingdom inflation. *Econometrica* 50: 987–1007.

Fisher, N. I. and P. Hall. 1998. Geoffrey Stuart Watson: Tributes and obituary (3 December 1921—3 January 1998). *Australian and New Zealand Journal of Statistics* 40: 257–267.

Godfrey, L. G. 1988. *Misspecification Tests in Econometrics*. Econometric Society Monographs, No. 16, Cambridge: Cambridge University Press.

Greene, W. H. 2003. *Econometric Analysis*. 5th ed. Upper Saddle River, NJ: Prentice Hall.

Johnston, J. and J. DiNardo. 1997. *Econometric Methods*. 4th ed. New York: McGraw–Hill.

Klein, L. 1950. *Economic Fluctuations in the United States 1921–1941*. New York: Wiley.

Phillips, P. C. B. 1988. The ET Interview: Professor James Durbin. *Econometric Theory* 4: 125–157.

Savin, E. and K. J. White. 1977. The Durbin–Watson test for serial correlation with extreme sample sizes or many regressors. *Econometrica* 45(8): 1989–1996.

Wooldridge, J. M. 2002. *Introductory Econometrics: A Modern Approach*. 2nd ed. Cincinnati, OH: South-Western.

James Durbin (1923– ) is a British statistician who was born in Wigan, near Manchester. He read mathematics at Cambridge and after military service and various research posts joined the London School of Economics in 1950. His many contributions to statistics have centered on serial correlation, time series, sample survey methodology, goodness-of-fit tests, and sample distribution functions, with emphasis on applications in the social sciences.

Geoffrey Stuart Watson (1921–1998) was born in Victoria, Australia, and earned degrees at Melbourne University and North Carolina State University. After a visit to the University of Cambridge, he returned to Australia, working at Melbourne and then the Australian National University. Following periods at Toronto and Johns Hopkins, he settled at Princeton. His best known contributions to statistics, in a very wide-ranging career, include the Durbin–Watson test for serial correlation, the Nadaraya–Watson estimator in nonparametric regression, and methods for directional data.

Leslie G. Godfrey (1946– ) was born in London and earned degrees at the Universities of Exeter and London. He now holds a chair in econometrics at the University of York. His interests center on implementation and interpretation of tests of econometric models, including non-nested models.

Trevor Stanley Breusch (1949– ) was born in Queensland and earned degrees at the University of Queensland and Australian National University (ANU). After a post at the University of Southampton, he returned to work at ANU. His background is in econometric methods and his recent interests include political values and social attitudes, earnings and income, and measurement of underground economic activity.

## Also See

**Complementary:** [TS] **tsset**

**Related:** [R] **regress postestimation**

# Title

**#review** — Review previous commands

# Syntax

`#review` [ $\#_1$ [ $\#_2$ ] ]

# Description

The `#review` command displays the last few lines typed at the terminal.

# Remarks

`#review` (pronounced *pound-review*) is a Stata preprocessor command. `#`*commands* do not generate a return code, or generate ordinary Stata errors. The only error message associated with `#`*commands* is "unrecognized #command".

The `#review` command displays the last few lines typed at the terminal. If no arguments follow `#review`, the last 5 lines typed at the terminal are displayed. The first argument specifies the number of lines to be reviewed, so `#review 10` displays the last 10 lines typed. The second argument specifies the number of lines to be displayed, so `#review 10 5` displays 5 lines, starting at the 10th previous line.

Stata reserves a buffer for `#review` lines and stores as many previous lines in the buffer as will fit, rolling out the oldest line to make room for the newest. Requests to `#review` lines no longer stored will be ignored. Only lines typed at the terminal are placed in the `#review` buffer. See [U] **10.5 Editing previous lines in Stata for all operating systems**.

## ▷ Example 1

Typing `#review` by itself will show the last five lines you typed at the terminal:

```
. #review
5 use mydata
4 * comments go into the #review buffer, too
3 describe
2 tabulate marriage educ [freq=number]
1 tabulate marriage educ [freq=number], chi2
. _
```

Typing `#review 15 2` shows the 15th and 14th previous lines:

```
. #review 15 2
15 replace x=. if x<200
14 summarize x
. _
```

◁

# Also See

**Background:** [U] **10.5 Editing previous lines in Stata for all operating systems**,
[U] **16.1.3 Long lines in do-files**

# Title

**roc** — Receiver-Operating-Characteristic (ROC) analysis

# Syntax

*Perform nonparametric ROC analysis*

`roctab` *refvar classvar* [*if*] [*in*] [*weight*] [, *roctab_options*]

*Test equality of ROC areas*

`roccomp` *refvar classvar* [*classvars*] [*if*] [*in*] [*weight*] [, *roccomp_options*]

*Test equality of ROC area against a standard ROC curve*

`rocgold` *refvar goldvar classvar* [*classvars*] [*if*] [*in*] [*weight*] [, *rocgold_options*]

| *roctab_options* | description |
|---|---|
| Main | |
| `lorenz` | report Gini and Pietra indices |
| `bamber` | calculate standard errors using the Bamber method |
| `hanley` | calculate standard errors using the Hanley method |
| `binomial` | calculate exact binomial confidence intervals |
| `table` | display the raw data in a $2 \times k$ contingency table |
| `detail` | show details on sensitivity/specificity for each cutpoint |
| `level(#)` | set confidence level; default is `level(95)` |
| `graph` | graph the ROC curve |
| `norefline` | suppress plotting the 45-degree reference line |
| `specificity` | graph sensitivity versus specificity |
| `summary` | report the area under the ROC curve |
| Plot | |
| `plotopts(`*plot_options*`)` | affect rendition of the ROC curve |
| Reference line | |
| `rlopts(`*cline_options*`)` | affect rendition of the reference line |
| Add plot | |
| `addplot(`*plot*`)` | add other plots to the generated graph |
| Y-Axis, X-Axis, Title, Caption, Legend, Overall | |
| *twoway_options* | any options other than `by()` documented in [G] ***twoway_options*** |

`fweight`s are allowed; see [U] **11.1.6 weight**.

| *roccomp_options* | description |
|---|---|
| Main | |
| by(*varname*) | split into groups by variable |
| binormal | estimate areas using binormal distribution assumption |
| test(*matname*) | use contrast matrix for comparing ROC areas |
| level(*#*) | set confidence level; default is level(95) |
| graph | graph the ROC curve |
| norefline | suppress plotting the 45-degree reference line |
| separate | place each ROC curve on its own graph |
| summary | report the area under the ROC curve |
| Plot 1, Plot 2, Plot 3 | |
| plot#opts(*plot_options*) | affect rendition of the #th ROC curve |
| Reference line | |
| rlopts(*cline_options*) | affect rendition of the reference line |
| Add plot | |
| addplot(*plot*) | add other plots to the generated graph |
| Y-Axis, X-Axis, Title, Caption, Legend, Overall | |
| *twoway_options* | any options documented in [G] ***twoway_options*** |
| † line#opts(*cline_options*) | affect rendition of the #th binormal fit line |

† line#opts(*cline_options*) does not appear in the dialog box.
fweights are allowed; see [U] **11.1.6 weight**.

(*Continued on next page*)

| *rocgold_options* | description |
|---|---|
| Main | |
| sidak | adjust the significance probability using Šidák's method |
| binormal | estimate areas using binormal distribution assumption |
| test(*matname*) | use contrast matrix for comparing ROC areas |
| level(*#*) | set confidence level; default is level(95) |
| graph | graph the ROC curve |
| norefline | suppress plotting the 45-degree reference line |
| separate | place each ROC curve on its own graph |
| summary | report the area under the ROC curve |
| Plot 1, Plot 2, Plot 3 | |
| plot#opts(*plot_options*) | affect rendition of the #th ROC curve; plot 1 is the gold standard |
| Reference line | |
| rlopts(*cline_options*) | affect rendition of the reference line |
| Y-Axis, X-Axis, Title, Caption, Legend, Overall | |
| *twoway_options* | any options documented in [G] ***twoway_options*** |
| †line#opts(*cline_options*) | affect rendition of the #th binormal fit line |

†line#opts(*cline_options*) does not appear in the dialog box.

fweights are allowed; see [U] **11.1.6 weight**.

| *plot_options* | description |
|---|---|
| *marker_options* | change look of markers (color, size, etc.) |
| *marker_label_options* | add marker labels; change look or position |
| *cline_options* | change the look of the line |

## Description

The above commands are used to perform Receiver Operating Characteristic (ROC) analyses with rating and discrete classification data.

The two variables *refvar* and *classvar* must be numeric. The reference variable indicates the true state of the observation, such as diseased and nondiseased or normal and abnormal, and must be coded as 0 and 1. The rating or outcome of the diagnostic test or test modality is recorded in *classvar*, which must be at least ordinal, with higher values indicating higher risk.

roctab performs nonparametric ROC analyses. By default, roctab calculates the area under the ROC curve. Optionally, roctab can plot the ROC curve, display the data in tabular form, and produce Lorenz-like plots.

roccomp tests the equality of two or more ROC areas obtained from applying two or more test modalities to the same sample or to independent samples. roccomp expects the data to be in wide form when comparing areas estimated from the same sample and in long form for areas estimated from independent samples.

rocgold independently tests the equality of the ROC area of each of several test modalities, specified by *classvar*, against a "gold" standard ROC curve, *goldvar*. For each comparison, rocgold reports the raw and the Bonferroni-adjusted significance probability. Optionally, Šidák's adjustment for multiple comparisons can be obtained.

See [R] **rocfit** for a command that fits maximum-likelihood ROC models.

## Options for roctab

Main

lorenz specifies that Gini and Pietra indices be reported. Optionally, graph will plot the Lorenz-like curve.

bamber specifies that the standard error for the area under the ROC curve be calculated using the method suggested by Bamber (1975). Otherwise, standard errors are obtained as suggested by DeLong, DeLong, and Clarke-Pearson (1988).

hanley specifies that the standard error for the area under the ROC curve be calculated using the method suggested by Hanley and McNeil (1982). Otherwise, standard errors are obtained as suggested by DeLong, DeLong, and Clarke-Pearson (1988).

binomial specifies that exact binomial confidence intervals be calculated.

table outputs a $2 \times k$ contingency table displaying the raw data.

detail outputs a table displaying the sensitivity, specificity, the percentage of subjects correctly classified, and two likelihood ratios for each possible cutpoint of *classvar*.

level(*#*) specifies the confidence level, as a percentage, for the confidence intervals. The default is level(95) or as set by set level; see [R] **level**.

graph produces graphical output of the ROC curve. If lorenz is specified, graphical output of a Lorenz-like curve will be produced.

norefline suppresses plotting the 45-degree reference line from the graphical output of the ROC curve.

specificity produces a graph of sensitivity versus specificity instead of sensitivity versus (1 − specificity). specificity implies graph.

summary reports the area under the ROC curve, its standard error, and its confidence interval. If lorenz is specified, Lorenz indices are reported. This option is only needed when also specifying graph.

Plot

plotopts(*plot_options*) affect the rendition of the plotted ROC curve—the curve's plotted points connected by lines. The *plot_options* can affect the size and color of markers, whether and how the markers are labeled, and whether and how the points are connected; see [G] ***marker_options***, [G] ***marker_label_options***, and [G] ***cline_options***.

Reference line

rlopts(*cline_options*) affect the rendition of the reference line; see [G] ***cline_options***.

Add plot

addplot(*plot*) provides a way to add other plots to the generated graph; see [G] ***addplot_option***.

Y-Axis, X-Axis, Title, Caption, Legend, Overall

*twoway_options* are any of the options documented in [G] ***twoway_options***, excluding by(). These include options for titling the graph (see [G] ***title_options***) and options for saving the graph to disk (see [G] ***saving_option***).

## Options for roccomp and rocgold

Main

by(*varname*) (roccomp only) is required when comparing independent ROC areas. The by() variable identifies the groups to be compared.

sidak (rocgold only) requests that the significance probability be adjusted for the effect of multiple comparisons using Šidák's method. Bonferroni's adjustment is reported by default.

binormal specifies that the areas under the ROC curves to be compared should be estimated using the binormal distribution assumption. By default, areas to be compared are computed using the trapezoidal rule.

test(*matname*) specifies the contrast matrix to be used when comparing ROC areas. By default, the null hypothesis that all areas are equal is tested.

level(*#*) specifies the confidence level, as a percentage, for the confidence intervals. The default is level(95) or as set by set level; see [R] **level**.

graph produces graphical output of the ROC curve.

norefline suppresses plotting the 45-degree reference line from the graphical output of the ROC curve.

separate is meaningful only with roccomp and specifies that each ROC curve be placed on its own graph rather than one curve on top of the other.

summary reports the area under the ROC curve, its standard error, and its confidence interval. This option is only needed when also specifying graph.

Plot 1, Plot 2, Plot 3

plot#opts(*plot_options*) affect the rendition of the #th ROC curve—the curve's plotted points connected by lines. The *plot_options* can affect the size and color of markers, whether and how the markers are labeled, and whether and how the points are connected; see [G] ***marker_options***, [G] ***marker_label_options***, and [G] ***cline_options***.

For rocgold, plot1opts() are applied to the ROC for the gold standard.

Reference line

rlopts(*cline_options*) affect the rendition of the reference line; see [G] ***cline_options***.

Add plot

addplot(*plot*) (roccomp only) provides a way to add other plots to the generated graph; see [G] ***addplot_option***.

Y-Axis, X-Axis, Title, Caption, Legend, Overall

*twoway_options* are any of the options documented in [G] ***twoway_options***. These include options for titling the graph (see [G] ***title_options***), options for saving the graph to disk (see [G] ***saving_option***), and the `by()` option (see [G] ***by_option***).

The following options are available with `roccomp` and `rocgold` but are not shown in the dialog boxes:

`line#opts(`*cline_options*`)` affect the rendition of the line representing the #th ROC curve drawn using the binormal distribution assumption; see [G] ***cline_options***. These lines are drawn only if the `binormal` option is specified.

# Remarks

Remarks are presented under the headings

*Introduction*
*Nonparametric ROC curves*
*Lorenz-like curves*
*Comparing areas under the ROC curve*
*Correlated data*
*Independent data*
*Comparing areas with a gold standard*

## Introduction

Receiver Operating Characteristic (ROC) analysis quantifies the accuracy of diagnostic tests or other evaluation modalities used to discriminate between two states or conditions, which are here referred to as normal and abnormal. The discriminatory accuracy of a diagnostic test is measured by its ability to correctly classify known normal and abnormal subjects. The analysis uses the ROC curve, a graph of the sensitivity versus $1 -$ specificity of the diagnostic test. The sensitivity is the fraction of positive cases that are correctly classified by the diagnostic test, while the specificity is the fraction of negative cases that are correctly classified. Thus the sensitivity is the true-positive rate, and the specificity is the true-negative rate.

The global performance of a diagnostic test is commonly summarized by the area under the ROC curve. This area can be interpreted as the probability that the result of a diagnostic test of a randomly selected abnormal subject will be greater than the result of the same diagnostic test from a randomly selected normal subject. The greater the area under the ROC curve, the better the global performance of the diagnostic test.

Both nonparametric methods and parametric (semiparametric) methods have been suggested for generating the ROC curve and for calculating its area. The following sections present these approaches, and the last section presents tests for comparing areas under ROC curves.

See Pepe (2003) for a discussion of ROC analysis. Pepe has posted Stata datasets and programs used to reproduce results presented in the book (*http://www.stata.com/bookstore/pepe.html*).

## Nonparametric ROC curves

The points on the nonparametric ROC curve are generated by using each possible outcome of the diagnostic test as a classification cutpoint and computing the corresponding sensitivity and $1 -$ specificity. These points are then connected by straight lines, and the area under the resulting ROC curve is computed using the trapezoidal rule.

## ▷ Example 1

Hanley and McNeil (1982) presented data from a study in which a reviewer was asked to classify, using a nine-point scale, a random sample of 109 tomographic images from patients with neurological problems. The rating scale was as follows: 1—definitely normal, 2—probably normal, 3—questionable, 4—probably abnormal, and 5—definitely abnormal. The true disease status was normal for 58 of the patients and abnormal for the remaining 51 patients.

Here we list 9 of the 109 observations.

```
. use http://www.stata-press.com/data/r9/hanley

. list disease rating in 1/9

     +------------------+
     | disease   rating |
     |------------------|
  1. |       1        5 |
  2. |       0        1 |
  3. |       1        5 |
  4. |       0        4 |
  5. |       0        1 |
     |------------------|
  6. |       0        3 |
  7. |       1        5 |
  8. |       0        5 |
  9. |       0        1 |
     +------------------+
```

For each observation, `disease` identifies the true disease status of the subject (0 = normal, 1 = abnormal), and `rating` contains the classification value assigned by the reviewer.

We can use `roctab` to calculate and plot the nonparametric ROC curve by specifying both the `summary` and `graph` options. By also specifying the `table` option, we obtain a contingency table summarizing our dataset.

(*Continued on next page*)

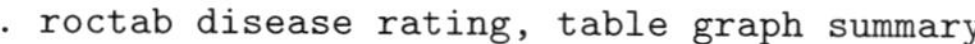

```
. roctab disease rating, table graph summary
```

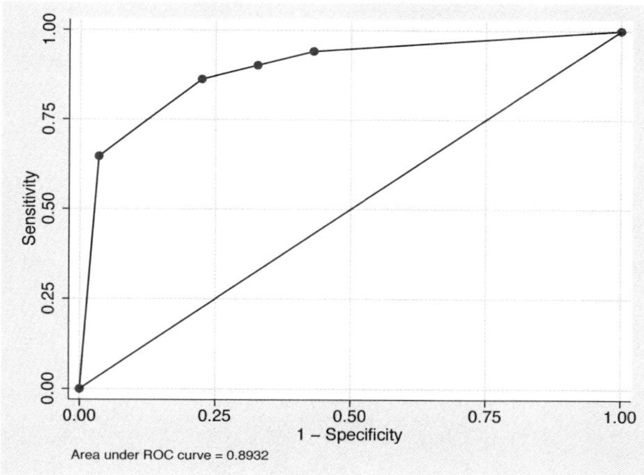


| disease | rating 1 | 2 | 3 | 4 | 5 | Total |
|---|---|---|---|---|---|---|
| 0 | 33 | 6 | 6 | 11 | 2 | 58 |
| 1 | 3 | 2 | 2 | 11 | 33 | 51 |
| Total | 36 | 8 | 8 | 22 | 35 | 109 |

| Obs | ROC Area | Std. Err. | —Asymptotic Normal— [95% Conf. | Interval] |
|---|---|---|---|---|
| 109 | 0.8932 | 0.0307 | 0.83295 | 0.95339 |

By default, `roctab` reports the area under the curve, its standard error, and its confidence interval. The `graph` option can be used to plot the ROC curve.

The ROC curve is plotted by computing the sensitivity and specificity using each value of the rating variable as a possible cutpoint. A point is plotted on the graph for each of the cutpoints. These plotted points are joined by straight lines to form the ROC curve, and the area under the ROC curve is computed using the trapezoidal rule.

We can tabulate the computed sensitivities and specificities for each of the possible cutpoints by specifying `detail`.

*(Continued on next page)*

```
. roctab disease rating, detail
Detailed report of Sensitivity and Specificity
```

| Cutpoint | Sensitivity | Specificity | Correctly Classified | LR+ | LR- |
|---|---|---|---|---|---|
| ( >= 1 ) | 100.00% | 0.00% | 46.79% | 1.0000 | |
| ( >= 2 ) | 94.12% | 56.90% | 74.31% | 2.1835 | 0.1034 |
| ( >= 3 ) | 90.20% | 67.24% | 77.98% | 2.7534 | 0.1458 |
| ( >= 4 ) | 86.27% | 77.59% | 81.65% | 3.8492 | 0.1769 |
| ( >= 5 ) | 64.71% | 96.55% | 81.65% | 18.7647 | 0.3655 |
| ( > 5 ) | 0.00% | 100.00% | 53.21% | | 1.0000 |

| Obs | ROC Area | Std. Err. | –Asymptotic Normal– [95% Conf. | Interval] |
|---|---|---|---|---|
| 109 | 0.8932 | 0.0307 | 0.83295 | 0.95339 |

Each cutpoint in the table indicates the ratings used to classify tomographs as being from an abnormal subject. For example, the first cutpoint $(>= 1)$ indicates that all tomographs rated as 1 or greater are classified as coming from abnormal subjects. Because all tomographs have a rating of 1 or greater, all are considered abnormal. Consequently, all abnormal cases are correctly classified (sensitivity = 100%), but none of the normal patients is classified correctly (specificity = 0%). For the second cutpoint $(>= 2)$, tomographs with ratings of 1 are classified as normal, and those with ratings of 2 or greater are classified as abnormal. The resulting sensitivity and specificity are 94.12% and 56.90%, respectively. Using this cutpoint, we correctly classified 74.31% of the 109 tomographs. Similar interpretations can be used on the remaining cutpoints. As mentioned, each cutpoint corresponds to a point on the nonparametric ROC curve. The first cutpoint $(>= 1)$ corresponds to the point at (1,1), and the last cutpoint $(> 5)$ corresponds to the point at (0,0).

`detail` also reports two likelihood ratios suggested by Choi (1998): the likelihood ratio for a positive test result (LR+) and the likelihood ratio for a negative test result (LR–). The likelihood ratio for a positive test result is the ratio of the probability of a positive test among the truly positive subjects to the probability of a positive test among the truly negative subjects. The likelihood ratio for a negative test result (LR–) is the ratio of the probability of a negative test among the truly positive subjects to the probability of a negative test among the truly negative subjects. Choi points out that LR+ corresponds to the slope of the line from the origin to the point on the ROC curve determined by the cutpoint. Similarly, LR– corresponds to the slope from the point (1,1) to the point on the ROC curve determined by the cutpoint.

By default, `roctab` calculates the standard error for the area under the curve using an algorithm suggested by DeLong, DeLong, and Clarke-Pearson (1988) and asymptotic normal confidence intervals. Optionally, standard errors based on methods suggested by Bamber (1975) or Hanley and McNeil (1982) can be computed by specifying `bamber` or `hanley`, respectively, and an exact binomial confidence interval can be obtained by specifying `binomial`.

```
. roctab disease rating, bamber
```

| Obs | ROC Area | Bamber Std. Err. | –Asymptotic Normal– [95% Conf. | Interval] |
|---|---|---|---|---|
| 109 | 0.8932 | 0.0306 | 0.83317 | 0.95317 |

```
. roctab disease rating, hanley binomial
                      ROC         Hanley           — Binomial Exact —
          Obs        Area        Std. Err.        [95% Conf. Interval]
       
          109      0.8932         0.0320          0.81559      0.94180
```

◁

## Lorenz-like curves

For applications where it is known that the risk status increases or decreases monotonically with increasing values of the diagnostic test, the ROC curve and associated indices are useful in assessing the overall performance of a diagnostic test. When the risk status does not vary monotonically with increasing values of the diagnostic test, however, the resulting ROC curve can be nonconvex and its indices can be unreliable. For these situations, Lee (1999) proposed an alternative to the ROC analysis based on Lorenz-like curves and the associated Pietra and Gini indices.

Lee (1999) mentions at least three specific situations where results from Lorenz curves are superior to those obtained from ROC curves: (1) a diagnostic test with similar means but very different standard deviations in the abnormal and normal populations, (2) a diagnostic test with bimodal distributions in either the normal or abnormal population, and (3) a diagnostic test distributed symmetrically in the normal population and skewed in the abnormal.

When the risk status increases or decreases monotonically with increasing values of the diagnostic test, the ROC and Lorenz curves yield interchangeable results.

### ▷ Example 2

To illustrate the use of the `lorenz` option, we constructed a fictitious dataset that yields results similar to those presented in Table III of Lee (1999). The data assume that a 12-point rating scale was used to classify 442 diseased and 442 healthy subjects. We list a few of the observations.

```
. use http://www.stata-press.com/data/r9/lorenz
. list in 1/7, noobs sep(0)
```

| disease | class | pop |
|---|---|---|
| 0 | 5 | 66 |
| 1 | 11 | 17 |
| 0 | 6 | 85 |
| 0 | 3 | 19 |
| 0 | 10 | 19 |
| 0 | 2 | 7 |
| 1 | 4 | 16 |

The data consist of 24 observations: 12 observations from diseased individuals and 12 from nondiseased individuals. Each observation corresponds to one of the 12 classification values of the rating-scale variable, `class`. The number of subjects represented by each observation is given by the `pop` variable, making this a frequency-weighted dataset. The data were generated assuming a binormal distribution of the latent variable with similar means for the normal and abnormal populations but with the standard deviation for the abnormal population 5 times greater than that of the normal population.

```
. roctab disease class [fweight=pop], graph summary
```

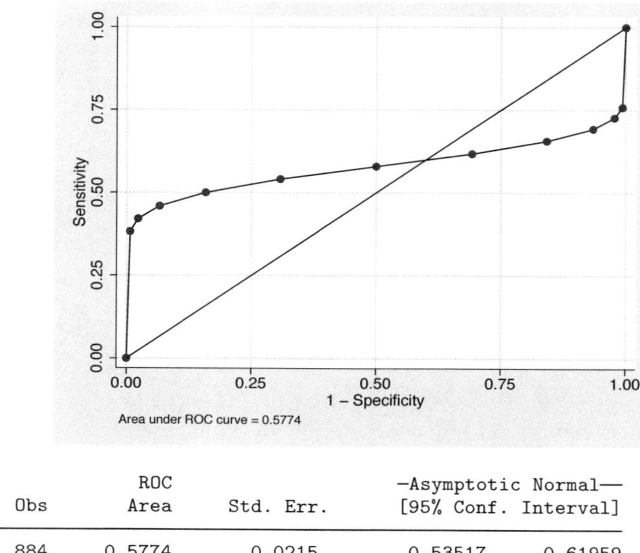


```
                      ROC                    -Asymptotic Normal--
           Obs       Area     Std. Err.      [95% Conf. Interval]
         ------------------------------------------------------------
           884     0.5774       0.0215        0.53517     0.61959
```

The resulting ROC curve is nonconvex or, as termed by Lee, "wiggly". Lee argues that for this and similar situations, the Lorenz curve and indices are preferred.

```
. roctab disease class [fweight=pop], lorenz graph summary
```

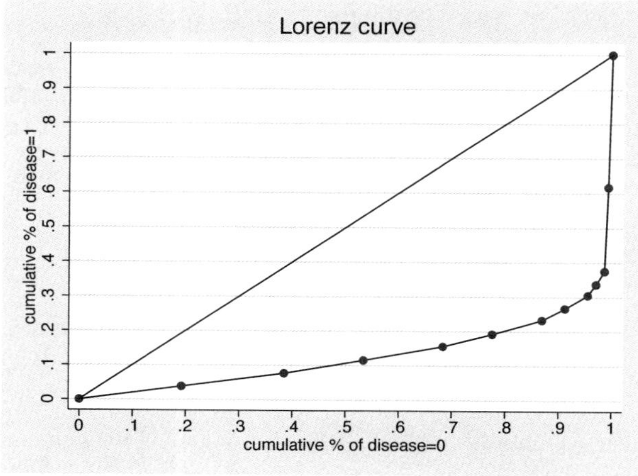


```
Lorenz curve
-----------------------------
 Pietra index =    0.6493
 Gini index   =    0.7441
```

Like ROC curves, a more bowed Lorenz curve suggests a better diagnostic test. This "bowedness" is quantified by the Pietra index, which is geometrically equivalent to twice the largest triangle that can be inscribed in the area between the curve and the diagonal line, and the Gini index, which is equivalent to twice the area between the Lorenz curve and the diagonal. Lee (1999) provides several additional interpretations for the Pietra and Gini indices.

◁

## Comparing areas under the ROC curve

The area under multiple ROC curves can be compared using `roccomp`. The command syntax is slightly different if the ROC curves are correlated (i.e., different diagnostic tests are applied to the same sample) or independent (i.e., diagnostic tests are applied to different samples).

## Correlated data

### ▷ Example 3

Hanley and McNeil (1983) presented data from an evaluation of two computer algorithms designed to reconstruct CT images from phantoms. We will call these two algorithms' modalities 1 and 2. A sample of 112 phantoms was selected; 58 phantoms were considered normal, and the remaining 54 were abnormal. Each of the two modalities was applied to each phantom, and the resulting images were rated by a reviewer using a six-point scale: 1—definitely normal, 2—probably normal, 3—possibly normal, 4—possibly abnormal, 5—probably abnormal, and 6—definitely abnormal. Because each modality was applied to the same sample of phantoms, the two sets of outcomes are correlated.

We list the first seven observations:

```
. use http://www.stata-press.com/data/r9/ct
. list in 1/7, sep(0)
```

| | mod1 | mod2 | status |
|---|---|---|---|
| 1. | 2 | 1 | 0 |
| 2. | 5 | 5 | 1 |
| 3. | 2 | 1 | 0 |
| 4. | 2 | 3 | 0 |
| 5. | 5 | 6 | 1 |
| 6. | 2 | 2 | 0 |
| 7. | 3 | 2 | 0 |

Note that the data are in wide form, which is required when dealing with correlated data. Each observation corresponds to one phantom. The variable `mod1` identifies the rating assigned for the first modality, and `mod2` identifies the rating assigned for the second modality. The true status of the phantoms is given by `status=0` if they are normal and `status=1` if they are abnormal. The observations with at least one missing rating were dropped from the analysis.

We plot the two ROC curves and compare their areas.

*(Continued on next page)*

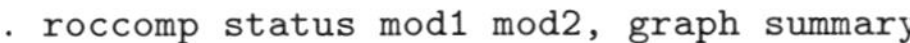

```
. roccomp status mod1 mod2, graph summary
```

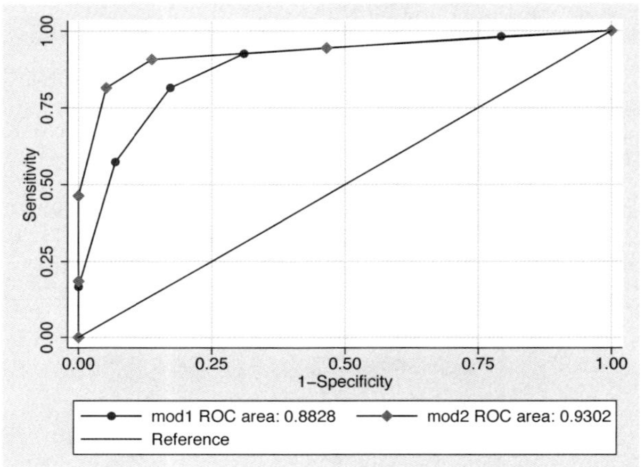


```
                              ROC                    -Asymptotic Normal--
                   Obs       Area     Std. Err.      [95% Conf. Interval]
-------------------------------------------------------------------------
mod1               112     0.8828        0.0317       0.82067     0.94498
mod2               112     0.9302        0.0256       0.88005     0.98042
-------------------------------------------------------------------------
Ho: area(mod1) = area(mod2)
    chi2(1) =      2.31       Prob>chi2 =    0.1282
```

By default, `roccomp`, with the **graph** option specified, plots the ROC curves on the same graph. Optionally the curves can be plotted side by side, each on its own graph, by also specifying **separate**.

For each curve, `roccomp` reports summary statistics and provides a test for the equality of the area under the curves using an algorithm suggested by DeLong, DeLong, and Clarke-Pearson (1988).

Although the area under the ROC curve for modality 2 is larger than that of modality 1, the chi-squared test yielded a significance probability of 0.1282, suggesting that there is no significant difference between these two areas.

The `roccomp` command can also be used to compare more than two ROC areas. To illustrate this, we modified the previous dataset by including a fictitious third modality.

*(Continued on next page)*

```
. use http://www.stata-press.com/data/r9/ct2
. roccomp status mod1 mod2 mod3, graph summary
```

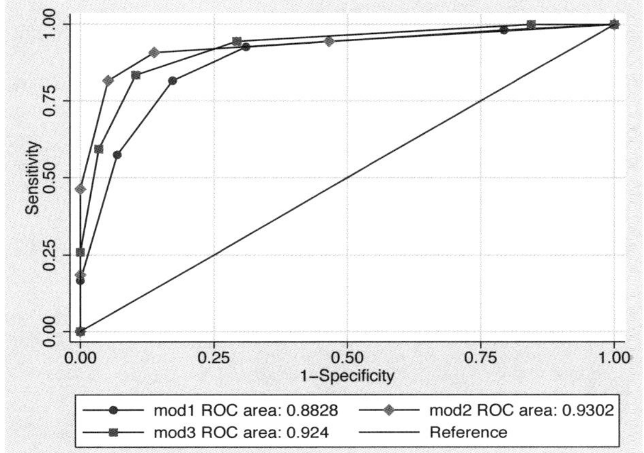


```
                               ROC                     -Asymptotic Normal--
                    Obs       Area     Std. Err.      [95% Conf. Interval]
-------------------------------------------------------------------------
mod1                112     0.8828       0.0317        0.82067     0.94498
mod2                112     0.9302       0.0256        0.88005     0.98042
mod3                112     0.9240       0.0241        0.87670     0.97132
-------------------------------------------------------------------------
Ho: area(mod1) = area(mod2) = area(mod3)
    chi2(2) =       6.54       Prob>chi2 =    0.0381
```

By default, `roccomp` tests whether the areas under the ROC curves are all equal. Other comparisons can be tested by creating a contrast matrix and specifying `test(`*matname*`)`, where *matname* is the name of the contrast matrix.

For example, assume that we are interested in testing whether the area under the ROC for `mod1` is equal to that of `mod3`. To do this, we can first create an appropriate contrast matrix and then specify its name with the `test()` option.

Of course, this is a trivial example because we could have just specified

```
. roccomp status mod1 mod3
```

without including `mod2` to obtain the same test results. However, for illustration, we will continue with this example.

The contrast matrix must have its number of columns equal to the number of *classvars* (i.e., the total number of ROC curves) and a number of rows less than or equal to the number of *classvars*, and the elements of each row must add to zero.

(*Continued on next page*)

```
. matrix C=(1,0,-1)

. roccomp status mod1 mod2 mod3, test(C)

                              ROC                    -Asymptotic Normal--
                   Obs       Area     Std. Err.      [95% Conf. Interval]
-------------------------------------------------------------------------
mod1               112     0.8828       0.0317        0.82067     0.94498
mod2               112     0.9302       0.0256        0.88005     0.98042
mod3               112     0.9240       0.0241        0.87670     0.97132
-------------------------------------------------------------------------
Ho: Comparison as defined by contrast matrix: C
    chi2(1) =     5.25       Prob>chi2 =   0.0220
```

Note that although all three areas are reported, the comparison is made using the specified contrast matrix.

Perhaps more interesting would be a comparison of the area from `mod1` and the average area of `mod2` and `mod3`.

```
. matrix C=(1,-.5,-.5)

. roccomp status mod1 mod2 mod3, test(C)

                              ROC                    -Asymptotic Normal--
                   Obs       Area     Std. Err.      [95% Conf. Interval]
-------------------------------------------------------------------------
mod1               112     0.8828       0.0317        0.82067     0.94498
mod2               112     0.9302       0.0256        0.88005     0.98042
mod3               112     0.9240       0.0241        0.87670     0.97132
-------------------------------------------------------------------------
Ho: Comparison as defined by contrast matrix: C
    chi2(1) =     3.43       Prob>chi2 =   0.0642
```

Other contrasts could be made. For example, we could test if `mod3` is different from at least one of the other two by first creating the following contrast matrix:

```
. matrix C=(-1, 0, 1 \ 0, -1, 1)

. matrix list C

C[2,3]
    c1  c2  c3
r1  -1   0   1
r2   0  -1   1
```

◁

## Independent data

### ▷ Example 4

In example 3, we noted that because each test modality was applied to the same sample of phantoms, the classification outcomes were correlated. Now assume that we have collected the same data presented by Hanley and McNeil (1983), except that we applied the first test modality to one sample of phantoms and the second test modality to a different sample of phantoms. The resulting measurements are now considered independent.

Here are a few of the observations.

```
. use http://www.stata-press.com/data/r9/ct3
```

```
. list in 1/7, sep(0)
```

|  | pop | status | rating | mod |
|---|---|---|---|---|
| 1. | 12 | 0 | 1 | 1 |
| 2. | 31 | 0 | 1 | 2 |
| 3. | 1 | 1 | 1 | 1 |
| 4. | 3 | 1 | 1 | 2 |
| 5. | 28 | 0 | 2 | 1 |
| 6. | 19 | 0 | 2 | 2 |
| 7. | 3 | 1 | 2 | 1 |

Note that the data are in long form, which is required when dealing with independent data. The data consist of 24 observations: 6 observations corresponding to abnormal phantoms and 6 to normal phantoms evaluated using the first modality, and similarly 6 observations corresponding to abnormal phantoms and 6 to normal phantoms evaluated using the second modality. The number of phantoms corresponding to each observation is given by the `pop` variable. Once again we have frequency-weighted data. The variable `mod` identifies the modality, and `rating` is the assigned classification.

We can better view our data by using the `table` command.

```
. table status rating [fw=pop], by(mod) row col
```

| mod and status | rating | | | | | | |
|---|---|---|---|---|---|---|---|
| | 1 | 2 | 3 | 4 | 5 | 6 | Total |
| 1 | | | | | | | |
| 0 | 12 | 28 | 8 | 6 | 4 | | 58 |
| 1 | 1 | 3 | 6 | 13 | 22 | 9 | 54 |
| Total | 13 | 31 | 14 | 19 | 26 | 9 | 112 |
| 2 | | | | | | | |
| 0 | 31 | 19 | 5 | 3 | | | 58 |
| 1 | 3 | 2 | 5 | 19 | 15 | 10 | 54 |
| Total | 34 | 21 | 10 | 22 | 15 | 10 | 112 |

The `status` variable indicates the true status of the phantoms: `status=0` if they are normal and `status=1` if they are abnormal.

We now compare the areas under the two ROC curves.

(*Continued on next page*)

```
. roccomp status rating [fw=pop], by(mod) graph summary
```

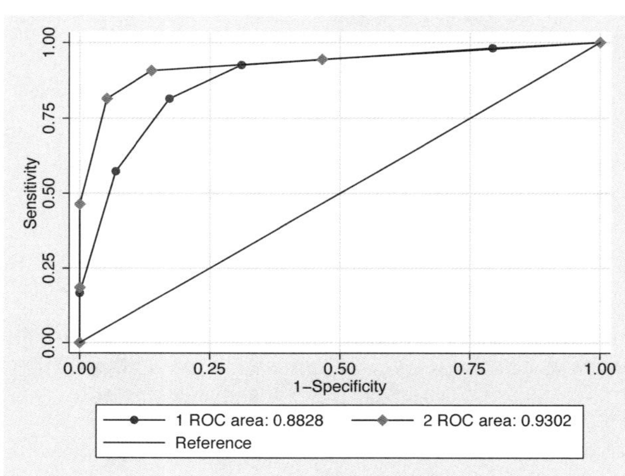


```
                             ROC                    -Asymptotic Normal--
mod              Obs        Area     Std. Err.      [95% Conf. Interval]
-------------------------------------------------------------------------
1                112      0.8828       0.0317        0.82067     0.94498
2                112      0.9302       0.0256        0.88005     0.98042
-------------------------------------------------------------------------
Ho: area(1) = area(2)
    chi2(1) =      1.35       Prob>chi2 =    0.2447
```

◁

## Comparing areas with a gold standard

The area under multiple ROC curves can be compared with a gold standard using `rocgold`. The command syntax is similar to that of `roccomp`. The tests are corrected for the effect of multiple comparisons.

### ▷ Example 5

We will use the same data (presented by Hanley and McNeil 1983) as in the `roccomp` examples. Let's assume that the first modality is considered to be the standard against which both the second and third modalities are compared.

We want to plot and compare both the areas of the ROC curves of `mod2` and `mod3` with `mod1`. Since we consider `mod1` to be the gold standard, it is listed first after the reference variable in the `rocgold` command line.

```
. use http://www.stata-press.com/data/r9/ct2
. rocgold status mod1 mod2 mod3, graph summary
```

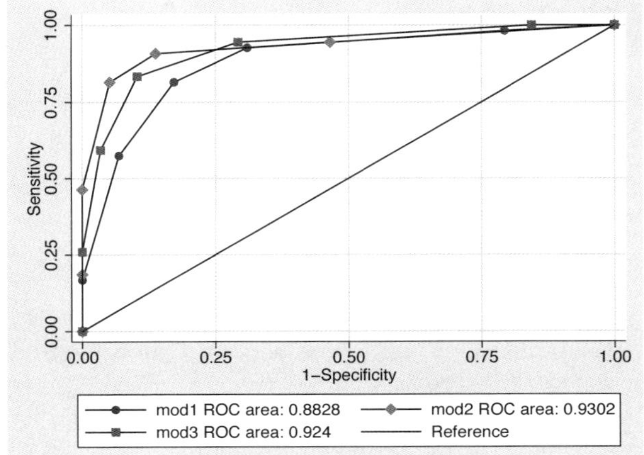


| | ROC Area | Std. Err. | chi2 | df | Pr>chi2 | Bonferroni Pr>chi2 |
|---|---|---|---|---|---|---|
| mod1 (standard) | 0.8828 | 0.0317 | | | | |
| mod2 | 0.9302 | 0.0256 | 2.3146 | 1 | 0.1282 | 0.2563 |
| mod3 | 0.9240 | 0.0241 | 5.2480 | 1 | 0.0220 | 0.0439 |

Equivalently, we could have done this in two steps using the roccomp command.

```
. roccomp status mod1 mod2, graph summary
. roccomp status mod1 mod3, graph summary msymbol(O S)
```

◁

## Saved Results

roctab saves in r():

Scalars

| | | | |
|---|---|---|---|
| r(N) | number of observations | r(area) | area under the ROC curve |
| r(se) | standard error for the area under the ROC curve | r(pietra) | Pietra index |
| r(lb) | lower bound of CI for the area under the ROC curve | r(gini) | Gini index |
| r(ub) | upper bound of CI for the area under the ROC curve | | |

roccomp saves in r():

Scalars

| | | | |
|---|---|---|---|
| r(N_g) | number of groups | r(df) | $\chi^2$ degrees of freedom |
| r(p) | significance probability | r(chi2) | $\chi^2$ |

Matrices

| | |
|---|---|
| r(V) | variance–covariance matrix |

rocgold saves in r():

Scalars

| | |
|---|---|
| r(N_g) | number of groups |

Matrices

| | | | |
|---|---|---|---|
| r(V) | variance–covariance matrix | r(p) | significance-probability vector |
| r(chi2) | $\chi^2$ vector | r(p_adj) | adjusted significance-probability vector |
| r(df) | $\chi^2$ degrees-of-freedom vector | | |

# Methods and Formulas

roctab, roccomp, and rocgold are implemented as ado-files.

Assume that we applied a diagnostic test to each of $N_n$ normal and $N_a$ abnormal subjects. Further assume that the higher the outcome value of the diagnostic test, the higher the risk of the subject being abnormal. Let $\widehat{\theta}$ be the estimated area under the curve, and let $X_i, i = 1, 2, \ldots, N_a$ and $Y_j, j = 1, 2, \ldots, N_n$ be the values of the diagnostic test for the abnormal and normal subjects, respectively.

## Nonparametric ROC

The points on the nonparametric ROC curve are generated by using each possible outcome of the diagnostic test as a classification cutpoint and computing the corresponding sensitivity and $1 -$ specificity. These points are then connected by straight lines, and the area under the resulting ROC curve is computed using the trapezoidal rule.

The default standard error for the area under the ROC curve is computed using the algorithm described by DeLong, DeLong, and Clarke-Pearson (1988). For each abnormal subject, $i$, define

$$V_{10}(X_i) = \frac{1}{N_n} \sum_{j=1}^{N_n} \psi(X_i, Y_j)$$

and for each normal subject, $j$, define

$$V_{01}(Y_j) = \frac{1}{N_a} \sum_{i=1}^{N_a} \psi(X_i, Y_j)$$

where

$$\psi(X, Y) = \begin{cases} 1 & Y < X \\ \frac{1}{2} & Y = X \\ 0 & Y > X \end{cases}$$

Define

$$S_{10} = \frac{1}{N_a - 1} \sum_{i=1}^{N_a} \{V_{10}(X_i) - \widehat{\theta}\}^2$$

and

$$S_{01} = \frac{1}{N_n - 1} \sum_{j=1}^{N_n} \{V_{01}(Y_j) - \widehat{\theta}\}^2$$

The variance of the estimated area under the ROC curve is given by

$$\text{var}(\widehat{\theta}) = \frac{1}{N_a}S_{10} + \frac{1}{N_n}S_{01}$$

The `hanley` standard error for the area under the ROC curve is computed using the algorithm described by Hanley and McNeil (1982). It requires the calculation of two quantities: $Q_1$ is Pr(two randomly selected abnormal subjects will both have a higher score than a randomly selected normal subject), and $Q_2$ is Pr(one randomly selected abnormal subject will have a higher score than any two randomly selected normal subjects). The Hanley and McNeil variance of the estimated area under the ROC curve is

$$\text{var}(\widehat{\theta}) = \frac{\widehat{\theta}(1-\widehat{\theta}) + (N_a-1)(Q_1-\widehat{\theta}^2) + (N_n-1)(Q_2-\widehat{\theta}^2)}{N_aN_n}$$

The `bamber` standard error for the area under the ROC curve is computed using the algorithm described by Bamber (1975). For any two $Y$ values, $Y_j$ and $Y_k$, and any $X_i$ value, define

$$b_{yyx} = p(Y_j, Y_k < X_i) + p(X_i < Y_j, Y_k) - 2p(Y_j < X_i < Y_k)$$

and similarly, for any two $X$ values, $X_i$ and $X_l$, and any $Y_j$ value, define

$$b_{xxy} = p(X_i, X_l < Y_j) + p(Y_j < X_i, X_l) - 2p(X_i < Y_j < X_l)$$

Bamber's unbiased estimate of the variance for the area under the ROC curve is

$$\text{var}(\widehat{\theta}) = \frac{1}{4}(N_a-1)(N_n-1)\{p(X \neq Y)+(N_a-1)b_{xxy}+(N_n-1)b_{yyx}-4(N_a+N_n-1)(\widehat{\theta}-0.5)^2\}$$

Asymptotic confidence intervals are constructed and reported by default, assuming a normal distribution for the area under the ROC curve.

Exact binomial confidence intervals are calculated as described in [R] **ci**, with $p$ equal to the area under the ROC curve.

## Comparing areas under the ROC curve

Areas under ROC curves are compared using an algorithm suggested by DeLong, DeLong, and Clarke-Pearson (1988). Let $\widehat{\theta} = (\widehat{\theta}^1, \widehat{\theta}^2, \ldots, \widehat{\theta}^k)$ be a vector representing the areas under $k$ ROC curves. For the $r$th area, define

$$V_{10}^r(X_i) = \frac{1}{N_n}\sum_{j=1}^{N_n}\psi(X_i^r, Y_j^r)$$

and for each normal subject, $j$, define

$$V_{01}^r(Y_j) = \frac{1}{N_a}\sum_{i=1}^{N_a}\psi(X_i^r, Y_j^r)$$

where

$$\psi(X^r, Y^r) = \begin{cases} 1 & Y^r < X^r \\ \frac{1}{2} & Y^r = X^r \\ 0 & Y^r > X^r \end{cases}$$

Define the $k \times k$ matrix $\mathbf{S_{10}}$ such that the $(r, s)$th element is

$$S_{10}^{r,s} = \frac{1}{N_a - 1} \sum_{i=1}^{N_a} \{V_{10}^r(X_i) - \widehat{\theta}^r\}\{V_{10}^s(X_i) - \widehat{\theta}^s\}$$

and $\mathbf{S_{01}}$ such that the $(r, s)$th element is

$$S_{01}^{r,s} = \frac{1}{N_n - 1} \sum_{j=1}^{N_n} \{V_{01}^r(Y_i) - \widehat{\theta}^r\}\{V_{01}^s(Y_i) - \widehat{\theta}^s\}$$

Then the covariance matrix is

$$S = \frac{1}{N_a} S_{10} + \frac{1}{N_n} S_{01}$$

Let $\mathbf{L}$ be a contrast matrix defining the comparison, so that

$$(\widehat{\theta} - \theta)'\mathbf{L}'\big(\mathbf{LSL}'\big)^{-1}\mathbf{L}(\widehat{\theta} - \theta)$$

has a chi-squared distribution with degrees of freedom equal to the rank of $\mathbf{LSL}'$.

# References

Bamber, D. 1975. The area above the ordinal dominance graph and the area below the receiver operating characteristic graph. *Journal of Mathematical Psychology* 12: 387–415.

Choi, B. C. K. 1998. Slopes of a receiver operating characteristic curve and likelihood ratios for a diagnostic test. *American Journal of Epidemiology* 148: 1127–1132.

Cleves, M. 1999. sg120: Receiver Operating Characteristic (ROC) analysis. *Stata Technical Bulletin* 52: 19–33. Reprinted in *Stata Technical Bulletin Reprints*, vol. 9, pp. 212–229.

——. 2000. sg120.2: Correction to roccomp command. *Stata Technical Bulletin* 54: 26. Reprinted in *Stata Technical Bulletin Reprints*, vol. 9, p. 231.

——. 2002a. Comparative assessment of three common algorithms for estimating the variance of the area under the nonparametric receiver operating characteristic curve. *Stata Journal* 2: 280–289.

——. 2002b. From the help desk: Comparing areas under receiver operating characteristic curves from two or more probit or logit models. *Stata Journal* 2: 301–313.

DeLong, E. R., D. M. DeLong, and D. L. Clarke-Pearson. 1988. Comparing the areas under two or more correlated receiver operating curves: A nonparametric approach. *Biometrics* 44: 837–845.

Erdreich, L. S. and E. T. Lee. 1981. Use of relative operating characteristic analysis in epidemiology: A method for dealing with subjective judgment. *American Journal of Epidemiology* 114: 649–662.

Hanley, J. A. and B. J. McNeil. 1982. The meaning and use of the area under a receiver operating characteristic (ROC) curve. *Radiology* 143: 26–36.

——. 1983. A method of comparing the areas under receiver operating characteristic curves derived from the same cases. *Radiology* 148: 839–843.

Lee, W. C. 1999. Probabilistic analysis of global performances of diagnostic test: Interpreting the Lorenz curve-based summary measures. *Statistics in Medicine* 18: 455–471.

Ma, G. and W. J. Hall. 1993. Confidence bands for the receiver operating characteristic curves. *Medical Decision Making* 13: 191–197.

Pepe, M. S. 2003. *The Statistical Evaluation of Medical Tests for Classification and Prediction.* New York: Oxford University Press.

Reichenheim, M. E. and A. Ponce de Leon. 2002. Estimation of sensitivity and specificity arising from validity studies with incomplete designs. *Stata Journal* 2: 267–279.

Seed, P. T. and A. Tobias. 2001. sbe36.1: Summary statistics for diagnostic tests. *Stata Technical Bulletin* 59: 25–27. Reprinted in *Stata Technical Bulletin Reprints*, vol. 10, pp. 90–93.

Tobias, A. 2000. sbe36: Summary statistics report for diagnostic tests. *Stata Technical Bulletin* 56: 16–18. Reprinted in *Stata Technical Bulletin Reprints*, vol. 10, pp. 87–90.

Working, H. and H. Hotelling. 1929. Application of the theory of error to the interpretation of trends. *Journal of the American Statistical Association* 24: 73–85.

# Also See

**Related:** [R] **logistic**, [R] **rocfit**

# Title

**rocfit** — Fit ROC models

# Syntax

`rocfit` *refvar classvar* [*if*] [*in*] [*weight*] [, *rocfit_options*]

| *rocfit_options* | description |
|---|---|
| Model | |
| `continuous(`*#*`)` | divide *classvar* into # groups of approximately equal length |
| `generate(`*newvar*`)` | create *newvar* containing classification groups |
| SE/Robust | |
| `vce(`*vcetype*`)` | *vcetype* may be `oim` or `opg` |
| Reporting | |
| `level(`*#*`)` | set confidence level; default is `level(95)` |
| Max options | |
| *maximize_options* | control the maximization process; seldom used |

`fweight`s are allowed; see [U] **11.1.6 weight**.

See [U] **20 Estimation and postestimation commands** for additional features of estimation commands.

# Description

`rocfit` fits maximum-likelihood ROC models assuming a binormal distribution of the latent variable.

The two variables *refvar* and *classvar* must be numeric. The reference variable indicates the true state of the observation, such as diseased and nondiseased or normal and abnormal, and must be coded as 0 and 1. The rating or outcome of the diagnostic test or test modality is recorded in *classvar*, which must be at least ordinal, with higher values indicating higher risk.

See [R] **roc** for other commands designed to perform Receiver Operating Characteristic (ROC) analyses with rating and discrete classification data.

# Options

Model

`continuous(`*#*`)` specifies that the continuous *classvar* be divided into # groups of approximately equal length. This option is required when *classvar* takes on more than 20 distinct values.

`continuous(.)` may be specified to indicate that *classvar* be used as it is, even though it could have more than 20 distinct values.

`generate(`*newvar*`)` specifies the new variable that is to contain the values indicating the groups produced by `continuous(`*#*`)`. `generate()` may only be specified with `continuous()`.

SE/Robust

vce(*vcetype*); see [R] ***vce_option***.

Reporting

level(*#*); see [R] **estimation options**.

Max options

*maximize_options*: difficult, technique(*algorithm_spec*), iterate(*#*), [no]log, trace, gradient, showstep, hessian, shownrtolerance, tolerance(*#*), ltolerance(*#*), gtolerance(*#*), nrtolerance(*#*), nonrtolerance, from(*init_specs*); see [R] **maximize**. These options are seldom used.

## Remarks

Dorfman and Alf (1969) developed a generalized approach for obtaining maximum likelihood estimates of the parameters for a smooth fitting ROC curve. The most commonly used method, and the one implemented here, is based upon the binormal model; see Pepe (2003) for a discussion of different approaches to ROC analysis.

The model assumes the existence of an unobserved, continuous, latent variable that is normally distributed (perhaps after a monotonic transformation) in both the normal and abnormal populations with means $\mu_n$ and $\mu_a$ and variances $\sigma_n^2$ and $\sigma_a^2$, respectively. The model further assumes that the $K$ categories of the rating variable result from partitioning the unobserved latent variable by $K-1$ fixed boundaries. The method fits a straight line to the empirical ROC points plotted using normal probability scales on both axes. Maximum likelihood estimates of the line's slope and intercept and the $K-1$ boundaries are obtained simultaneously. See *Methods and Formulas* for details.

The intercept from the fitted line is a measurement of $(\mu_a - \mu_n)/\sigma_a$, and the slope measures $\sigma_n/\sigma_a$.

Thus the intercept is the standardized difference between the two latent population means, and the slope is the ratio of the two standard deviations. The null hypothesis that there is no difference between the two population means is evaluated by testing that the intercept $=0$, and the null hypothesis that the variances in the two populations are equal is evaluated by testing that the slope $=1$.

### ▷ Example 1

We use Hanley and McNeil's (1982) dataset, described in example 1 of [R] **roc**, to fit a smooth ROC curve assuming a binormal model.

(*Continued on next page*)

```
.use http://www.stata-press.com/data/r9/hanley
. rocfit disease rating
Fitting binormal model:
Iteration 0:   log likelihood = -123.68069
Iteration 1:   log likelihood = -123.64867
Iteration 2:   log likelihood = -123.64855
Iteration 3:   log likelihood = -123.64855
Binormal model of disease on rating                 Number of obs   =        109
Goodness-of-fit chi2(2) =        0.21
Prob > chi2             =      0.9006
Log likelihood          =  -123.64855

------------------------------------------------------------------------------
             |      Coef.   Std. Err.        z    P>|z|    [95% Conf. Interval]
-------------+----------------------------------------------------------------
   intercept |   1.656782   0.310456      5.34   0.000     1.048300    2.265265
   slope (*) |   0.713002   0.215882     -1.33   0.092     0.289881    1.136123
-------------+----------------------------------------------------------------
       /cut1 |   0.169768   0.165307      1.03   0.152    -0.154227    0.493764
       /cut2 |   0.463215   0.167235      2.77   0.003     0.135441    0.790990
       /cut3 |   0.766860   0.174808      4.39   0.000     0.424243    1.109477
       /cut4 |   1.797938   0.299581      6.00   0.000     1.210770    2.385106
------------------------------------------------------------------------------

------------------------------------------------------------------------------
             |                 Indices from binormal fit
       Index |   Estimate    Std. Err.                     [95% Conf. Interval]
-------------+----------------------------------------------------------------
    ROC area |   0.911331    0.029506                      0.853501    0.969161
    delta(m) |   2.323671    0.502370                      1.339044    3.308298
        d(e) |   1.934361    0.257187                      1.430284    2.438438
        d(a) |   1.907771    0.259822                      1.398530    2.417012
------------------------------------------------------------------------------
(*) z test for slope==1
```

`rocfit` outputs the MLE for the intercept and slope of the fitted regression line along with, in this case, 4 boundaries (because there are 5 ratings) labeled `/cut1` through `/cut4`. In addition, `rocfit` also computes and reports 4 indices based on the fitted ROC curve: the area under the curve (labeled `ROC area`), $\delta(m)$ (labeled `delta(m)`), $d_e$ (labeled `d(e)`), and $d_a$ (labeled `d(a)`). More information about these indices can be found in *Methods and Formulas* and in Erdreich and Lee (1981). ◁

## Saved Results

`rocfit` saves in `e()`:

Scalars

| | | | |
|---|---|---|---|
| `e(N)` | number of observations | `e(df_gf)` | goodness-of-fit degrees of freedom |
| `e(k)` | number of parameters | `e(p_gf)` | $\chi^2$ goodness-of-fit significance probability |
| `e(k_eq)` | number of equations | `e(area)` | area under the ROC curve |
| `e(k_dv)` | number of dependent variables | `e(se_area)` | standard error for the area under the ROC curve |
| `e(df_m)` | model degrees of freedom | | |
| `e(ll)` | log likelihood | `e(deltam)` | $\delta(m)$ |
| `e(chi2_gf)` | goodness-of-fit $\chi^2$ | `e(se_delm)` | standard area for $\delta(m)$ |
| `e(rank)` | rank of `e(V)` | `e(de)` | $d_e$ index |
| `e(ic)` | number of iterations | `e(se_de)` | standard error for $d_e$ index |
| `e(rc)` | return code | `e(da)` | $d_a$ index |
| `e(converged)` | 1 if converged, 0 otherwise | `e(se_da)` | standard error for $d_a$ index |

Macros

| | | | |
|---|---|---|---|
| `e(cmd)` | `rocfit` | `e(opt)` | type of optimization |
| `e(depvar)` | names of dependent variables | `e(ml_method)` | type of `ml` method |
| `e(title)` | title in estimation output | `e(user)` | name of likelihood-evaluator program |
| `e(wtype)` | weight type | `e(technique)` | maximization technique |
| `e(wexp)` | weight expression | `e(crittype)` | optimization criterion |
| `e(chi2type)` | `GOF`; type of model $\chi^2$ test | `e(properties)` | `b V` |

Matrices

| | | | |
|---|---|---|---|
| `e(b)` | coefficient vector | `e(V)` | variance–covariance matrix of the estimators |
| `e(ilog)` | iteration log (up to 20 iterations) | | |
| `e(gradient)` | gradient vector | | |

Functions

| | |
|---|---|
| `e(sample)` | marks estimation sample |

# Methods and Formulas

`rocfit` is implemented as an ado-file.

Dorfman and Alf (1969) developed a general procedure for obtaining maximum likelihood estimates of the parameters of a smooth-fitting ROC curve. The most common method, and the one implemented in Stata, is based upon the binormal model.

The model assumes that there is an unobserved continuous latent variable that is normally distributed in both the normal and abnormal populations. The idea is better explained with the following illustration:

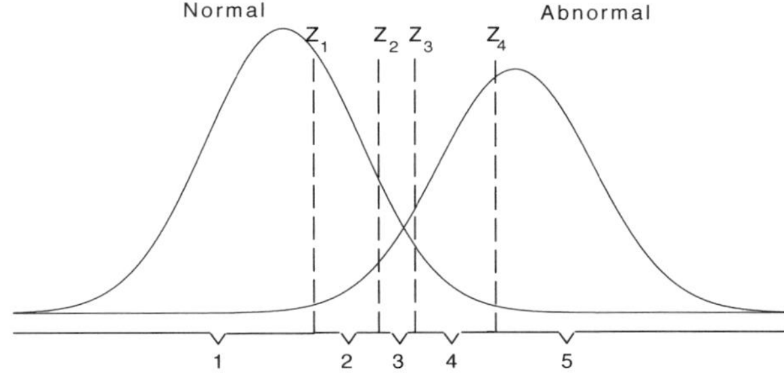


The latent variable is assumed to be normally distributed for both the normal and abnormal subjects, perhaps after a monotonic transformation, with means $\mu_n$ and $\mu_a$ and variances $\sigma_n^2$ and $\sigma_a^2$, respectively.

This latent variable is assumed to be partitioned into the $k$ categories of the rating variable by $k-1$ fixed boundaries. In the above figure, the $k=5$ categories of the rating variable identified on the bottom result from the partition of the four boundaries $Z_1$ through $Z_4$.

Let $R_j$ for $j = 1, 2, \ldots, k$ indicate the categories of the rating variable, let $i = 1$ if the subject belongs to the normal group, and let $i = 2$ if the subject belongs to the abnormal group.

Then

$$p(R_j|i=1) = F(Z_j) - F(Z_{j-1})$$

where $Z_k = (x_k - \mu_n)/\sigma_n$, $F$ is the cumulative normal distribution, $F(Z_0) = 0$ and $F(Z_k) = 1$. Also,

$$p(R_j|i=2) = F(bZ_j - a) - F(bZ_{j-1} - a)$$

where $b = \sigma_n/\sigma_a$ and $a = (\mu_a - \mu_n)/\sigma_a$.

The parameters $a$, $b$ and the $k-1$ fixed boundaries $Z_j$ are simultaneously estimated by maximizing the log-likelihood function

$$\log L = \sum_{i=1}^{2}\sum_{j=1}^{k} r_{ij}\log\{p(R_j|i)\}$$

where $r_{ij}$ is the number of $R_j$s in group $i$.

The area under the fitted ROC curve is computed as

$$\Phi\left(\frac{a}{\sqrt{1+b^2}}\right)$$

where $\Phi$ is the standard normal cumulative distribution function.

Point estimates for the ROC curve indices are as follows:

$$\delta(m) = \frac{a}{b} \qquad d_e = \frac{2a}{b+1} \qquad d_a = \frac{a\sqrt{2}}{\sqrt{1+b^2}}$$

Variances for these indices are computed using the delta method.

The $\delta(m)$ estimates $(\mu_a - \mu_n)/\sigma_n$, $d_e$ estimates $2(\mu_a - \mu_n)/(\sigma_a - \sigma_n)$, and $d_a$ estimates $\sqrt{2}(\mu_a - \mu_n)/(\sigma_a^2 - \sigma_n^2)^2$.

Simultaneous confidence bands for the entire curve are obtained, as suggested by Ma and Hall (1993), by first obtaining Working–Hotelling (1929) confidence bands for the fitted straight line in normal probability coordinates and then transforming them back to ROC coordinates.

## References


Bamber, D. 1975. The area above the ordinal dominance graph and the area below the receiver operating characteristic graph. *Journal of Mathematical Psychology* 12: 387–415.

Choi, B. C. K. 1998. Slopes of a receiver operating characteristic curve and likelihood ratios for a diagnostic test. *American Journal of Epidemiology* 148: 1127–1132.

Cleves, M. 1999. sg120: Receiver Operating Characteristic (ROC) analysis. *Stata Technical Bulletin* 52: 19–33. Reprinted in *Stata Technical Bulletin Reprints*, vol. 9, pp. 212–229.

——. 2000a. sg120.1: Two new options added to rocfit command. *Stata Technical Bulletin* 53: 18–19. Reprinted in *Stata Technical Bulletin Reprints*, vol. 9, pp. 230–231.

Dorfman, D. D. and E. Alf. 1969. Maximum likelihood estimation of parameters of signal detection theory and determination of confidence intervals-rating method data. *Journal of Mathematical Psychology* 6: 487–496.

Erdreich, L. S. and E. T. Lee. 1981. Use of relative operating characteristic analysis in epidemiology: A method for dealing with subjective judgment. *American Journal of Epidemiology* 114: 649–662.

Hanley, J. A. and B. J. McNeil. 1982. The meaning and use of the area under a receiver operating characteristic (ROC) curve. *Radiology* 143: 26–36.

Ma, G. and W. J. Hall. 1993. Confidence bands for the receiver operating characteristic curves. *Medical Decision Making* 13: 191–197.

Pepe, M. S. 2003. *The Statistical Evaluation of Medical Tests for Classification and Prediction*. New York: Oxford University Press.

Working, H. and H. Hotelling. 1929. Application of the theory of error to the interpretation of trends. *Journal of the American Statistical Association* 24: 73–85.

## Also See

**Complementary:** [R] **rocfit postestimation**; [R] **roc**

**Background:** [U] **20 Estimation and postestimation commands**,
[R] **maximize**, [R] ***vce_option***

# Title

**rocfit postestimation** — Postestimation tools for rocfit

# Description

The following command is of special interest after `rocfit`:

| command | description |
|---|---|
| `rocplot` | plot the fitted ROC curve and simultaneous confidence bands |

For information about `rocplot`, see below.

In addition, the following standard postestimation commands are available:

| command | description |
|---|---|
| `estat` | AIC, BIC, VCE, and estimation sample summary |
| `estimates` | cataloging estimation results |
| `lincom`[1] | point estimates, standard errors, testing, and inference for linear combinations of coefficients |
| `test`[1] | Wald tests for simple and composite linear hypotheses |

[1] See *Using lincom and test after rocfit* below.
See the corresponding entries in the *Stata Base Reference Manual* for details.

## Special-interest postestimation command

`rocplot` plots the fitted ROC curve and simultaneous confidence bands.

(*Continued on next page*)

## Syntax for rocplot

`rocplot` [ , *rocplot_options* ]

| *rocplot_options* | description |
|---|---|
| Main | |
| `confband` | display confidence bands |
| `norefline` | suppress plotting the reference line |
| `level(#)` | set confidence level; default is `level(95)` |
| Plot | |
| `plotopts(`*plot_options*`)` | affect rendition of the ROC points |
| Fit line | |
| `lineopts(`*cline_options*`)` | affect rendition of the fitted ROC line |
| CI plot | |
| `ciopts(`*area_options*`)` | affect rendition of the confidence bands |
| Reference line | |
| `rlopts(`*cline_options*`)` | affect rendition of the reference line |
| Add plot | |
| `addplot(`*plot*`)` | add other plots to the generated graph |
| Y-Axis, X-Axis, Title, Caption, Legend, Overall | |
| *twoway_options* | any options other than `by()` documented in [G] ***twoway_options*** |

| *plot_options* | description |
|---|---|
| *marker_options* | change look of markers (color, size, etc.) |
| *marker_label_options* | add marker labels; change look or position |
| *cline_options* | change the look of the line |

## Options for rocplot

Main

`confband` specifies that simultaneous confidence bands be plotted around the ROC curve.

`norefline` suppresses plotting the 45-degree reference line from the graphical output of the ROC curve.

`level(#)` specifies the confidence level, as a percentage, for the confidence bands. The default is `level(95)` or as set by `set level`; see [R] **level**.

Plot

`plotopts(`*plot_options*`)` affect the rendition of the plotted ROC points, including the size and color of markers, whether and how the markers are labeled, and whether and how the points are connected. For the full list of available *plot_options*, see [G] ***marker_options***, [G] ***marker_label_options***, and [G] ***cline_options***.

Fit line

`fitopts(`*cline_options*`)` affect the rendition of the fitted ROC line; see [G] ***cline_options***.

CI plot

`ciopts(`*area_options*`)` affect the rendition of the confidence bands; see [G] ***area_options***.

Reference line

`rlopts(`*cline_options*`)` affect the rendition of the reference line; see [G] ***cline_options***.

Add plot

`addplot(`*plot*`)` provides a way to add other plots to the generated graph. See [G] ***addplot_option***.

Y-Axis, X-Axis, Title, Caption, Legend, Overall

*twoway_options* are any of the options documented in [G] ***twoway_options***, excluding `by()`. These include options for titling the graph (see [G] ***title_options***) and options for saving the graph to disk (see [G] ***saving_option***).

# Remarks

Remarks are presented under the headings

*Using lincom and test*
*Using rocplot*

## Using lincom and test

Commands such as `lincom` and `test` permit reference to *name* instead of `_b[`*name*`]`, but this is not the case when `lincom` and `test` are used after `rocfit`. You need to specify full names, such as `_b[intercept]` and `_b[slope]`. For instance, you could type

```
. test _b[intercept] = 0

 ( 1)  intercept = 0

           chi2(  1) =   28.48
         Prob > chi2 =    0.0000
```

For the shorthand notation to work, you need a variable named *name* in the data. In `rocfit`, however, *name* is just a coefficient label that does not necessarily correspond to any variable in the data.

## Using rocplot

▷ Example 1

In [R] **rocfit**, we fitted a ROC curve by typing `rocfit disease rating`.

Note that in the output table for our model, we are testing whether or not the variances of the two latent populations are equal by testing that the slope = 1.

We plot the fitted ROC curve.

```
. rocplot, confband
```

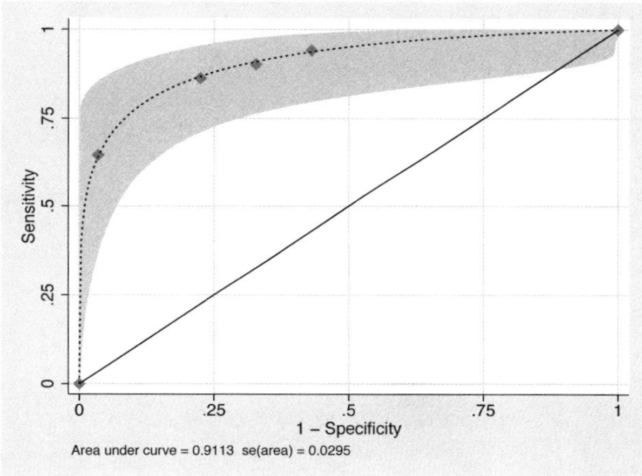


◁

# Methods and Formulas

All postestimation commands listed above are implemented as ado-files.

# Also See

**Complementary:** [R] **rocfit**; [R] **roc**; [R] **estimates**, [R] **lincom**, [R] **test**

**Background:** [U] **13.5 Accessing coefficients and standard errors**, [U] **20 Estimation and postestimation commands**, [R] **estat**

# Title

**rologit** — Rank-ordered logistic regression

# Syntax

`rologit` *depvar indepvars* [*if*] [*in*] [*weight*] , `group(`*varname*`)` [*options*]

| *options* | description |
|---|---|
| Model | |
| * `group(`*varname*`)` | identifier variable that links the alternatives |
| `offset(`*varname*`)` | include *varname* in model with coefficient constrained to 1 |
| `incomplete(`*#*`)` | use # to code unranked alternatives; default is `incomplete(0)` |
| `reverse` | reverse the preference order |
| `notestrhs` | keep RHS variables that do not vary within group |
| `ties(`*spec*`)` | method to handle ties: `exactm`, `breslow`, `efron`, or `none` |
| SE/Robust | |
| `robust` | compute standard errors using the robust/sandwich estimator |
| `cluster(`*varname*`)` | adjust standard errors for intragroup correlation |
| Reporting | |
| `level(`*#*`)` | set confidence level; default is `level(95)` |
| Max options | |
| *maximize_options* | control the maximization process; seldom used |

*`group(`*varname*`)` is required.

`bootstrap`, `by`, `jackknife`, `rolling`, `statsby`, and `xi` are allowed; see [U] **11.1.10 Prefix commands**.

`fweight`s, `iweight`s, and `pweight`s are allowed; see [U] **11.1.6 weight**. No weights are allowed in the case of ties.

See [U] **20 Estimation and postestimation commands** for additional capabilities of estimation commands.

# Description

`rologit` fits the rank-ordered logistic regression model by maximum likelihood (Beggs, Cardell, and Hausman 1981). This model is also known as the Plackett–Luce model (Marden 1995), as the exploded logit model (Punj and Staelin 1978), and as the choice-based method of conjoint analysis (Hair et al. 1995).

`rologit` expects the data to be in long form, similar to `clogit`, in which each of the ranked alternatives forms an observation; all observations related to an individual are linked together by the variable that you specify in the `group()` option. The distinction from `clogit` is that *depvar* in `rologit` records the rankings of the alternatives, whereas for `clogit`, *depvar* only marks the best alternative by a value not equal to zero. `rologit` interprets equal scores of *depvar* as ties. The ranking information may be incomplete "at the bottom" (least preferred alternatives). That is, unranked alternatives may be coded as 0 or as a common value that may be specified with the `incomplete()` option.

If your data record only the unique best alternative, `rologit` fits the same model as `clogit`.

# Options

Model

group(*varname*) is required, and it specifies the identifier variable (numeric or string) that links the alternatives for an individual, which have been compared and rank-ordered with respect to one another.

offset(*varname*); see [R] **estimation options**.

incomplete(*#*) specifies the numeric value used to code alternatives that are not ranked. It is assumed that unranked alternatives are less preferred than the ranked alternatives (i.e., the data record the ranking of the most preferred alternatives). It is not assumed that subjects are indifferent between the unranked alternatives. # defaults to 0.

reverse specifies that in the preference order, a higher number means a less attractive alternative. The default is that higher values indicate more attractive alternatives. The rank-ordered logit model is not symmetric in the sense that reversing the ordering simply leads to a change in the signs of the coefficients.

notestrhs suppresses the test that the independent variables vary within (at least some of) the groups. Effects of variables that are always constant are not identified. For instance, a rater's gender cannot directly affect his or her rankings; it could affect the rankings only via an interaction with a variable that does vary over alternatives.

ties(*spec*) specifies the method for handling ties (indifference between alternatives) (see [ST] **stcox** for details):

| | |
|---|---|
| exactm | exact marginal likelihood (default) |
| breslow | Breslow's approximation |
| efron | Efron's approximation |
| none | no ties allowed |

SE/Robust

robust, cluster(*varname*); see [R] **estimation options**. Note that if robust or cluster() are specified and there are tied rankings in the data, ties(efron) is imposed.

Reporting

level(*#*); see [R] **estimation options**.

Max options

*maximize_options*: iterate(*#*), trace, [no]log, tolerance(*#*), ltolerance(*#*); see [R] **maximize**. These options are seldom used.

# Remarks

The rank-ordered logit model can be applied to analyze how decision-makers combine attributes of alternatives into overall evaluations of the attractiveness of these alternatives. The model generalizes McFadden's choice model, as fitted by the clogit command. It employs richer information about the comparison of alternatives, namely, how decision-makers rank the alternatives rather than just specifying the alternative that they like best.

Remarks are presented under the headings

*Examples*
*Comparing respondents*
*Incomplete rankings and ties*
*Clustered choice data*
*Comparison of rologit and clogit*
*On reversals of rankings*

## Examples

A popular way to study employer preferences for characteristics of employees is the quasi-experimental "vignette method". As an example, we consider the research by De Wolf on the labor market position of social science graduates (De Wolf 2000). This study addresses how the educational portfolio (e.g., general skills versus specific knowledge) affects short term and long-term labor-market opportunities. De Wolf asked 22 human resource managers (the respondents) to rank order the 6 most suitable candidates out of 20 fictitious applicants, and to rank order these 6 candidates for 3 jobs, namely, (1) researcher, (2) management trainee, and (3) policy adviser. Applicants were described by 10 attributes, including their age, gender, details of their portfolio, and work experience. In this example, we analyze a subset of the data. Also, to simplify the output, we drop, at random, 10 nonselected applicants per case. The resulting dataset is comprised of 29 cases, consisting of 10 applicants each. The data are in long form: observations correspond to alternatives (the applications), and alternatives that figured in one decision task are identified by the variable `caseid`. We list the observations for `caseid==7`, in which the respondent considered applicants for a social science research position.

```
. use http://www.stata-press.com/data/r9/evignet
(Vignet study employer prefs (Inge de Wolf 2000))
. list pref female age grades edufit workexp boardexp if caseid==7, noobs
```

| pref | female | age | grades | edufit | workexp | boardexp |
|---|---|---|---|---|---|---|
| 0 | yes | 28 | A/B | no | none | no |
| 0 | no | 25 | C/D | yes | one year | no |
| 0 | no | 25 | C/D | yes | none | yes |
| 0 | yes | 25 | C/D | no | internship | yes |
| 1 | no | 25 | C/D | yes | one year | yes |
| 2 | no | 25 | A/B | yes | none | no |
| 3 | yes | 25 | A/B | yes | one year | no |
| 4 | yes | 25 | A/B | yes | none | yes |
| 5 | no | 25 | A/B | yes | internship | no |
| 6 | yes | 28 | A/B | yes | one year | yes |

In this case, 6 applicants were selected. The rankings are stored in the variable `pref`, where a value of 6 corresponds to "best among the candidates", a value of 5 corresponds to "second best among the candidates", etc. The applicants with a ranking of 0 were not among the best 6 candidates for the job. The respondent was not asked to express his preferences among these 4 applicants, but by the elicitation procedure, it is known that he ranks these 4 applicants below the 6 selected applicants. The best candidate was a female, 28 years old, with education fitting the job, with good grades (A/B), with one year of work experience, and with experience being a board member of a fraternity, a sports club, etc. The profiles of the other candidates read similarly. In this case, the respondent completed the task, that is, he selected and rank ordered the 6 most suitable applicants. In some cases, the respondent performed only part of the task.

```
. list pref female age grades edufit workexp boardexp if caseid==18, noobs
```

| pref | female | age | grades | edufit | workexp | boardexp |
|---|---|---|---|---|---|---|
| 0 | no | 25 | C/D | yes | none | yes |
| 0 | no | 25 | C/D | no | internship | yes |
| 0 | no | 28 | C/D | no | internship | yes |
| 0 | yes | 25 | A/B | no | one year | no |
| 2 | yes | 25 | A/B | no | none | yes |
| 2 | no | 25 | A/B | no | none | yes |
| 2 | no | 25 | A/B | no | one year | yes |
| 5 | no | 25 | A/B | no | none | yes |
| 5 | no | 25 | A/B | no | none | yes |
| 5 | yes | 25 | A/B | no | none | no |

The respondent selected the six best candidates and segmented these six candidates into two groups: one group with the three best candidates, and a second group of three candidates that were "still acceptable". The numbers 2 and 5, indicating these two groups, are arbitrary apart from the implied ranking of the groups. The ties between the candidates in a group indicate that the respondent was not able to rank the candidates within the group.

The purpose of the vignette experiment was to explore and test hypotheses about which of the employees' attributes are valued by employers, how these attributes are weighted depending on the type of job (described by variable `job` in these data), etc. In the psychometric tradition of Thurstone (1927), *value* is assumed to be linear in the attributes, with the coefficients expressing the direction and weight of the attributes. In addition, it is assumed that *valuation* is to some extent a random procedure, captured by an additive random term. For instance, if value only depends on an applicant's age and gender, we would have

$$\text{value}(\texttt{female}_i, \texttt{age}_i) = \beta_1 \texttt{female}_i + \beta_2 \texttt{age}_i + \epsilon_i$$

where the random residual $\epsilon_i$ captures all omitted attributes. Thus $\beta_1 > 0$ means that the employer assigns higher value to a woman than to a man. Given this conceptualization of value, it is straightforward to model the decision (selection) among alternatives or the ranking of alternatives: The alternative with the highest value is selected (chosen), or the alternatives are ranked according to their value. To complete the specification of a model of choice and of ranking, we assume that the random residual $\epsilon_i$ follows an "extreme value distribution of type I", introduced in this context by Luce (1959). This specific assumption is made mostly for computational convenience.

This model is known by many names. Among others, it is known as the rank-ordered logit model in economics (Beggs, Cardell, and Hausman 1981), as the exploded logit model in marketing research (Punj and Staelin 1978), as the choice-based conjoint analysis model (Hair et al. 1995), and as the Plackett–Luce model (Marden 1995). The model coefficients are estimated using the method of maximum likelihood. The implementation in `rologit` uses an analogy between the rank-ordered logit model and the Cox regression model observed by Allison and Christakis (1994); see *Methods and Formulas*. The `rologit` command implements this method for rankings, while `clogit` deals with the variant of choices, that is, only the most highly valued alternative is recorded. In the latter case, the model is also known as the Luce–McFadden choice model. In fact, when the data record the most preferred (unique) alternative and no additional ranking information about preferences is available, `rologit` and `clogit` return the same information, though formatted somewhat differently.

```
. rologit pref female age grades edufit workexp boardexp if job==1, group(caseid)
Iteration 0:   log likelihood =   -95.41087
Iteration 1:   log likelihood = -71.180903
Iteration 2:   log likelihood =   -68.47734
Iteration 3:   log likelihood = -68.345918
Iteration 4:   log likelihood = -68.345389
Refining estimates:
Iteration 0:   log likelihood = -68.345389

Rank-ordered logistic regression                Number of obs      =        80
Group variable: caseid                          Number of groups   =         8

no ties in data                                 Obs per group: min =        10
                                                               avg =     10.00
                                                               max =        10

                                                LR chi2(6)         =     54.13
Log likelihood = -68.34539                      Prob > chi2        =    0.0000

------------------------------------------------------------------------------
        pref |      Coef.   Std. Err.      z    P>|z|     [95% Conf. Interval]
-------------+----------------------------------------------------------------
      female |  -.4487287   .3671307    -1.22   0.222    -1.168292    .2708343
         age |  -.0984926   .0820473    -1.20   0.230    -.2593024    .0623172
      grades |   3.064534   .6148245     4.98   0.000       1.8595    4.269568
      edufit |   .7658064   .3602366     2.13   0.034     .0597556    1.471857
     workexp |   1.386427    .292553     4.74   0.000     .8130341    1.959821
    boardexp |   .6944377   .3762596     1.85   0.065    -.0430176    1.431893
------------------------------------------------------------------------------
```

Focusing only on the variables whose coefficients are significant at the 10% level (we are analyzing 8 respondents only!), the estimated value of an applicant for a job of type 1 (research positions) can be written as

```
value = 3.06*grades + 0.77*edufit + 1.39*workexp + 0.69*boardexp
```

Thus employers prefer applicants for a research position (`job==1`) whose educational portfolio fits the job, who have better grades, who have more relevant work experience, and who have (extracurricular) board experience. They don't seem to care much about the sex and age of applicants, which is comforting.

Given these estimates of the valuation by employers, we consider the probabilities that each of the applications is ranked first. Under the assumption that the $\epsilon_i$ are independent and follow an extreme value type I distribution, Luce (1959) showed that the probability, $\pi_i$, that alternative $i$ is valued higher than alternatives $2, \ldots, k$ can be written in the multinomial logit form

$$\pi_i = \Pr\left\{\text{value}_1 > \max(\text{value}_2, \ldots, \text{value}_m)\right\} = \frac{\exp(\text{value}_i)}{\sum_{j=1}^{k} \exp(\text{value}_i)}$$

The probability of observing a specific ranking can be written as the *product* of such terms, representing a sequential decision interpretation in which the rater first chooses the most preferred alternative, then the most preferred alternative among the rest, etc.

The probabilities for alternatives to be ranked first are conveniently computed by `predict`.

```
. predict p if e(sample)
(option pr assumed; conditional probability that alternative is ranked first)
(210 missing values generated)
. sort caseid pref p
. list pref p grades edufit workexp boardexp if caseid==7, noobs
```

| pref | p | grades | edufit | workexp | boardexp |
|---|---|---|---|---|---|
| 0 | .0027178 | C/D | yes | none | yes |
| 0 | .0032275 | C/D | no | internship | yes |
| 0 | .0064231 | A/B | no | none | no |
| 0 | .0217202 | C/D | yes | one year | no |
| 1 | .0434964 | C/D | yes | one year | yes |
| 2 | .0290762 | A/B | yes | none | no |
| 3 | .2970933 | A/B | yes | one year | no |
| 4 | .0371747 | A/B | yes | none | yes |
| 5 | .1163203 | A/B | yes | internship | no |
| 6 | .4427504 | A/B | yes | one year | yes |

There clearly is a positive relation between the stated ranking and the predicted probabilities for alternatives to be ranked first, but the association is not perfect. In fact, we would not have expected a perfect association, as the model specifies a (nondegenerate) probability distribution over the possible rankings of the alternatives. These predictions for sets of 10 candidates can also be used to make predictions for subsets of the alternatives. For instance, suppose that only the last three candidates listed in this table would be available. According to parameter estimates of the rank-ordered logit model, the probability that the last of these candidates is selected equals $0.443/(0.037+0.116+0.443) = 0.743$.

## Comparing respondents

The `rologit` model assumes that all respondents, HR managers in large public-sector organizations in the Netherlands, use the *same* valuation function; i.e., they apply the same decision weights. This is the substantive interpretation of the assumption that the $\beta$s are constant between the respondents. To probe this assumption, we could test whether the coefficients vary between different groups of respondents. For a metric characteristic of the HR manager, such as `firmsize`, we can consider a trend-model in the valuation weights,

$$\beta_{ij} = \alpha_{i0} + \alpha_{i1}\texttt{firmsize}_j$$

and we can test that the slopes $\alpha_{i1}$ of `firmsize` are zero.

```
. generate firmsize = employer
. generate size_edufit   = firmsize * edufit
. generate size_grades   = firmsize * grades
. generate size_workexp  = firmsize * workexp
. generate size_boardexp = firmsize * boardexp
```

(*Continued on next page*)

```
. rologit pref edufit grades workexp size* if job==1, group(caseid) nolog
Rank-ordered logistic regression                Number of obs      =        80
Group variable: caseid                          Number of groups   =         8
no ties in data                                 Obs per group: min =        10
                                                               avg =     10.00
                                                               max =        10
                                                LR chi2(7)         =     57.17
Log likelihood = -66.82346                      Prob > chi2        =    0.0000

        pref       Coef.   Std. Err.      z    P>|z|     [95% Conf. Interval]

      edufit     1.29122     1.13764    1.13   0.256    -.9385127    3.520953
      grades    6.439776    2.288056    2.81   0.005     1.955267    10.92428
     workexp     1.23342    .8065067    1.53   0.126     -.347304    2.814144
 size_edufit   -.0173333    .0711942   -0.24   0.808    -.1568714    .1222048
 size_grades   -.2099279    .1218251   -1.72   0.085    -.4487008     .028845
size_workexp    .0097508    .0525081    0.19   0.853    -.0931632    .1126649
size_board~p    .0382304    .0227545    1.68   0.093    -.0063676    .0828284

. testparm size*
 ( 1)  size_edufit = 0
 ( 2)  size_grades = 0
 ( 3)  size_workexp = 0
 ( 4)  size_boardexp = 0
           chi2(  4) =    7.14
         Prob > chi2 =    0.1288
```

The Wald test that the slopes `size_*` are jointly zero provides no evidence upon which we would reject the null hypothesis; i.e., we do not find evidence against the assumption of constant valuation weights of the attributes by firms of different size. Note that we did not enter `firmsize` as a predictor variable. Characteristics of the decision-making agent do not vary between alternatives. Thus an additive effect of these characteristics on the valuation of alternatives does *not* affect the agent's ranking of alternatives and his choice. Consequently the coefficient of `firmsize` is not identified. `rologit` would in fact have diagnosed the problem and dropped `firmsize` from the analysis. Diagnosing this problem can slow the estimation considerably; the test may be suppressed by specifying the option `notestrhs`.

## Incomplete rankings and ties

`rologit` allows incomplete rankings and ties in the rankings as proposed by Allison and Christakis (1994). `rologit` only permits rankings to be incomplete "at the bottom"; namely, that the ranking of the least attractive alternatives for subjects may not be known—do not confuse this with the situation that a subject is indifferent between these alternatives. This form of incompleteness occurred in the example discussed here, as the respondents were instructed to select and rank only the top 6 alternatives. It may also be that respondents refused to rank the alternatives that are very unattractive. `rologit` does not allow other forms of incompleteness, for instance, data in which respondents indicate which of four cars they like best, and which one they like least, but not how they rank the two intermediate cars. Another example of incompleteness that cannot be analyzed with `rologit` is data in which respondents select the three alternatives they like best but are not requested to express their preferences among the three selected alternatives.

`rologit` also permits ties in rankings. `rologit` assumes that if a subject expresses a tie between two or more alternatives, he or she actually holds one particular strict preference ordering, but with all possibilities of a strict ordering consistent with the expressed weak ordering being equally probable.

For instance, suppose that a respondent ranks alternative 1 highest. He prefers alternatives 2 and 3 over alternative 4, and he is indifferent between alternatives 2 and 3. We assume that this respondent either has the strict preference ordering $1 > 2 > 3 > 4$ or $1 > 3 > 2 > 4$, with both possibilities being equally likely. From a psychometric perspective, it may actually be more appropriate to also assume that the alternatives 2 and 3 are close; for instance, the difference between the associated valuations (utilities) is less than some threshold or minimally discernible difference. Computationally, however, this is a more demanding model.

## Clustered choice data

We have seen that applicants with work experience are in a relatively favorable position. To test whether the effects of work experience vary between the jobs, we can include interactions between the type of job and the attributes of applicants. Such interactions can be obtained via `xi`. To suppress the meaningless (and unidentified) main effect of `job`, we use `xi` specifications of the form `i.job|`*varname*.

Since some HR managers contributed data for more than one job, we cannot assume that their selection decisions for different jobs are independent. We can account for this by specifying the `cluster()` option. By treating choice data as incomplete ranking data with only the most preferred alternative marked, `rologit` may be used to estimate the model parameters for clustered choice data.

```
. xi: rologit pref i.job|female i.job|grades i.job|edufit i.job|workexp,
> group(caseid) cluster(employer) ties(efron) nolog
i.job             _Ijob_1-3          (naturally coded; _Ijob_1 omitted)
i.job|female      _IjobXfemal_#      (coded as above)
i.job|grades      _IjobXgrade_#      (coded as above)
i.job|edufit      _IjobXedufi_#      (coded as above)
i.job|workexp     _IjobXworke_#      (coded as above)

Rank-ordered logistic regression                 Number of obs      =       290
Group variable: caseid                           Number of groups   =        29

ties handled via the efron method                Obs per group: min =        10
                                                                avg =     10.00
                                                                max =        10

                                                 Wald chi2(12)      =     79.57
Log pseudolikelihood=-296.3855                   Prob > chi2        =    0.0000

                            (Std. Err. adjusted for 22 clusters in employer)
------------------------------------------------------------------------------
             |               Robust
        pref |      Coef.   Std. Err.      z    P>|z|     [95% Conf. Interval]
-------------+----------------------------------------------------------------
      female |  -.2286609   .2519883    -0.91   0.364    -.7225489    .2652272
_IjobXfema~2 |   .0293815   .4829166     0.06   0.951    -.9171177    .9758808
_IjobXfema~3 |   .1195538   .3688844     0.32   0.746    -.6034463    .8425538
      grades |   2.812555   .8517878     3.30   0.001     1.143081    4.482028
_IjobXgrad~2 |  -2.364247   1.005963    -2.35   0.019    -4.335898   -.3925961
_IjobXgrad~3 |   -1.88232   .8995277    -2.09   0.036    -3.645362   -.1192782
      edufit |   .7027757   .2398396     2.93   0.003     .2326987    1.172853
_IjobXeduf~2 |   -.267475   .4244964    -0.63   0.529    -1.099473    .5645226
_IjobXeduf~3 |  -.3182995   .3689972    -0.86   0.388    -1.041521    .4049217
     workexp |   1.224453   .3396773     3.60   0.000     .5586978    1.890208
_IjobXwork~2 |  -.6870077   .3692946    -1.86   0.063    -1.410812    .0367964
_IjobXwork~3 |  -.4656993   .4515712    -1.03   0.302    -1.350763    .4193639
------------------------------------------------------------------------------
```

The parameter estimates for the first job type are very similar to those that would have been obtained from an analysis isolated to these data. Differences are due only to an implied change in the method of handling ties. With clustered observations, `rologit` uses Efron's method. If we had specified the option `ties(efron)` with the separate analyses, then the parameter estimates would have been identical to the simultaneous results. Another difference is that `rologit` now reports robust standard errors, adjusted for clustering within respondents. These could have been obtained for the separate analyses, as well by specifying the `robust` option. In fact, this option would also have forced `rologit` to switch to Efron's approximation as well.

Given the combined results for the three types of jobs, we can test easily whether the weights for the attributes of applicants vary between the jobs, in other words, whether employers are looking for different qualifications in applicants for different jobs. A Wald test for the equality hypothesis of no difference can be obtained with the `testparm` command:

```
. testparm _I*
 ( 1)  _IjobXfemal_2 = 0
 ( 2)  _IjobXfemal_3 = 0
 ( 3)  _IjobXgrade_2 = 0
 ( 4)  _IjobXgrade_3 = 0
 ( 5)  _IjobXedufi_2 = 0
 ( 6)  _IjobXedufi_3 = 0
 ( 7)  _IjobXworke_2 = 0
 ( 8)  _IjobXworke_3 = 0

           chi2(  8) =   14.96
         Prob > chi2 =    0.0599
```

We find only mild evidence that employers look for different qualities in candidates according to the job for which they are being considered.

### ❑ Technical Note

Allison (1999) stressed that the comparison between groups of the coefficients of logistic regression is problematic, especially in its latent-variable interpretation. In many common latent-variable models, only the regression coefficients divided by the scale of the latent variable are identified. Thus a comparison of logit regression coefficients between, say, men and women is only meaningful if one is willing to argue that the standard deviation of the latent residual does not differ between the sexes. The rank-ordered logit model is also affected by this problem. While we formulated the model with a scale-free residual, we can actually think of the model for the value of an alternative as being scaled by the standard deviation of the random term, representing other relevant attributes of alternatives. Again comparing attribute weights between jobs is meaningful to the extent that we are willing to defend that "all omitted attributes" are equally important for different kinds of jobs.

❑

## Comparison of rologit and clogit

The rank-ordered logit model also has a sequential interpretation. A subject first chooses the best among the alternatives. Next, he or she selects the best alternative among the remaining alternatives, etc. The decisions at each of the subsequent stages are described by McFadden's choice model, and a subject is assumed to apply the same decision weights at each stage. Some authors have expressed concern that later choices may well be made in a more random way than the first few decisions. A formalization of this idea is a heteroskedastic version of the rank-ordered logit model in which the scale of the random term increases with the number of decisions made (e.g., Hausman and Ruud 1987). This extended model is currently not supported by `rologit`. However, the hypothesis that the same decision weights are applied at the first stage and at later stages can be tested by applying a Hausman test.

First, we fit the rank-ordered logit model on the full ranking data for the first type of job,

```
. rologit pref age female edufit grades workexp boardexp if job==1, group(caseid) nolog
Rank-ordered logistic regression                 Number of obs      =        80
Group variable: caseid                           Number of groups   =         8

no ties in data                                  Obs per group: min =        10
                                                                avg =     10.00
                                                                max =        10

                                                 LR chi2(6)         =     54.13
Log likelihood = -68.34539                       Prob > chi2        =    0.0000

------------------------------------------------------------------------------
        pref |      Coef.   Std. Err.      z    P>|z|     [95% Conf. Interval]
-------------+----------------------------------------------------------------
         age |  -.0984926   .0820473    -1.20   0.230    -.2593024    .0623172
      female |  -.4487287   .3671307    -1.22   0.222    -1.168292    .2708343
      edufit |   .7658064   .3602366     2.13   0.034     .0597556    1.471857
      grades |   3.064534   .6148245     4.98   0.000       1.8595    4.269568
     workexp |   1.386427    .292553     4.74   0.000     .8130341    1.959821
    boardexp |   .6944377   .3762596     1.85   0.065    -.0430176    1.431893
------------------------------------------------------------------------------
```

and we save the estimates for later use with the `estimates` command.

```
. estimates store Ranking
```

To estimate the decision weights based on the most preferred alternatives only, we create a variable `best` that is 1 for the best alternatives, and 0 otherwise. The `by` prefix is useful here.

```
. by caseid (pref), sort: gen best = pref == pref[_N] if job==1
(210 missing values generated)
```

Note that by specifying `(pref)` with `by caseid`, we ensured that the data were sorted in increasing order on `pref` within `caseid`. Hence, the most preferred alternatives are last in the sort order. The expression `pref == pref[_N]` is true (1) for the most preferred alternatives, even if the alternative is not unique, and false (0) otherwise. If the most preferred alternatives were sometimes tied, we could still fit the model for the based-alternatives-only data via `rologit`, but `clogit` would yield different results since it deals with ties in a way less appropriate in the case of continuous valuations. To ascertain whether there are ties in the selected data regarding applicants for research positions, we can combine `by` with `assert`:

```
. by caseid (pref), sort: assert pref[_N-1] != pref[_N] if job==1
```

There are no ties. We can now fit the model on the choice data by either `clogit` or `rologit`.

*(Continued on next page)*

```
. rologit best age edufit grades workexp boardexp if job==1, group(caseid) nolog
Rank-ordered logistic regression                Number of obs      =        80
Group variable: caseid                          Number of groups   =         8
no ties in data                                 Obs per group: min =        10
                                                               avg =     10.00
                                                               max =        10
                                                LR chi2(5)         =     17.27
Log likelihood = -9.783205                      Prob > chi2        =    0.0040

------------------------------------------------------------------------------
        best |      Coef.   Std. Err.      z    P>|z|     [95% Conf. Interval]
-------------+----------------------------------------------------------------
         age |  -.1048959   .2017068    -0.52   0.603    -.5002339    .2904421
      edufit |   .4558387   .9336775     0.49   0.625    -1.374136    2.285813
      grades |   3.443851   1.969002     1.75   0.080    -.4153223    7.303025
     workexp |   2.545648   1.099513     2.32   0.021     .3906422    4.700655
    boardexp |   1.765176   1.112763     1.59   0.113    -.4157988    3.946152
------------------------------------------------------------------------------
. estimates store Choice
```

The same results, though with a slightly different formatted header, would have been obtained by using `clogit` on these data.

```
. clogit best age edufit grades workexp boardexp if job==1, group(caseid) nolog
Conditional (fixed-effects) logistic regression   Number of obs   =         80
                                                  LR chi2(5)      =      17.27
                                                  Prob > chi2     =     0.0040
Log likelihood = -9.7832046                       Pseudo R2       =     0.4689

------------------------------------------------------------------------------
        best |      Coef.   Std. Err.      z    P>|z|     [95% Conf. Interval]
-------------+----------------------------------------------------------------
         age |  -.1048959   .2017068    -0.52   0.603    -.5002339    .2904421
      edufit |   .4558387   .9336775     0.49   0.625    -1.374136    2.285813
      grades |   3.443851   1.969002     1.75   0.080    -.4153223    7.303025
     workexp |   2.545648   1.099513     2.32   0.021     .3906422    4.700655
    boardexp |   1.765176   1.112763     1.59   0.113    -.4157988    3.946152
------------------------------------------------------------------------------
```

The parameters of the ranking and choice models look different, but the standard errors based on the choice data are much larger. Are we estimating different parameters with the ranking data than with the choice data? A Hausman test compares two estimators of a parameter. One of the estimators should be efficient under the null hypothesis; namely, that choosing the second-best alternative is determined with the same decision weights as the best, etc. In our case, the efficient estimator of the decision weights uses the ranking information. The other estimator should be consistent, even if the null hypothesis is false. In our application, this is the estimator that uses the first-choice data only.

```
. hausman Choice Ranking
                 ---- Coefficients ----
             |      (b)          (B)            (b-B)     sqrt(diag(V_b-V_B))
             |     Choice      Ranking       Difference          S.E.
-------------+----------------------------------------------------------------
         age |   -.1048959    -.0984926       -.0064033        .1842657
      edufit |    .4558387     .7658064       -.3099676        .8613846
      grades |    3.443851     3.064534        .3793169        1.870551
     workexp |    2.545648     1.386427        1.159221        1.059878
    boardexp |    1.765176     .6944377        1.070739         1.04722
------------------------------------------------------------------------------
                          b = consistent under Ho and Ha; obtained from rologit
           B = inconsistent under Ha, efficient under Ho; obtained from rologit

    Test:  Ho:  difference in coefficients not systematic

                  chi2(5) = (b-B)'[(V_b-V_B)^(-1)](b-B)
                          =        3.05
                Prob>chi2 =      0.6918
```

We do not find evidence for misspecification. We have to be cautious, though, since Hausman-type tests are often not very powerful, and the number of observations in our example is very small, which makes the quality of the approximation of the null distribution by a chi-square rather uncertain.

## On reversals of rankings

The rank-ordered logit model has a property that you may find unexpected and even unfortunate. Compare two analyses with the rank-ordered logit model, one in which alternatives are ranked from "most attractive" to "least attractive", the other a reversed analysis in which these alternatives are ranked from "most unattractive" to "least unattractive". Likely, you mean by unattractiveness just the opposite of attractiveness, and you expect that the weights of the attributes in predicting "attractiveness" to be minus the weights in predicting "unattractiveness". This is, however, *not* true for the rank-ordered logit model. The assumed distribution of the random residual takes the form $F(\epsilon) = 1 - \exp\{\exp(-\epsilon)\}$. This distribution is right-skewed. Therefore, slightly different models result from adding and subtracting the random residual, corresponding with high-to-low and low-to-high rankings. Thus the estimated coefficients will differ between the two specifications, though usually not in an important way. You may observe the difference by specifying the option **reverse** of **rologit**. Note that reversing the rank order makes rankings that are incomplete at the bottom to become incomplete at the top. Only the first kind of incompleteness is supported by **rologit**. Thus for this comparison, we exclude the alternatives that are not ranked, omitting the information that ranked alternatives are preferred over excluded ones.

```
. rologit pref grades edufit workexp boardexp if job==1 & pref!=0, group(caseid)
  (output omitted )
. estimates store Original
. rologit pref grades edufit workexp boardexp if job==1 & pref!=0, group(caseid) reverse
  (output omitted )
. estimates store Reversed
```

*(Continued on next page)*

```
. estimates table Original Reversed, stats(aic bic)
```

| Variable | Original | Reversed |
|---|---|---|
| grades | 2.0032332 | -1.0955335 |
| edufit | -.13111006 | -.05710681 |
| workexp | 1.2805373 | -1.2096383 |
| boardexp | .46213212 | -.27200317 |
| aic | 96.750452 | 99.665642 |
| bic | 104.23526 | 107.15045 |

Thus, while the weights of the attributes in the case of reversed rankings are indeed mostly of opposite signs, the magnitudes of the weights and their standard errors differ. Which one is more appropriate? We have no advice to offer here. The specific science of the problem will determine what is appropriate, though we would be surprised indeed if this helps in this particular case. Formal testing does not help much either, as the models for the original and reversed rankings are not nested. The model selection indices, such as the AIC and BIC, however, suggest that you stick to the rank-ordered logit model applied to the original ranking rather than to the reversed ranking.

## Saved Results

`rologit` saves in `e()`:

Scalars

| | | | |
|---|---|---|---|
| `e(N)` | number of observations | `e(N_g)` | number of groups |
| `e(ll_0)` | log likelihood of the null model ("all rankings are equiprobable") | `e(g_min)` | minimum group size |
| | | `e(g_avg)` | average group size |
| `e(ll)` | log likelihood | `e(g_max)` | maximum group size |
| `e(df_m)` | model degrees of freedom | `e(code_inc)` | value for incomplete preferences |
| `e(chi2)` | $\chi^2$ | `e(N_clust)` | number of clusters |
| `e(r2_p)` | pseudo-$R^2$ | | |

Macros

| | | | |
|---|---|---|---|
| `e(cmd)` | `rologit` | `e(offset)` | offset |
| `e(depvar)` | name of dependent variable | `e(chi2type)` | Wald or LR; type of model $\chi^2$ test |
| `e(group)` | name of `group()` variable | `e(ties)` | `breslow`, `efron`, `exactm` |
| `e(wtype)` | weight type | `e(vcetype)` | title used to label Std. Err. |
| `e(wexp)` | weight expression | `e(crittype)` | optimization criterion |
| `e(title)` | title in estimation output | `e(properties)` | `b V` |
| `e(clustvar)` | name of cluster variable | `e(predict)` | program used to implement `predict` |

Matrices

| | | | |
|---|---|---|---|
| `e(b)` | coefficient vector | `e(V)` | variance–covariance matrix of the estimators |

Functions

| | |
|---|---|
| `e(sample)` | marks estimation sample |

# Methods and Formulas

rologit is implemented as an ado-file.

Allison and Christakis (1994) demonstrate that maximum likelihood estimates for the rank-ordered logit model can be obtained as the maximum partial-likelihood estimates of an appropriately specified Cox regression model for waiting time ([ST] **stcox**). In this analogy, a higher value for an alternative is formally equivalent to a higher hazard rate of failure. rologit uses stcox to fit the rank-ordered logit model based on such a specification of the data in Cox terms. A higher stated preference is represented by a shorter waiting time until failure. Incomplete rankings are dealt with via censoring. Moreover, decision situations (subjects) are to be treated as strata. Finally, as proposed by Allison and Christakis, ties in rankings are handled by the marginal-likelihood method, specifying that all strict preference orderings consistent with the stated weak preference ordering are equally likely. The marginal-likelihood estimator is available in stcox via the option exactm. The approximations of the marginal likelihood due to Breslow and Efron are also appropriate for the analysis of rank-ordered logit models. Since in most applications the number of ranked alternatives by a single subject will be fairly small (at most, say, 20), the number of ties is small as well, and so you rarely will need to turn to approximations to restrict computer time. Since the marginal-likelihood estimator in stcox does not support the cluster adjustment or pweights, you should use the Efron approximation in such cases.

# Acknowledgment

The rologit command was written by Jeroen Weesie, Department of Sociology, Utrecht University, the Netherlands.

# References

Allison, P. 1999. Comparing logit and probit coefficients across groups. *Sociological Methods and Research* 28: 186–208.

Allison, P. and N. Christakis. 1994. Logit models for sets of ranked items. In *Sociological methodology*, Volume 24, ed. Peter Marsden, 199–228. Oxford: Blackwell.

Beggs, S., S. Cardell, and J. Hausman. 1981. Assessing the potential demand for electric cars. *Journal of Econometrics* 16: 1–19.

Hair, J. E. Jr., R. E. Anderson, R. L. Tatham, and W. C. Black. 1995. *Multivariate Data Analysis with Readings*. 4th ed. London: Prentice Hall.

Luce, R. D. 1959. *Individual Choice Behavior*. New York: Wiley.

Marden, J. I. 1995. *Analyzing and Modeling Rank Data*. London: Chapman & Hall.

McCullagh, P. 1993. Permutations and regression models. In *Probability Models and Statistical Analysis for Ranking Data*, ed. M. A. Fligner and J. S. Verducci, 196–215. New York: Springer.

Punj, G. N. and R. Staelin. 1978. The choice process for graduate business schools. *Journal of Marketing Research* 15: 588–598.

Ruud, B. and J. Hausman. 1987. Specifying and testing econometric models for rank-ordered data. *Journal of Econometrics* 34: 83–104.

Thurstone, L. L. 1927. A law of comparative judgment. *Psychological Reviews* 34: 273–286.

Wolf, I. de 2000. *Opleidingsspecialisatie en arbeidsmarktsucces van sociale wetenschappers*. Amsterdam: ThelaThesis.

Yellott, J. I., Jr. 1977. The relationship between Luce's choice axiom, Thurstone's theory of comparative judgment, and the double-exponential distribution. *Journal of Mathematical Psychology* 15: 109–144.

## Also See

**Complementary:** [R] **rologit postestimation**

**Related:** [R] **clogit**, [R] **logit**, [R] **mlogit**, [R] **nlogit**

**Background:** [U] **11.1.10 Prefix commands**,
[U] **20 Estimation and postestimation commands**,
[R] **estimation options**, [R] **maximize**

# Title

**rologit postestimation** — Postestimation tools for rologit

# Description

The following postestimation commands are available for `rologit`:

| command | description |
|---|---|
| `adjust` | adjusted predictions of $\mathbf{x}\boldsymbol{\beta}$ or $\exp(\mathbf{x}\boldsymbol{\beta})$ |
| `estat` | AIC, BIC, VCE, and estimation sample summary |
| `estimates` | cataloging estimation results |
| `hausman` | Hausman's specification test |
| `lincom` | point estimates, standard errors, testing, and inference for linear combinations of coefficients |
| `linktest` | link test for model specification |
| `lrtest` | likelihood-ratio test |
| `mfx` | marginal effects or elasticities |
| `nlcom` | point estimates, standard errors, testing, and inference for nonlinear combinations of coefficients |
| `predict` | predictions, residuals, influence statistics, and other diagnostic measures |
| `predictnl` | point estimates, standard errors, testing, and inference for generalized predictions |
| `test` | Wald tests for simple and composite linear hypotheses |
| `testnl` | Wald tests of nonlinear hypotheses |

See the corresponding entries in the *Stata Base Reference Manual* for details.

# Syntax for predict

`predict` [*type*] *newvar* [*if*] [*in*] [, *statistic* `nooffset`]

| *statistic* | description |
|---|---|
| `pr` | probability that alternatives are ranked first; the default |
| `xb` | linear prediction |
| `stdp` | standard error of the prediction |

These statistics are available both in and out of sample; type `predict ... if e(sample) ...` if wanted only for the estimation sample.

# Options for predict

`pr`, the default, calculates the probability that alternatives are ranked first.

`xb` calculates the linear prediction.

`stdp` calculates the standard error of the linear prediction.

`nooffset` is relevant only if you specified `offset(`*varname*`)` for `rologit`. It modifies the calculations made by `predict` so that they ignore the offset variable; the linear prediction is treated as $\mathbf{x}_j\mathbf{b}$ rather than as $\mathbf{x}_j\mathbf{b} + \text{offset}_j$.

# Remarks

See *Comparing respondents* and *Comparing types of jobs* in [R] **rologit** for examples of the use of `testparm`, an alternative to the `test` command.

See *Comparison of rologit and clogit* and *On reversals of rankings* in [R] **rologit** for examples of the use of `estimates`.

See *Comparison of rologit and clogit* in [R] **rologit** for an example of the use of `hausman`.

# Methods and Formulas

All postestimation commands listed above are implemented as ado-files.

# Also See

**Complementary:** [R] **rologit**; [R] **adjust**, [R] **estimates**, [R] **hausman**, [R] **lincom**, [R] **linktest**, [R] **lrtest**, [R] **mfx**, [R] **nlcom**, [R] **predictnl**, [R] **test**, [R] **testnl**

**Background:** [U] **13.5 Accessing coefficients and standard errors**, [U] **20 Estimation and postestimation commands**, [R] **estat**, [R] **predict**

# Title

**rreg** — Robust regression

# Syntax

`rreg` *depvar* [*indepvars*] [*if*] [*in*] [, *options*]

| *options* | description |
|---|---|
| Model | |
| `tune(#)` | use # as the biweight tuning constant; default is `tune(7)` |
| Reporting | |
| `level(#)` | set confidence level; default is `level(95)` |
| `genwt(newvar)` | create *newvar* containing the weights assigned to each observation |
| Opt options | |
| *optimization_options* | control the optimization process; seldom used |
| `graph` | graph weights during convergence |

*depvar* and *indepvars* may contain time-series operators; see [U] **11.4.3 Time-series varlists**.
`by`, `rolling`, `statsby`, and `xi` are allowed; see [U] **11.1.10 Prefix commands**.
See [U] **20 Estimation and postestimation commands** for additional capabilities of estimation commands.

# Description

`rreg` performs one version of robust regression of *depvar* on *indepvars*.

Also see *Robust standard errors* in [R] **regress** for standard regression with robust variance estimates and [R] **qreg** for quantile (including median or least-absolute-residual) regression.

# Options

Model

`tune(#)` is the biweight tuning constant. The default is 7, meaning 7 times the median absolute deviation from the median residual (MAD); see *Methods and Formulas*. Lower tuning constants downweight outliers rapidly but may lead to unstable estimates (below 6 is not recommended). Higher tuning constants produce milder downweighting.

Reporting

`level(#)`; see [R] **estimation options**.

`genwt(newvar)` creates the new variable *newvar* containing the weights assigned to each observation.

Opt options

*optimization_options*: `iterate(#)`, `tolerance(#)`, [`no`]`log`. `iterate()` specifies the maximum number of iterations; iterations stop when the maximum change in weights drops below `tolerance()`; and `log`/`nolog` specifies whether or not to show the iteration log. These options are seldom used.

`graph` allows you to graphically watch the convergence of the iterative technique. The weights obtained from the most recent round of estimation are graphed against the weights obtained from the previous round.

## Remarks

`rreg` first performs an initial screening based on Cook's distance > 1 to eliminate gross outliers before calculating starting values and then performs Huber iterations followed by biweight iterations, as suggested by Li (1985).

### ▷ Example 1

We wish to examine the relationship between mileage rating, weight, and location of manufacture for the 74 cars in our automobile data. As a point of comparison, we begin by fitting an ordinary regression:

```
. use http://www.stata-press.com/data/r9/auto
(1978 Automobile Data)

. regress mpg weight foreign

      Source |       SS       df       MS              Number of obs =      74
-------------+------------------------------           F(  2,    71) =   69.75
       Model |  1619.2877     2  809.643849           Prob > F      =  0.0000
    Residual |  824.171761    71   11.608053           R-squared     =  0.6627
-------------+------------------------------           Adj R-squared =  0.6532
       Total |  2443.45946    73  33.4720474           Root MSE      =  3.4071

------------------------------------------------------------------------------
         mpg |      Coef.   Std. Err.      t    P>|t|     [95% Conf. Interval]
-------------+----------------------------------------------------------------
      weight |  -.0065879   .0006371   -10.34   0.000    -.0078583   -.0053175
     foreign |  -1.650029   1.075994    -1.53   0.130      -3.7955    .4954422
       _cons |    41.6797   2.165547    19.25   0.000     37.36172    45.99768
------------------------------------------------------------------------------
```

We now compare this with the results from `rreg`:

```
. rreg mpg weight foreign

   Huber iteration 1:  maximum difference in weights = .80280176
   Huber iteration 2:  maximum difference in weights = .2915438
   Huber iteration 3:  maximum difference in weights = .08911171
   Huber iteration 4:  maximum difference in weights = .02697328
Biweight iteration 5:  maximum difference in weights = .29186818
Biweight iteration 6:  maximum difference in weights = .11988101
Biweight iteration 7:  maximum difference in weights = .03315872
Biweight iteration 8:  maximum difference in weights = .00721325

Robust regression                                      Number of obs =      74
                                                       F(  2,    71) =  168.32
                                                       Prob > F      =  0.0000

------------------------------------------------------------------------------
         mpg |      Coef.   Std. Err.      t    P>|t|     [95% Conf. Interval]
-------------+----------------------------------------------------------------
      weight |  -.0063976   .0003718   -17.21   0.000     -.007139   -.0056562
     foreign |  -3.182639    .627964    -5.07   0.000    -4.434763   -1.930514
       _cons |   40.64022   1.263841    32.16   0.000      38.1202    43.16025
------------------------------------------------------------------------------
```

Note the large change in the `foreign` coefficient.

◁

## ❑ Technical Note

It would have been better if we had fitted the previous robust regression by typing **rreg mpg weight foreign, genwt(w)**. The new variable **w** would then contain the estimated weights. Let's pretend that we did this:

```
. rreg mpg weight foreign, genwt(w)
  (output omitted)
. summarize w, detail
                    Robust Regression Weight

      Percentiles      Smallest
 1%            0              0
 5%     .0442957              0
10%     .4674935              0       Obs                  74
25%     .8894815       .0442957       Sum of Wgt.          74

50%     .9690193                      Mean           .8509966
                        Largest       Std. Dev.      .2746451
75%     .9949395       .9996715
90%     .9989245       .9996953       Variance       .0754299
95%     .9996715       .9997343       Skewness      -2.287952
99%     .9998585       .9998585       Kurtosis       6.874605
```

We discover that three observations in our data were dropped altogether (they have weight 0). We could further explore our data:

```
. sort w
. list make mpg weight w if w<.467, sep(0)
```

| | make | mpg | weight | w |
|---|---|---|---|---|
| 1. | Datsun 210 | 35 | 2,020 | 0 |
| 2. | VW Diesel | 41 | 2,040 | 0 |
| 3. | Subaru | 35 | 2,050 | 0 |
| 4. | Plym. Arrow | 28 | 3,260 | .04429567 |
| 5. | Cad. Seville | 21 | 4,290 | .08241943 |
| 6. | Toyota Corolla | 31 | 2,200 | .10443129 |
| 7. | Olds 98 | 21 | 4,060 | .28141296 |

Being familiar with the automobile data, we immediately spotted two things: The VW is the only diesel in our data, and the weight recorded for the Plymouth Arrow is incorrect.

❑

## ▷ Example 2

If we do not specify any explanatory variables, **rreg** produces a robust estimate of the mean:

```
. rreg mpg
   Huber iteration 1:  maximum difference in weights = .64471879
   Huber iteration 2:  maximum difference in weights = .05098336
   Huber iteration 3:  maximum difference in weights = .0099887
Biweight iteration 4:  maximum difference in weights = .25197391
Biweight iteration 5:  maximum difference in weights = .00358606
```

```
Robust regression                                      Number of obs =      74
                                                       F(  0,    73) =    0.00
                                                       Prob > F      =       .

------------------------------------------------------------------------------
         mpg |      Coef.   Std. Err.      t    P>|t|     [95% Conf. Interval]
-------------+----------------------------------------------------------------
       _cons |   20.68825    .641813    32.23   0.000     19.40912    21.96738
------------------------------------------------------------------------------
```

The estimate is given by the coefficient on `_cons`. The mean is 20.69 with an estimated standard error of .6418. The 95% confidence interval is [ 19.4, 22.0 ]. By comparison, `ci` gives us the standard calculation:

```
. ci mpg
    Variable |        Obs        Mean    Std. Err.       [95% Conf. Interval]
-------------+---------------------------------------------------------------
         mpg |         74     21.2973    .6725511        19.9569    22.63769
```

◁

## Saved Results

`rreg` saves in `e()`:

Scalars

| | | | |
|---|---|---|---|
| `e(N)` | number of observations | `e(r2)` | $R$-squared |
| `e(mss)` | model sum of squares | `e(r2_a)` | adjusted $R$-squared |
| `e(df_m)` | model degrees of freedom | `e(F)` | $F$ statistic |
| `e(rss)` | residual sum of squares | `e(rmse)` | root mean squared error |
| `e(df_r)` | residual degrees of freedom | | |

Macros

| | | | |
|---|---|---|---|
| `e(cmd)` | `rreg` | `e(model)` | `ols` |
| `e(depvar)` | name of dependent variable | `e(properties)` | `b V` |
| `e(genwt)` | variable containing the weights | `e(predict)` | program used to implement `predict` |
| `e(title)` | title in estimation output | | |

Matrices

| | | | |
|---|---|---|---|
| `e(b)` | coefficient vector | `e(V)` | variance–covariance matrix of the estimators |

Functions

| | |
|---|---|
| `e(sample)` | marks estimation sample |

## Methods and Formulas

`rreg` is implemented as an ado-file.

See Berk (1990), Goodall (1983), and Rousseeuw and Leroy (1987) for a general description of the issues and methods. Hamilton (1991, 1992) provides a more detailed description of `rreg` and some Monte Carlo evaluations.

`rreg` begins by fitting the regression (see [R] **regress**), calculating Cook's $D$ (see [R] **predict** and [R] **regress postestimation**), and excluding any observation for which $D > 1$.

Thereafter `rreg` works iteratively: it performs a regression, calculates case weights based on absolute residuals, and regresses again using those weights. Iterations stop when the maximum change in weights drops below `tolerance()`. Weights derive from one of two weight functions, Huber weights and biweights. Huber weights (Huber 1964) are used until convergence, and then, based on that result, biweights are used until convergence. The biweight was proposed by A. E. Beaton and J. W. Tukey (1974, 151–152) after the Princeton robustness study (Andrews et al. 1972) had compared various estimators. Both weighting functions are used because Huber weights have problems dealing with severe outliers, while biweights sometimes fail to converge or have multiple solutions. The initial Huber weighting should improve the behavior of the biweight estimator.

In Huber weighting, cases with small residuals receive weights of 1; cases with larger residuals receive gradually smaller weights. Let $e_i = y_i - \mathbf{X}_i\mathbf{b}$ represent the $i$th-case residual. The $i$th scaled residual $u_i = e_i/s$ is calculated, where $s = M/.6745$ is the residual scale estimate and $M = \text{med}(|e_i - \text{med}(e_i)|)$ is the median absolute deviation from the median residual. Huber estimation obtains case weights:

$$w_i = \begin{cases} 1 & \text{if } |u_i| \le c_h \\ c_h/|u_i| & \text{otherwise} \end{cases}$$

`rreg` defines $c_h = 1.345$, so downweighting begins with cases whose absolute residual exceeds $(1.345/.6745)M \approx 2M$.

With biweights, all cases with nonzero residuals receive some downweighting, according to the smoothly decreasing biweight function

$$w_i = \begin{cases} \{1 - (u_i/c_b)^2\}^2 & \text{if } |u_i| \le c_b \\ 0 & \text{otherwise} \end{cases}$$

where $c_b = 4.685 \cdot \texttt{tune()}/7$. Thus when $\texttt{tune()} = 7$, cases with absolute residuals of $(4.685/.6745)M \approx 7M$ or more are assigned 0 weight and thus are effectively dropped. Goodall (1983, 377) suggests using a value between 6 and 9, inclusive, for `tune()` in the biweight case and states that performance is good between 6 and 12, inclusive.

The tuning constants $c_h = 1.345$ and $c_b = 4.685$ (assuming `tune()` is set at the default 7) give `rreg` about 95% of the efficiency of OLS when applied to data with normally distributed errors (Hamilton 1991). Lower tuning constants downweight outliers more drastically (but give up Gaussian efficiency); higher tuning constants make the estimator more like OLS.

Standard errors are calculated using the pseudovalues approach described in Street, Carroll, and Ruppert (1988).

## Acknowledgment

The current version of `rreg` is due to the work of Lawrence Hamilton, Department of Sociology, University of New Hampshire.

## References

Andrews, D. F., P. J. Bickel, F. R. Hampel, P. J. Huber, W. H. Rogers, and J. W. Tukey. 1972. *Robust Estimates of Location: Survey and Advances*. Princeton: Princeton University Press.

Beaton, A. E. and J. W. Tukey. 1974. The fitting of power series, meaning polynomials, illustrated on band-spectroscopic data. *Technometrics* 16: 146–185.

Berk, R. A. 1990. A primer on robust regression. In *Modern Methods of Data Analysis*, ed. J. Fox and J. S. Long, 292–324. Newbury Park, CA: Sage.

Goodall, C. 1983. M-estimators of location: An outline of the theory. In *Understanding Robust and Exploratory Data Analysis*, ed. D. C. Hoaglin, F. Mosteller, and J. W. Tukey, 339–431. New York: Wiley.

Gould, W. W. and W. H. Rogers. 1994. Quantile regression as an alternative to robust regression. *1994 Proceedings of the Statistical Computing Section*. Alexandria, VA: American Statistical Association.

Hamilton, L. C. 1991. srd1: How robust is robust regression? *Stata Technical Bulletin* 2: 21–26. Reprinted in *Stata Technical Bulletin Reprints*, vol. 1, pp. 169–175.

——. 1992. *Regression with Graphics*. Pacific Grove, CA: Brooks/Cole.

——. 2004. *Statistics with Stata*. Belmont, CA: Brooks/Cole.

Huber, P. J. 1964. Robust estimation of a location parameter. *Annals of Mathematical Statistics* 35: 73–101.

Li, G. 1985. Robust regression. In *Exploring Data Tables, Trends, and Shapes*, ed. D. C. Hoaglin, F. Mosteller, and J. W. Tukey, 281–340. New York: Wiley.

Mosteller, F. and J. W. Tukey. 1977. *Data Analysis and Regression*. Reading, MA: Addison–Wesley.

Relles, D. A. and W. H. Rogers. 1977. Statisticians are fairly robust estimators of location. *Journal of the American Statistical Association* 72: 107–111.

Rousseeuw, P. J. and A. M. Leroy. 1987. *Robust Regression and Outlier Detection*. New York: Wiley.

Street, J. O., R. J. Carroll, and D. Ruppert. 1988. A note on computing robust regression estimates via iteratively reweighted least squares. *The American Statistician* 42: 152–154.

## Also See

**Complementary:** [R] **rreg postestimation**

**Related:** [R] **qreg**, [R] **regress**, [R] **regress postestimation**

**Background:** [U] **11.1.10 Prefix commands**,
[U] **20 Estimation and postestimation commands**,
[R] **estimation options**

# Title

**rreg postestimation** — Postestimation tools for rreg

# Description

The following postestimation commands are available for `rreg`:

| command | description |
|---|---|
| `adjust`[1] | adjusted predictions of $\mathbf{x}\beta$ or $\exp(\mathbf{x}\beta)$ |
| `estat` | VCE and estimation sample summary |
| `estimates` | cataloging estimation results |
| `lincom` | point estimates, standard errors, testing, and inference for linear combinations of coefficients |
| `mfx` | marginal effects or elasticities |
| `nlcom` | point estimates, standard errors, testing, and inference for nonlinear combinations of coefficients |
| `predict` | predictions, residuals, influence statistics, and other diagnostic measures |
| `predictnl` | point estimates, standard errors, testing, and inference for generalized predictions |
| `test` | Wald tests for simple and composite linear hypotheses |
| `testnl` | Wald tests of nonlinear hypotheses |

[1] `adjust` does not work with time-series operators.

See the corresponding entries in the *Stata Base Reference Manual* for details.

# Syntax for predict

`predict` [*type*] *newvar* [*if*] [*in*] [, *statistic*]

| *statistic* | description |
|---|---|
| `xb` | linear prediction; the default |
| `stdp` | standard error of the linear prediction |
| `residuals` | residuals |
| `hat` | diagonal elements of the hat matrix |

These statistics are available both in and out of sample; type `predict ... if e(sample) ...` if wanted only for the estimation sample.

# Options for predict

`xb`, the default, calculates the linear prediction.

`stdp` calculates the standard error of the linear prediction.

`residuals` calculates the residuals.

`hat` calculates the diagonal elements of the hat matrix. Note that you must have run the `rreg` command with the `genwt()` option.

## Methods and Formulas

All postestimation commands listed above are implemented as ado-files.

## Also See

**Complementary:** [R] **rreg**; [R] **adjust**, [R] **estimates**, [R] **lincom**, [R] **mfx**, [R] **nlcom**, [R] **predictnl**, [R] **test**, [R] **testnl**

**Background:** [U] **13.5 Accessing coefficients and standard errors**, [U] **20 Estimation and postestimation commands**, [R] **estat**, [R] **predict**

## Title

**runtest** — Test for random order

## Syntax

`runtest` *varname* [*in*] [, *options*]

| *options* | description |
|---|---|
| `continuity` | continuity correction |
| `drop` | ignore values equal to the threshold |
| `split` | randomly split values equal to the threshold as above or below the threshold; default is to count as below |
| `mean` | use mean as threshold; default is median |
| `threshold(#)` | assign arbitrary threshold; default is median |

## Description

`runtest` tests whether the observations of *varname* are serially independent—that is, whether they occur in a random order—by counting how many runs there are above and below a threshold. By default, the median is used as the threshold. A small number of runs indicates positive serial correlation; a large number indicates negative serial correlation.

## Options

`continuity` specifies a continuity correction that may be helpful in small samples. If there are fewer than 10 observations either above or below the threshold, however, the tables in Swed and Eisenhart (1943) provide more reliable critical values. By default, no continuity correction is used.

`drop` directs `runtest` to ignore any values of *varname* that are equal to the threshold value when counting runs and tabulating observations. By default, `runtest` counts a value as being above the threshold when it is strictly above the threshold and as being below the threshold when it is less than or equal to the threshold.

`split` directs `runtest` to randomly split values of *varname* that are equal to the threshold. In other words, when *varname* is equal to threshold, a "coin" is flipped. If it comes up heads, the value is counted as above the threshold. If it comes up tails, the value is counted as below the threshold.

`mean` directs `runtest` to tabulate runs above and below the mean rather than the median.

`threshold(#)` specifies an arbitrary threshold to use in counting runs. For example, if *varname* has already been coded as a 0/1 variable, the median generally will not be a meaningful separating value.

## Remarks

`runtest` performs a nonparametric test of the hypothesis that the observations of *varname* occur in a random order by counting how many runs there are above and below a threshold. If *varname* is positively serially correlated, it will tend to remain above or below its median for several observations in a row; that is, there will be relatively few runs. If, on the other hand, *varname* is negatively serially correlated, observations above the median will tend to be followed by observations below the median and vice versa; that is, there will be relatively many runs.

By default, `runtest` uses the median for the threshold, and this is not necessarily the best choice. If `mean` is specified, the mean is used instead of the median. If `threshold(#)` is specified, # is used. Since `runtest` divides the data into two states—above and below the threshold—it is appropriate for data that are already binary; for example, win or lose, live or die, rich or poor, etc. Such variables are often coded as 0 for one state and 1 for the other. In this case, you should specify `threshold(0)` since, by default, `runtest` separates the observations into those that are greater than the threshold and those that are less than *or equal* to the threshold.

As with most nonparametric procedures, the treatment of ties complicates the test. Observations equal to the threshold value are ties and can be treated in one of three ways. By default, they are treated as if they were below the threshold. If `drop` is specified, they are omitted from the calculation and the total number of observations is adjusted. If `split` is specified, each is randomly assigned to the above- and below-threshold groups. The random assignment is different each time the procedure is run unless you specify the random-number seed; see [D] **generate**.

### ▷ Example 1

We can use `runtest` to check regression residuals for serial correlation.

```
. regress ...
  (output omitted )
. predict resid, resid
. scatter resid year, connect(l) yline(0) title(Regression residuals)
```

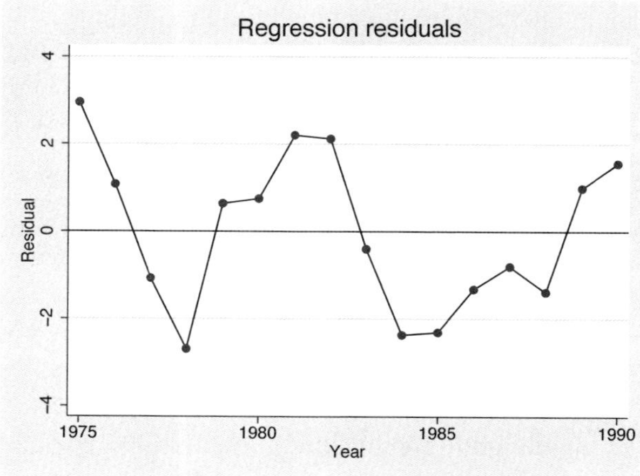


The graph gives the impression that these residuals are positively correlated. Excursions above or below zero—the natural threshold for regression residuals—tend to last for several observations. `runtest` can evaluate the statistical significance of this impression.

```
. runtest resid, thresh(0)
 N(resid <= 0) = 8
 N(resid >  0) = 8
           obs = 16
       N(runs) = 5
           z   = -2.07
      Prob>|z| = .04
```

There are 5 runs in these sixteen observations. Using the normal approximation to the true distribution of the number of runs, the 5 runs in this series are fewer than would be expected if the residuals were serially independent. The $p$-value is 0.03, indicating a two-sided significant result at the 5 percent level. If the alternative hypothesis is positive serial correlation, rather than any deviation from randomness, then the one-sided $p$-value is $.03/2 = .015$. With so few observations, however, the normal approximation may be inaccurate. (Tables compiled by Swed and Eisenhart list 5 runs as the 5 percent critical value for a one-sided test.)

`runtest` is a nonparametric test. It ignores the magnitudes of the observations and notes only whether the values are above or below the threshold. We can demonstrate this feature by reducing the information about the regression residuals in this example to a 0/1 variable that indicates only whether a residual is positive or negative.

```
. generate byte sign = resid>0

. runtest sign, thresh(0)
 N(sign <= 0) = 8
 N(sign >  0) = 8
          obs = 16
      N(runs) = 5
          z   = -2.07
     Prob>|z| = .04
```

As expected, `runtest` produces the same answer as before.

◁

## ❑ Technical Note

The run test can also be used to test the null hypothesis that two samples are drawn from the same underlying distribution. The run test is sensitive to differences in the shapes, as well as the locations, of the empirical distributions.

Suppose, for example, that two different additives are added to the oil in ten different cars during an oil change. The cars are run until a viscosity test determines that another oil change is needed, and the number of miles traveled between oil changes is recorded. The data are

*(Continued on next page)*

```
. use http://www.stata-press.com/data/r9/additive, clear
. list

     +--------------------+
     | additive   miles   |
     |--------------------|
  1. |        1    4024   |
  2. |        1    4756   |
  3. |        1    7993   |
  4. |        1    5025   |
  5. |        1    4188   |
     |--------------------|
  6. |        2    3007   |
  7. |        2    1988   |
  8. |        2    1051   |
  9. |        2    4478   |
 10. |        2    4232   |
     +--------------------+
```

To test whether the additives generate different distributions of miles between oil changes, we sort the data by `miles` and then use `runtest` to see whether the marker for each additive occurs in random order:

```
. sort miles
. runtest additive, thresh(1)
 N(additive <= 1) = 5
 N(additive >  1) = 5
              obs = 10
          N(runs) = 4
              z   = -1.34
         Prob>|z| = .18
```

In this example, the additives do not produce statistically different results.

❑

## ❑ Technical Note

A test that is related to the run test is the runs up-and-down test. In the latter test, the data are classified not by whether they lie above or below a threshold, but by whether they are steadily increasing or decreasing. Thus an unbroken string of increases in the variable of interest is counted as one run, as is an unbroken string of decreases. According to Madansky (1988), the run test is superior to the runs up-and-down test for detecting trends in the data, but the runs up-and-down test is superior for detecting autocorrelation.

`runtest` can be used to perform a runs up-and-down test. Using the regression residuals from the example above, we can perform a `runtest` on their first differences:

```
. use http://www.stata-press.com/data/r9/run1
. generate resid_D = resid - resid[_n-1]
(1 missing value generated)
. runtest resid_D, thresh(0)
 N(resid_D <= 0) = 7
 N(resid_D >  0) = 8
             obs = 15
         N(runs) = 6
             z   = -1.33
        Prob>|z| = .18
```

Edgington (1961) has compiled a table of the small sample distribution of the runs up-and-down statistic, and this table is reprinted in Madansky (1988). For large samples, the $z$ statistic reported by `runtest` is incorrect for the runs up-and-down test. Let $N$ be the number of observations (15 in this example), and let $r$ be the number of runs (6). The expected number of runs in the runs up-and-down test is

$$\mu_r = \frac{2N-1}{3}$$

the variance is

$$\sigma_r^2 = \frac{16N-29}{90}$$

and the correct $z$ statistic is

$$\widehat{z} = \frac{r-\mu_r}{\sigma_r}$$

❑

❑ Technical Note

`runtest` will tolerate missing values at the beginning or end of a series, as occurred in the technical note above (generating first differences resulted in a missing value for the first observation). `runtest`, however, will issue an error message if there are any missing observations in the interior of the series (in the portion covered by the `in` *range* modifier). To perform the test anyway, simply `drop` the missing observations before using `runtest`.

❑

## Saved Results

`runtest` saves in `r()`:

Scalars

| | | | |
|---|---|---|---|
| `r(N)` | number of observations | `r(p)` | $p$-value of $z$ |
| `r(N_below)` | number below the threshold | `r(z)` | $z$ statistic |
| `r(N_above)` | number above the threshold | `r(n_runs)` | number of runs |
| `r(mean)` | expected number of runs | `r(Var)` | variance of the number of runs |

## Methods and Formulas

`runtest` is implemented as an ado-file.

`runtest` begins by calculating the number of observations below the threshold, $n_0$; the number of observations above the threshold, $n_1$; the total number of observations, $N = n_0 + n_1$; and the number of runs, $r$. These statistics are always reported, so the exact tables of critical values in Swed and Eisenhart (1943) may be consulted if necessary.

The expected number of runs under the null is

$$\mu_r = \frac{2n_0 n_1}{N} + 1$$

the variance is

$$\sigma_r^2 = \frac{2n_0n_1\,(2n_0n_1 - N)}{N^2\,(N-1)}$$

and the normal approximation test statistic is

$$\widehat{z} = \frac{r - \mu_r}{\sigma_r}$$

## Acknowledgment

`runtest` was written by Sean Becketti, a past editor of the *Stata Technical Bulletin*.

## References

Edgington, E. S. 1961. Probability table for number of runs of signs of first differences in ordered series. *Journal of the American Statistical Association* 56: 156–159.

Madansky, A. 1988. *Prescriptions for Working Statisticians.* New York: Springer.

Swed, F. S. and C. Eisenhart. 1943. Tables for testing randomness of grouping in a sequence of alternatives. *Annals of Mathematical Statistics* 14: 83–86.

## Also See

**Related:** [R] **ksmirnov**, [R] **kwallis**, [R] **signrank**

# Title

**sampsi** — Sample size and power determination

# Syntax

`sampsi` $\#_1$ $\#_2$ [ , *options* ]

| *options* | description |
|---|---|
| Main | |
| `onesample` | one-sample test; default is two-sample |
| `sd1(#)` | standard deviation of sample 1 |
| `sd2(#)` | standard deviation of sample 2 |
| Options | |
| `alpha(#)` | significance level of test; default is `alpha(0.05)` |
| `power(#)` | power of test; default is `power(0.90)` |
| `n1(#)` | size of sample 1 |
| `n2(#)` | size of sample 2 |
| `ratio(#)` | ratio of sample sizes; default is `ratio(1)` |
| `pre(#)` | number of baseline measurements; default is `pre(0)` |
| `post(#)` | number of follow-up measurements; default is `post(1)` |
| `nocontinuity` | do not use continuity correction for two-sample test on proportions |
| `r0(#)` | correlation between baseline measurements; default is `r0()=r1()` |
| `r1(#)` | correlation between follow-up measurements |
| `r01(#)` | correlation between baseline and follow-up measurements |
| `onesided` | one-sided test; default is two-sided |
| `method(`*method*`)` | analysis method where *method* is `post`, `change`, `ancova`, or `all`; default is `method(all)` |

# Description

`sampsi` estimates required sample size or power of tests for studies comparing two groups. `sampsi` can be used when comparing means or proportions for simple studies where only one measurement of the outcome is planned and for comparing mean summary statistics for more complex studies where repeated measurements of the outcome on each experimental unit are planned.

If `n1(#)` or `n2(#)` is specified, `sampsi` computes power; otherwise, it computes sample size. For simple studies, if `sd1(#)` or `sd2(#)` is specified, `sampsi` assumes a comparison of means; otherwise, it assumes a comparison of proportions. In the case of repeated measurements, `sd1(#)` or `sd2(#)` must be specified. `sampsi` is an immediate command; all its arguments are numbers.

(*Continued on next page*)

# Options

Main

onesample indicates a one-sample test. The default is two-sample.

sd1(#) and sd2(#) are the standard deviations of population 1 and population 2, respectively. One or both must be specified when doing a comparison of means. When the onesample option is used, sd1(#) is the standard deviation of the single sample (note that it can be abbreviated as sd(#)). If only one of sd1(#) or sd2(#) is specified, sampsi assumes that sd1 = sd2. If neither sd1(#) nor sd2(#) is specified, sampsi assumes a test of proportions. In the case of repeated measurements, sd1(#) and/or sd2(#) must be specified.

Options

alpha(#) is the significance level of the test. The default is alpha(0.05) unless set level has been used to reset the default significance level for confidence intervals. If a set level $\#_{\mathrm{lev}}$ command has been issued, the default value is alpha($1 - \#_{\mathrm{lev}}/100$). See [R] **level**.

power(#) $= 1 - \beta$ is the power of the test. The default is power(0.90).

n1(#) and n2(#) are the sizes of sample 1 and sample 2, respectively. One or both must be specified when computing power. If neither n1(#) nor n2(#) is specified, sampsi computes sample size. When the onesample option is used, n1(#) is the size of the single sample (note that it can be abbreviated as n(#)). If only one of n1(#) or n2(#) is specified, the unspecified one is computed using the formula ratio = n2()/n1().

ratio(#) is the ratio of sample sizes for two-sample tests: ratio = n2()/n1(). The default is ratio(1).

pre(#) specifies the number of baseline measurements (prerandomization) planned in a repeated-measure study. The default is pre(0).

post(#) specifies the number of follow-up measurements (postrandomization) planned in a repeated-measure study. The default is post(1).

nocontinuity requests power and sample size calculations without the continuity correction for the two-sample test on proportions. If not specified, the continuity correction is used.

r0(#) specifies the correlation between baseline measurements in repeated-measurement studies. If r0(#) is not specified, sampsi assumes that r0() = r1().

r1(#) specifies the correlation between follow-up measurements in repeated-measurement studies. For repeated-measurement studies, either r1(#) or r01(#) must be specified. If r1(#) is not specified, sampsi assumes that r1() = r01().

r01(#) specifies the correlation between baseline and follow-up measurements in repeated-measurement studies. For repeated-measurement studies, either r01(#) or r1(#) must be specified. If r01(#) is not specified, sampsi assumes that r01() = r1().

onesided indicates a one-sided test. The default is two-sided.

method(post | change | ancova | all) specifies the analysis method to be used with repeated measures. change and ancova can be used only if baseline measurements are planned. The default is method(all), which means to use all three methods. Each method is described in *Methods and Formulas*.

# Remarks

Remarks are presented under the headings

*Studies with a single measurement of the outcome*
*Two-sample test of equality of means*
*One-sample test of mean*
*Two-sample test of equality of proportions*
*One-sample test of proportion*
*Clinical trials with repeated measures*

## Studies with a single measurement of the outcome

For simple studies, where only one measurement of the outcome is planned, `sampsi` computes sample size or power for four types of tests:

1. two-sample comparison of mean $\mu_1$ of population 1 with mean $\mu_2$ of population 2. The null hypothesis is $\mu_1 = \mu_2$, and normality is assumed. The postulated values of the means are $\mu_1 = \#_1$ and $\mu_2 = \#_2$, and the postulated standard deviations are `sd1(#)` and `sd2(#)`.
2. one-sample comparison of the mean $\mu$ of a population with a hypothesized value $\mu_0$. The null hypothesis is $\mu = \mu_0$, and normality is assumed. The first argument, $\#_1$, to `sampsi` is $\mu_0$. The second argument, $\#_2$, is the postulated value of $\mu$, i.e., the alternative hypothesis is $\mu = \#_2$. The postulated standard deviation is `sd1(#)`. To get this test, the `onesample` option must be specified.
3. two-sample comparison of proportion $p_1$ with proportion $p_2$. The null hypothesis is $p_1 = p_2$, and the postulated values are $p_1 = \#_1$ and $p_2 = \#_2$.
4. one-sample comparison of a proportion $p$ with a hypothesized value $p_0$. The null hypothesis is $p = p_0$, where $p_0 = \#_1$. The alternative hypothesis is $p = \#_2$. To get this test, the `onesample` option must be specified.

   Examples of these follow.

## Two-sample test of equality of means

### ▷ Example 1

We are doing a study of the relationship of oral contraceptives (OC) and blood pressure (BP) level for women ages 35–39 (Rosner 2000, 307–308). From a pilot study, it was determined that the mean and standard deviation BP of OC users were 132.86 and 15.34, respectively. The mean and standard deviation of OC nonusers in the plot study were found to be 127.44 and 18.23. Since it is easier to find OC nonusers than users, we decide that $n_2$, the size of the sample of OC nonusers, should be twice $n_1$, the size of the sample of OC users; that is, $r = n_2/n_1 = 2$. To compute the sample sizes for $\alpha = 0.05$ (two-sided) and the power of 0.80, we issue the following command:

(*Continued on next page*)

```
. sampsi 132.86 127.44, p(0.8) r(2) sd1(15.34) sd2(18.23)
Estimated sample size for two-sample comparison of means
Test Ho: m1 = m2, where m1 is the mean in population 1
                    and m2 is the mean in population 2
Assumptions:
         alpha =   0.0500  (two-sided)
         power =   0.8000
            m1 =   132.86
            m2 =   127.44
           sd1 =    15.34
           sd2 =    18.23
         n2/n1 =     2.00
Estimated required sample sizes:
            n1 =      108
            n2 =      216
```

We now find out that we only have enough money to study 100 subjects from each group. We can compute the power for $n_1 = n_2 = 100$ by typing

```
. sampsi 132.86 127.44, n1(100) sd1(15.34) sd2(18.23)
Estimated power for two-sample comparison of means
Test Ho: m1 = m2, where m1 is the mean in population 1
                    and m2 is the mean in population 2
Assumptions:
         alpha =   0.0500  (two-sided)
            m1 =   132.86
            m2 =   127.44
           sd1 =    15.34
           sd2 =    18.23
sample size n1 =      100
            n2 =      100
         n2/n1 =     1.00
Estimated power:
         power =   0.6236
```

Note that we did not have to specify `n2(#)` or `ratio(#)` since `ratio(1)` is the default.

◁

## One-sample test of mean

▷ Example 2

Suppose that we wish to test the effects of a low-fat diet on serum cholesterol levels. We will measure the difference in cholesterol level for each subject before and after being on the diet. Since there is only one group of subjects, all on the diet, this is a one-sample test, and we must use the `onesample` option with `sampsi`.

Our null hypothesis is that the mean of individual differences in cholesterol level will be zero, i.e., $\mu = 0\,\mathrm{mg}/100\,\mathrm{ml}$. If the effect of the diet is as large as a mean difference of $-10\,\mathrm{mg}/100\,\mathrm{ml}$, then we wish to have power of 0.95 for rejecting the null hypothesis. Since we expect a reduction in levels, we want to use a one-sided test with $\alpha = 0.025$. Based on past studies, we estimate that the standard deviation of the difference in cholesterol levels will be about $20\,\mathrm{mg}/100\,\mathrm{ml}$. To compute the required sample size, we type

```
. sampsi 0 -10, sd(20) onesam a(0.025) onesided p(0.95)
Estimated sample size for one-sample comparison of mean
  to hypothesized value
Test Ho: m =      0, where m is the mean in the population
Assumptions:
         alpha =   0.0250  (one-sided)
         power =   0.9500
 alternative m =      -10
            sd =       20
Estimated required sample size:
             n =       52
```

We decide to conduct the study with $n = 60$ subjects, and we wonder what the power will be at a one-sided significance level of $\alpha = 0.01$:

```
. sampsi 0 -10, sd(20) onesam a(0.01) onesided n(60)
Estimated power for one-sample comparison of mean
  to hypothesized value
Test Ho: m =      0, where m is the mean in the population
Assumptions:
         alpha =   0.0100  (one-sided)
 alternative m =      -10
            sd =       20
 sample size n =       60
Estimated power:
         power =   0.9390
```

◁

## Two-sample test of equality of proportions

### ▷ Example 3

We want to conduct a survey on people's opinions of the president's performance. Specifically, we want to determine whether members of the president's party have a different opinion from people with another party affiliation. Using past surveys as a guide, we estimate that only 25 percent of members of the president's party will say that the president is doing a poor job, whereas 40 percent of members of other parties will rate the president's performance as poor. We compute the required sample sizes for $\alpha = 0.05$ (two-sided) and the power of 0.90 by typing

```
. sampsi 0.25 0.4
Estimated sample size for two-sample comparison of proportions
Test Ho: p1 = p2, where p1 is the proportion in population 1
                  and p2 is the proportion in population 2
Assumptions:
         alpha =   0.0500  (two-sided)
         power =   0.9000
            p1 =   0.2500
            p2 =   0.4000
         n2/n1 =   1.00
Estimated required sample sizes:
            n1 =      216
            n2 =      216
```

To compute the power for a survey with a sample of $n_1 = 300$ members of the president's party and a sample of $n_2 = 150$ members of other parties, we type

```
. sampsi 0.25 0.4, n1(300) r(0.5)
Estimated power for two-sample comparison of proportions
Test Ho: p1 = p2, where p1 is the proportion in population 1
                  and p2 is the proportion in population 2
Assumptions:
         alpha =   0.0500  (two-sided)
            p1 =   0.2500
            p2 =   0.4000
sample size n1 =      300
            n2 =      150
         n2/n1 =     0.50
Estimated power:
         power =   0.8790
```

◁

## One-sample test of proportion

### ▷ Example 4

Someone claims that females are more likely than males to study French. Our null hypothesis is that the proportion of female French students is 0.5. We wish to compute the sample size that will give us 80% power to reject the null hypothesis if the true proportion of female French students is 0.75:

```
. sampsi 0.5 0.75, power(0.8) onesample
Estimated sample size for one-sample comparison of proportion
  to hypothesized value
Test Ho: p = 0.5000, where p is the proportion in the population
Assumptions:
         alpha =   0.0500  (two-sided)
         power =   0.8000
 alternative p =   0.7500
Estimated required sample size:
             n =       29
```

What is the power if the true proportion of female French students is only 0.6 and the biggest sample of French students we can survey is $n = 200$?

```
. sampsi 0.5 0.6, n(200) onesample
Estimated power for one-sample comparison of proportion
  to hypothesized value
Test Ho: p = 0.5000, where p is the proportion in the population
Assumptions:
         alpha =   0.0500  (two-sided)
 alternative p =   0.6000
 sample size n =      200
Estimated power:
         power =   0.8123
```

◁

❑ Technical Note

`r(warning)` is saved only for power calculations for one- and two-sample tests on proportions. If sample sizes are not large enough (Tamhane 2000, 300, 307) for sample proportions to be approximately normally distributed, `r(warning)` is set to 1. Otherwise, a note is displayed in the output, and `r(warning)` is set to 0.

❑

## Clinical trials with repeated measures

In randomized controlled trials (RCTs), when comparing a standard treatment with an experimental therapy, it is not unusual for the study design to allow for repeated measurements of the outcome. Typically, one or more measurements are taken at baseline immediately before randomization, and additional measurements are taken at regular intervals during follow-up. Depending on the analysis method planned and the correlations between measurements at different time points, there can be a great increase in efficiency (variance reduction) from such designs over a simple study with only one measurement of the outcome.

Frison and Pocock (1992) discuss three methods used in RCTs to compare two treatments using a continuous outcome measured at different times on each patient.

Post-treatment means (POST): uses the mean of each patient's follow-up measurements as the summary measurement. It compares the two groups using a simple $t$ test. This method ignores any baseline measurements.

Mean changes (CHANGE): uses each patient's difference between the mean of the follow-up measurements and the mean of baseline measurements as the summary measurement. It compares the two groups using a simple $t$ test.

Analysis of covariance (ANCOVA): uses the mean baseline measurement for each patient as a covariate in a linear model for treatment comparisons of follow-up means.

`method()` specifies which of these three analyses is planned to be used. `sampsi` will calculate the decrease in variance of the estimate of treatment effect based on the number of measurements at baseline, the number of measurements during follow-up, and the correlations between measurements at different times, and use it to estimate power and/or sample size.

▷ Example 5

We are designing a clinical trial comparing a new medication for the treatment of angina to a placebo. We are planning on performing an exercise stress test on each patient four times during the study: once at time of treatment randomization and 3 more times at 4, 6 and 8 weeks after randomization. From each test, we will measure the time in seconds from the beginning of the test until the patient is unable to continue due to angina pain. From a previous pilot study, we estimated the means (sds) for the new drug and the placebo group to be 498 seconds (20.2) and 485 seconds (19.5), respectively, and an overall correlation at follow-up of 0.7. We will analyze these data by comparing each patient's difference between the mean of post-treatment measurements and the mean of baseline measurements, i.e., the change method. To compute the number of subjects needed for allocation to each treatment group for $\alpha = 0.05$ (two-sided) and power of 90%, we issue the following command:

```
. sampsi 498 485, sd1(20.2) sd2(19.5) method(change) pre(1) post(3) r1(.7)

Estimated sample size for two samples with repeated measures

Assumptions:
         alpha =   0.0500  (two-sided)
                                                 power =    0.9000
                                                    m1 =       498
                                                    m2 =       485
                                                   sd1 =      20.2
                                                   sd2 =      19.5
                                                 n2/n1 =      1.00
           number of follow-up measurements =         3
 correlation between follow-up measurements =     0.700
            number of baseline measurements =         1
   correlation between baseline & follow-up =     0.700

Method: CHANGE
 relative efficiency =    2.500
    adjustment to sd =    0.632
        adjusted sd1 =   12.776
        adjusted sd2 =   12.333

 Estimated required sample sizes:
                  n1 =       20
                  n2 =       20
```

The output from `sampsi` for repeated measurements includes the specified parameters used to estimate the sample sizes or power, the relative efficiency of the design, and the adjustment to the standard deviation. These last two are the inverse and the square root of the calculated improvement in the variance compared to a similar study where only one measurement is planned.

We see that we need to allocate 20 subjects to each treatment group. Assume that we only have funds to enroll 30 patients into our study. If we randomly assigned 15 patients to each treatment group, what would be the expected power of our study, assuming that all other parameters remain the same?

```
. sampsi 498 485, sd1(20.2) sd2(19.5) meth(change) pre(1) post(3) r1(.7) n1(15) n2(15)

Estimated power for two samples with repeated measures

Assumptions:
                                                 alpha =    0.0500  (two-sided)
                                                    m1 =       498
                                                    m2 =       485
                                                   sd1 =      20.2
                                                   sd2 =      19.5
                                        sample size n1 =        15
                                                    n2 =        15
                                                 n2/n1 =      1.00
           number of follow-up measurements =         3
 correlation between follow-up measurements =     0.700
            number of baseline measurements =         1
   correlation between baseline & follow-up =     0.700

Method: CHANGE
 relative efficiency =    2.500
    adjustment to sd =    0.632
        adjusted sd1 =   12.776
        adjusted sd2 =   12.333

 Estimated power:
               power =    0.809
```

If we enroll 30 patients into our study instead of the recommended 40, the power of the study decreases from 90% to approximately 81%.

◁

## Saved Results

`sampsi` saves in `r()`:

Scalars

| | |
|---|---|
| `r(N_1)` | sample size $n_1$ |
| `r(N_2)` | sample size $n_2$ |
| `r(power)` | power |
| `r(adj)` | adjustment to the SE |
| `r(warning)` | 0 if assumptions are satisfied and 1 otherwise |

## Methods and Formulas

`sampsi` is implemented as an ado-file.

In the following formulas, $\alpha$ is the significance level, $1-\beta$ is the power, $z_{1-\alpha/2}$ is the $(1-\alpha/2)$ quantile of the normal distribution, and $r = n_2/n_1$ is the ratio of sample sizes. The formulas below are for two-sided tests. The formulas for one-sided tests can be obtained by replacing $z_{1-\alpha/2}$ with $z_{1-\alpha}$.

1. The required sample sizes for a two-sample test of equality of means (assuming normality) are

$$n_1 = \frac{(\sigma_1^2 + \sigma_2^2/r)(z_{1-\alpha/2} + z_{1-\beta})^2}{(\mu_1 - \mu_2)^2}$$

   and $n_2 = rn_1$ (Rosner 2000, 308).

2. For a one-sample test of a mean where the null hypothesis is $\mu = \mu_0$ and the alternative hypothesis is $\mu = \mu_{\rm A}$, the required sample size (assuming normality) is

$$n = \left\{ \frac{(z_{1-\alpha/2} + z_{1-\beta})\sigma}{\mu_{\rm A} - \mu_0} \right\}^2$$

   (Pagano and Gauvreau 2000, 247–248).

3. The required sample sizes for a two-sample test of equality of proportions (using a normal approximation with a continuity correction) are

$$n_1 = \frac{n'}{4}\left[1 + \left\{1 + \frac{2(r+1)}{n'r\,|\,p_1 - p_2|}\right\}^{1/2}\right]^2$$

$$n_2 = rn_1$$

   where

$$n' = \frac{\left[z_{1-\alpha/2}\left\{(r+1)\overline{pq}\right\}^{1/2} + z_{1-\beta}\left(rp_1q_1 + p_2q_2\right)^{1/2}\right]^2}{r(p_1 - p_2)^2}$$

   and $\overline{p} = (p_1 + rp_2)/(r+1)$ and $\overline{q} = 1 - \overline{p}$ (Fleiss, Levin, and Paik 76).

   Without a continuity correction, the sample sizes are

$$n_1 = n'$$

$$n_2 = rn_1$$

   where $n'$ is defined above.

4. For a one-sample test of proportion where the null hypothesis is $p = p_0$ and the alternative hypothesis is $p = p_A$, the required sample size (using a normal approximation) is

$$n = \left[ \frac{z_{1-\alpha/2}\{p_0(1-p_0)\}^{1/2} + z_{1-\beta}\{p_A(1-p_A)\}^{1/2}}{p_A - p_0} \right]^2$$

(Pagano and Gauvreau 2000, 332).

5. For repeated measurements, Frison and Pocock (1992) discuss three methods for use in randomized clinical trials to compare two treatments using a continuous outcome measured at different times on each patient. Each uses the average of baseline measurements, $\bar{x}_0$, and follow-up measurements, $\bar{x}_1$:

POST outcome is $\bar{x}_1$, where the analysis is by simple $t$ test.

CHANGE outcome is $\bar{x}_1 - \bar{x}_0$, where the analysis is by simple $t$ test.

ANCOVA outcome is $\bar{x}_1 - \beta\bar{x}_0$, where the $\beta$ is estimated by analysis of covariance, correcting for the average at baseline.

ANCOVA will always be the most efficient of the three approaches. $\beta$ is set so that $\beta\bar{x}_0$ accounts for the largest possible variation of $\bar{x}_1$.

For a study with one measurement each at baseline and follow-up, CHANGE will be more efficient than POST, provided that the correlation between measurements at baseline and measurements at follow-up is more than 0.5. POST ignores all baseline measurements, which tends to make it unpopular. CHANGE is the method most commonly used. With more than one baseline measurement, CHANGE and ANCOVA tend to produce similar sample sizes and power.

The improvements in variance of the estimate of treatment effect over a study with only one measurement depend on the number of measurements $p$ at baseline; the number of measurements during follow-up; and the correlations between measurements at baseline $\bar{\rho}_{\text{pre}}$, between measurements at follow-up $\bar{\rho}_{\text{post}}$, and between measurements at baseline and measurements at follow-up $\bar{\rho}_{\text{mix}}$. The improvements in variance for the POST method are given by

$$\frac{1 + (r-1)\bar{\rho}_{\text{post}}}{r}$$

for the CHANGE method by

$$\frac{1 + (r-1)\bar{\rho}_{\text{post}}}{r} + \frac{1 + (p-1)\bar{\rho}_{\text{pre}}}{p} - 2\bar{\rho}_{\text{mix}}$$

and for the ANCOVA method by

$$\frac{1 + (r-1)\bar{\rho}_{\text{post}}}{r} - \frac{\bar{\rho}^2_{\text{mix}}p}{1 + (p-1)\bar{\rho}_{\text{pre}}}$$

Often the three correlations are assumed equal. In data from a number of trials, Frison and Pocock found that $\bar{\rho}_{\text{pre}}$ and $\bar{\rho}_{\text{post}}$ typically had values around 0.7, while $\bar{\rho}_{\text{mix}}$ was nearer 0.5. This is consistent with the common finding that measurements closer in time are more strongly correlated.

Power calculations are based on estimates of a single variance at all time points.

## Acknowledgments

`sampsi` is based on the `sampsiz` command written by Joseph Hilbe of Arizona State University (Hilbe 1993). Paul Seed of Queen Mary's School of Medicine and Dentistry, University of London (Seed 1997, 1998), expanded the command to allow for repeated measurements.

## References

Fleiss, J. L., B. Levin, and M. C. Paik. 2003. *Statistical Methods for Rates and Proportions*. 3rd ed. New York: Wiley.

Frison, L. and S. Pocock. 1992. Repeated measures in clinical trials: Analysis using mean summary statistics and its implications for design. *Statistics in Medicine* 11: 1685-1704.

Hilbe, J. 1993. sg15: Sample size determination for means and proportions. *Stata Technical Bulletin* 11: 17–20. Reprinted in *Stata Technical Bulletin Reprints*, vol. 2, pp. 145–149.

Pagano, M. and K. Gauvreau. 2000. *Principles of Biostatistics*. 2nd ed. Pacific Grove, CA: Brooks/Cole.

Rosner, B. 2000. *Fundamentals of Biostatistics*. 5th ed. Pacific Grove, CA: Duxbury.

Royston, J. P. and A. Babiker. 2002. A menu-drive facility for complex sample size calculation in randomized controlled trials with a survival or a binary outcome. *Stata Journal* 2: 151–163.

Seed, P. 1997. sbe18: Sample size calculations for clinical trials with repeated measures data. *Stata Technical Bulletin* 40: 16–18. Reprinted in *Stata Technical Bulletin Reprints*, vol. 7, pp. 121–125.

——. 1998. sbe18.1: Update of sampsi. *Stata Technical Bulletin* 45: 21. Reprinted in *Stata Technical Bulletin Reprints*, vol. 8, p. 84.

Tamhane, A. and D. Dunlop. 2000. *Statistics and Data Analysis: From Elementary to Intermediate*. Upper Saddle River, NJ: Prentice Hall.

## Also See

**Background:** [U] **19 Immediate commands**

# Title

**saved results** — Saved results

# Syntax

*List results stored in r()*

`return list`

*List results stored in e()*

`ereturn list`

*List results stored in s()*

`sreturn list`

# Description

Results of calculations are saved by many Stata commands so that they can be easily accessed and substituted into subsequent commands.

`return list` lists results stored in `r()`.

`ereturn list` lists results stored in `e()`.

`sreturn list` lists results stored in `s()`.

This entry discusses using saved results. Programmers wishing to save results should see [P] **return** and [P] **ereturn**.

# Remarks

Stata commands are classified as being

| | |
|---|---|
| r-class | general commands that save results in `r()` |
| e-class | estimation commands that save results in `e()` |
| s-class | parsing commands that save results in `s()` |
| n-class | commands that do not save in `r()`, `e()`, or `s()` |

There is also a c-class, `c()`, containing the values of system parameters and settings, along with certain constants, such as the value of pi; see [P] **creturn**. A program, however, cannot be c-class.

You can look at the *Saved Results* section of the manual entry of a command to determine whether it is r-, e-, s-, or n-class, but it is easy enough to guess.

Commands producing statistical results are either r-class or e-class. They are e-class if they present estimation results and r-class otherwise. No commands are s-class, which is a class used by programmers. n-class commands explicitly state where the result is to go. For instance, `generate` and `replace` are n-class because their syntax is `generate` *varname* `=` ... and `replace` *varname* `=` ....

After executing a command, you can type `return list`, `ereturn list`, or `sreturn list` to see what has been saved:

```
. use http://www.stata-press.com/data/r9/auto
(1978 Automobile Data)

. summarize mpg

    Variable |       Obs        Mean    Std. Dev.       Min        Max
-------------+--------------------------------------------------------
         mpg |        74     21.2973    5.785503         12         41

. return list

scalars:
                  r(N) =  74
              r(sum_w) =  74
               r(mean) =  21.2972972972973
                r(Var) =  33.47204738985561
                 r(sd) =  5.785503209735141
                r(min) =  12
                r(max) =  41
                r(sum) =  1576
```

Following **summarize**, you can use **r(N)**, **r(mean)**, **r(sd)**, etc., in expressions:

```
. gen double mpgstd = (mpg-r(mean))/r(sd)

. summarize mpgstd

    Variable |       Obs        Mean    Std. Dev.       Min        Max
-------------+--------------------------------------------------------
      mpgstd |        74   -1.64e-16           1  -1.606999    3.40553
```

Be careful to use results stored in **r()** soon because they will be replaced the next time you execute another r-class command. For instance, although **r(mean)** was 21.3 (approximately) after **summarize mpg**, it is $-1e-16$ now because you just ran **summarize** again.

e-class is really no different from r-class, except for where results are stored and that, when an estimation command stores results, it tends to store a lot of them:

```
. regress mpg weight displ
  (output omitted)

. ereturn list

scalars:
                  e(N) =  74
               e(df_m) =  2
               e(df_r) =  71
                  e(F) =  66.78504752026517
                 e(r2) =  .6529306984682528
               e(rmse) =  3.45606176570828
                e(mss) =  1595.409691543724
                e(rss) =  848.0497679157352
               e(r2_a) =  .643154098425105
                 e(ll) =  -195.2397979466294
               e(ll_0) =  -234.3943376482347

macros:
              e(title) : "Linear regression"
             e(depvar) : "mpg"
                e(cmd) : "regress"
         e(properties) : "b V"
            e(predict) : "regres_p"
              e(model) : "ols"
          e(estat_cmd) : "regress_estat"

matrices:
                  e(b) :  1 x 3
                  e(V) :  3 x 3

functions:
             e(sample)
```

These e-class results will stick around until you run another estimation command. Typing `return list` and `ereturn list` is the easy way to find out what a command stores.

Both r- and e-class results come in four flavors: scalars, macros, matrices, and functions. (s-class results come in only one flavor—macros—and as earlier noted, s class is used solely by programmers, so ignore it.)

Scalars are just that—numbers by any other name. You can subsequently refer to `r(mean)` or `e(rmse)` in numeric expressions and obtain the result to full precision.

Macros are strings. For instance, `e(depvar)` contains "`mpg`". You can refer to it, too, in subsequent expressions, but really that would be of most use to programmers, who will refer to it using constructs like `"`e(depvar)'"`. In any case, macros are macros, and you obtain their contents just as you would a local macro, by enclosing their name in single quotes. The name in this case is the full name, so `` `e(depvar)' `` is `mpg`.

Matrices are matrices, and all estimation commands store `e(b)` and `e(V)` containing the coefficient vector and variance–covariance matrix of the estimates (VCE).

Functions are saved by e-class commands only, and the only function existing is `e(sample)`. `e(sample)` evaluates to 1 (meaning true) if the observation was used in the previous estimation and to 0 (meaning false) otherwise.

## ❑ Technical Note

Say that some command set `r(scalar)` and `r(macro)`, the first being stored as a scalar and the second as a macro. In theory, in subsequent use you are supposed to refer to `r(scalar)` and `` `r(macro)' ``. In fact, however, you can refer to either one with or without quotes, so you could refer to `` `r(scalar)' `` and `r(macro)`. Programmers sometimes do this.

In the case of `r(scalar)`, when you refer to `r(scalar)`, you are referring to the full double-precision saved result. Think of `r(scalar)` without quotes as a function returning the value of the saved result `scalar`. When you refer to `r(scalar)` in quotes, Stata understands `` `r(scalar)' `` to mean "substitute the printed result of evaluating `r(scalar)`". Pretend that `r(scalar)` equals the number 23. Then `` `r(scalar)' `` is `23`, the character `2` followed by `3`.

Referring to `r(scalar)` in quotes is sometimes useful. For instance, say that you want to use the immediate command `ci` with `r(scalar)`. The immediate command `ci` requires its arguments to be numbers—numeric literals in programmer's jargon—and it will not take an expression. Thus you could not type '`ci r(scalar) ...`'. You could, however, type '`` ci `r(scalar)' ... ``' because `` `r(scalar)' `` is just a numeric literal.

In the case of `r(macro)`, you are supposed to refer to it in quotes: `` `r(macro)' ``. If, however, you omit the quotes in an expression context, Stata evaluates the macro and then pretends that it is the result of function-returning-string. There are side effects of this, the most important being that the result is trimmed to 80 characters.

Referring to `r(macro)` without quotes is never a good idea; the feature was included merely for completeness.

You can even refer to `r(matrix)` in quotes (assume that `r(matrix)` is a matrix). `` `r(matrix)' `` does not result in the matrix being substituted; it returns the word `matrix`. Programmers sometimes find that useful.

❑

## Also See

**Related:** [P] **ereturn**, [P] **return**

**Background:** [U] **18.8 Accessing results calculated by other programs**,
[U] **18.9 Accessing results calculated by estimation commands**

# Title

**scobit** — Skewed logistic regression

# Syntax

`scobit` *depvar* [*indepvars*] [*if*] [*in*] [*weight*] [, *options*]

| *options* | description |
|---|---|
| Model | |
| `noconstant` | suppress constant term |
| `offset(`*varname*`)` | include *varname* in model with coefficient constrained to 1 |
| `asis` | retain perfect predictor variables |
| `constraints(`*constraints*`)` | apply specified linear constraints |
| SE/Robust | |
| `vce(`*vcetype*`)` | *vcetype* may be `oim`, `robust`, `opg`, `bootstrap`, or `jackknife` |
| `robust` | synonym for `vce(robust)` |
| `cluster(`*varname*`)` | adjust standard errors for intragroup correlation |
| Reporting | |
| `level(#)` | set confidence level; default is `level(95)` |
| `or` | report odds ratios |
| Max options | |
| *maximize_options* | control the maximization process; seldom used |

`bootstrap`, `by`, `jackknife`, `rolling`, `statsby`, `stepwise`, and `xi` are allowed; see [U] **11.1.10 Prefix commands**.

`fweight`s, `iweight`s, and `pweight`s are allowed; see [U] **11.1.6 weight**.

See [U] **20 Estimation and postestimation commands** for additional capabilities of estimation commands.

# Description

`scobit` fits a maximum-likelihood skewed logit model.

See [R] **logistic** for a list of related estimation commands.

# Options

Model

`noconstant`, `offset(`*varname*`)`, `constraints(`*constraints*`)`; see [R] **estimation options.**

`asis` forces retention of perfect predictor variables and their associated perfectly predicted observations and may produce instabilities in maximization; see [R] **probit.**

SE/Robust

`vce(`*vcetype*`)`; see [R] ***vce_option***.

`robust`, `cluster(`*varname*`)`; see [R] **estimation options.**

Reporting

level(#); see [R] **estimation options**.

or reports the estimated coefficients transformed to odds ratios, i.e., $e^b$ rather than $b$. Standard errors and confidence intervals are similarly transformed. This option affects how results are displayed, not how they are estimated. or may be specified at estimation or when replaying previously estimated results.

Max options

*maximize_options*: difficult, technique(*algorithm_spec*), iterate(#), [no]log, trace, gradient, showstep, hessian, shownrtolerance, tolerance(#), ltolerance(#), gtolerance(#), nrtolerance(#), nonrtolerance, from(*init_specs*); see [R] **maximize**.

# Remarks

Remarks are presented under the headings

*Skewed logistic model*
*Robust standard errors*

## Skewed logistic model

scobit fits maximum likelihood models with dichotomous dependent variables coded as 0/1 (or, more precisely, coded as 0 and not 0).

### ▷ Example 1

We have data on the make, weight, and mileage rating of 22 foreign and 52 domestic automobiles. We wish to fit a model explaining whether a car is foreign based on its mileage. Here is an overview of our data:

```
. use http://www.stata-press.com/data/r9/auto
(1978 Automobile Data)

. keep make mpg weight foreign

. describe

Contains data from http://www.stata-press.com/data/r9/auto.dta
  obs:            74                          1978 Automobile Data
 vars:             4                          13 Apr 2005 17:45
 size:         1,998 (99.9% of memory free)  (_dta has notes)
-------------------------------------------------------------------------------
              storage  display     value
variable name   type   format      label      variable label
-------------------------------------------------------------------------------
make            str18  %-18s                  Make and Model
mpg             int    %8.0g                  Mileage (mpg)
weight          int    %8.0gc                 Weight (lbs.)
foreign         byte   %8.0g       origin     Car type
-------------------------------------------------------------------------------
Sorted by:  foreign
     Note:  dataset has changed since last saved
```

```
. inspect foreign
foreign:  Car type                           Number of Observations
                                                                Non-
                                             Total    Integers    Integers
 |  #                        Negative           -          -           -
 |  #                        Zero              52         52           -
 |  #                        Positive          22         22           -
 |  #                                       -----      -----       -----
 |  #   #                    Total             74         74           -
 |  #   #                    Missing            -
 +-----------------------                   -----
0                     1                        74
   (2 unique values)
      foreign is labeled and all values are documented in the label.
```

The variable **foreign** takes on two unique values, 0 and 1. The value 0 denotes a domestic car, and 1 denotes a foreign car.

The model that we wish to fit is

$$\Pr(\texttt{foreign} = 1) = F(\beta_0 + \beta_1 \texttt{mpg})$$

where $F(z) = 1 - 1/\{1 + \exp(z)\}^{\alpha}$.

To fit this model, we type

```
. scobit foreign mpg
Fitting logistic model:
Iteration 0:   log likelihood =  -45.03321
Iteration 1:   log likelihood = -39.380959
Iteration 2:   log likelihood = -39.288802
Iteration 3:   log likelihood =  -39.28864
Fitting full model:
Iteration 0:   log likelihood =  -39.28864
Iteration 1:   log likelihood = -39.286393
Iteration 2:   log likelihood = -39.284415
Iteration 3:   log likelihood = -39.284234
Iteration 4:   log likelihood = -39.284197
Iteration 5:   log likelihood = -39.284196
Skewed logistic regression                        Number of obs     =         74
                                                  Zero outcomes     =         52
Log likelihood =  -39.2842                        Nonzero outcomes  =         22

------------------------------------------------------------------------------
     foreign |      Coef.   Std. Err.      z    P>|z|     [95% Conf. Interval]
-------------+----------------------------------------------------------------
         mpg |   .1813879   .2407362     0.75   0.451    -.2904463    .6532222
       _cons |  -4.274883   1.399305    -3.06   0.002    -7.017471   -1.532295
-------------+----------------------------------------------------------------
    /lnalpha |  -.4450405   3.879885    -0.11   0.909    -8.049476    7.159395
-------------+----------------------------------------------------------------
       alpha |   .6407983   2.486224                      .0003193    1286.133
------------------------------------------------------------------------------
Likelihood-ratio test of alpha=1:    chi2(1) =     0.01    Prob > chi2 = 0.9249
note: Likelihood-ratio tests are recommended for inference with scobit models.
```

We find that cars yielding better gas mileage are less likely to be foreign. The likelihood-ratio test at the bottom of the output indicates that the model is not significantly different from a logit model. Therefore, we should use the more parsimonious model.

◁

### ❑ Technical Note

Stata interprets a value of 0 as a negative outcome (failure) and treats all other values (except missing) as positive outcomes (successes). Thus if the dependent variable takes on the values 0 and 1, 0 is interpreted as failure and 1 as success. If the dependent variable takes on the values 0, 1, and 2, 0 is still interpreted as failure, but both 1 and 2 are treated as successes.

Formally, when we type `scobit` $y$ $x$, Stata fits the model

$$\Pr(y_j \neq 0 \mid \mathbf{x}_j) = 1 - 1\Big/\Big\{1 + \exp(\mathbf{x}_j\boldsymbol{\beta})\Big\}^{\alpha}$$

❑

## Robust standard errors

If you specify the `robust` option, `scobit` reports robust standard errors as described in [U] **20.14 Obtaining robust variance estimates**. In the case of the model of `foreign` on `mpg`, the robust calculation increases the standard error of the coefficient on `mpg` by around 25 percent:

```
. scobit foreign mpg, robust nolog

Skewed logistic regression                        Number of obs     =         74
                                                  Zero outcomes     =         52
Log pseudolikelihood =  -39.2842                  Nonzero outcomes  =         22

------------------------------------------------------------------------------
             |               Robust
     foreign |      Coef.   Std. Err.      z    P>|z|     [95% Conf. Interval]
-------------+----------------------------------------------------------------
         mpg |   .1813879   .3028487     0.60   0.549    -.4121847    .7749606
       _cons |  -4.274883   1.335521    -3.20   0.001    -6.892455   -1.657311
-------------+----------------------------------------------------------------
    /lnalpha |  -.4450405    4.71561    -0.09   0.925    -9.687466    8.797385
-------------+----------------------------------------------------------------
       alpha |   .6407983   3.021755                      .0000621    6616.919
------------------------------------------------------------------------------
```

Without `robust`, the standard error for the coefficient on `mpg` was reported to be .241, with a resulting confidence interval of $[-.29, .65]$.

Specifying the `cluster()` option relaxes the independence assumption required by the skewed logit estimator to being just independence between clusters. To demonstrate this, we will switch to a different dataset.

### ▷ Example 2

We are studying the unionization of women in the United States and have a dataset with 26,200 observations on 4,434 women between 1970 and 1988. For our purposes, we will use the variables `age` (the women were 14–26 in 1968 and the data thus span the age range of 16–46), `grade` (years of schooling completed, ranging from 0 to 18), `not_smsa` (28% of the person-time was spent living outside an SMSA—standard metropolitan statistical area), `south` (41% of the person-time was in the South), and `southXt` (`south` interacted with year, treating 1970 as year 0). We also have variable `union`. Overall, 22% of the person-time is marked as time under union membership and 44% of these women have belonged to a union.

We fit the following model, ignoring that women are observed an average of 5.9 times each in these data:

```
. use http://www.stata-press.com/data/r9/union
(NLS Women 14-24 in 1968)

. scobit union age grade not_smsa south southXt, nolog nonrtol

Skewed logistic regression                  Number of obs     =      26200
                                            Zero outcomes     =      20389
Log likelihood =  -13544.2                  Nonzero outcomes  =       5811

------------------------------------------------------------------------------
       union |      Coef.   Std. Err.      z    P>|z|     [95% Conf. Interval]
-------------+----------------------------------------------------------------
         age |   .0085889   .0023036     3.73   0.000      .004074    .0131039
       grade |   .0447166   .0057073     7.83   0.000     .0335304    .0559027
    not_smsa |  -.1906374   .0317694    -6.00   0.000    -.2529042   -.1283707
       south |  -.6446248   .0557704   -11.56   0.000    -.7539328   -.5353169
     southXt |   .0068271   .0047299     1.44   0.149    -.0024433    .0160976
       _cons |  -10.82928   63.79149    -0.17   0.865    -135.8583    114.1997
-------------+----------------------------------------------------------------
    /lnalpha |   8.862483   63.79072     0.14   0.890     -116.165      133.89
-------------+----------------------------------------------------------------
       alpha |   7061.993   450489.6                      3.55e-51    1.41e+58
------------------------------------------------------------------------------
Likelihood-ratio test of alpha=1:    chi2(1) =     3.07    Prob > chi2 = 0.0799
note: Likelihood-ratio tests are recommended for inference with scobit models.
```

The reported standard errors in this model are probably meaningless. Women are observed repeatedly, so the observations are not independent. Looking at the coefficients, we find a large southern effect against unionization and little time trend. The **robust** and **cluster()** options provide a way to fit this model and obtain correct standard errors:

```
. scobit union age grade not_smsa south southXt, robust cluster(id) nolog nonrtol

Skewed logistic regression                  Number of obs     =      26200
                                            Zero outcomes     =      20389
Log pseudolikelihood =  -13544.2            Nonzero outcomes  =       5811

                           (Std. Err. adjusted for 4434 clusters in idcode)
------------------------------------------------------------------------------
             |               Robust
       union |      Coef.   Std. Err.      z    P>|z|     [95% Conf. Interval]
-------------+----------------------------------------------------------------
         age |   .0085889   .0033835     2.54   0.011     .0019575    .0152204
       grade |   .0447166   .0125938     3.55   0.000     .0200332    .0693999
    not_smsa |  -.1906374   .0641961    -2.97   0.003    -.3164594   -.0648155
       south |  -.6446248   .0833872    -7.73   0.000    -.8080608   -.4811889
     southXt |   .0068271   .0063044     1.08   0.279    -.0055292    .0191834
       _cons |  -10.82928   .9169089   -11.81   0.000    -12.62639   -9.032174
-------------+----------------------------------------------------------------
    /lnalpha |   8.862483   .7787163    11.38   0.000     7.336227    10.38874
-------------+----------------------------------------------------------------
       alpha |   7061.993   5499.289                      1534.909    32491.65
------------------------------------------------------------------------------
```

What is important to understand is that **scobit, robust cluster()** is robust to assumptions about within-cluster correlation. That is, it inefficiently sums within cluster for the standard error calculation rather than attempting to exploit what might be assumed about the within-cluster correlation (as do the **xtgee** population-averaged models; see [XT] **xtgee**).

◁

## ❑ Technical Note

The scobit model can be very difficult to fit because of the functional form. Often it requires many iterations, or the optimizer prints out warning and informative messages during the optimization. For example, without the `nrtolerance` option, the model using the `union` dataset will not converge. See [R] **maximize** for details about the optimizer.

❑

## ❑ Technical Note

The primary reason for using `scobit` rather that `logit` is that the effects of the regressors on the probability of success are not constrained to be the largest when the probability is 0.5. Rather, the independent variables might show their largest impact when the probability of success is 0.3 or 0.6. This added flexibility results because the scobit function, unlike the logit function, can be skewed and is not constrained to be mirror symmetric about the 0.5 probability of success.

As Nagler (1994) pointed out, the point of maximum impact is constrained under the scobit model to fall within the interval $(0, 1 - e^{(-1)})$ or approximately (0, 0.63). Christopher Achen (2002) notes that if we believe the maximum impact to be outside that range, we can instead estimate the "power logit" model by simply reversing the 0s and 1s of our outcome variable and estimating a scobit model on failure, rather than success. We would need to reverse the signs of the coefficients if we wanted to interpret them in terms of impact on success, or we could leave them as they are and interpret them in terms of impact on failure. The important thing to remember is that the scobit model, unlike the logit model, is not invariant to the choice of which result is assigned to success.

❑

*(Continued on next page)*

# Saved Results

scobit saves in e():

Scalars

| | | | |
|---|---|---|---|
| e(N) | number of observations | e(alpha) | alpha |
| e(k) | number of parameters | e(N_clust) | number of clusters |
| e(k_eq) | number of equations | e(chi2) | $\chi^2$ |
| e(k_dv) | number of dependent variables | e(chi2_c) | $\chi^2$ for comparison test |
| e(df_m) | model degrees of freedom | e(p) | significance |
| e(ll) | log likelihood | e(rank) | rank of e(V) |
| e(ll_0) | log likelihood, constant-only model | e(ic) | number of iterations |
| e(ll_c) | log likelihood, comparison model | e(rc) | return code |
| e(N_f) | number of failure (zero) outcomes | e(converged) | 1 if converged, 0 otherwise |
| e(N_s) | number of success (nonzero) outcomes | | |

Macros

| | | | |
|---|---|---|---|
| e(cmd) | scobit | e(vce) | *vcetype* specified in vce() |
| e(depvar) | name of dependent variable | e(vcetype) | title used to label Std. Err. |
| e(wtype) | weight type | e(opt) | type of optimization |
| e(wexp) | weight expression | e(ml_method) | type of ml method |
| e(title) | title in estimation output | e(user) | name of likelihood-evaluator program |
| e(clustvar) | name of cluster variable | e(crittype) | optimization criterion |
| e(offset) | offset | e(technique) | maximization technique |
| e(chi2type) | Wald or LR; type of model $\chi^2$ test | e(properties) | b V |
| e(chi2_ct) | Wald or LR; type of model $\chi^2$ test corresponding to e(chi2_c) | e(predict) | program used to implement predict |

Matrices

| | | | |
|---|---|---|---|
| e(b) | coefficient vector | e(V) | variance–covariance matrix of the estimators |
| e(ilog) | iteration log (up to 20 iterations) | | |
| e(gradient) | gradient vector | | |

Functions

| | |
|---|---|
| e(sample) | marks estimation sample |

# Methods and Formulas

scobit is implemented as an ado-file.

Skewed logit analysis is an alternative to logit, which relaxes the assumption that individuals with initial probability of .5 are most sensitive to changes in independent variables.

The log-likelihood function for skewed logit is

$$\ln L = \sum_{j \in S} w_j \ln F(\mathbf{x}_j \mathbf{b}) + \sum_{j \notin S} w_j \ln \big\{1 - F(\mathbf{x}_j \mathbf{b})\big\}$$

where $S$ is the set of all observations $j$ such that $y_j \neq 0$, $F(z) = 1 - 1/\big\{1 + \exp(z)\big\}^{\alpha}$, and $w_j$ denotes the optional weights. $\ln L$ is maximized as described in [R] **maximize**.

If robust standard errors are requested, the calculation described in *Methods and Formulas* of [R] **regress** is carried forward with

$$\mathbf{u}_j^1 = \mathbf{x}_j \frac{\alpha \exp(\mathbf{x}_j\mathbf{b})}{\{1+\exp(\mathbf{x}_j\mathbf{b})\}[\{1+\exp(\mathbf{x}_j\mathbf{b})\}^\alpha - 1]}$$

$$\mathbf{u}_j^2 = \frac{\alpha \ln\{1+\exp(\mathbf{x}_j\mathbf{b})\}}{\{1+\exp(\mathbf{x}_j\mathbf{b})\}[\{1+\exp(\mathbf{x}_j\mathbf{b})\}^\alpha - 1]}$$

for the positive outcomes and

$$\mathbf{u}_j^1 = -\mathbf{x}_j \frac{\alpha \exp(\mathbf{x}_j\mathbf{b})}{1+\exp(\mathbf{x}_j\mathbf{b})}$$

$$\mathbf{u}_j^2 = \alpha \ln\{1+\exp(\mathbf{x}_j\mathbf{b})\}$$

for the negative outcomes.

## References

Achen, C. H. 2002. Toward a new political methodology: Microfoundations and ART. *Annual Review of Political Science* 5: 423–450.

Nagler, J. 1994. Scobit: An alternative estimator to logit and probit. *American Journal of Political Science* 38: 230–255.

## Also See

**Complementary:** [R] **scobit postestimation**; [R] **constraint**

**Related:** [R] **biprobit**, [R] **cloglog**, [R] **cusum**, [R] **glm**, [R] **glogit**, [R] **logistic**, [R] **logit**, [R] **probit**

**Background:** [U] **11.1.10 Prefix commands**, [U] **20 Estimation and postestimation commands**, [R] **estimation options**, [R] **maximize**, [R] ***vce_option***

# Title

**scobit postestimation** — Postestimation tools for scobit

# Description

The following postestimation commands are available for `scobit`:

| command | description |
|---|---|
| `adjust` | adjusted predictions of $\mathbf{x}\beta$ |
| `estat` | AIC, BIC, VCE, and estimation sample summary |
| `estimates` | cataloging estimation results |
| `lincom` | point estimates, standard errors, testing, and inference for linear combinations of coefficients |
| `lrtest` | likelihood-ratio test |
| `mfx` | marginal effects or elasticities |
| `nlcom` | point estimates, standard errors, testing, and inference for nonlinear combinations of coefficients |
| `predict` | predictions, residuals, influence statistics, and other diagnostic measures |
| `predictnl` | point estimates, standard errors, testing, and inference for generalized predictions |
| `suest` | seemingly unrelated estimation |
| `test` | Wald tests for simple and composite linear hypotheses |
| `testnl` | Wald tests of nonlinear hypotheses |

See the corresponding entries in the *Stata Base Reference Manual* for details.

# Syntax for predict

`predict` [*type*] *newvar* [*if*] [*in*] [, *statistic* `nooffset`]

`predict` [*type*] { *stub** | $newvar_{\text{reg}}$ $newvar_{\text{lnalpha}}$ } [*if*] [*in*] , `scores`

| *statistic* | description |
|---|---|
| `pr` | probability of a positive outcome; the default |
| `xb` | $\mathbf{x}_j\mathbf{b}$, linear prediction |
| `stdp` | standard error of the prediction |

These statistics are available both in and out of sample; type `predict ... if e(sample) ...` if wanted only for the estimation sample.

# Options for predict

`pr`, the default, calculates the probability of a positive outcome.

`xb` calculates the linear prediction.

`stdp` calculates the standard error of the linear prediction.

`nooffset` is relevant only if you specified `offset(`*varname*`)` for `scobit`. It modifies the calculations made by `predict` so that they ignore the offset variable; the linear prediction is treated as $\mathbf{x}_j\mathbf{b}$ rather than as $\mathbf{x}_j\mathbf{b} + \text{offset}_j$.

`scores` calculates equation-level score variables.

The first new variable will contain $\partial \ln L/\partial(\mathbf{x}_j\boldsymbol{\beta})$.

The second new variable will contain $\partial \ln L/\partial \ln \alpha$.

## Remarks

Once you have fitted a model, you can obtain the predicted probabilities using the `predict` command for both the estimation sample and other samples; see [U] **20 Estimation and postestimation commands** and [R] **predict**. Here we will make only a few additional comments.

`predict` without arguments calculates the predicted probability of a positive outcome. With the `xb` option, it calculates the linear combination $\mathbf{x}_j\mathbf{b}$, where $\mathbf{x}_j$ are the independent variables in the $j$th observation and $\mathbf{b}$ is the estimated parameter vector.

With the `stdp` option, `predict` calculates the standard error of the prediction, which is *not* adjusted for replicated covariate patterns in the data.

▷ Example 1

In example 1 of [R] **scobit**, we fitted the model `scobit foreign mpg`. To obtain predicted probabilities, we type

```
. use http://www.stata-press.com/data/r9/auto
(1978 Automobile Data)
. keep make mpg weight foreign
. scobit foreign mpg
  (output omitted)
. predict p
(option p assumed; Pr(foreign))
. summarize foreign p
    Variable |       Obs        Mean    Std. Dev.       Min        Max
-------------+--------------------------------------------------------
     foreign |        74    .2972973    .4601885          0          1
           p |        74    .2974049     .182352   .0714664    .871624
```

◁

## Methods and Formulas

All postestimation commands listed above are implemented as ado-files.

(Continued on next page)

# Also See

**Complementary:** [R] **scobit**; [R] **adjust**, [R] **estimates**, [R] **lincom**, [R] **lrtest**, [R] **mfx**, [R] **nlcom**, [R] **predictnl**, [R] **suest**, [R] **test**, [R] **testnl**

**Background:** [U] **13.5 Accessing coefficients and standard errors**,
[U] **20 Estimation and postestimation commands**,
[R] **estat**, [R] **predict**

# Title

**sdtest** — Variance-comparison tests

# Syntax

*One-sample variance-comparison test*

sdtest *varname* == # [*if*] [*in*] [, level(#)]

*Two-sample variance-comparison test*

sdtest $varname_1$ == $varname_2$ [*if*] [*in*] [, level(#)]

*Test that two groups have the same variance*

sdtest *varname* [*if*] [*in*] , by(*groupvar*) [level(#)]

*Immediate form of one-sample variance-comparison test*

sdtesti $\#_{obs}$ $\{\#_{mean} \mid .\}$ $\#_{sd}$ $\#_{val}$ [, level(#)]

*Immediate form of two-sample variance-comparison test*

sdtesti $\#_{obs,1}$ $\{\#_{mean,1} \mid .\}$ $\#_{sd,1}$ $\#_{obs,2}$ $\{\#_{mean,2} \mid .\}$ $\#_{sd,2}$ [, level(#)]

*Robust tests for equality of variance*

robvar *varname* [*if*] [*in*] , by(*groupvar*)

by may be used with sdtest (but not with sdtesti) and robvar; see [D] **by**.

# Description

sdtest performs tests on the equality of standard deviations (variances). In the first form, sdtest tests that the standard deviation of *varname* is #. In the second form, sdtest tests that $varname_1$ and $varname_2$ have the same standard deviation. In the third form, sdtest performs the same test, using the standard deviations of the two groups defined by *groupvar*.

sdtesti is the immediate form of sdtest; see [U] **19 Immediate commands**.

Both the traditional $F$ test for the homogeneity of variances and Bartlett's generalization of this test to $K$ samples are very sensitive to the assumption that the data are drawn from an underlying Gaussian distribution. See, for example, the cautionary results discussed by Markowski and Markowski (1990). Levene (1960) proposed a test statistic for equality of variance that was found to be robust under non-normality. Subsequently, Brown and Forsythe (1974) proposed alternative formulations of Levene's test statistic that use more robust estimators of central tendency in place of the mean. These reformulations were demonstrated to be more robust than Levene's test when dealing with skewed populations.

robvar reports Levene's robust test statistic ($W_0$) for the equality of variances between the groups defined by *groupvar* and the two statistics proposed by Brown and Forsythe that replace the mean in Levene's formula with alternative location estimators. The first alternative ($W_{50}$) replaces the mean with the median. The second alternative replaces the mean with the 10 percent trimmed mean ($W_{10}$).

# Options

level(*#*) specifies the confidence level, as a percentage, for confidence intervals of the means. The default is level(95) or as set by set level; see [U] **20.6 Specifying the width of confidence intervals**.

by(*groupvar*) specifies the *groupvar* that defines the groups to be compared. For sdtest, there should be two groups, but for robvar there may be more than two groups. Do not confuse the by() option with the by prefix; both may be specified.

# Remarks

Remarks are presented under the headings

*Basic form*
*Immediate form*
*Robust test*

## Basic form

sdtest performs two different statistical tests: one testing equality of variances and the other testing that the standard deviation is equal to a known constant. Which test it performs is determined by whether you type a variable name or a number to the right of the equal sign.

### ▷ Example 1: One-sample test of variance

We have a sample of 74 automobiles. For each automobile, we know the mileage rating. We wish to test whether the overall standard deviation is 5:

```
. use http://www.stata-press.com/data/r9/auto
(1978 Automobile Data)

. sdtest mpg == 5

One-sample test of variance

------------------------------------------------------------------------------
Variable |     Obs        Mean    Std. Err.   Std. Dev.   [95% Conf. Interval]
---------+--------------------------------------------------------------------
     mpg |      74     21.2973    .6725511    5.785503    19.9569    22.63769
------------------------------------------------------------------------------
    sd = sd(mpg)                                          c = chi2 =  97.7384
Ho: sd = 5                                       degrees of freedom =       73

     Ha: sd < 5                 Ha: sd != 5                   Ha: sd > 5
 Pr(C < c) = 0.9717       2*Pr(C > c) = 0.0565           Pr(C > c) = 0.0283
```

◁

### ▷ Example 2: Variance ratio test

We are testing the effectiveness of a new fuel additive. We run an experiment on 12 cars, running each without and with the additive. The data can be found in [R] **ttest**. The results for each car are stored in the variables `mpg1` and `mpg2`:

```
. use http://www.stata-press.com/data/r9/fuel
. sdtest mpg1==mpg2
Variance ratio test
------------------------------------------------------------------------------
Variable |     Obs        Mean    Std. Err.   Std. Dev.   [95% Conf. Interval]
---------+--------------------------------------------------------------------
    mpg1 |      12          21    .7881701    2.730301    19.26525    22.73475
    mpg2 |      12       22.75    .9384465    3.250874    20.68449    24.81551
---------+--------------------------------------------------------------------
combined |      24      21.875    .6264476    3.068954    20.57909    23.17091
------------------------------------------------------------------------------
    ratio = sd(mpg1) / sd(mpg2)                                   f =   0.7054
Ho: ratio = 1                                    degrees of freedom =   11, 11

    Ha: ratio < 1               Ha: ratio != 1                 Ha: ratio > 1
  Pr(F < f) = 0.2862         2*Pr(F < f) = 0.5725           Pr(F > f) = 0.7138
```

We cannot reject the hypothesis that the standard deviations are the same.

In [R] **ttest**, we draw an important distinction between paired and unpaired data, which, in this example, means whether there are 12 cars in a before-and-after experiment or 24 different cars. For `sdtest`, on the other hand, there is no distinction. If the data had been unpaired and stored as described in [R] **ttest**, we could have typed `sdtest mpg, by(treated)`, and the results would have been the same.

◁

## Immediate form

### ▷ Example 3: sdtesti

Immediate commands are used not with data, but with reported summary statistics. For instance, to test whether a variable on which we have 75 observations and a reported standard deviation of 6.5 comes from a population with underlying standard deviation 6, we would type

```
. sdtesti 75 . 6.5 6
One-sample test of variance
------------------------------------------------------------------------------
         |     Obs        Mean    Std. Err.   Std. Dev.   [95% Conf. Interval]
---------+--------------------------------------------------------------------
       x |      75           .    .7505553         6.5           .           .
------------------------------------------------------------------------------
    sd = sd(x)                                            c = chi2 =  86.8472
Ho: sd = 6                                       degrees of freedom =       74

     Ha: sd < 6                 Ha: sd != 6                   Ha: sd > 6
  Pr(C < c) = 0.8542         2*Pr(C > c) = 0.2916           Pr(C > c) = 0.1458
```

The mean plays no role in the calculation, so it may be omitted.

To test whether the variable comes from a population with the same standard deviation as another for which we have a calculated standard deviation of 7.5 over 65 observations, we would type

```
. sdtesti 75 . 6.5  65 . 7.5
Variance ratio test
------------------------------------------------------------------------------
         |     Obs        Mean    Std. Err.   Std. Dev.   [95% Conf. Interval]
---------+--------------------------------------------------------------------
       x |      75           .    .7505553         6.5           .           .
       y |      65           .    .9302605         7.5           .           .
---------+--------------------------------------------------------------------
combined |     140           .           .           .           .           .
------------------------------------------------------------------------------
    ratio = sd(x) / sd(y)                                         f =   0.7511
Ho: ratio = 1                                    degrees of freedom =   74, 64

    Ha: ratio < 1               Ha: ratio != 1                 Ha: ratio > 1
  Pr(F < f) = 0.1172         2*Pr(F < f) = 0.2344           Pr(F > f) = 0.8828
```

◁

## Robust test

### ▷ Example 4: robvar

We wish to test whether the standard deviation of the length of stay for patients hospitalized for a given medical procedure differs by gender. Our data consist of observations on the length of hospital stay for 1778 patients: 884 males and 894 females. Length of stay, `lengthstay`, is highly skewed (Skewness coefficient = 4.912591) and thus violates Bartlett's normality assumption. Therefore, we use `robvar` to compare the variances.

```
. use http://www.stata-press.com/data/r9/stay
. robvar lengthstay, by(sex)
            |  Summary of Length of stay in days
        sex |        Mean   Std. Dev.       Freq.
------------+------------------------------------
       male |   9.0874434   9.7884747         884
     female |    8.800671   9.1081478         894
------------+------------------------------------
      Total |   8.9432508   9.4509466        1778
W0  = .55505315   df(1, 1776)     Pr > F = .45635888
W50 = .42714734   df(1, 1776)     Pr > F = .51347664
W10 = .44577674   df(1, 1776)     Pr > F = .50443411
```

For these data, we cannot reject the null hypothesis that the variances are equal. However, Bartlett's test yields a significance probability of 0.0319 due to the pronounced skewness of the data.

◁

### ❑ Technical Note

`robvar` implements both the conventional Levene's test centered at the mean and a median-centered test. In a simulation study, Conover, Johnson, and Johnson (1981) compare the properties of the two tests and recommend using the median test for asymmetric data, although for small sample sizes the test is somewhat conservative. See Carroll and Schneider (1985) for an explanation of why both mean- and median-centered tests have approximately the same level for symmetric distributions, but for asymmetric distributions the median test is closer to the correct level.

❑

## Saved Results

`sdtest` and `sdtesti` save in `r()`:

Scalars

| | | | |
|---|---|---|---|
| `r(N)` | number of observations | `r(sd)` | standard deviation |
| `r(p_l)` | lower one-sided $p$-value | `r(sd_1)` | standard deviation for first variable |
| `r(p_u)` | upper one-sided $p$-value | `r(sd_2)` | standard deviation for second variable |
| `r(p)` | two-sided $p$-value | `r(df_1)` | numerator degrees of freedom |
| `r(F)` | $F$ statistic | `r(df_2)` | denominator degrees of freedom |

`robvar` saves in `r()`:

Scalars

| | | | |
|---|---|---|---|
| `r(N)` | number of observations | `r(w10)` | Brown and Forsythe's $F$ statistic (trimmed mean) |
| `r(w50)` | Brown and Forsythe's $F$ statistic (median) | | |
| `r(p_w50)` | Brown and Forsythe's $p$-value | `r(p_w10)` | Brown and Forsythe's $p$-value (trimmed mean) |
| `r(w0)` | Levene's $F$ statistic | | |
| `r(p_w0)` | Levene's $p$-value | `r(df_1)` | numerator degrees of freedom |
| | | `r(df_2)` | denominator degrees of freedom |

## Methods and Formulas

`sdtest` and `sdtesti` are implemented as ado-files.

See Hoel (1984; 279–284, 298–300) or Pagano and Gauvreau (2000, chapter 11) for an introduction and explanation of the calculation of these tests.

The test for $\sigma = \sigma_0$ is given by

$$\chi^2 = \frac{(n-1)s^2}{\sigma_0^2}$$

which is distributed as $\chi^2$ with $n-1$ degrees of freedom.

The test for $\sigma_x^2 = \sigma_y^2$ is given by

$$F = \frac{s_x^2}{s_y^2}$$

which is distributed as $F$ with $n_x - 1$ and $n_y - 1$ degrees of freedom.

`robvar` is also implemented as an ado-file.

Let $X_{ij}$ be the $j$th observation of $X$ for the $i$th group. Let $Z_{ij} = |X_{ij} - \overline{X}_i|$, where $\overline{X}_i$ is the mean of $X$ in the $i$th group. Levene's test statistic is

$$W_0 = \frac{\sum_i n_i(\overline{Z}_i - \overline{Z})^2/(g-1)}{\sum_i \sum_j (Z_{ij} - \overline{Z}_i)^2 / \sum_i (n_i - 1)}$$

where $n_i$ is the number of observations in group $i$ and $g$ is the number of groups. $W_{50}$ is obtained by replacing $\overline{X}_i$ with the $i$th group median of $X_{ij}$, while $W_{10}$ is obtained by replacing $\overline{X}_i$ with the 10 percent trimmed mean for group $i$.

## References

Brown, M. B. and A. B. Forsythe. 1974. Robust test for the equality of variances. *Journal of the American Statistical Association* 69: 364–367.

Carroll, R. J. and H. Schneider. 1985. A note on Levene's test for equality of variances. *Statistics & Probability Letters* 3(4): 191–194.

Cleves, M. A. 1995. sg35: Robust tests for the equality of variances. *Stata Technical Bulletin* 25: 13–15. Reprinted in *Stata Technical Bulletin Reprints*, vol. 5, pp. 91–93.

——. 2000. sg35.2: Robust tests for the equality of variances update to Stata 6. *Stata Technical Bulletin* 53: 17–18. Reprinted in *Stata Technical Bulletin Reprints*, vol. 9, pp. 158–159.

Conover, W. J., M. E. Johnson, and M. M. Johnson. 1981. A comparative study of tests for homogeneity of variances, with applications to the outer continental shelf bidding data. *Technometrics* 23: 351–361.

Hoel, P. G. 1984. *Introduction to Mathematical Statistics*. 5th ed. New York: Wiley.

Levene, H. 1960. Robust tests for equality of variances. *Contributions to Probability and Statistics: Essays in honor of Harold Hotelling*. ed. I. Olkin, S. G. Ghurye, W. Hoeffding, W. G. Madow, and H. B. Mann, 278–292. Menlo Park, CA: Stanford University Press.

Markowski, C. A. and E. P. Markowski. 1990. Conditions for the effectiveness of a preliminary test of variance. *The American Statistician* 44: 322–326.

Pagano, M. and K. Gauvreau. 2000. *Principles of Biostatistics*. 2nd ed. Pacific Grove, CA: Brooks/Cole.

Seed, P. 2000. sbe33: Comparing several methods of measuring the same quantity. *Stata Technical Bulletin* 55: 2–9. Reprinted in *Stata Technical Bulletin Reprints*, vol. 10, pp. 73–82.

Tobias, A. 1998. gr28: A graphical procedure to test equality of variances. *Stata Technical Bulletin* 42: 4–6. Reprinted in *Stata Technical Bulletin Reprints*, vol. 7, pp. 68–70.

## Also See

**Related:** [R] **ttest**

**Background:** [U] **19 Immediate commands**

# Title

**search** — Search Stata documentation

# Syntax

`search` *word* [*word* ...] [, *search_options*]

`set searchdefault` { `local` | `net` | `all` } [, `permanently`]

`findit` *word* [*word* ...]

| *search_options* | description |
|---|---|
| `local` | search using Stata's keyword database; the default |
| `net` | search across materials available via Stata's `net` command |
| `all` | search across both the local keyword database and the `net` material |
| `author` | search by author's name |
| `entry` | search by entry ID |
| `exact` | search across both the local keyword database and the `net` materials; prevents matching on abbreviations |
| `faq` | search the FAQs posted to the Stata web site |
| `historical` | search entries that are of historical interest only |
| `or` | list an entry if *any* of the words typed after `search` are associated with the entry |
| `manual` | search the entries in the *Stata Documentation* |
| `sj` | search the entries in the *Stata Journal* and the STB |

# Description

`search` searches a keyword database and the Internet.

Capitalization of the words following `search` is irrelevant, as is the inclusion or exclusion of special characters such as commas, hyphens, and the like.

`set searchdefault` affects the default behavior of the `search` command. `local` is the default.

`findit` is equivalent to `search` *word* [*word* ...], `all`. `findit` results are displayed in the Viewer. `findit` is the best way to search for information on a topic across all sources, including the online help, the FAQs at the Stata web site, the *Stata Journal*, and all Stata-related Internet sources including user-written additions. From `findit`, you can click to go to a source or to install additions.

# Options for search

`local`, the default (unless changed by `set searchdefault`), specifies that the search be performed using only Stata's keyword database.

`net` specifies that the search be performed across the materials available via Stata's `net` command. Using `search` *word* [*word* ...], `net` is equivalent to typing `net search` *word* [*word* ...] (without options); see [R] **net search**.

all specifies that the search be performed across both the local keyword database and the net materials.

author specifies that the search be performed on the basis of author's name rather than keywords. A search with the author option is performed on the local keyword database only.

entry specifies that the search be performed on the basis of entry IDs rather than keywords. A search with the entry option is performed on the local keyword database only.

exact prevents matching on abbreviations. A search with the exact option is performed across both the local keyword database and the net materials.

faq limits the search to the FAQs posted on the Stata web site: *http://www.stata.com*. A search with the faq option is performed on the local keyword database only.

historical adds to the search entries that are of historical interest only. By default, such entries are not listed. Past entries are classified as historical if they discuss a feature that later became an official part of Stata. Updates to historical entries will always be found, even if historical is not specified. A search with the historical option is performed on the local keyword database only.

or specifies that an entry be listed if any of the words typed after search are associated with the entry. The default is to list the entry only if all the words specified are associated with the entry. A search with the or option is performed on the local keyword database only.

manual limits the search to entries in the *Stata Documentation*; that is, the search is limited to the *Stata User's Guide* and all the *Stata Reference* manuals. A search with the manual option is performed on the local keyword database only.

sj limits the search to entries in the *Stata Journal* and its predecessor, the *Stata Technical Bulletin*. A search with the sj option is performed on the local keyword database only.

## Option for set searchdefault

permanently specifies that, in addition to making the change right now, the searchdefault setting be remembered and become the default setting when you invoke Stata.

## Remarks

Remarks are presented under the headings

*Introduction*
*Internet searches*
*Author searches*
*Entry ID searches*
*Return codes*

### Introduction

See [U] **4 Stata's online help and search facilities** for a tutorial introduction to search. search is one of Stata's most useful commands. To understand the advanced features of search, you need to know how it works.

search has a database—files—containing the titles, etc., of every entry in the *Stata User's Guide*, the *Stata Base Reference Manual*, the *Stata Graphics Reference Manual*, the *Stata Longitudinal/Panel Data Reference Manual*, the *Stata Multivariate Statistics Reference Manual*, the *Stata Programming Reference Manual*, the *Stata Survey Data Reference Manual*, the *Stata Survival Analysis and Epidemiological Tables Reference Manual*, the *Stata Time-Series Reference Manual*, the *Mata Reference Manual*, the articles in the *Stata Journal* and the *Stata Technical Bulletin*, and the FAQs posted on the Stata web site. In these files is a list of words, called keywords, associated with each entry.

When you type search *xyz*, search reads the database and compares the list of keywords with *xyz*. If it finds *xyz* in the list or a keyword that allows an abbreviation of *xyz*, it displays the entry.

When you type search *xyz abc*, search does the same thing but displays an entry only if it contains both keywords. The order does not matter, so you can search linear regression or search regression linear.

Obviously, how many entries search finds depends on how the search database was constructed. We have included a plethora of keywords under the theory that, for a given request, it is better to list too much rather than risk listing nothing at all. Still, you are in the position of guessing the keywords. Do you look up normality test, normality tests, or tests of normality? Answer: normality test would be best, but all would work. In general, use the singular, and strike the unnecessary words. For guidelines for specifying keywords, see [U] **4.5 More on search**.

set searchdefault allows you to specify where search searches. set searchdefault local, the default, restricts search to using only Stata's keyword database. set searchdefault net restricts search to searching only the Internet. set searchdefault all indicates that both the keyword database and the Internet are to be searched.

## Internet searches

search with the net option searches the Internet for user-written additions to Stata, including, but not limited to, user-written additions published in the *Stata Journal* (SJ) and the *Stata Technical Bulletin* (STB). search *keywords*, net performs the same search as the command net search (with no options); see [R] **net search**.

```
. search random effect, net
Keyword search
        Keywords:  random effect
          Search:  (1) Web resources from Stata and from other users

Web resources from Stata and other users
(contacting http://www.stata.com)
63 packages found (Stata Journal and STB listed first)
------------------------------------------------------

st0046 from http://www.stata-journal.com/software/sj3-3
    SJ3-3 st0046.  From the help desk: Swamy's random ... / From the help
    desk: Swamy's random coefficients model / by Brian P. Poi, Stata
    Corporation / Support:  bpoi@stata.com / After installation, type help
    xtrchh2
st0031 from http://www.stata-journal.com/software/sj3-1
    SJ3-1 st0031.  Intraclass correlation in random-effects models ... /
    Intraclass correlation in random-effects models for binary data / by
    German Rodriguez, Princeton University / Irma Elo, University of
    Pennsylvania / Support:  grodri@princeton.edu, popelo@pop.upenn.edu /
```

```
st0012 from http://www.stata-journal.com/software/sj2-2
    SJ2-2 st0012.  Randomization-based efficacy estimator / by Ian R. White,
    MRC Biostatistics Unit, Cambridge, UK / Sarah Walker, MRC Clinical Trials
    Unit, London, UK / Abdel Babiker, MRC Clinical Trials Unit, London, UK /
    Support:  ian.white@mrc-bsu.cam.ac.uk, abdel.babiker@ctu.mrc.ac.uk /
```

*(output omitted)*

```
contents from http://www.socio.ethz.ch/people/jannb/stuff/stata
    Packages by Ben Jann, ETH Zurich, jannsoz.gess.ethz.ch / Ben Jann is
    author or coauthor of the following packages / in the SSC Archive.  /
    PACKAGES you could cmd:-ssc describe-: / net "describe
    http://fmwww.bc.edu/repec/bocode/a/alphawgt":alphawgtcol 23Cronbach's

permutation from http://www-gene.cimr.cam.ac.uk/clayton/software/stata
    Routines for generalized permutation testing / David Clayton / / Diabetes
    and Inflammation Laboratory Tel: 44 (0)1223 762669 / Cambridge Institute
    for Medical Research Fax: 44 (0)1223 762102 / Wellcome Trust/MRC Building
    david.claytoncimr.cam.ac.uk / Addenbrooke's Hospital, Cambridge, CB2 2XY

hammersley from http://www.sun.rhbnc.ac.uk/ uhss021/stata
    hammersley.  Hammersley sequence (nonrandom sampling). / Program by
    Kenneth L. Simons. / hammersley creates hammersley sequence points, which
    can be used as an / alternative to random numbers for simulation, or as an
    alternative to grid / search to explore a parameter space.  Hammersley

rndallo from http://biostat-resources.com/stata
    rndallo.  Randomly allocate subjects to treatment groups / Program by
    Mario A. Cleves / rndallo generates a list of randomly assign subjects to
    treatment / groups using permuted blocks of fix size.
```

*(output omitted)*

```
25 references found in tables of contents
-----------------------------------------

http://www.stata-journal.com/software/sj3-3/
    Stata Journal volume 3, issue 3 / Lean mainstream schemes for Stata 8
    graphics / B-splines and splines parameterized by their values / at
    reference points on the x-axis / somersd -- Confidence intervals for
    nonparametric / statistics and their differences / Robust confidence

http://www.stata-journal.com/software/sj3-1/
    Stata Journal volume 3, issue 1 / Instrumental variables and GMM:
    Estimation and testing / Intraclass correlation in random-effects models
    for / binary data / Sample size calculations for main effects and /
    interactions in case-control studies using Stata's nchi2 / and npnchi2
```

*(output omitted)*

```
http://fmwww.bc.edu/RePEc/bocode/x/
    module to make data set of summary statistics on disk or in memory /
    module to make a data set of frequencies and percents on disk or in memory
    / module to calculate and graph cross-correlation function / module to
    produce tabulation using categories defined by fractions of a cut-off
(end of search)
```

## Author searches

`search` ordinarily compares the words following `search` with the keywords for the entry. If you specify the `author` option, however, it compares the words with the author's name. In the search database, we have filled in author names for all SJ and STB inserts.

For instance, in [R] **kdensity** in this manual you will discover that Isaías H. Salgado-Ugarte wrote the first version of Stata's `kdensity` command and published it in the STB. Assume that you read his original insert and found the discussion useful. You might now wonder what else he has written in the SJ or STB. To find out, you type

```
. search Salgado-Ugarte, author
  (output omitted)
```

Names like Salgado-Ugarte are confusing to many people. `search` does not require you to specify the entire name; what you type is compared with each "word" of the name and, if any part matches, the entry is listed. The dash is a special character, and you can omit it. Thus you can obtain the same list by looking up Salgado, Ugarte, or Salgado Ugarte without the dash.

Actually, to find all entries written by Salgado-Ugarte, you need to type

```
. search Salgado-Ugarte, author historical
  (output omitted)
```

Prior inserts in the SJ or STB that provide a feature that later was superseded by a built-in feature of Stata are marked as historical in the search database and, by default, are not listed. The `historical` option ensures that all entries are listed.

## Entry ID searches

If you specify the `entry` option, `search` compares what you have typed with the entry ID. The entry ID is not the title—it is the reference listed to the left of the title that tells you where to look. For instance, in

```
[R]      regress . . . . . . . . . . . . . . . . . . . . . . . Linear regression
         (help regress)
```

[R] **regress** is the entry ID. This is a reference, of course, to this manual. In

```
FAQ        . . . . . . . . . . . . Analysis of multiple failure-time survival data
           . . . . . . . . . . . . . . . . . . . . . . . . . . . . . . . . M. Cleves
           11/99   How do I analyze multiple failure-time data using Stata?
                   http://www.stata.com/support/faqs/stat/stmfail.html
```

"FAQ" is the entry ID. In

```
STB-40   dm51 . . . . . . . . . . . . . Defining and recording variable orderings
         (help vorder if installed) . . . . . . . . . . . . . . . . John R. Gleason
         11/97   pp.10--12; STB Reprints Vol 7, pp.49--52
         tool for managing varlists to be processed by order
```

"STB-40" is the entry ID.

`search` with the `entry` option searches these entry ids.

Thus you could generate a table of contents for the *User's Guide* by typing

```
. search [U], entry
  (output omitted)
```

You could generate a table of contents for *Stata Journal*, Volume 1, Issue 1, by typing

```
. search sj-1-1, entry
  (output omitted)
```

To generate a table of contents for the 26th issue of the STB, you would type

```
. search STB-26, entry historical
  (output omitted)
```

The `historical` option in this case is possibly important. STB-26 was published in July 1995, and perhaps some of its inserts have already been marked historical.

You could obtain a complete list of all inserts associated with sg53 by typing

```
. search sg53, entry historical
  (output omitted )
```

Again we include the `historical` option in case any of the relevant inserts have been marked historical.

### Return codes

In addition to indexing the entries in the *User's Guide* and all the *Reference* manuals, `search` also can be used to search return codes.

To see information on return code 131, type

```
. search rc 131
[R]     . . . . . . . . . . . . . . . . . . . . . . . . . . Return code 131
        not possible with test;
        You requested a test of a hypothesis that is nonlinear in the
        variables.  test tests only linear hypotheses.  Use testnl.
```

If you want a list of all Stata return codes, type

```
. search error, entry
  (output omitted )
```

## Acknowledgment

`findit` grew out of a suggestion by Nicholas J. Cox, University of Durham.

## Also See

**Complementary:** [R] **help**, [R] **net search**

**Background:** [U] **4 Stata's online help and search facilities**

# Title

**serrbar** — Graph standard error bar chart

# Syntax

`serrbar` *mvar svar xvar* [*if*] [*in*] [, *options*]

| *options* | description |
|---|---|
| Main | |
| `scale(#)` | scale length of graph bars; default is `scale(1)` |
| Error bars | |
| *rcap_options* | affect rendition of capped spikes |
| Plotted points | |
| `mvopts(`*scatter_options*`)` | affect rendition of plotted points |
| Add plot | |
| `addplot(`*plot*`)` | add other plots to generated graph |
| Y-Axis, X-Axis, Title, Caption, Legend, Overall | |
| *twoway_options* | any options other than `by()` documented in [G] ***twoway_options*** |

# Description

`serrbar` graphs *mvar* ± `scale()` × *svar* against *xvar*. Usually, but not necessarily, *mvar* and *svar* will contain means and standard errors or standard deviations of some variable so that a standard error bar chart is produced.

# Options

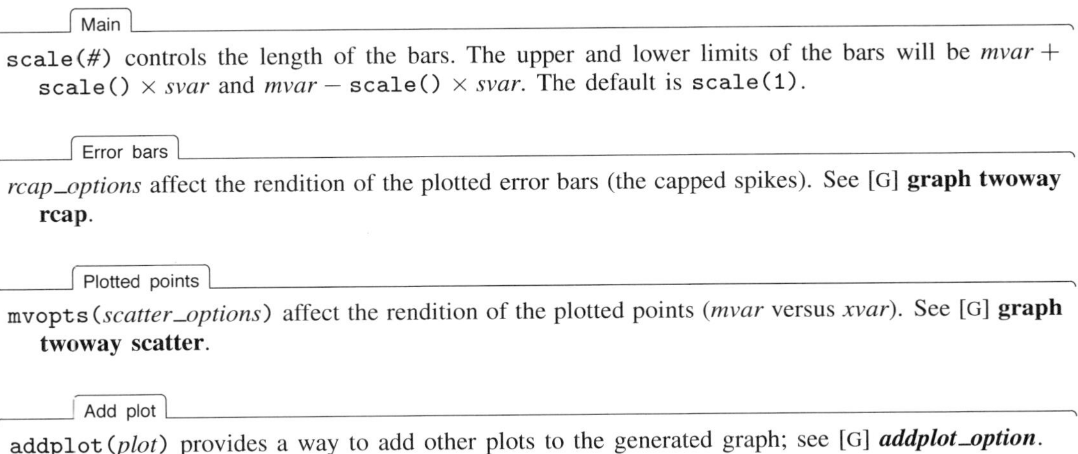

Main

`scale(#)` controls the length of the bars. The upper and lower limits of the bars will be *mvar* + `scale()` × *svar* and *mvar* − `scale()` × *svar*. The default is `scale(1)`.

Error bars

*rcap_options* affect the rendition of the plotted error bars (the capped spikes). See [G] **graph twoway rcap**.

Plotted points

`mvopts(`*scatter_options*`)` affect the rendition of the plotted points (*mvar* versus *xvar*). See [G] **graph twoway scatter**.

Add plot

`addplot(`*plot*`)` provides a way to add other plots to the generated graph; see [G] ***addplot_option***.

Y-Axis, X-Axis, Title, Caption, Legend, Overall

*twoway_options* are any of the options documented in [G] ***twoway_options***, excluding `by()`. These include options for titling the graph (see [G] ***title_options***) and options for saving the graph to disk (see [G] ***saving_option***).

# Remarks

## ▷ Example 1

In quality-control applications, the three most commonly used variables with this command are the process mean, process standard deviation, and time. For instance, we have data on the average weights and standard deviations from an assembly line in San Francisco for the period January 8 to January 16. Our data are

```
. use http://www.stata-press.com/data/r9/assembly
. list, sep(0) divider
```

| | date | mean | std |
|---|---|---|---|
| 1. | 108 | 192.22 | 3.94 |
| 2. | 109 | 192.64 | 2.83 |
| 3. | 110 | 192.37 | 4.58 |
| 4. | 113 | 194.76 | 3.25 |
| 5. | 114 | 192.69 | 2.89 |
| 6. | 115 | 195.02 | 1.73 |
| 7. | 116 | 193.40 | 2.62 |

We type `serrbar mean std date, scale(2)` but, after seeing the result, decide to make it fancier:

```
. serrbar mean std date, scale(2) title("Observed Weight Variation")
> sub("San Francisco plant, 1/8 to 1/16") yline(195) yaxis(1 2)
> ylab(195, axis(2)) ytitle("", axis(2))
```

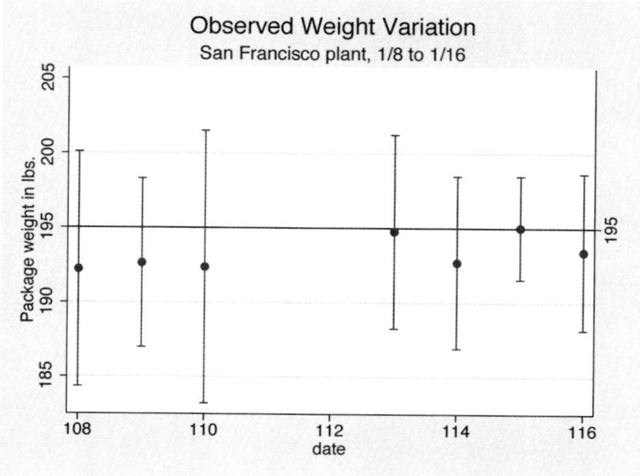


◁

## Methods and Formulas

serrbar is implemented as an ado-file.

## Acknowledgment

serrbar was written by Nicholas J. Cox of the University of Durham.

## Also See

**Related:** [R] **qc**

**Complementary:** *Stata Graphics Reference Manual*

## Title

**set** — Overview of system parameters

## Syntax

`set` [ *setcommand* ... ]

`set` typed without arguments is equivalent to `query` typed without arguments.

## Description

This entry provides a reference to Stata's `set` commands. For many entries, more thorough information is provided elsewhere; see the Reference field in each entry below for the location of this information.

To reset system parameters to factory defaults, see [R] **set_defaults**.

## Remarks

`set adosize`

Syntax: `set adosize` # [ , `permanently` ]
Default: 500
Description: sets the maximum amount of memory that automatically loaded do-files may consume. $10 \le \# \le 1000$.
Reference: [P] **sysdir**

`set checksum`

Syntax: `set checksum` { `on` | `off` } [ , `permanently` ]
Default: `off`
Description: determines whether files should be prevented from being downloaded from the Internet if checksums do not match.
Reference: [D] **checksum**

(*Continued on next page*)

set conren (Unix console only)

Syntax 1: set conren

Syntax 2: set conren clear

Syntax 3: set conren [sf | bf | it] {result | [txt | text] | input | error | link | hilite} [char[char...]]

Syntax 4: set conren {ulon | uloff} [char [char ...]]

Syntax 5: set conren reset [char [char ...]]

Description: this can possibly make the output on your screen appear prettier.
set conren displays a list of the currently defined display codes.
set conren clear clears all codes.
set conren followed by a font type (bf, sf, or it) and display context (result, error, link or hilite) and then followed by a series of space-separated characters sets the code for the specified font type and display context. If the font type is omitted, the code is set to the same specified code for all three font types.
set conren ulon and set conren uloff set the codes for turning on and off underlining.
set conren reset sets the code that will turn off all display and underlining codes.

Reference: [GSU] **conren**

set copycolor (Macintosh and Windows only)

Syntax: set copycolor {automatic|asis|gs1|gs2|gs3} [, permanently]

Default: automatic

Description: determines how colors are handled when graphs are copied to the clipboard.

Reference: [G] **set printcolor**

set dockable (Windows only)

Syntax: set dockable {on|off} [, permanently]

Default: on

Description: determines whether to enable the use of dockable window characteristics, including the ability to dock, pin, or tab a window into another window.

set dockingguides (Windows only)

Syntax: set dockingguides {on|off} [, permanently]

Default: on

Description: determines whether to enable the use of dockable guides when repositioning a dockable window.

set dp

Syntax: set dp {comma|period} [, permanently]

Default: period

Description: determines whether a period or a comma is to be used as the decimal point.

Reference: [D] **format**

set eolchar (Macintosh only)

Syntax: set eolchar {mac|unix} [, permanently]

Default: mac

Description: sets the default end-of-line delimiter for text files created in Stata.

set fastscroll
Syntax: set fastscroll {on | off} [, permanently]
Default: on
Description: sets the scrolling method for new output in the Results window. Setting fastscroll to on is faster but can be jumpy. Setting fastscroll to off is slower but smoother.

set graphics
Syntax: set graphics {on | off}
Default: on; default is off for console Stata
Description: determines whether graphs are displayed on your monitor.
Reference: [G] **set graphics**

set httpproxy
Syntax: set httpproxy {on | off} [, init]
Default: off
Description: turns on/off the use of a proxy server. There is no permanently option because permanently is implied.
Reference: [R] **netio**

set httpproxyauth
Syntax: set httpproxyauth {on | off}
Default: off
Description: turns on/off whether or not authorization is required for the proxy server. There is no permanently option because permanently is implied.
Reference: [R] **netio**

set httpproxyhost
Syntax: set httpproxyhost ["]*name*["]
Description: sets the name of a host to be used as a proxy server. There is no permanently option because permanently is implied.
Reference: [R] **netio**

set httpproxyport
Syntax: set httpproxyport #
Default: 8080 if Stata cannot auto-detect the proper setting for your computer.
Description: sets the port number for a proxy server. There is no permanently option because permanently is implied.
Reference: [R] **netio**

set httpproxypw
Syntax: set httpproxypw ["]*password*["]
Description: sets the appropriate password. There is no permanently option because permanently is implied.
Reference: [R] **netio**

`set httpproxyuser`

Syntax: `set httpproxyuser [''']`*name*`[''']`
Description: sets the appropriate user ID. There is no `permanently` option because `permanently` is implied.
Reference: [R] **netio**

`set icmap` (Macintosh only)

Syntax: `set icmap {on|off}`
Default: `on`
Description: turns on/off Internet Config file mapping for files created in Stata. There is no `permanently` option because `permanently` is implied.

`set level`

Syntax: `set level` # [`, permanently`]
Default: 95
Description: sets the default significance level for confidence intervals for all commands that report confidence intervals. $10.00 \leq \# \leq 99.99$, and # can have at most two digits after the decimal point.
Reference: [R] **level**

`set linegap`

Syntax: `set linegap` #
Default: 1
Description: sets the space between lines, in pixels, in the Results window. There is no `permanently` option because `permanently` is implied.

`set linesize`

Syntax: `set linesize` #
Default: 1 less than the full width of the screen
Description: sets the line width, in characters, for both the screen and the log file.
Reference: [R] **log**

`set locksplitters` (Windows only)

Syntax: `set locksplitters {on|off}` [`, permanently`]
Default: `off`
Description: determines whether splitters should be locked so that docked windows cannot be resized.

`set logtype`

Syntax: `set logtype {text|smcl}` [`, permanently`]
Default: `smcl`
Description: sets the default log filetype.
Reference: [R] **log**

*(Continued on next page)*

`set macgphengine` (Macintosh only)

Syntax: `set macgphengine {quartz|quickdraw} [, permanently]`

Default: `quartz`

Description: determines whether Quartz or QuickDraw is used to draw graphs. Quartz is the primary graphics engine for OS X and supports smooth lines, curves, and text. QuickDraw was the graphics engine for Mac OS prior to OS X and lacks many of the advanced features of Quartz. Graphs drawn by Quartz are visibly superior to those drawn by QuickDraw and closely resemble printed output.

`set matsize`

Syntax: `set matsize # [, permanently]`

Default: `400` for Stata/SE; `200` for Intercooled Stata; `40` for Small Stata

Description: sets the maximum number of variables that can be included in any estimation command. This cannot be changed in Small Stata.

Reference: [R] **matsize**

`set maxdb`

Syntax: `set maxdb # [, permanently]`

Default: `50`

Description: sets the maximum number of dialog boxes whose contents are remembered from one invocation to the next during a session.

Reference: [R] **db**

`set maxiter`

Syntax: `set maxiter # [, permanently]`

Default: `16000`

Description: sets the default maximum number of iterations for estimation commands. $0 \le \# \le 16000$.

Reference: [R] **maximize**

`set maxvar`

Syntax: `set maxvar # [, permanently]`

Default: `5000` for Stata/SE; `2048` for Intercooled Stata; `99` for Small Stata

Description: sets the maximum number of variables. This can only be changed in Stata/SE.

Reference: [D] **memory**

`set memory`

Syntax: `set memory #[b|k|m|g] [, permanently]`

Default: `10m` for Stata/SE; `1m` for Intercooled Stata; fixed data area of 99 variables on roughly 1,000 observations for Small Stata

Description: set memory allocated to Stata's data areas. In Small Stata this cannot be changed.

Reference: [D] **memory**

`set more`

Syntax: `set more {on|off} [, permanently]`

Default: `on`

Description: pause when `—more—` is displayed, continuing only when the user presses a key.

Reference: [R] **more**

`set obs`

Syntax: `set obs #`
Default: current number of observations
Description: changes the number of observations in the current dataset. # must be at least as large as the current number of observations. If there are variables in memory, the values of all new observations are set to *missing*.
Reference: [D] **obs**

`set output`

Syntax: `set output {proc|inform|error}`
Default: `proc`
Description: specifies the output to be displayed. `proc` means display all output; `inform` suppresses procedure output but displays informative messages and error messages; `error` suppresses all output except error messages. `set output` is seldom used.
Reference: [P] **quietly**

`set pagesize`

Syntax: `set pagesize #`
Default: 2 less than the physical number of lines on the screen
Description: sets the number of lines between —`more`— messages.
Reference: [R] **more**

`set piccomments` (Macintosh only)

Syntax: `set piccomments {on|off} [, permanently]`
Default: `on`
Description: determines whether PICT files use PICCOMMENTS for drawing true rotated text.

`set persistfv` (Windows only)

Syntax: `set persistfv {on|off}`
Default: `off`
Description: determines whether floating viewers are to persist between Stata sessions or when saving and loading a windowing preference.

`set persistvtopic` (Windows only)

Syntax: `set persistvtopic {on|off}`
Default: `off`
Description: determines whether viewer topics are to persist between Stata sessions or when saving and loading a windowing preference.

`set printcolor`

Syntax: `set printcolor { automatic|asis|gs1|gs2|gs3} [, permanently]`
Default: `automatic`
Description: determines how colors are handled when graphs are printed.
Reference: [G] **set printcolor**

`set reventries` (Windows only)
Syntax: `set reventries # [, permanently]`
Default: 100
Description: sets the number of scrollback lines available in the Review window.

`set revwindow` (Macintosh only)
Syntax: `set revwindow {float|nofloat} [, permanently]`
Default: `float`
Description: specifies whether the Review window should be placed behind other windows. `float` indicates that the window should stay in front. Stata must be restarted for a new setting to take effect.

`set rmsg`
Syntax: `set rmsg {on|off} [, permanently]`
Default: `off`
Description: indicates whether a return message telling the execution time is to be displayed at the completion of each command.
Reference: [P] **rmsg**

`set scheme`
Syntax: `set scheme` *schemename* `[, permanently]`
Default: `s2color`
Description: determines the overall look for graphs.
Reference: [G] **set scheme**

`set scrollbufsize`
Syntax: `set scrollbufsize #`
Default: `32000`
Description: sets the scrollback buffer size, in bytes, for the Results window; may be set between 10k and 500k.

`set searchdefault`
Syntax: `set searchdefault {local|net|all} [, permanently]`
Default: `local`
Description: sets the default behavior of the `search` command. `set searchdefault local` restricts `search` to use only Stata's keyword database. `set searchdefault net` restricts `search` to searching only the Internet. `set searchdefault all` indicates that both the keyword database and the Internet are to be searched.
Reference: [R] **search**

`set seed`
Syntax: `set seed {#|`*code*`}`
Default: `123456789`
Description: specifies initial value of the random-number seed used by the `uniform()` function.
Reference: [D] **generate**

`set smalldlg` (Windows 98/ME only)
Syntax: `set smalldlg {on|off} [, permanently]`
Default: `off`
Description: determines whether to use smaller versions of some dialog boxes–those dialog boxes with fewer controls.

`set smoothfonts` (Macintosh only)
Syntax: `set smoothfonts {on|off} [, permanently]`
Default: `on`
Description: determines whether to use the Quartz graphics engine to draw smooth text edges in all Stata windows except the Graph windows, which use the setting in `macgphengine`.

`set smoothsize` (Macintosh only)
Syntax: `set smoothsize # [, permanently]`
Default: `10`
Description: sets the minimum font size for font smoothing; see `set smoothfonts`.

`set timeout1`
Syntax: `set timeout1` *#seconds* `[, permanently]`
Default: `120`
Description: sets the number of seconds Stata will wait for a remote host to respond to an initial contact before giving up. In general, users should not modify this value unless instructed to do so by Stata Technical Services.
Reference: [R] **netio**

`set timeout2`
Syntax: `set timeout2` *#seconds* `[, permanently]`
Default: `300`
Description: sets the number of seconds Stata will keep trying to get information from a remote host after initial contact before giving up. In general, users should not modify this value unless instructed to do so by Stata Technical Services.
Reference: [R] **netio**

`set trace`
Syntax: `set trace {on|off}`
Default: `off`
Description: indicates whether to trace the execution of programs for debugging.
Reference: [P] **trace**

`set tracedepth`
Syntax: `set tracedepth #`
Default: `32000` (equivalent to $\infty$)
Description: if `trace` is set on, trace execution of programs and nested programs up to `tracedepth`. For example, if `tracedepth` is 2, the current program and any subroutine called would be traced, but subroutines of subroutines would not be traced.
Reference: [P] **trace**

`set traceexpand`

Syntax: `set traceexpand {on|off} [, permanently]`
Default: `on`
Description: if `trace` is set on, show lines both before and after macro expansion. If `traceexpand` is set off, only the line before macro expansion is shown.
Reference: [P] **trace**

`set tracehilite`

Syntax: `set tracehilite "`*pattern*`" [, word]`
Default: `""`
Description: if *pattern* is specified, it will be highlighted in the trace output.
Reference: [P] **trace**

`set traceindent`

Syntax: `set traceindent {on|off} [, permanently]`
Default: `on`
Description: if `trace` is set on, indent displayed lines according to their nesting level. The lines of the main program are not indented. Two spaces of indentation are used for each level of nested subroutine.
Reference: [P] **trace**

`set tracenumber`

Syntax: `set tracenumber {on|off} [, permanently]`
Default: `off`
Description: if `trace` is set on, show the nesting level numerically in front of the line. Lines of the main program are preceded by 01, lines of subroutines called by the main program are preceded by 02, etc.
Reference: [P] **trace**

`set tracesep`

Syntax: `set tracesep {on|off} [, permanently]`
Default: `on`
Description: if `trace` is set on, display a horizontal separator line that displays the name of the subroutine whenever a subroutine is called or exits.
Reference: [P] **trace**

`set type`

Syntax: `set type {float|double} [, permanently]`
Default: `float`
Description: specifies the default type assigned to new variables.
Reference: [D] **generate**

`set update_interval` (Macintosh and Windows only)

Syntax: `set update_interval #`
Default: 7
Description: sets the number of days to elapse before performing the next automatic `update query`.
Reference: [R] **update**

set update_prompt (Macintosh and Windows only)
Syntax: set update_prompt {on|off} [, init]
Default: on
Description: determines wheter a dialog is to be displayed before performing an automatic update query. There is no permanently option because permanently is implied.
Reference: [R] **update**

set update_query (Macintosh and Windows only)
Syntax: set update_query {on|off} [, init]
Default: on
Description: determines whether update query is to be automatically performed when Stata is launched. There is no permanently option because permanently is implied.
Reference: [R] **update**

set varabbrev
Syntax: set varabbrev {on|off} [, permanently]
Default: on
Description: indicates whether Stata should allow variable abbreviations

set varlabelpos
Syntax: set varlabelpos #
Default: 32
Description: sets the maximum number of characters to display for variable names in the Variables window.

set varwindow (Macintosh only)
Syntax: set varwindow {float|nofloat} [, permanently]
Default: float
Description: specifies whether the Variables window should be placed behind other windows. float indicates that the window should stay in front. Stata must be restarted for a new setting to take effect.

set virtual
Syntax: set virtual {on|off}
Default: off
Description: indicates whether Stata should work to arrange its memory to keep objects close together.
Reference: [D] **memory**

set xptheme (Windows only)
Syntax: set xptheme {on|off} [, permanently]
Default: on
Description: indicates whether the XP or the Classic visual style be applied to Stata. on, the Classic style, allows programmable dialogs to load faster. Stata must be restarted for a new setting to take effect.

# Also See

**Complementary:** [R] **query**, [R] **set_defaults**

**Related:** [M-3] **mata set**

# Title

**set_defaults** — Reset system parameters to original Stata defaults

# Syntax

set_defaults { *category* | _all } [ , <u>perm</u>anently ]

where *category* is one of <u>mem</u>ory | <u>out</u>put | <u>inter</u>face | <u>graph</u>ics | <u>eff</u>iciency |

<u>net</u>work | <u>up</u>date | trace | mata | <u>oth</u>er

# Description

`set_defaults` resets settings made by `set` to the original default settings that were shipped with Stata.

`set_defaults _all` resets all the categories, whereas `set defaults` *category* resets only the settings for the specified category.

# Option

`permanently` specifies that, in addition to making the change right now, the settings be remembered and become the default settings when you invoke Stata.

# Remarks

## ▷ Example 1

To assist us in debugging a new command, we modified some of the `trace` settings. To return them to their original values, we type

```
. set_defaults trace
-> set trace off
-> set tracedepth 32000
-> set traceexpand on
-> set tracesep on
-> set traceindent on
-> set tracenumber off
-> set tracehilite ""
(preferences reset)
```

◁

# Also See

**Complementary:** [R] **query**, [R] **set**

**Related:** [M-3] **mata set**

## Title

**signrank** — Equality tests on matched data

## Syntax

*Wilcoxon matched-pairs signed-ranks test*

`signrank` *varname* = *exp* [*if*] [*in*]

*Sign test of matched pairs*

`signtest` *varname* = *exp* [*if*] [*in*]

`by` may be used with `signrank` and `signtest`; see [D] **by**.

## Description

`signrank` tests the equality of matched pairs of observations using the Wilcoxon matched-pairs signed-ranks test (Wilcoxon 1945). The null hypothesis is that both distributions are the same.

`signtest` also tests the equality of matched pairs of observations (Arbuthnott 1712, but better explained by Snedecor and Cochran 1989) by calculating the differences between *varname* and the expression. The null hypothesis is that the median of the differences is zero; no further assumptions are made about the distributions. This, in turn, is equivalent to the hypothesis that the true proportion of positive (negative) signs is one-half.

For equality tests on unmatched data, see [R] **ranksum**.

## Remarks

### ▷ Example 1: signrank

We are testing the effectiveness of a new fuel additive. We run an experiment with 12 cars. We first run each car without the fuel treatment and measure the mileage. We then add the fuel treatment and repeat the experiment. The results of the experiment are

| Without Treatment | With Treatment | Without Treatment | With Treatment |
|---|---|---|---|
| 20 | 24 | 18 | 17 |
| 23 | 25 | 24 | 28 |
| 21 | 21 | 20 | 24 |
| 25 | 22 | 24 | 27 |
| 18 | 23 | 23 | 21 |
| 17 | 18 | 19 | 23 |

We create two variables called `mpg1` and `mpg2` representing mileage without and with the treatment, respectively. We can test the null hypothesis that the treatment had no effect by typing

```
. use http://www.stata-press.com/data/r9/fuel
. signrank mpg1=mpg2
Wilcoxon signed-rank test
        sign |      obs   sum ranks    expected
-------------+---------------------------------
    positive |        3        13.5        38.5
    negative |        8        63.5        38.5
        zero |        1           1           1
-------------+---------------------------------
         all |       12          78          78
unadjusted variance       162.50
adjustment for ties        -1.62
adjustment for zeros       -0.25
                        ----------
adjusted variance         160.62
Ho: mpg1 = mpg2
             z =  -1.973
    Prob > |z| =   0.0485
```

The output indicates that we can reject the null hypothesis at any level above 4.85%. ◁

### ▷ Example 2: signtest

**signtest** tests that the median of the differences is zero, making no further assumptions, whereas **signrank** assumed that the distributions are equal as well. Using the data above,

```
. signtest mpg1=mpg2
Sign test
        sign |    observed    expected
-------------+------------------------
    positive |           3         5.5
    negative |           8         5.5
        zero |           1           1
-------------+------------------------
         all |          12          12
One-sided tests:
  Ho: median of mpg1 - mpg2 = 0 vs.
  Ha: median of mpg1 - mpg2 > 0
      Pr(#positive >= 3) =
         Binomial(n = 11, x >= 3, p = 0.5) =  0.9673
  Ho: median of mpg1 - mpg2 = 0 vs.
  Ha: median of mpg1 - mpg2 < 0
      Pr(#negative >= 8) =
         Binomial(n = 11, x >= 8, p = 0.5) =  0.1133
Two-sided test:
  Ho: median of mpg1 - mpg2 = 0 vs.
  Ha: median of mpg1 - mpg2 != 0
      Pr(#positive >= 8 or #negative >= 8) =
         min(1, 2*Binomial(n = 11, x >= 8, p = 0.5)) =  0.2266
```

The summary table indicates that there were 3 comparisons for which `mpg1` exceeded `mpg2`, 8 comparisons for which `mpg2` exceeded `mpg1`, and one comparison for which they were the same.

The output below the summary table is based on the binomial distribution. The significance of the one-sided test, where the alternative hypothesis is that the median of `mpg2` − `mpg1` is greater than zero, is 0.1133. The significance of the two-sided test, where the alternative hypothesis is simply that the median of the differences is different from zero, is $0.2266 = 2 \times 0.1133$. ◁

## Saved Results

`signrank` saves in `r()`:

Scalars

| | | | |
|---|---|---|---|
| `r(N_neg)` | number of negative comparisons | `r(sum_neg)` | sum of the negative ranks |
| `r(N_pos)` | number of positive comparisons | `r(z)` | $z$ statistic |
| `r(N_tie)` | number of tied comparisons | `r(Var_a)` | adjusted variance |
| `r(sum_pos)` | sum of the positive ranks | | |

`signtest` saves in `r()`:

Scalars

| | | | |
|---|---|---|---|
| `r(N_neg)` | number of negative comparisons | `r(p_2)` | two-sided probability |
| `r(N_pos)` | number of positive comparisons | `r(p_neg)` | one-sided probability of negative comparison |
| `r(N_tie)` | number of tied comparisons | `r(p_pos)` | one-sided probability of positive comparison |

## Methods and Formulas

`signrank` and `signtest` are implemented as ado-files.

For a practical introduction to these techniques with an emphasis on examples rather than theory, see Bland (2000) or Sprent (1993). For a summary of these tests, see Snedecor and Cochran (1989).

### signrank

Both the sign test and Wilcoxon signed-rank tests test the null hypothesis that the distribution of a random variable $D = \textit{varname} - \textit{exp}$ has median zero. The sign test makes no additional assumptions, but the Wilcoxon signed-rank test makes the additional assumption that the distribution of $D$ is symmetric. If $D = X_1 - X_2$, where $X_1$ and $X_2$ have the same distribution, then it follows that the distribution of $D$ is symmetric about zero. Thus the Wilcoxon signed-rank test is often described as a test of the hypothesis that two distributions are the same, i.e., $X_1 \sim X_2$.

Let $d_j$ denote the difference for any matched pair of observations,

$$d_j = x_{1j} - x_{2j} = \textit{varname} - \textit{exp}$$

for $j = 1, 2, \ldots, n$.

Rank the absolute values of the differences $|d_j|$, and assign any tied values the average rank. Consider the signs of $d_j$, and let

$$r_j = \text{sign}(d_j)\,\text{rank}(|d_j|)$$

be the signed ranks. The test statistic is

$$T_{\text{obs}} = \sum_{j=1}^{n} r_j = (\text{sum of ranks for + signs}) - (\text{sum of ranks for } - \text{ signs})$$

The null hypothesis is that the distribution of $d_j$ is symmetric about 0. Hence the likelihood is unchanged if we flip signs on the $d_j$, and thus the randomization datasets are the $2^n$ possible sign changes for the $d_j$. Thus the randomization distribution of our test statistic $T$ can be computed by considering all the $2^n$ possible values of

$$T = \sum_{j=1}^{n} S_j r_j$$

where the $r_j$ are the observed signed-ranks (considered fixed) and $S_j$ is either $+1$ or $-1$.

With this distribution, the mean and variance of $T$ are given by

$$E(T) = 0 \qquad \text{and} \qquad \text{Var}_{\text{adj}}(T) = \sum_{j=1}^{n} r_j^2$$

Note that the test statistic for the Wilcoxon signed-rank test is often expressed (equivalently) as the sum of the positive signed-ranks, $T_+$, where

$$E(T_+) = \frac{n(n+1)}{4} \qquad \text{and} \qquad \text{Var}_{\text{adj}}(T_+) = \frac{1}{4}\sum_{j=1}^{n} r_j^2$$

Zeros and ties do not affect the theory above, and the exact variance is still given by the above formula for $\text{Var}_{\text{adj}}(T_+)$. When $d_j = 0$ is observed, $d_j$ will always be zero in each of the randomization datasets (using $\text{sign}(0) = 0$). When there are ties, you can assign averaged ranks for each group of ties and then treat them the same as the other ranks.

The "unadjusted variance" reported by `signrank` is the variance that the randomization distribution would have had if there had been no ties or zeros:

$$\text{Var}_{\text{unadj}}(T_+) = \frac{1}{4}\sum_{j=1}^{n} j^2 = \frac{n(n+1)(2n+1)}{24}$$

The adjustment for zeros is the change in the variance when the ranks for the zeros are signed to make $r_j = 0$,

$$\Delta\text{Var}_{\text{zero adj}}(T_+) = -\frac{1}{4}\sum_{j=1}^{n_0} j^2 = -\frac{n_0(n_0+1)(2n_0+1)}{24}$$

where $n_0$ is the number of zeros. The adjustment for ties is the change in the variance when the ranks (for nonzero observations) are replaced by averaged ranks:

$$\Delta\text{Var}_{\text{ties adj}}(T_+) = \text{Var}_{\text{adj}}(T_+) - \text{Var}_{\text{unadj}}(T_+) - \Delta\text{Var}_{\text{zero adj}}(T_+)$$

A normal approximation is used to calculate

$$z = \frac{T_+ - E(T_+)}{\sqrt{\text{Var}_{\text{adj}}(T_+)}}$$

## signtest

The test statistic for the sign test is the number $n_+$ of differences

$$d_j = x_{1j} - x_{2j} = \mathit{varname} - \mathit{exp}$$

greater than zero. Assuming that the probability of a difference being equal to zero is exactly zero, then, under the null hypothesis, $n_+ \sim \text{Binomial}(n, p = \frac{1}{2})$, where $n$ is the total number of observations.

But what if some differences are zero? This question has a ready answer if you view the test from the perspective of Fisher's Principle of Randomization (Fisher 1935). Fisher's idea (stated in a modern way) was to look at a family of transformations of the observed data such that the *a priori* likelihood (under the null hypothesis) of the transformed data is the same as the likelihood of the observed data. The distribution of the test statistic is then produced by calculating its value for each of the transformed "randomization" datasets, assuming that each dataset is equally likely.

For the sign test, the "data" are simply the set of signs of the differences. Under the null hypothesis of the sign test, the probability that $d_j$ is less than zero is equal to the probability that $d_j$ is greater than zero. Thus you can transform the observed signs by flipping any number of them, and the set of signs will have the same likelihood. The $2^n$ possible sign changes form the family of randomization datasets. If you have no zeros, this procedure again leads to $n_+ \sim \text{Binomial}(n, p = \frac{1}{2})$.

If you do have zeros, changing their signs leaves them as zeros. So, if you observe $n_0$ zeros, each of the $2^n$ sign-change datasets will also have $n_0$ zeros. Hence, the values of $n_+$ calculated over the sign-change datasets range from 0 to $n - n_0$, and the "randomization" distribution of $n_+$ is $\text{Binomial}(n - n_0, p = \frac{1}{2})$.

The work of Arbuthnott (1712) and later 18th-century contributions is discussed by Hald (1990, chapter 17).

Frank Wilcoxon (1892–1965) was born in Ireland to American parents. After working in various occupations (including merchant seaman, oil-well pump attendant and tree surgeon), he settled in chemistry, gaining degrees from Rutgers and Cornell and employment from various companies. Working mainly on the development of fungicides and insecticides, Wilcoxon became interested in statistics in 1925 and made several key contributions to nonparametric methods. After retiring from industry, he taught statistics at Florida State until his death.

## References

Arbuthnott, J. 1712. An argument for divine providence, taken from the constant regularity observ'd in the births of both sexes. *Philosophical Transactions* 27: 186–190.

Bland, M. 2000. *An Introduction to Medical Statistics*. 3rd ed. Oxford: Oxford University Press.

Bradley, R. A. 2001. Frank Wilcoxon. In *Statisticians of the Centuries*, ed. C. C. Heyde and E. Seneta, 420–424. New York: Springer.

Fisher, R. A. 1935. *Design of Experiments*. Edinburgh: Oliver & Boyd.

Hald, A. 1990. *A History of Probability and Statistics and Their Applications before 1750*. New York: Wiley.

Snedecor, G. W. and W. G. Cochran. 1989. *Statistical Methods*. 8th ed. Ames, IA: Iowa State University Press.

Sprent, P. and N. C. Smeeton. 2001. *Applied Nonparametric Statistical Methods*. 3rd ed. Boca Raton, FL: Chapman & Hall/CRC.

Sribney, W. M. 1995. crc40: Correcting for ties and zeros in sign and rank tests. *Stata Technical Bulletin* 26: 2–4. Reprinted in *Stata Technical Bulletin Reprints*, vol. 5, pp. 5–8.

Wilcoxon, F. 1945. Individual comparisons by ranking methods. *Biometrics* 1: 80–83.

## Also See

**Related:** [R] **kwallis**, [R] **nptrend**, [R] **ranksum**, [R] **runtest**, [R] **ttest**

## Title

**simulate** — Monte Carlo simulations

## Syntax

simulate [*exp_list*] , reps(*#*) [*options*] : *command*

| *options* | description |
|---|---|
| nodots | suppress the replication dots |
| noisily | display any output from *command* |
| trace | trace the *command* |
| saving(*filename*, ...) | save results to *filename* |
| nolegend | suppress the table legend |
| verbose | display the full table legend |
| seed(*#*) | set random-number seed to *#* |

fweights, aweights, iweights, and pweights are allowed in *command*.

| | |
|---|---|
| *exp_list* contains | (*name*: *elist*) |
| | *elist* |
| | *eexp* |
| *elist* contains | *newvar* = (*exp*) |
| | (*exp*) |
| *eexp* is | *specname* |
| | [*eqno*]*specname* |
| *specname* is | _b |
| | _b[] |
| | _se |
| | _se[] |
| *eqno* is | ## |
| | *name* |

*exp* is a standard Stata expression; see [U] **13 Functions and expressions**.

Distinguish between [], which are to be typed, and [ ], which indicate optional arguments.

## Description

simulate eases the programming task of performing Monte Carlo-type simulations. Typing

```
. simulate exp_list, reps(#): command
```

runs *command* for # replications and collects the results in *exp_list*.

*command* defines the command that performs a single simulation. Most Stata commands and user-written programs can be used with simulate, as long as they follow standard Stata syntax; see [U] **11 Language syntax**. The by prefix may not be part of *command*.

*exp_list* specifies the expression to be calculated from the execution of *command*. If no expressions are given, *exp_list* assumes a default, depending upon whether *command* changes results in `e()` or `r()`. If *command* changes results in `e()`, the default is `_b`. If *command* changes results in `r()` (but not `e()`), the default is all the scalars posted to `r()`. It is an error not to specify an expression in *exp_list* otherwise.

## Options

`reps(`*#*`)` is required—it specifies the number of replications to be performed.

`nodots` suppresses display of the replication dots. By default, a single dot character is displayed for each successful replication. A single red 'x' is displayed if *command* returns an error or if one of the values in *exp_list* is missing.

`noisily` requests that any output from *command* be displayed. This option implies the `nodots` option.

`trace` causes a trace of the execution of *command* to be displayed. This option implies the `noisily` option.

`saving(`*filename* [`,` *suboptions* ]`)` creates a Stata data file (`.dta` file) consisting of, for each statistic in *exp_list*, a variable containing the simulated values.

`double` specifies that the results for each replication be stored as `doubles`, meaning 8-byte reals. By default, they are stored as `floats`, meaning 4-byte reals.

`every(`*#*`)` specifies that results be written to disk every #th replication. `every()` should only be specified in conjunction with `saving()` when *command* takes a long time for each replication. This will allow recovery of partial results should some other software crash your computer. See [P] **postfile**.

`replace` specifies that *filename* be overwritten if it exists.

`nolegend` suppresses display of the table legend. The table legend identifies the rows of the table with the expressions they represent.

`verbose` requests that the full table legend be displayed. By default, coefficients and standard errors are not displayed.

`seed(`*#*`)` sets the random-number seed. Specifying this option is equivalent to typing the following command before calling `simulate`:

```
. set seed #
```

## Remarks

For an introduction to Monte Carlo methods, see Johnston and DiNardo (1997, 348–358).

### ▷ Example 1

We have a dataset containing means and variances of 100-observation samples from a lognormal distribution (as a first step in evaluating, say, the coverage of a 95%, $t$-based confidence interval). Then we perform the experiment 1,000 times.

The following command definition will generate 100 independent observations from a lognormal distribution and compute the summary statistics for this sample.

```
program lnsim, rclass
        version 9
        drop _all
        set obs 100
        gen z = exp(invnormal(uniform()))
        summarize z
        return scalar mean = r(mean)
        return scalar Var  = r(Var)
end
```

We can save 1,000 simulated means and variances from `lnsim` by typing

```
. set seed 1234
. simulate mean=r(mean) var=r(Var), reps(1000) nodots: lnsim
      command:  lnsim
         mean:  r(mean)
          var:  r(Var)
. describe *
              storage  display     value
variable name   type   format      label      variable label
-------------------------------------------------------------------------------
mean            float  %9.0g                  r(mean)
var             float  %9.0g                  r(Var)
. summarize
    Variable |       Obs        Mean    Std. Dev.       Min        Max
-------------+--------------------------------------------------------
        mean |      1000    1.641109    .2230346   1.053379    2.47622
         var |      1000    4.794263    4.597844   .8202238   50.07779
```

◁

## ❑ Technical Note

Before executing our `lnsim` simulator, we can verify that it works by executing it interactively.

```
. set seed 1234
. lnsim
obs was 0, now 100
    Variable |       Obs        Mean    Std. Dev.       Min        Max
-------------+--------------------------------------------------------
           z |       100    1.581671    1.908433   .0338387   11.18932
. return list
scalars:
                 r(Var) =  3.642116429498976
                r(mean) =  1.581670569628477
```

❑

## ▷ Example 2

Consider a more complicated problem. Let's experiment with fitting $y_j = a + bx_j + u_j$ when the true model has $a = 1$, $b = 2$, $u_j = z_j + cx_j$, and when $z_j$ is $\mathrm{N}(0,1)$. We will save the parameter estimates and standard errors and experiment with varying $c$. $x_j$ will be fixed across experiments but will originally be generated as $\mathrm{N}(0,1)$. We begin by interactively making the true data:

```
. drop _all
. set obs 100
obs was 0, now 100
```

```
. set seed 54321
. gen x = invnormal(uniform())
. gen true_y = 1+2*x
. save truth
file truth.dta saved
```

Our program is

```
program hetero1
        version 9
        args c
        use truth, clear
        gen y = true_y + (invnormal(uniform()) + `c'*x)
        regress y x
end
```

Note the use of `` `c' `` in our statement for generating y. c is a local macro generated from args c and thus refers to the first argument supplied to hetero1. If we want $c = 3$ for our experiment, we type

```
. simulate _b _se, reps(10000): hetero1 3
  (output omitted)
```

Our program hetero1 could, however, be more efficient because it rereads the file truth once every replication. It would be better if we could read the data just once. In fact, if we read in the data right before running simulate, we really shouldn't have to reread for each subsequent replication. A faster version reads

```
program hetero2
        version 9
        args c
        capture drop y
        gen y = true_y + (invnormal(uniform()) + `c'*x)
        regress y x
end
```

Requiring that the current dataset has the variables true_y and x may become inconvenient. Another improvement would be to require that the user supply variable names, such as in

```
program hetero3
        version 9
        args truey x c
        capture drop y
        gen y = `truey' + (invnormal(uniform()) + `c'*`x')
        regress y x
end
```

Thus we can type

```
. simulate _b _se, reps(10000): hetero3 true_y x 3
  (output omitted)
```

◁

## ▷ Example 3

Now let's consider the problem of simulating the ratio of two medians. Suppose that each sample of size $n_i$ comes from a normal population with a mean $\mu_i$ and standard deviation $\sigma_i$, where $i = 1, 2$. We write the program below and save it as a text file called myratio.ado (see [U] **17 Ado-files**). Our program is an rclass command that requires six arguments as input, identified by the local macros n1, mu1, sigma1, n2, mu2, sigma2, which correspond to $n_1$, $\mu_1$, $\sigma_1$, $n_2$, $\mu_2$, $\sigma_2$, respectively. With these arguments, myratio will generate the data for the two samples, use summarize to compute the two medians, and save the ratio of the medians in r(ratio).

```
program myratio, rclass
        version 9
        args n1 mu1 sigma1 n2 mu2 sigma2
        // generate the data
        drop _all
        local N = `n1'+`n2'
        set obs `N'
        tempvar y
        generate `y' = invnormal(uniform())
        replace `y' = cond(_n<=`n1',`mu1'+`y'*`sigma1',`mu2'+`y'*`sigma2')
        // calculate the medians
        tempname m1
        summarize `y' if _n<=`n1', detail
        scalar `m1' = r(p50)
        summarize `y' if _n>`n1', detail
        // save the results
        return scalar ratio = `m1' / r(p50)
end
```

The result of running our simulation is

```
. set seed 19192

. simulate ratio=r(ratio), reps(1000) nodots: myratio 5 3 1 10 3 2

      command:  myratio 5 3 1 10 3 2
        ratio:  r(ratio)

. summarize

    Variable |       Obs        Mean    Std. Dev.       Min        Max
-------------+--------------------------------------------------------
       ratio |      1000    1.085591    .4203416   .3174837   5.936007
```

◁

## ❑ Technical Note

Stata lets us do simulations of simulations and simulations of bootstraps. Stata's **bootstrap** command (see [R] **bootstrap**) works much like **simulate**, except that it feeds the user-written program a bootstrap sample. Say that we want to evaluate the bootstrap estimator of the standard error of the median when applied to lognormally distributed data. We want to perform a simulation, resulting in a dataset of medians and bootstrap estimated standard errors.

As background, **summarize** calculates summary statistics, leaving the mean in **r(mean)** and the variance in **r(Var)**. **summarize** with the **detail** option also calculates summary statistics, but more of them, and leaves the median in **r(p50)**.

Thus our plan is to perform simulations by randomly drawing a dataset: we calculate the median of our random sample, we use **bootstrap** to obtain a dataset of medians calculated from bootstrap samples of our random sample, the standard deviation of those medians is our estimate of the standard error, and the summary statistics are saved in the results of **summarize**.

Our simulator is

*(Continued on next page)*

```
program define bsse, rclass
        version 9
        drop _all
        set obs 100
        gen x = invnormal(uniform())
        tempfile bsfile
        bootstrap midp=r(p50), rep(100) saving(`bsfile'): summarize x, detail
        use `bsfile', clear
        summarize midp
        return scalar mean = r(mean)
        return scalar sd   = r(sd)
end
```

We can obtain final results, running our simulation 1,000 times, by typing

```
. set seed 48901
. simulate med=r(mean) bs_se=r(sd), reps(1000): bsse
     command:  bsse
         med:  r(mean)
       bs_se:  r(sd)
Simulations (1000)
────┼─── 1 ───┼─── 2 ───┼─── 3 ───┼─── 4 ───┼─── 5
..................................................    50
..................................................   100
..................................................   150
..................................................   200
  (output omitted)
..................................................   850
..................................................   900
..................................................   950
..................................................  1000
. summarize
    Variable |       Obs        Mean    Std. Dev.       Min        Max
─────────────┼────────────────────────────────────────────────────────
         med |      1000    .0002238    .1175051  -.3625843   .3817343
       bs_se |      1000    .1252704    .0294073   .0552211   .2785387
```

This is a case where the simulation dots (drawn by default, unless the `nodots` option is specified) will give us an idea of how long this simulation will take to finish as it runs. ❑

# Methods and Formulas

`simulate` is implemented as an ado-file.

# References

Gould, W. W. 1994. ssi6.1: Simplified Monte Carlo simulations. *Stata Technical Bulletin* 20: 22–24. Reprinted in *Stata Technical Bulletin Reprints*, vol. 4, pp. 207–210.

Hamilton, L. C. 2004. *Statistics with Stata*. Belmont, CA: Brooks/Cole.

Johnston, J. and J. DiNardo. 1997. *Econometric Methods*. 4th ed. New York: McGraw–Hill.

Weesie, J. 1998. ip25: Parameterized Monte Carlo simulations: An enhancement to the simulation command. *Stata Technical Bulletin* 43: 13–15. Reprinted in *Stata Technical Bulletin Reprints*, vol. 8, pp. 75–77.

## Also See

**Complementary:** [P] **postfile**

**Related:** [R] **bootstrap**, [R] **jackknife**, [R] **permute**,
[D] **statsby**

## Title

**sj** — Stata Journal and STB installation instructions

## Description

The *Stata Journal* (SJ) is a quarterly journal containing articles about statistics, data analysis, teaching methods, and effective use of Stata's language. The SJ publishes reviewed papers together with shorter notes and comments, regular columns, tips, book reviews, and other material of interest to researchers applying statistics in a variety of disciplines. You can read all about the *Stata Journal* at *http://www.stata-journal.com.*

The *Stata Journal* is a printed and electronic journal with corresponding software. If you want the journal, you must subscribe, but the software is available for free from our web site at *http://www.stata-journal.com.*

The predecessor to the *Stata Journal* was the *Stata Technical Bulletin* (STB). The STB was also a printed and electronic journal with corresponding software. Individual STB journals may still be purchased. The STB software is available for free from our web site at *http://www.stata.com.*

Below are instructions for installing the *Stata Journal* and the *Stata Technical Bulletin* software from our web site.

## Remarks

Remarks are presented under the headings



### Installing the Stata Journal software

Each issue of the *Stata Journal* is labeled Volume #, Number #. Volume 1 refers to the first year of publication, Volume 2 to the second, and so on. Issues are numbered 1, 2, 3, and 4 within each year. The first issue of the *Journal* was published fourth quarter, 2001, and that issue is numbered Volume 1, Number 1. For installation purposes, we refer to this issue as `sj1-1`.

The articles, columns, notes, and comments that make up the *Stata Journal* are assigned a letter-and-number code, called an insert tag, such as st0001, an0034, or ds0011. The letters represent a category: st is the statistics category, an is the announcements category, etc. The numbers are assigned sequentially, so st0001 is the first article in the statistics category.

Sometimes inserts are subsequently updated, either to fix bugs or to add new features. A number such as st0001_1 indicates that this article, column, note, or comment is an update to the original st0001 article. Updates are complete; that is, installing st0001_1 provides all the features of the original article and more.

The *Stata Journal* software may be obtained by pointing and clicking or by using command mode.

The sections below detail how to install an insert. In all cases, pretend that you wish to install insert st0001_1 from sj2-2.

### Obtaining from the Internet by pointing and clicking

1. Select **Help > SJ and User-written Programs**.
2. Click on *Stata Journal*.
3. Click on *sj2-2*.
4. Click on *st0001_1*.
5. Click on *(click here to install)*.

### Obtaining from the Internet via command mode

Type the following:

```
. net from http://www.stata-journal.com/software
. net cd sj2-2
. net describe st0001_1
. net install st0001_1
```

The above could be shortened to

```
. net from http://www.stata-journal.com/software/sj2-2
. net describe st0001_1
. net install st0001_1
```

Alternatively, you could type

```
. net sj 2-2
. net describe st0001_1
. net install st0001_1
```

but going about it the long way around is more entertaining, at least the first time.

## Installing the STB software

Each issue of the STB is numbered. STB-1 refers to the first issue (published May 1991), STB-2 to the second (published July 1991), and so on.

An issue of the STB consists of inserts—articles—and these are assigned letter-and-number combinations, such as sg84, dm80, sbe26.1, etc. The letters represent a category: sg is the general statistics category and dm the data-management category. The numbers are assigned sequentially, so sbe39 is the 39th insert in the biostatistics and epidemiology series.

Insert sbe39, it turns out, provides a method of accounting for publication bias in meta-analysis; it adds a new command called `metatrim` to Stata. If you installed sbe39, you would have that command and its online help. Insert sbe39 was published in STB-57 (September 2000). Obtaining `metatrim` simply requires going to STB-57 and getting sbe39.

Sometimes inserts were subsequently updated, either to fix bugs or to add new features. sbe39 was updated: the first update is sbe39.1 and the second is sbe39.2. You could install insert sbe39.2, and it would not matter whether you had previously installed sbe39.1. Updates are complete: installing sbe39.2 provides all the features of the original insert and more.

For computer naming purposes, insert sbe39.2 is referred to as `sbe39_2`. When referred to in normal text, however, the insert is still called sbe39.2 because that looks nicer.

Inserts are easily available from the Internet. Inserts may be obtained by pointing and clicking or by using command mode.

The sections below detail how to install an insert. In all cases, pretend that you wish to install insert sbe39.2 from STB-61.

### Obtaining from the Internet by pointing and clicking

1. Select **Help > SJ and User-written Programs**.
2. Click on *STB*.
3. Click on *stb61*.
4. Click on *sbe39_2*.
5. Click on *(click here to install)*.

### Obtaining from the Internet via command mode

Type the following:

```
. net from http://www.stata.com
. net cd stb
. net cd stb61
. net describe sbe39_2
. net install sbe39_2
```

The above could be shortened to

```
. net from http://www.stata.com/stb/stb61
. net describe sbe39_2
. net install sbe39_2
```

but going about it the long way around is more entertaining, at least the first time.

## Also See

**Complementary:** [R] **search**

**Related:** [R] **net**, [R] **net search**, [R] **update**

**Background:** [U] **3.5 The Stata Journal and the Stata Technical Bulletin**,
[U] **28 Using the Internet to keep up to date**,
[GSM] **19 Using the Internet**,
[GSU] **19 Using the Internet**,
[GSW] **19 Using the Internet**

# Title

**sktest** — Skewness and kurtosis test for normality

# Syntax

`sktest` *varlist* [*if*] [*in*] [*weight*] [, `noadjust`]

`aweight`s and `fweight`s are allowed; see [U] **11.1.6 weight**.

# Description

For each variable in *varlist*, `sktest` presents a test for normality based on skewness and another based on kurtosis and then combines the two tests into an overall test statistic. `sktest` requires a minimum of 8 observations to make its calculations.

# Option

Main

`noadjust` suppresses the empirical adjustment made by Royston (1991c) to the overall $\chi^2$ and its significance level and presents the unaltered test as described by D'Agostino, Balanger, and D'Agostino, Jr. (1990).

# Remarks

Also see [R] **swilk** for the Shapiro–Wilk and Shapiro–Francia tests for normality. Those tests are, in general, preferred for nonaggregated data (Gould and Rogers 1991, Gould 1992, Royston 1991c). Moreover, a normal quantile plot should be used in conjunction with any test for normality; see [R] **diagnostic plots** for more information.

## ▷ Example 1

Using our automobile dataset, we will test whether the variables `mpg` and `trunk` are normally distributed:

```
. use http://www.stata-press.com/data/r9/auto
(1978 Automobile Data)

. sktest mpg trunk

                   Skewness/Kurtosis tests for Normality
                                                     ------- joint ------
    Variable | Pr(Skewness)   Pr(Kurtosis)  adj chi2(2)     Prob>chi2
-------------+-------------------------------------------------------
         mpg |      0.002         0.080          10.95         0.0042
       trunk |      0.912         0.044           4.19         0.1228
```

We can reject the hypothesis that mpg is normally distributed, but we cannot reject the hypothesis that trunk is normally distributed, at least at the 12% level. The kurtosis for trunk is 2.19, as can be verified by issuing the command

```
. summarize trunk, detail
  (output omitted)
```

and the $p$-value of 0.044 shown in the table above indicates that it is significantly different from the kurtosis of a normal distribution at the 5% significance level. However, based on skewness alone, we cannot reject the hypothesis that trunk is normally distributed.

◁

❑ Technical Note

sktest implements the test as described by D'Agostino et al. (1990) but with the adjustment made by Royston (1991c). In the above example, if we had specified the noadjust option, the $\chi^2$ values would have been 13.13 for mpg and 4.05 for trunk. With the adjustment, the $\chi^2$ value might show as '.'. This should be interpreted as an absurdly large number; the data are most certainly not normal.

❑

## Saved Results

sktest saves in r():

Scalars

| | |
|---|---|
| r(chi2) | $\chi^2$ |
| r(P_skew) | Pr(skewness) |
| r(P_kurt) | Pr(kurtosis) |
| r(P_chi2) | Pr($\chi^2$) |

## Methods and Formulas

sktest is implemented as an ado-file.

sktest implements the test described by D'Agostino, Balanger, and D'Agostino, Jr. (1990) with the empirical correction developed by Royston (1991c).

Let $g_1$ denote the coefficient of skewness and $b_2$ denote the coefficient of kurtosis as calculated by summarize, and let $n$ denote the sample size. If weights are specified, then $g_1$, $b_2$, and $n$ denote the weighted coefficients of skewness and kurtosis and weighted sample size, respectively. See [R] **summarize** for the formulas for skewness and kurtosis.

To perform the test of skewness, we compute

$$Y = g_1 \left\{ \frac{(n+1)(n+3)}{6(n-2)} \right\}^{1/2}$$

$$\beta_2(g_1) = \frac{3(n^2+27n-70)(n+1)(n+3)}{(n-2)(n+5)(n+7)(n+9)}$$

$$W^2 = -1 + [2\{\beta_2(g_1) - 1\}]^{1/2}$$

and

$$\alpha = \left\{2/(W^2-1)\right\}^{1/2}$$

Then the distribution of the test statistic

$$Z_1 = \frac{1}{\sqrt{\ln W}} \ln \left[ Y/\alpha + \left\{ (Y/\alpha)^2 + 1 \right\}^{1/2} \right]$$

is approximately standard normal under the null hypothesis that the data are distributed normally.

To perform the test of kurtosis, we compute

$$\begin{aligned}
\mathrm{E}(b_2) &= \frac{3(n-1)}{n+1} \\
\mathrm{var}(b_2) &= \frac{24n(n-2)(n-3)}{(n+1)^2(n+3)(n+5)} \\
X &= \{b_2 - \mathrm{E}(b_2)\} / \sqrt{\mathrm{var}(b_2)} \\
\sqrt{\beta_1(b_2)} &= \frac{6(n^2-5n+2)}{(n+7)(n+9)} \left\{ \frac{6(n+3)(n+5)}{n(n-2)(n-3)} \right\}^{1/2}
\end{aligned}$$

and

$$A = 6 + \frac{8}{\sqrt{\beta_1(b_2)}} \left[ \frac{2}{\sqrt{\beta_1(b_2)}} + \left\{ 1 + \frac{4}{\beta_1(b_2)} \right\}^{1/2} \right]$$

Then the distribution of the test statistic

$$Z_2 = \frac{1}{\sqrt{2/(9A)}} \left[ \left( 1 - \frac{2}{9A} \right) - \left\{ \frac{1 - 2/A}{1 + X\sqrt{2/(A-4)}} \right\}^{1/3} \right]$$

is approximately standard normal under the null hypothesis that the data are distributed normally.

D'Agostino, Balanger, and D'Agostino, Jr.'s omnibus test of normality uses the statistic

$$K^2 = Z_1^2 + Z_2^2$$

which has approximately a $\chi^2$ distribution with two degrees of freedom under the null of normality.

Royston (1991c) proposed the following adjustment to the test of normality, which `sktest` uses by default. Let $\Phi(x)$ denote the cumulative standard normal distribution function for $x$, and let $\Phi^{-1}(p)$ denote the inverse cumulative standard normal function [i.e. $x = \Phi^{-1}\{\Phi(x)\}$]. Define the following terms:

$$\begin{aligned}
Z_c &= -\Phi^{-1} \left\{ \exp\left( -\frac{1}{2} K^2 \right) \right\} \\
Z_t &= 0.55n^{0.2} - 0.21 \\
a_1 &= (-5 + 3.46 \ln n) \exp(-1.37 \ln n) \\
b_1 &= 1 + (0.854 - 0.148 \ln n) \exp(-0.55 \ln n) \\
a_2 &= a_1 - \{2.13/(1 - 2.37 \ln n)\} Z_t
\end{aligned}$$

and

$$b_2 = 2.13/(1 - 2.37 \ln n) + b1$$

If $Z_c < -1$ set $Z = Z_c$; else if $Z_c < Z_t$ set $Z = a_1 + b_1 Z_c$; else set $Z = a_2 + b_2 Z_c$. Define $P = 1 - \Phi(Z)$. Then $K^2 = -2 \ln P$ is approximately distributed $\chi^2$ with two degrees of freedom.

The relative merits of the skewness and kurtosis test versus the Shapiro–Wilk and Shapiro–Francia tests have been a subject of debate. The interested reader is directed to the articles in the *Stata Technical Bulletin*. Our recommendation is to use the Shapiro–Francia test whenever possible, that is, whenever dealing with nonaggregated or ungrouped data (Gould and Rogers 1991, Gould 1992); see [R] **swilk**. If normality is rejected, use `sktest` to determine the source of the problem.

As both D'Agostino, Balanger, and D'Agostino, Jr. (1990) and Royston (1991d) mention, researchers should also examine the normal quantile plot to determine normality rather than blindly relying on a few test statistics. See the `qnorm` command documented in [R] **diagnostic plots** for more information on normal quantile plots.

Note that `sktest` is similar in spirit to the Jarque–Bera (1987) test of normality. The Jarque–Bera test statistic is also calculated from the sample skewness and kurtosis, though it is based on asymptotic standard errors without any corrections for sample size. In effect, `sktest` offers two adjustments for sample size, that of Royston (1991c) and that of D'Agostino, Balanger, and D'Agostino, Jr. (1990).

# Acknowledgments

`sktest` has benefited greatly by the comments and work of Patrick Royston of the MRC Clinical Trials Unit, London; at this point, the program should be viewed as due as much to Royston as to us, except, of course, for any errors. We are also indebted to Nicholas J. Cox, University of Durham, for helpful comments.

# References

D'Agostino, R. B., A. Balanger and R. B. D'Agostino, Jr. 1990. A suggestion for using powerful and informative tests of normality. *The American Statistician* 44(4): 316–321.

——. 1991. sg3.3: Comment on tests of normality. *Stata Technical Bulletin* 3: 20. Reprinted in *Stata Technical Bulletin Reprints*, vol. 1, pp. 105–106.

Gould, W. W. 1991. sg3: Skewness and kurtosis tests of normality. *Stata Technical Bulletin* 1: 20–21. Reprinted in *Stata Technical Bulletin Reprints*, vol. 1, pp. 99–101.

——. 1992. sg3.7: Final summary of tests of normality. *Stata Technical Bulletin* 5: 10–11. Reprinted in *Stata Technical Bulletin Reprints*, vol. 1, pp. 114–115.

Gould, W. W. and W. H. Rogers. 1991. sg3.4: Summary of tests of normality. *Stata Technical Bulletin* 3: 20–23. Reprinted in *Stata Technical Bulletin Reprints*, vol. 1, pp. 106–110.

Jarque, C. M. and A. K. Bera. 1987. A test for normality of observations and regression residuals. *International Statistical Review* 2: 163–172.

Royston, P. 1991a. sg3.1: Tests for departure from normality. *Stata Technical Bulletin* 2: 16–17. Reprinted in *Stata Technical Bulletin Reprints*, vol. 1, pp. 101–104.

——. 1991b. sg3.2: Shapiro–Wilk and Shapiro–Francia tests. *Stata Technical Bulletin* 3: 19. Reprinted in *Stata Technical Bulletin Reprints*, vol. 1, p. 105.

——. 1991c. sg3.5: Comment on sg3.4 and an improved D'Agostino test. *Stata Technical Bulletin* 3: 23–24. Reprinted in *Stata Technical Bulletin Reprints*, vol. 1, pp. 110–112.

——. 1991d. sg3.6: A response to sg3.3: Comment on tests of normality. *Stata Technical Bulletin* 4: 8–9. Reprinted in *Stata Technical Bulletin Reprints*, vol. 1, pp. 112–114.

# Also See

**Related:** [R] **diagnostic plots**, [R] **ladder**, [R] **lv**, [R] **swilk**

# Title

**slogit** — Stereotype logistic regression

# Syntax

`slogit` *depvar* [*indepvars*] [*if*] [*in*] [*weight*] [, *options*]

| *options* | description |
|---|---|
| Model | |
| `dimension(#)` | dimension of the model; default is `dimension(1)` |
| `baseoutcome(#)` | set the base outcome to level #; default is the last outcome |
| `constraints(numlist)` | apply specified linear constraints |
| `nocorner` | do not generate the corner constraints |
| SE/Robust | |
| `vce(vcetype)` | *vcetype* may be `oim`, `robust`, `opg`, `bootstrap`, or `jackknife` |
| `robust` | synonym for `vce(robust)` |
| `cluster(varname)` | adjust standard errors for intragroup correlation |
| Reporting | |
| `level(#)` | set confidence level; default is `level(95)` |
| Max options | |
| *maximize_options* | control the maximization process; seldom used |
| `initialize(initype)` | method of initializing scale parameters; *initype* can be `constant`, `random`, or `svd`; see *Options* for details |
| `nonormalize` | do not normalize the numeric variables |

`bootstrap`, `by`, `jackknife`, `rolling`, `statsby`, and `xi` are allowed; see [U] **11.1.10 Prefix commands**.

`fweights`, `iweights`, and `pweights` are allowed; see [U] **11.1.6 weight**.

See [U] **20 Estimation and postestimation commands** for additional capabilities of estimation commands.

# Description

`slogit` fits maximum-likelihood stereotype regression models as developed by Anderson (1984). Like multinomial logistic and ordered logistic models, stereotype logistic models are for use with categorical dependent variables. In a multinomial logistic model, the categories cannot be ranked, while in an ordered logistic model the categories follow a natural ranking scheme. You can view stereotype logistic models as a compromise between those two models. You can use them when you are unsure of the relevance of the ordering, as is often the case when subjects are asked to assess or judge something. You can also use them in place of multinomial logistic models when you suspect some of the alternatives are similar. Unlike ordered logistic models, stereotype logistic models do not impose the proportional odds assumption.

# Options

Model

`dimension(#)` specifies the dimension of the model, which is the number of equations required to describe the relationship between the dependent variable and the independent variables. The maximum dimension is $\min(m-1,p)$, where $m$ is the number of categories of the dependent variable and $p$ is the number of independent variables in the model. The stereotype model with maximum dimension is a reparameterization of the multinomial logistic model.

`baseoutcome(#)` specifies the outcome level whose scale parameters and intercept are constrained to be zero. By default, `slogit` assumes the outcome levels are ordered and uses the largest level of the dependent variable as the base outcome.

`constraints(`*numlist*`)` specifies the linear constraints to be applied during estimation. Constraints are defined using the `constraint` command (see [R] **constraint**) and are numbered.

By default, the linear equality constraints suggested by Anderson (1984), termed the corner constraints, are generated for you. You can add constraints to these, as needed, or you can turn off the corner constraints altogether by specifying `nocorner`. Note that these constraints are in addition to the constraints placed on the $\phi$ parameters corresponding to `baseoutcome(#)`.

`nocorner` specifies that `slogit` not generate the corner constraints. If you specify `nocorner`, you must specify at least `dimension()` × `dimension()` constraints for the model to be identified.

SE/Robust

`vce(`*vcetype*`)`; see [R] ***vce_option***.

`robust`, `cluster(`*varname*`)`; see [R] **estimation options**.

Reporting

`level(#)`; see [R] **estimation options**.

Max options

*maximize_options*: `difficult`, `technique(`*algorithm_spec*`)`, `iterate(#)`, `[no]log`, `trace`, `gradient`, `showstep`, `hessian`, `shownrtolerance`, `tolerance(#)`, `ltolerance(#)`, `gtolerance(#)`, `nrtolerance(#)`, `nonrtolerance`, `from(`*init_specs*`)`; see [R] **maximize**. These options are seldom used.

`initialize(constant|random|svd)` specifies how initial estimates are computed. The default, `initialize(constant)`, is to set the scale parameters to the constant $\min(1/2, 1/d)$, where $d$ is the dimension specified in `dimension()`.

`initialize(random)` requests that uniformly distributed random numbers between 0 and 1 be used as initial values for the scale parameters. If you specify this option, you should also use `set seed` to ensure that you can replicate your results (see [D] **generate**).

`initialize(svd)` requests that a singular value decomposition be performed on the matrix of regression estimates from `mlogit` to reduce its rank to the dimension specified in `dimension()`. `slogit` uses the reduced-rank components of the SVD as initial estimates for the scale and regression coefficients. For details, see *Methods and Formulas*.

`nonormalize` requests that the numeric variables not be normalized. Normalization of the numeric variables improves numerical stability but consumes more memory in generating temporary double-precision variables. Variables that are of type `byte` are not normalized, and if initial estimates are specified using the `from()` option, normalization of variables is not performed. See *Methods and Formulas* for more information.

# Remarks

Remarks are presented under the headings

*Introduction*
*One-dimensional model*
*Higher-dimension models*

## Introduction

Stereotype logistic models are frequently used when subjects are requested to assess or judge something. For example, consider a survey in which consumers may be asked to rate the quality of a product on a scale from one to five, one indicating poor quality and five indicating excellent quality. If the categories are monotonically related to a single underlying latent variable, the ordered logistic model is appropriate. However, suppose that consumers assess quality not just along a single dimension, but rather weigh two or three latent factors. Stereotype logistic regression allows you to specify multiple equations to capture the effects of those latent variables, which you then parameterize in terms of observable characteristics. Unlike with multinomial logit, the number of equations you specify could be less than $m-1$, where $m$ is the number of categories of the dependent variable.

Stereotype logistic models are also used when categories may be indistinguishable. Suppose that a consumer must choose among A, B, C, or D. Multinomial logistic modeling assumes that the four choices are distinct in the sense that a consumer choosing one of the goods can distinguish its characteristics from the others. If goods A and B are in fact quite similar, consumers may be randomly picking between the two. One alternative is to combine the two categories and fit a three-category multinomial logistic model. A more flexible alternative is to use a stereotype logistic model.

In the multinomial logistic model, you estimate $m-1$ parameter vectors $\widetilde{\boldsymbol{\beta}}_k$, $k=1\dots m-1$, where $m$ is the number of categories of the dependent variable. The stereotype logistic model is a restriction on the multinomial model in the sense that there are $d$ parameter vectors, where $d$ is between one and $\min(m-1,p)$, and $p$ is the number of regressors. The relationship between the stereotype model's coefficients $\boldsymbol{\beta}_j$, $j=1,\dots,d$, and the multinomial model's coefficients is $\widetilde{\boldsymbol{\beta}}_k = -\sum_{j=1}^{d}\phi_{jk}\boldsymbol{\beta}_j$. The $\phi$s are scale parameters to be estimated along with the $\boldsymbol{\beta}_j$s.

Given a row vector of covariates $\mathbf{x}$, let $\eta_k = \theta_k - \sum_{j=1}^{d}\phi_{jk}\mathbf{x}\boldsymbol{\beta}_j$. The probability of observing outcome $k$ is

$$\Pr(Y_i=k) = \begin{cases} \dfrac{\exp(\eta_k)}{1+\sum_{l=1}^{m-1}\exp(\eta_l)} & k<m \\ \dfrac{1}{1+\sum_{l=1}^{m-1}\exp(\eta_l)} & k=m. \end{cases}$$

Notice that this model includes a set of $\theta$ parameters so that each equation has an unrestricted constant term. If $d=m-1$, the stereotype model is just a reparameterization of the multinomial logistic model. To identify the $\phi$s and the $\boldsymbol{\beta}$s, you must place at least $d^2$ restrictions on the parameters. By default, `slogit` uses the "corner constraints" $\phi_{jj}=1$ and $\phi_{jk}=0$ for $j\neq k$, $k\leq d$, and $j\leq d$.

## One-dimensional model

### ▷ Example 1

We have two years of repair rating data on the make, price, mileage rating, and gear ratio of 104 foreign and 44 domestic automobiles (with 13 missing values on repair rating). We wish to fit a stereotype logistic model to discriminate between the levels of repair rating using mileage, price, gear ratio, and origin of the manufacturer. Here is an overview of our data:

```
. use http://www.stata-press.com/data/r9/auto2yr
(Automobile Models)

. tabulate repair

     repair |      Freq.     Percent        Cum.
------------+-----------------------------------
       Poor |          5        3.70        3.70
       Fair |         19       14.07       17.78
    Average |         57       42.22       60.00
       Good |         38       28.15       88.15
  Excellent |         16       11.85      100.00
------------+-----------------------------------
      Total |        135      100.00
```

The variable `repair` can take five values, 1, …, 5, which represent the subjective rating of the car model's repair record as *Poor*, *Fair*, *Average*, *Good*, and *Excellent*.

We wish to fit the following one-dimensional stereotype logistic model

$$\eta_k = \theta_k - \phi_k\left(\beta_1\texttt{foreign} + \beta_2\texttt{mpg} + \beta_3\texttt{price} + \beta_4\texttt{gratio}\right)$$

for $k < 5$ and $\eta_5 = 0$. To fit this model, we type

(*Continued on next page*)

```
. slogit repair foreign mpg price gratio
Iteration 0:   log likelihood =  -173.7818  (not concave)
Iteration 1:   log likelihood = -164.77318
Iteration 2:   log likelihood =  -161.7069
Iteration 3:   log likelihood = -159.76139
Iteration 4:   log likelihood = -159.34328
Iteration 5:   log likelihood = -159.25914
Iteration 6:   log likelihood = -159.25691
Iteration 7:   log likelihood = -159.25691
Stereotype logistic regression                    Number of obs   =        135
                                                  Wald chi2(4)    =       9.33
Log likelihood = -159.25691                       Prob > chi2     =     0.0535
 ( 1)  [phi1_1]_cons = 1

------------------------------------------------------------------------------
      repair |      Coef.   Std. Err.      z    P>|z|     [95% Conf. Interval]
-------------+----------------------------------------------------------------
     foreign |   5.947382   2.094126     2.84   0.005      1.84297    10.05179
         mpg |   .1911968     .08554     2.24   0.025     .0235414    .3588521
       price |  -.0000576   .0001357    -0.42   0.671    -.0003236    .0002083
      gratio |  -4.307571   1.884713    -2.29   0.022     -8.00154   -.6136017
-------------+----------------------------------------------------------------
     /phi1_1 |          1          .        .       .            .           .
     /phi1_2 |   1.262268   .3530565     3.58   0.000     .5702904    1.954247
     /phi1_3 |    1.17593   .3169397     3.71   0.000     .5547394     1.79712
     /phi1_4 |   .8657195   .2411228     3.59   0.000     .3931275    1.338311
     /phi1_5 |          0  (base outcome)
-------------+----------------------------------------------------------------
     /theta1 |  -6.864749    4.21252    -1.63   0.103    -15.12114    1.391639
     /theta2 |  -7.613977   4.861803    -1.57   0.117    -17.14294    1.914981
     /theta3 |   -5.80655   4.987508    -1.16   0.244    -15.58189    3.968786
     /theta4 |   -3.85724   3.824132    -1.01   0.313     -11.3524    3.637922
     /theta5 |          0  (base outcome)
------------------------------------------------------------------------------
(repair=Excellent is the base outcome)
```

Notice that the coefficient associated with the first scale parameter, $\phi_{11}$, is 1, and its standard error and other statistics are missing. This is the corner constraint applied to the one-dimensional model; in the header, this constraint is listed as `[phi1_1]_cons = 1`. Notice also that the $\phi$ and $\theta$ parameters that are associated with the base outcome are identified. Keep in mind, though, there are no coefficient estimates `[phi1_5]_cons` or `[theta5]_cons` in the `ereturn` matrix `e(b)`. The Wald statistic is for a test of the joint significance of the regression coefficients on `foreign`, `mpg`, `price`, and `gratio`.

The one-dimensional stereotype model restricts the multinomial logistic regression coefficients $\widetilde{\boldsymbol{\beta}}_k$, $k = 1, \ldots, m-1$ to be parallel; that is, $\widetilde{\boldsymbol{\beta}}_k = -\phi_k\boldsymbol{\beta}$. As Lunt (2001) discusses, in the one-dimensional stereotype model a single linear combination $\mathbf{x}_i\boldsymbol{\beta}$ best discriminates the outcomes of the dependent variable, and the scale parameters $\phi_k$ measure the distance between the outcome levels and the linear predictor. If $\phi_1 \geq \phi_2 \geq \cdots \phi_{m-1} \geq \phi_m \equiv 0$, the model suggests that the subjective assessment of the dependent variable is indeed ordered. In this example, the maximum likelihood estimates of the $\phi$s are not monotonic, as would be assumed in an ordered logit model.

We test $\phi_1 = \phi_2$ by typing

```
. test [phi1_2]_cons = [phi1_1]_cons
 ( 1)  - [phi1_1]_cons + [phi1_2]_cons = 0
           chi2(  1) =    0.55
         Prob > chi2 =    0.4576
```

Since the two parameters are not statistically different, we decide to add a constraint to force $\phi_1 = \phi_2$:

```
. constraint define 1 [phi1_2]_cons = [phi1_1]_cons
. slogit repair foreign mpg price gratio, constraint(1) nolog
Stereotype logistic regression                    Number of obs   =        135
                                                  Wald chi2(4)    =      21.28
Log likelihood = -159.65769                       Prob > chi2     =     0.0003
 ( 1)  [phi1_1]_cons = 1
 ( 2) - [phi1_1]_cons + [phi1_2]_cons = 0

      repair        Coef.   Std. Err.      z    P>|z|     [95% Conf. Interval]

     foreign     7.166515   1.690177     4.24   0.000     3.853829     10.4792
         mpg     .2340043   .0807042     2.90   0.004     .0758271    .3921816
       price     -.000041   .0001618    -0.25   0.800    -.0003581     .000276
      gratio    -5.218107   1.798717    -2.90   0.004    -8.743528   -1.692686

     /phi1_1            1          .        .       .            .           .
     /phi1_2            1          .        .       .            .           .
     /phi1_3     .9751096   .1286563     7.58   0.000     .7229478    1.227271
     /phi1_4     .7209343   .1220353     5.91   0.000     .4817494    .9601191
     /phi1_5            0  (base outcome)

     /theta1    -8.293452   4.645182    -1.79   0.074    -17.39784    .8109368
     /theta2    -6.958451   4.629292    -1.50   0.133     -16.0317    2.114795
     /theta3    -5.620232   4.953981    -1.13   0.257    -15.32986    4.089392
     /theta4    -3.745624   3.809189    -0.98   0.325     -11.2115    3.720249
     /theta5            0  (base outcome)

(repair=Excellent is the base outcome)
```

The $\phi$ estimates are now monotonically decreasing and the standard errors of the $\phi$s are small relative to the size of the estimates, so we conclude that, with the exception of outcomes *Poor* and *Fair*, the groups are distinguishable for the one-dimensional model and that the quality assessment can be ordered.

◁

## Higher-dimension models

The stereotype logistic model is not limited to ordered categorical dependent variables; you can use it on nominal data to reduce the dimension of the regressions. Recall that a multinomial model fitted to a categorical dependent variable with $m$ levels will have $m-1$ sets of regression coefficients. However, a model with fewer dimensions may fit the data equally well, suggesting that some of the categories are indistinguishable.

▷ Example 2

As discussed in [R] **mlogit**, we have data on the type of health insurance available to 616 psychologically depressed subjects in the U.S. (Tarlov et al. 1989; Wells et al. 1989). Patients may have either an indemnity (fee-for-service) plan or a prepaid plan, such as an HMO, or may be uninsured. Demographic variables include age, gender, race, and site.

First we fit the saturated, two-dimensional model that is equivalent to a multinomial logistic model. We choose the base outcome to be 1 (indemnity insurance) since that is the default for `mlogit`.

```
. use http://www.stata-press.com/data/r9/sysdsn3, clear
(Health insurance data)

. slogit insure age male nonwhite site2 site3, dim(2) base(1) nolog

Stereotype logistic regression                    Number of obs   =        615
                                                  Wald chi2(10)   =      38.17
Log likelihood = -534.36165                       Prob > chi2     =     0.0000

 ( 1)  [phi1_2]_cons = 1
 ( 2)  [phi1_3]_cons = 0
 ( 3)  [phi2_2]_cons = 0
 ( 4)  [phi2_3]_cons = 1

------------------------------------------------------------------------------
      insure |      Coef.   Std. Err.      z    P>|z|     [95% Conf. Interval]
-------------+----------------------------------------------------------------
dim1         |
         age |    .011745   .0061946     1.90   0.058    -.0003962    .0238862
        male |  -.5616934   .2027465    -2.77   0.006    -.9590693   -.1643175
    nonwhite |  -.9747768   .2363213    -4.12   0.000    -1.437953   -.5115955
       site2 |  -.1130359   .2101903    -0.54   0.591    -.5250013    .2989296
       site3 |   .5879879   .2279351     2.58   0.010     .1412433    1.034733
-------------+----------------------------------------------------------------
dim2         |
         age |   .0077961   .0114418     0.68   0.496    -.0146294    .0302217
        male |  -.4518496   .3674867    -1.23   0.219     -1.17211     .268411
    nonwhite |  -.2170589   .4256361    -0.51   0.610     -1.05129    .6171725
       site2 |   1.211563   .4705127     2.57   0.010     .2893747    2.133751
       site3 |   .2078123   .3662926     0.57   0.570     -.510108    .9257327
-------------+----------------------------------------------------------------
     /phi1_1 |          0  (base outcome)
     /phi1_2 |          1          .        .       .            .           .
     /phi1_3 |          0          .        .       .            .           .
-------------+----------------------------------------------------------------
     /phi2_1 |          0  (base outcome)
     /phi2_2 |          0          .        .       .            .           .
     /phi2_3 |          1          .        .       .            .           .
-------------+----------------------------------------------------------------
     /theta1 |          0  (base outcome)
     /theta2 |   .2697127   .3284422     0.82   0.412    -.3740222    .9134476
     /theta3 |  -1.286943   .5923219    -2.17   0.030    -2.447872   -.1260134
------------------------------------------------------------------------------
(insure=Indemnity is the base outcome)
```

For comparison, we also fit the model using `mlogit`:

(*Continued on next page*)

```
. mlogit insure age male nonwhite site2 site3, nolog
Multinomial logistic regression                   Number of obs   =        615
                                                  LR chi2(10)     =      42.99
                                                  Prob > chi2     =     0.0000
Log likelihood = -534.36165                       Pseudo R2       =     0.0387

------------------------------------------------------------------------------
      insure |      Coef.   Std. Err.      z    P>|z|     [95% Conf. Interval]
-------------+----------------------------------------------------------------
Prepaid      |
         age |   -.011745   .0061946    -1.90   0.058    -.0238862    .0003962
        male |   .5616934   .2027465     2.77   0.006     .1643175    .9590693
    nonwhite |   .9747768   .2363213     4.12   0.000     .5115955    1.437958
       site2 |   .1130359   .2101903     0.54   0.591    -.2989296    .5250013
       site3 |  -.5879879   .2279351    -2.58   0.010    -1.034733   -.1412433
       _cons |   .2697127   .3284422     0.82   0.412    -.3740222    .9134476
-------------+----------------------------------------------------------------
Uninsure     |
         age |  -.0077961   .0114418    -0.68   0.496    -.0302217    .0146294
        male |   .4518496   .3674867     1.23   0.219     -.268411     1.17211
    nonwhite |   .2170589   .4256361     0.51   0.610    -.6171725     1.05129
       site2 |  -1.211563   .4705127    -2.57   0.010    -2.133751   -.2893747
       site3 |  -.2078123   .3662926    -0.57   0.570    -.9257327     .510108
       _cons |  -1.286943   .5923219    -2.17   0.030    -2.447872   -.1260135
------------------------------------------------------------------------------
(insure==Indemnity is the base outcome)
```

Notice that, apart from having opposite signs, the coefficients from the stereotype logistic model are identical to those from the multinomial logit model. Recall the definition of $\eta_k$ given in the *Remarks*, particularly the minus sign in front of the summation. One other difference in the output is that the constant estimates labeled `/theta` in the `slogit` output are the constants labeled `_cons` in the `mlogit` output.

Next we examine the one-dimensional model.

```
. slogit insure age male nonwhite site2 site3, dim(1) base(1) nolog
Stereotype logistic regression                    Number of obs   =        615
                                                  Wald chi2(5)    =      28.20
Log likelihood = -539.75205                       Prob > chi2     =     0.0000
 ( 1)  [phi1_2]_cons = 1

------------------------------------------------------------------------------
      insure |      Coef.   Std. Err.      z    P>|z|     [95% Conf. Interval]
-------------+----------------------------------------------------------------
         age |   .0108366   .0061918     1.75   0.080    -.0012992    .0229723
        male |  -.5032537   .2078171    -2.42   0.015    -.9105678   -.0959396
    nonwhite |  -.9480351   .2340604    -4.05   0.000    -1.406785    -.489285
       site2 |  -.2444316   .2246366    -1.09   0.277    -.6847113    .1958481
       site3 |    .556665   .2243799     2.48   0.013     .1168886    .9964415
-------------+----------------------------------------------------------------
     /phi1_1 |          0  (base outcome)
     /phi1_2 |          1          .        .       .            .           .
     /phi1_3 |   .0383539   .4079705     0.09   0.925    -.7612535    .8379613
-------------+----------------------------------------------------------------
     /theta1 |          0  (base outcome)
     /theta2 |    .187542   .3303847     0.57   0.570    -.4600001     .835084
     /theta3 |  -1.860134   .2158898    -8.62   0.000     -2.28327   -1.436997
------------------------------------------------------------------------------
(insure=Indemnity is the base outcome)
```

We have reduced a two-dimensional multinomial model to one dimension, reducing the number of estimated parameters by four and decreasing the model likelihood by $\approx 5.4$.

`slogit` does not report a model likelihood-ratio test. The test of $d = 1$ (a one-dimensional model) versus $d = 0$ (the null model) does not have an asymptotic $\chi^2$ distribution because the unconstrained $\phi$ parameters (`/phi1_3` in the previous example) cannot be identified if $\boldsymbol{\beta} = 0$. More generally, this problem precludes testing any hierarchical model of dimension $d$ versus $d-1$. Of course, the likelihood-ratio test of a full-dimension model versus $d = 0$ is valid because the full model is just multinomial logistic, and all of the $\phi$ parameters are fixed at 0 or 1.

◁

## ❑ Technical Note

The stereotype model is a special case of the reduced-rank vector generalized linear model discussed by Yee and Hastie (2003). If we define $\eta_{ik} = \theta_k - \sum_{j=1}^{d} \phi_{jk} \mathbf{x}_i \boldsymbol{\beta}_j$, for $k = 1, \ldots, m-1$, we can write the expression in matrix notation as

$$\boldsymbol{\eta}_i = \boldsymbol{\theta} + \boldsymbol{\Phi} \left( \mathbf{x}_i \mathbf{B} \right)'$$

where $\boldsymbol{\Phi}$ is a $(m-1) \times d$ matrix containing the $\phi_{jk}$ parameters and $\mathbf{B}$ is a $p \times d$ matrix with columns containing the $\boldsymbol{\beta}_j$ parameters, $j = 1, \ldots, d$. The factorization $\boldsymbol{\Phi}\mathbf{B}'$ is not unique since $\boldsymbol{\Phi}\mathbf{B}'$ = $\boldsymbol{\Phi}\mathbf{M}\mathbf{M}^{-1}\mathbf{B}'$ for any nonsingular $d \times d$ matrix $\mathbf{M}$. To avoid this identifiability problem, we choose $\mathbf{M} = \boldsymbol{\Phi}_1^{-1}$, where

$$\boldsymbol{\Phi} = \begin{pmatrix} \boldsymbol{\Phi}_1 \\ \boldsymbol{\Phi}_2 \end{pmatrix}$$

and $\boldsymbol{\Phi}_1$ is $d \times d$ of rank $d$ so that

$$\boldsymbol{\Phi}\mathbf{M} = \begin{pmatrix} \mathbf{I}_d \\ \boldsymbol{\Phi}_2 \boldsymbol{\Phi}_1^{-1} \end{pmatrix}$$

and $\mathbf{I}_d$ is a $d \times d$ identity matrix. Thus the corner constraints used by `slogit` are $\phi_{jj} \equiv 1$ and $\phi_{jk} \equiv 0$ for $j \neq k$ and $k, j \leq d$.

❑

(*Continued on next page*)

# Saved Results

`slogit` saves in `e()`:

Scalars

| | | | |
|---|---|---|---|
| `e(N)` | number of observations | `e(ll)` | log likelihood |
| `e(k)` | number of parameters | `e(ll_0)` | null model log likelihood |
| `e(k_indvars)` | number of independent variables | `e(N_clust)` | number of clusters |
| `e(k_out)` | number of outcomes | `e(chi2)` | $\chi^2$, if `e(chi2type)=Wald` |
| `e(df_m)` | Wald test degrees of freedom, if `e(chi2type)=Wald` | `e(p)` | significance |
| | | `e(ic)` | number of iterations |
| `e(df_0)` | null model degrees of freedom, | `e(rank)` | rank of `e(V)` |
| `e(k_dim)` | model dimension | `e(rc)` | return code |
| `e(i_base)` | base outcome index | `e(converged)` | `1` if converged, `0` otherwise |

Macros

| | | | |
|---|---|---|---|
| `e(cmd)` | `slogit` | `e(vce)` | *vcetype* specified in `vce()` |
| `e(depvar)` | name of dependent variable | `e(vcetype)` | title used to label Std. Err. |
| `e(indvars)` | independent variables | `e(opt)` | type of optimization |
| `e(wtype)` | weight type | `e(ml_method)` | type of `ml` method |
| `e(wexp)` | weight expression | `e(user)` | name of likelihood-evaluator program |
| `e(title)` | title in estimation output | `e(technique)` | maximization technique |
| `e(clustvar)` | name of cluster variable | `e(crittype)` | optimization criterion |
| `e(out`$i$`)` | outcome labels, $i$ = `1...e(k_out)` | `e(properties)` | `b V` |
| `e(chi2type)` | `Wald` | `e(predict)` | program implementing `predict` |
| `e(labels)` | outcome labels or numeric levels | | |

Matrices

| | | | |
|---|---|---|---|
| `e(b)` | coefficient vector | `e(V)` | variance–covariance matrix of the estimators |
| `e(outcomes)` | outcome values | | |
| `e(ilog)` | iteration log (up to 20 iterations) | `e(gradient)` | gradient vector |

Functions

| | |
|---|---|
| `e(sample)` | marks estimation sample |

# Methods and Formulas

`slogit` obtains the maximum likelihood estimates for the stereotype logistic model using `ml`; see [R] **ml**. Each set of regression estimates, one set of $\boldsymbol{\beta}_j$s for each dimension, constitutes one `ml model` equation. The $d \times (m-1)$ $\phi$s and the $(m-1)$ $\theta$s are `ml` ancillary parameters.

Without loss of generality, let the base outcome level be the $m$th level of the dependent variable. Define the row vector $\boldsymbol{\phi}_k = (\phi_{1k}, \ldots, \phi_{dk})$ for $k = 1, \ldots, m-1$, and define the $p \times d$ matrix $\mathbf{B} = (\boldsymbol{\beta}_1, \ldots, \boldsymbol{\beta}_d)$. For observation $i$, the log odds of outcome level $k$ relative to level $m$, $k = 1, \ldots, m-1$ is the index

$$\begin{aligned}\ln\left\{\frac{\Pr(Y_i = k)}{\Pr(Y_i = m)}\right\} = \eta_{ik} &= \theta_k - \boldsymbol{\phi}_k\left(\mathbf{x}_i\mathbf{B}\right)' \\ &= \theta_k - \boldsymbol{\phi}_k\boldsymbol{\nu}_i'\end{aligned}$$

The row vector $\boldsymbol{\nu}_i$ can be interpreted as a latent variable reducing the $p$-dimensional vector of covariates to a more interpretable $d < p$ dimensions.

The probability of the $i$th observation having outcome level $k$ is then

$$\Pr(Y_i = k) = p_{ik} = \begin{cases} \dfrac{e^{\eta_{ik}}}{1 + \sum_{j=1}^{m-1} e^{\eta_{ij}}}, \text{ if } & k < m \\ \dfrac{1}{1 + \sum_{j=1}^{m-1} e^{\eta_{ij}}}, \text{ if } & k = m \end{cases}$$

from which the log-likelihood function is computed as

$$L = \sum_{i=1}^{n} w_i \sum_{k=1}^{m} I_k(y_i) \ln(p_{ik}) \tag{1}$$

Here $w_i$ is the weight for observation $i$ and

$$I_k(y_i) = \begin{cases} 1, \text{ if observation } y_i \text{ has outcome } k \\ 0, \text{ otherwise} \end{cases}$$

Numeric variables are normalized for numerical stability during optimization where a new double-precision variable $\widetilde{x}_j$ is created from variable $x_j$, $j = 1, \ldots, p$, such that $\widetilde{x}_j = (x_j - \bar{x}_j)/s_j$. This feature is turned off if you specify `nonormalize`, or if you use the `from()` option for initial estimates. Normalization is not performed on byte variables, including the indicator variables generated by [R] **xi**. The linear equality constraints for regression parameters, if specified, must be scaled also. Assume that a constraint is applied to the regression parameter associated with variable $j$ and dimension $i$, $\beta_{ji}$, and the corresponding element of the constraint matrix (see [P] **makecns**) is divided by $s_j$.

After convergence, the parameter estimates for variable $j$ and dimension $i$—$\widetilde{\beta}_{ji}$ say—are transformed back to their original scale, $\beta_{ji} = \widetilde{\beta}_{ji}/s_j$. For the intercepts, you compute

$$\theta_k = \widetilde{\theta}_k + \sum_{i=1}^{d} \phi_{ik} \sum_{j=1}^{p} \frac{\widetilde{\beta}_{ji}\bar{x}_j}{s_j}$$

Initial values are computed using estimates obtained by using `mlogit` to fit a multinomial logistic model. Let the $p \times (m-1)$ matrix $\widetilde{\mathbf{B}}$ contain the multinomial logistic regression parameters less the $m-1$ intercepts. Each $\phi$ is initialized with constant values $\min(1/2, 1/d)$, option `initialize(constant)` (the default), or, with uniform random numbers, option `initialize(random)`. Constraints are then applied to the starting values so that the structure of the $(m-1) \times d$ matrix $\mathbf{\Phi}$ is

$$\mathbf{\Phi} = \begin{pmatrix} \phi_1 \\ \phi_2 \\ \vdots \\ \phi_{m-1} \end{pmatrix} = \begin{pmatrix} \mathbf{I}_d \\ \widetilde{\mathbf{\Phi}} \end{pmatrix}$$

where $\mathbf{I}_d$ is a $d \times d$ identity matrix. Assume that only the corner constraints are used, but any constraints you place on the scale parameters are also applied to the initial scale estimates, so the structure of $\mathbf{\Phi}$ will change accordingly. Note that the $\phi$ parameters are invariant to the scale of the covariates, so initial estimates in $[0, 1]$ are reasonable. The constraints guarantee that the rank of $\mathbf{\Phi}$ is at least $d$, so the initial estimates for the stereotype regression parameters are obtained from $\mathbf{B} = \widetilde{\mathbf{B}}\mathbf{\Phi}(\mathbf{\Phi}'\mathbf{\Phi})^{-1}$.

One other approach for initial estimates is provided: `initialize(svd)`. It starts with the `mlogit` estimates and computes $\widetilde{\mathbf{B}}' = \mathbf{UDV}'$, where $\mathbf{U}_{m-1\times p}$ and $\mathbf{V}_{p\times p}$ are orthonormal matrices and $\mathbf{D}_{p\times p}$ is a diagonal matrix containing the singular values of $\widetilde{\mathbf{B}}$. The estimates for $\mathbf{\Phi}$ and $\mathbf{B}$ are the first $d$ columns of $\mathbf{U}$ and $\mathbf{VD}$, respectively (Yee and Hastie 2003).

The score for regression coefficients is

$$\mathbf{u}_i(\boldsymbol{\beta}_j) = \frac{\partial L_{ik}}{\partial \boldsymbol{\beta}_j} = \mathbf{x}_i \left( \sum_{l=1}^{m-1} \phi_{jl} p_{il} - \phi_{jk} \right)$$

the score for the scale parameters is

$$u_i(\phi_{jl}) = \frac{\partial L_{ik}}{\partial \phi_{jl}} = \begin{cases} \mathbf{x}_i \boldsymbol{\beta}_j (p_{ik} - 1), & \text{if } l = k \\ \mathbf{x}_i \boldsymbol{\beta}_j p_{il}, & \text{if } l \neq k \end{cases}$$

for $l = 1, \ldots, m-1$; and the score for the intercepts is

$$u_i(\theta_l) = \frac{\partial L_{ik}}{\partial \theta_l} = \begin{cases} 1 - p_{ik}, & \text{if } l = k \\ -p_{il}, & \text{if } l \neq k \end{cases}$$

# References


Anderson, J. A. 1984. Regression and ordered categorical variables. *Journal of the Royal Statistical Society, B.* 46: 1–30

Lunt, M. 2001. sg163: Stereotype ordinal regression. *Stata Technical Bulletin* 61: 12–18. Reprinted in *Stata Technical Bulletin Reprints*, vol. 10, pp. 298–307.

Tarlov, A. R., J. E. Ware, Jr., S. Greenfield, E. C. Nelson, E. Perrin, and M. Zubkoff. 1989. The medical outcomes study. *Journal of the American Medical Association*, 262: 925–930

Wells, K. E., R. D. Hays, M. A. Burnam, W. H. Rogers, S. Greenfield, and J. E. Ware, Jr. 1989. Detection of depressive disorder for patients receiving prepaid or fee-for-service care. *Journal of the American Medical Association*, 262: 3298–3302

Yee, T. W. and T. J. Hastie. 2003. Reduced-rank vector generalized linear models. *Statistical Modelling*, 3: 1–41


# Also See

**Complementary:** [R] **slogit postestimation**; [R] **constraint**, [R] **roc**

**Related:** [R] **mlogit**, [R] **ologit**, [R] **oprobit**

**Background:** [U] **11.1.10 Prefix commands**,
[U] **20 Estimation and postestimation commands**,
[R] **estimation options**, [R] **maximize**, [R] ***vce_option***

## Title

**slogit postestimation** — Postestimation tools for slogit

## Description

The following postestimation commands are available for `slogit`:

| command | description |
|---|---|
| `estat` | AIC, BIC, VCE, and estimation sample summary |
| `estimates` | cataloging estimation results |
| `lincom` | point estimates, standard errors, testing, and inference for linear combinations of coefficients |
| `lrtest` | likelihood-ratio test |
| `mfx` | marginal effects or elasticities |
| `nlcom` | point estimates, standard errors, testing, and inference for nonlinear combinations of coefficients |
| `predict` | predicted probabilities, estimated index and its approximate standard error |
| `predictnl` | point estimates, standard errors, testing, and inference for generalized predictions |
| `suest` | seemingly unrelated estimation |
| `test` | Wald tests for simple and composite linear hypotheses |
| `testnl` | Wald tests of nonlinear hypotheses |

See the corresponding entries in the *Stata Base Reference Manual* for details.

## Syntax for predict

`predict` [*type*] *newvarlist* [*if*] [*in*] [, *statistic* `outcome(`*#*`)`]

`predict` [*type*] *stub** [*if*] [*in*] , `scores`

| *statistic* | description |
|---|---|
| `pr` | probability of one or all of the dependent variable outcomes; the default |
| `xb` | index for the $k$th outcome |
| `stdp` | standard error of the index for the $k$th outcome |

Note that you specify one new variable with `xb` and `stdp` and specify either one or $m$ new variables with `pr`, where $m$ is the number of categories.

Statistics are available both in and out of sample; type `predict ... if e(sample) ...` if wanted only for the estimation sample.

## Options for predict

`pr`, the default, calculates the probability of each of the categories of the dependent variable or the probability of the level specified in `outcome(`*#*`)`. If you specify the `outcome(`*#*`)` option, you only need to specify one new variable; otherwise, you must specify a new variable for each category of the dependent variable.

`xb` calculates the index, $\theta_k - \sum_{j=1}^{d} \phi_{\rm jk} \mathbf{x}_i \boldsymbol{\beta}_j$, for outcome level $k \neq$ `e(i_base)` and dimension $d =$`e(k_dim)`. It returns a vector of zeros if $k =$`e(i_base)`. A synonym for `xb` is `index`. This option requires the `outcome(#)` option.

`stdp` calculates the standard error of the index. A synonym for `stdp` is `seindex`. This option requires the `outcome(#)` option.

`outcome(#)` specifies the outcome for which the statistic is to be calculated.

`scores` calculates the equation-level score variables. For models with $d$ dimensions and $m$ levels, $d + (d+1)(m-1)$ new variables are created. Assume $j = 1, \ldots, d$ and $k = 1, \ldots, m$ in the following.

The first $d$ new variables will contain $\partial \ln L / \partial (\mathbf{x}\boldsymbol{\beta}_j)$.

The next $d(m-1)$ new variables will contain $\partial \ln L / \partial \phi_{jk}$.

The last $m-1$ new variables will contain $\partial \ln L / \partial \theta_k$.

## Remarks

Once you have fitted a stereotype logistic model, you can obtain the predicted probabilities using the `predict` command for both the estimation sample and other samples; see [U] **20 Estimation and postestimation commands** and [R] **predict**.

`predict` without arguments (or with the `pr` option) calculates the predicted probability of each outcome of the dependent variable. You must, therefore, give a new variable name for each of the outcomes. To compute the estimated probability of a single outcome, you use the `outcome(#)` option where # is the level encoding the outcome. Note that if the dependent variable's levels are labeled, the outcomes can also be identified by the label values (see [D] **label**).

The option `xb` in conjunction with `outcome(#)` specifies that the index be computed for the outcome encoded by level #. Its approximate standard error is computed if the `stdp` option is specified. Only one of the options `pr`, `xb`, or `stdp` can be specified with a call to `predict`.

### ▷ Example 1

In the second example in [R] **slogit**, we fitted the one-dimensional stereotype model, where the dependent variable is `insure` with levels $k = 1$ for outcome *Indemnity*, $k = 2$ for *Prepaid*, and $k = 3$ for *Uninsure*. The base outcome for the model is *Indemnity*, so for $k \neq 1$ the vector of indices for the $k$th level is

$$\eta_k = \theta_k - \phi_k \left( \beta_1 \texttt{age} + \beta_2 \texttt{male} + \beta_3 \texttt{nonwhite} + \beta_4 \texttt{site2} + \beta_5 \texttt{site3} \right)$$

We estimate the group probabilities by calling `predict` after `slogit`.

```
. predict pIndemnity pPrepaid pUninsure, p
. list pIndemnity pPrepaid pUninsure insure in 1/10
```

| | pIndemn~y | pPrepaid | pUninsure | insure |
|---|---|---|---|---|
| 1. | .54193437 | .37548746 | .08257817 | Indemnity |
| 2. | .43596386 | .496328 | .06770813 | Prepaid |
| 3. | .51115833 | .41051074 | .07833093 | Indemnity |
| 4. | .39411326 | .54422342 | .06166332 | Prepaid |
| 5. | .46556509 | .46250642 | .07192849 | . |
| 6. | .44017793 | .49151025 | .06831183 | Prepaid |
| 7. | .46321218 | .46519304 | .07159478 | Prepaid |
| 8. | .37723017 | .5635696 | .05920023 | . |
| 9. | .48677573 | .43830181 | .07492246 | Uninsure |
| 10. | .58236678 | .32958016 | .08805306 | Prepaid |

Note here that observations 5 and 8 are not used to fit the model since `insure` is missing at these points, but `predict` estimates the probabilities for these observations since none of the independent variables is missing. You can use `if e(sample)` in the call to `predict` to use only those observations that are used to fit the model.

◁

# Methods and Formulas

All postestimation commands listed above are implemented as ado-files.

## predict

Let level $b$ be the base outcome that is used to fit the stereotype logistic regression model of dimension $d$. The index for observation $i$ and level $k \neq b$ is $\eta_{ik} = \theta_k - \sum_{j=1}^{d} \phi_{jk}\mathbf{x}_i\boldsymbol{\beta}_j$. This is the log odds of outcome encoded as level $k$ relative to that of $b$ so that we define $\eta_{ib} \equiv 0$. The outcome probabilities for this model are defined as $\Pr(Y_i = k) = e^{\eta_{ik}} / \sum_{j=1}^{m} e^{\eta_{ij}}$. Note that, unlike in `mlogit`, `ologit`, and `oprobit`, the index is no longer a linear function of the parameters. The standard error of index $\eta_{ik}$ is thus computed using the delta method (see also [R] **predictnl**).

The equation-level score for regression coefficients is

$$\frac{\partial \ln L_{ik}}{\partial \mathbf{x}_i\boldsymbol{\beta}_j} = \left( \sum_{l=1}^{m-1} \phi_{jl} p_{il} - \phi_{jk} \right)$$

the equation-level score for the scale parameters is

$$\frac{\partial \ln L_{ik}}{\partial \phi_{jl}} = \begin{cases} \mathbf{x}_i\boldsymbol{\beta}_j(p_{ik} - 1), & \text{if } l = k \\ \mathbf{x}_i\boldsymbol{\beta}_j p_{il}, & \text{if } l \neq k \end{cases}$$

for $l = 1, \ldots, m-1$; and the equation-level score for the intercepts is

$$\frac{\partial \ln L_{ik}}{\partial \theta_l} = \begin{cases} 1 - p_{ik}, & \text{if } l = k \\ -p_{il}, & \text{if } l \neq k \end{cases}$$

# Also See

**Complementary:** [R] **slogit**; [R] **estimates**, [R] **lincom**, [R] **lrtest**, [R] **mfx**, [R] **nlcom**, [R] **predictnl**, [R] **suest**, [R] **test**, [R] **testnl**

**Background:** [U] **13.5 Accessing coefficients and standard errors**, [U] **20 Estimation and postestimation commands**, [R] **estat**, [R] **predict**

# Title

**smooth** — Robust nonlinear smoother

# Syntax

`smooth` *smoother*[`,twice`] *varname* [*if*] [*in*], `generate(`*newvar*`)`

where *smoother* is specified as *Sm*[*Sm*[...]] and *Sm* is one of

```
{1|2|3|4|5|6|7|8|9}[R]
3[R]S[S|R][S|R]...
E
H
```

Letters may be specified in lowercase if preferred. Examples of *smoother*[`,twice`] include

| | | | | |
|---|---|---|---|---|
| `3RSSH` | `3RSSH,twice` | `4253H` | `4253H,twice` | `43RSR2H,twice` |
| `3rssh` | `3rssh,twice` | `4253h` | `4253h,twice` | `43rsr2h,twice` |

# Description

`smooth` applies the specified resistant, nonlinear smoother to *varname* and stores the smoothed series in *newvar*.

# Option

`generate(`*newvar*`)` is required; it specifies the name of the new variable that will contain the smoothed values.

# Remarks

Smoothing is an exploratory data-analysis technique for making the general shape of a series apparent. In this approach (Tukey 1977), the observed data series is assumed to be the sum of an underlying process that evolves smoothly (the smooth) and of an unsystematic noise component (the rough); that is,

$$\text{data} = \text{smooth} + \text{rough}$$

Smoothed values $z_t$ are obtained by taking medians (or some other location estimate) of each point in the original data $y_t$ and a few of the points around it. The number of points used is called the span of the smoother. Thus a span-3 smoother produces $z_t$ by taking the median of $y_{t-1}$, $y_t$, and $y_{t+1}$. `smooth` provides running median smoothers of spans 1 to 9—indicated by the digit that specifies their span. Median smoothers are resistant to isolated outliers, so they provide robustness to spikes in the data. Since the median is also a nonlinear operator, such smoothers are known as robust (or resistant) nonlinear smoothers.

smooth also provides the Hanning linear, nonrobust smoother, indicated by the letter H. Hanning is a span-3 smoother with binomial weights. Repeated applications of H—for example, HH, HHH, etc.—provide binomial smoothers of span 5, 7, etc. See Cox (2004, 1997) for a graphical application of this fact.

Because a single smoother usually cannot adequately separate the smooth from the rough, compound smoothers—multiple smoothers applied in sequence—are used. The smoother 35H, for instance, smooths the data with a span-3 median smoother, then smooths the result with a span-5 median smoother, and finally smooths that result with the Hanning smoother. smooth allows you to specify any number of smoothers in any sequence.

Three refinements can be combined with the running median and Hanning smoothers. First, the end points of a smooth can be given special treatment. This is specified by the E operator. Second, smoothing by 3, the span-3 running median, tends to produce flat-topped hills and valleys. The splitting operator, S, "splits" these repeated values, applies the end-point operator to them, and then "rejoins" the series. Finally, it is sometimes useful to repeat an odd-span median smoother or the splitting operator until the smooth no longer changes. Following a digit or an S with an R specifies this type of repetition.

Even the best smoother may fail to separate the smooth from the rough adequately. To guard against losing any systematic components of the data series, after smoothing, the smoother can be reapplied to the resulting rough, and any recovered signal can be added back to the original smooth. The twice operator specifies this procedure. More generally, an arbitrary smoother can be applied to the rough (using a second smooth command), and the recovered signal can be added back to the smooth. This more general procedure is called reroughing (Tukey 1977).

The details of each of the smoothers and operators are explained in *Methods and Formulas* below.

## ▷ Example 1

smooth is designed to recover the general features of a series that has been contaminated with noise. To demonstrate this, we construct a series, add noise to it, and then smooth the noisy version to recover an estimate of the original data. First, we construct and display the data:

```
. drop _all
. set obs 10
. set seed 123456789
. generate time = _n
. label variable time "Time"
. generate x = _n^3 - 10*_n^2 + 5*_n
. label variable x "Signal"
. generate z = x + 50*invnormal(uniform())
. label variable z "Observed series"
```

```
. scatter x z time, c(l .) m(i o) ytitle("")
```

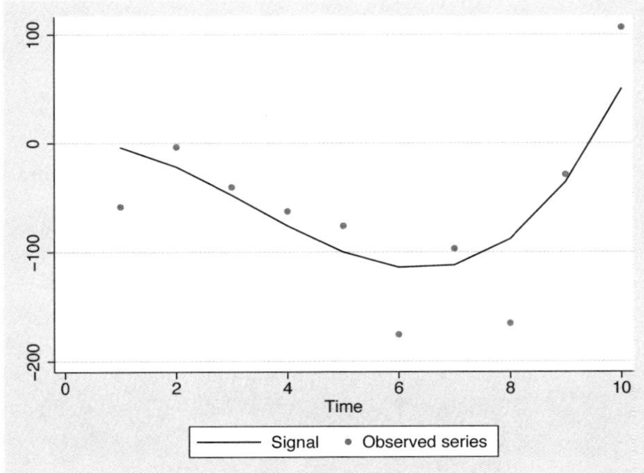


Now we smooth the noisy series, z, assumed to be the only data we would observe:

```
. smooth 4253eh,twice z, gen(sz)
. label variable sz "Smoothed series"
. scatter x z sz time, c(l . l) m(i o i) ytitle("") || scatter sz time,
> c(l . l) m(i o i) ytitle("") clpatter(dash_dot)
```

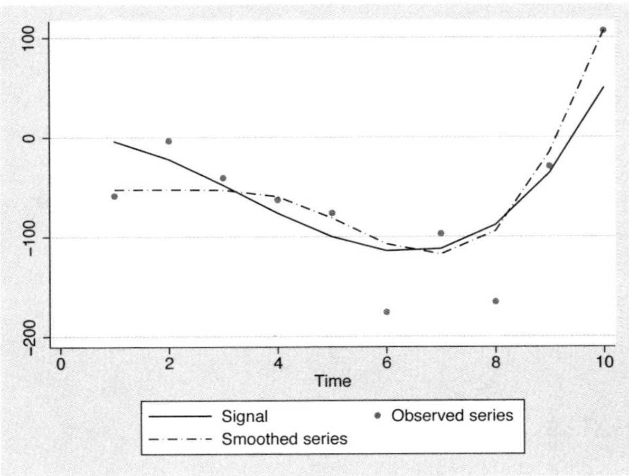


◁

## ▷ Example 2

Salgado-Ugarte and Curts-García (1993) provide data on the frequencies of observed fish lengths. In this example, the series to be smoothed—the frequencies—is ordered by fish length rather than by time.

```
. use http://www.stata-press.com/data/r9/fishdata, clear
. smooth 4253eh,twice freq, gen(sfreq)
. label var sfreq "4253EH,twice of frequencies"
```

```
. scatter sfreq freq length, c(l .) m(i o) title("Smoothed frequencies of fish lengths")
ytitle("") xlabel(#4)
```

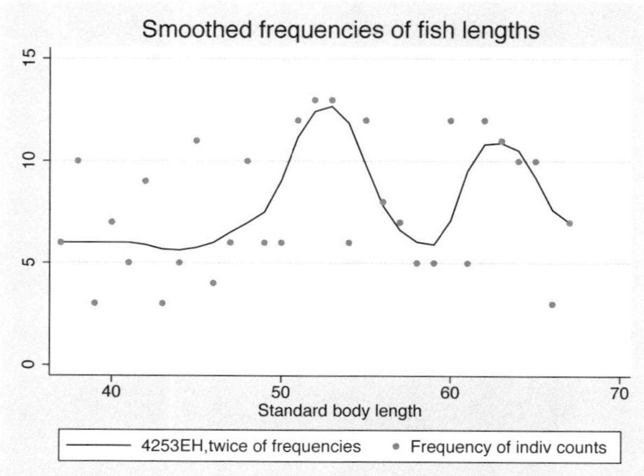


◁

❑ Technical Note

`smooth` allows missing values at the beginning and end of the series, but missing values in the middle are not allowed. Leading and trailing missing values are ignored. If you wish to ignore missing values in the middle of the series, you must `drop` the missing observations before using `smooth`. Doing so, of course, would violate `smooth`'s assumption that observations are equally spaced—each observation represents a year, a quarter, or a month (or a one-year birth-rate category). In practice, `smooth` produces good results as long as the spaces between adjacent observations do not vary too much.

Smoothing is usually applied to time series, but any variable with a natural order can be smoothed. For example, a smoother might be applied to the birth rate recorded by the age of the mothers (birth rate for 17-year-olds, birth rate for 18-year-olds, and so on).

❑

## Methods and Formulas

`smooth` is implemented as an ado-file.

### Running median smoothers of odd span

The smoother `3` defines

$$z_t = \text{median}(y_{t-1}, y_t, y_{t+1})$$

The smoother `5` defines

$$z_t = \text{median}(y_{t-2}, y_{t-1}, y_t, y_{t+1}, y_{t+2})$$

and so on. The smoother `1` defines $z_t = \text{median}(y_t)$, so it does nothing.

End points are handled by using smoothers of shorter, odd span. Thus in the case of 3,

$$
\begin{aligned}
z_1 &= y_1 \\
z_2 &= \text{median}(y_1, y_2, y_3) \\
&\vdots \\
z_{N-1} &= \text{median}(y_{N-2}, y_{N-1}, y_N) \\
Z_N &= y_N
\end{aligned}
$$

In the case of 5,

$$
\begin{aligned}
z_1 &= y_1 \\
z_2 &= \text{median}(y_1, y_2, y_3) \\
z_3 &= \text{median}(y_1, y_2, y_3, y_4, y_5) \\
z_4 &= \text{median}(y_2, y_3, y_4, y_5, y_6) \\
&\vdots \\
z_{N-2} &= \text{median}(y_{N-4}, y_{N-3}, y_{N-2}, y_{N-1}, y_N) \\
z_{N-1} &= \text{median}(y_{N-2}, y_{N-1}, y_N) \\
Z_N &= y_N
\end{aligned}
$$

and so on.

## Running median smoothers of even span

Define the median() function as returning the linearly interpolated value when given an even number of arguments. Thus the smoother 2 defines

$$z_{t+.5} = (y_t + y_{t+1})/2$$

The smoother 4 defines $z_{t+.5}$ as the linearly interpolated median of $(y_{t-1}, y_t, y_{t+1}, y_{t+2})$, and so on. In all cases, end points are handled by using smoothers of shorter, even span. Thus in the case of 4,

$$
\begin{aligned}
z_{.5} &= y_1 \\
z_{1.5} &= \text{median}(y_1, y_2) = (y_1 + y_2)/2 \\
z_{2.5} &= \text{median}(y_1, y_2, y_3, y_4) \\
&\vdots \\
z_{N-2.5} &= \text{median}(y_{N-4}, y_{N-3}, y_{N-2}, y_N) \\
z_{N-1.5} &= \text{median}(y_{N-2}, y_{N-1}) \\
z_{N-.5} &= \text{median}(y_{N-1}, y_N) \\
z_{N+.5} &= y_N
\end{aligned}
$$

As defined above, an even-span smoother increases the length of the series by one observation. However, the series can be recentered on the original observation numbers, and the "extra" observation can be eliminated by smoothing the series again with another even-span smoother. For instance, the smooth of 4 illustrated above could be followed by a smooth of 2 to obtain

$$
\begin{aligned}
z_1^* &= (z_{.5} + z_{1.5})/2 \\
z_2^* &= (z_{1.5} + z_{2.5})/2 \\
z_3^* &= (z_{2.5} + z_{3.5})/2 \\
&\vdots \\
z_{N-2}^* &= (z_{N-2.5} + z_{N-1.5})/2 \\
z_{N-1}^* &= (z_{N-1.5} + z_{N-.5})/2 \\
z_N^* &= (z_{N-.5} + z_{N+.5})/2
\end{aligned}
$$

`smooth` keeps track of the number of even smoothers applied to the data and expands and shrinks the length of the series accordingly. To ensure that the final smooth has the same number of observations as *varname*, `smooth` requires you to specify an even number of even-span smoothers. However, the pairs of even-span smoothers need not be contiguous; for instance, `4253` and `4523` are both allowed.

### Repeat operator

`R` indicates that a smoother is to be repeated until convergence, that is, until repeated applications of the smoother produce the same series. Thus `3` applies the smoother of running medians of span 3. `33` applies the smoother twice. `3R` produces the result of repeating `3` an infinite number of times. `R` should only be used with odd-span smoothers since even-span smoothers are not guaranteed to converge.

The smoother `453R2` applies a span-4 smoother, followed by a span-5 smoother, followed by repeated applications of a span-3 smoother, followed by a span-2 smoother.

### End-point rule

The end-point rule `E` modifies the values $z_1$ and $z_N$ according to the following formulas:

$$
\begin{aligned}
z_1 &= \text{median}(3z_2 - 2z_3, z_1, z_2) \\
z_N &= \text{median}(3z_{N-2} - 2z_{N-1}, z_N, z_{N-1})
\end{aligned}
$$

When the end-point rule is not applied, end points are typically "copied in"; that is, $z_1 = y_1$ and $z_N = y_N$.

### Splitting operator

The smoothers `3` and `3R` can produce flat-topped hills and valleys. The split operator attempts to eliminate such hills and valleys by splitting the sequence, applying the end-point rule `E`, rejoining the series, and then resmoothing by `3R`.

The `S` operator may be applied only after `3`, `3R`, or `S`.

We recommend that the `S` operator be repeated once (`SS`) or until no further changes take place (`SR`).

### Hanning smoother

`H` is the Hanning linear smoother:

$$z_t = (y_{t-1} + 2y_t + y_{t+1})/4$$

End points are copied in: $z_1 = y_1$ and $z_N = y_N$. `H` should be applied only after all nonlinear smoothers.

### Twicing

A smoother divides the data into a smooth and a rough:

$$\text{data} = \text{smooth} + \text{rough}$$

If the smoothing is successful, the rough should exhibit no pattern. Twicing refers to applying the smoother to the observed, calculating the rough, and then applying the smoother to the rough. The resulting "smoothed rough" is then added back to the smooth from the first step.

## Acknowledgments

`smooth` was originally written by William Gould (1992)—at which time it was named `nlsm`—and was inspired by Salgado-Ugarte and Curts-García (1992). Salgado-Ugarte and Curts-García (1993) subsequently reported anomalies in `nlsm`'s treatment of even-span median smoothers. `smooth` corrects these problems and incorporates other improvements but otherwise is essentially the same as originally published.

## References

Cox, N. J. 2004. Software update for bsmplot. *Stata Journal* 4: 490.

——. 1997. gr22: Binomial smoothing plot. *Stata Technical Bulletin* 35: 7–9. Reprinted in *Stata Technical Bulletin Reprints*, vol. 6, pp. 36–38.

Gould, W. W. 1992. sed7.1: Resistant nonlinear smoothing using Stata. *Stata Technical Bulletin* 8: 9–12. Reprinted in *Stata Technical Bulletin Reprints*, vol. 2, pp. 104–107.

Salgado-Ugarte, I. and J. Curts-García. 1992. sed7: Resistant smoothing using Stata. *Stata Technical Bulletin* 7: 8–11. Reprinted in *Stata Technical Bulletin Reprints*, vol. 2, pp. 99–103.

——. 1993. sed7.2: Twice reroughing procedure for resistant nonlinear smoothing. *Stata Technical Bulletin* 11: 14–16. Reprinted in *Stata Technical Bulletin Reprints*, vol. 2, pp. 108–111.

Sasieni, P. 1998. gr27: An adaptive variable span running line smoother. *Stata Technical Bulletin* 41: 4–7. Reprinted in *Stata Technical Bulletin Reprints*, vol. 7, pp. 63–68.

Tukey, J. W. 1977. *Exploratory Data Analysis*. Reading, MA: Addison–Wesley.

Velleman, P. F. 1977. Robust nonlinear data smoothers: Definitions and recommendations. *Proceedings of the National Academy of Sciences USA* 74(2): 434–436.

——. 1980. Definition and comparison of robust nonlinear data smoothing algorithms. *Journal of the American Statistical Association* 75(371): 609–615.

Velleman, P. F. and D. C. Hoaglin. 1981. *Applications, Basics, and Computing of Exploratory Data Analysis*. Boston: Duxbury.

## Also See

**Related:** [R] **lowess**,
[TS] **tssmooth**

# Title

**spearman** — Spearman's and Kendall's correlations

# Syntax

*Spearman's rank correlation coefficients*

spearman [*varlist*] [*if*] [*in*] [, *spearman_options*]

*Kendall's rank correlation coefficients*

ktau [*varlist*] [*if*] [*in*] [, *ktau_options*]

| *spearman_options* | description |
|---|---|
| Main | |
| stats(*spearman_list*) | list of statistics; select up to 3 statistics; default is stats(rho) |
| print(#) | significance level for displaying coefficients |
| star(#) | significance level for displaying with a star |
| bonferroni | use Bonferroni-adjusted significance level |
| sidak | use Šidák-adjusted significance level |
| pw | calculate all the pairwise correlation coefficients between the variables |
| matrix | display output in matrix form |

| *ktau_options* | description |
|---|---|
| Main | |
| stats(*ktau_list*) | list of statistics; select up to 6 statistics; default is stats(taua) |
| print(#) | significance level for displaying coefficients |
| star(#) | significance level for displaying with a star |
| bonferroni | use Bonferroni-adjusted significance level |
| sidak | use Šidák-adjusted significance level |
| pw | calculate all the pairwise correlation coefficients between the variables |
| matrix | display output in matrix form |

by may be used with spearman and ktau; see [D] **by**.

where the elements of *spearman_list* may be

| | |
|---|---|
| rho | correlation coefficient |
| obs | number of observations |
| p | significance level |

and the elements of *ktau_list* may be

| | |
|---|---|
| taua | correlation coefficient $\tau_a$ |
| taub | correlation coefficient $\tau_b$ |
| score | score |
| se | standard error of score |
| obs | number of observations |
| p | significance level |

# Description

spearman displays Spearman's rank correlation coefficients for all pairs of variables in *varlist* or, if *varlist* is not specified, all the variables in the dataset.

ktau displays Kendall's rank correlation coefficients between the variables in *varlist* or, if *varlist* is not specified, all the variables in the dataset. ktau is intended for use on small and moderate-sized datasets; it requires considerable computation time for larger datasets.

# Options for spearman

Main

stats(*spearman_list*) specifies the statistics to be displayed in the matrix of output. stats(rho) is the default. Up to 3 statistics may be specified. stats(rho obs p) would display the correlation coefficient, number of observations, and significance level. If *varlist* contains only two variables, all statistics are shown in tabular form, and stats(), print(), and star() have no effect unless the matrix option is specified.

print(*#*) specifies the significance level of correlation coefficients to be printed. Correlation coefficients with larger significance levels are left blank in the matrix. Typing spearman, print(.10) would list only those correlation coefficients that are significant at the 10% level or lower.

star(*#*) specifies the significance level of correlation coefficients to be marked with a star. Typing spearman, star(.05) would "star" all correlation coefficients significant at the 5% level or lower.

bonferroni makes the Bonferroni adjustment to calculated significance levels. This affects printed significance levels and the print() and star() options. Thus spearman, print(.05) bonferroni prints coefficients with Bonferroni-adjusted significance levels of .05 or less.

sidak makes the Šidák adjustment to calculated significance levels. This affects printed significance levels and the print() and star() options. Thus spearman, print(.05) sidak prints coefficients with Šidák-adjusted significance levels of .05 or less.

pw specifies that correlations be calculated using pairwise deletion of observations with missing values. By default, spearman uses casewise deletion, where observations are ignored if any of the variables in *varlist* are missing.

matrix forces spearman to display the statistics as a matrix, even if *varlist* contains only two variables. matrix is implied if more than two variables are specified.

# Options for ktau

Main

stats(*ktau_list*) specifies the statistics to be displayed in the matrix of output. stats(taua) is the default. Up to 6 statistics may be specified. stats(taua taub score se obs p) would display the correlation coefficients $\tau_a$, $\tau_b$, score, standard error of score, number of observations, and significance level. If *varlist* contains only two variables, all statistics are shown in tabular form and stats(), print(), and star() have no effect unless the matrix option is specified.

print(*#*) specifies the significance level of correlation coefficients to be printed. Correlation coefficients with larger significance levels are left blank in the matrix. Typing ktau, print(.10) would list only those correlation coefficients that are significant at the 10% level or lower.

`star(#)` specifies the significance level of correlation coefficients to be marked with a star. Typing `ktau, star(.05)` would "star" all correlation coefficients significant at the 5% level or lower.

`bonferroni` makes the Bonferroni adjustment to calculated significance levels. This affects printed significance levels and the `print()` and `star()` options. Thus `ktau, print(.05) bonferroni` prints coefficients with Bonferroni-adjusted significance levels of .05 or less.

`sidak` makes the Šidák adjustment to calculated significance levels. This affects printed significance levels and the `print()` and `star()` options. Thus `ktau, print(.05) sidak` prints coefficients with Šidák-adjusted significance levels of .05 or less.

`pw` specifies that correlations be calculated using pairwise deletion of observations with missing values. By default, `ktau` uses casewise deletion, where observations are ignored if any of the variables in *varlist* are missing.

`matrix` forces `ktau` to display the statistics as a matrix, even if *varlist* contains only two variables. `matrix` is implied if more than two variables are specified.

# Remarks

## ▷ Example 1

We wish to calculate the correlation coefficients among marriage rate (`mrgrate`), divorce rate (`divorce_rate`), and median age (`medage`) in state data. We can calculate the standard Pearson correlation coefficients and significance by typing

```
.use http://www.stata-press.com/data/r9/states2
(State data)

. pwcorr mrgrate divorce_rate medage, sig

             |  mrgrate divorc~e   medage
-------------+---------------------------
     mrgrate |   1.0000
             |
             |
divorce_rate |   0.7895   1.0000
             |   0.0000
             |
      medage |   0.0011  -0.1526   1.0000
             |   0.9941   0.2900
             |
```

(Continued on next page)

We can calculate Spearman's rank correlation coefficients by typing

```
. spearman mrgrate divorce_rate medage, stats(rho p)
(obs=50)

  +-----------------+
  | Key             |
  |-----------------|
  |   rho           |
  |   Sig. level    |
  +-----------------+

             |  mrgrate divorc~e   medage
-------------+---------------------------
     mrgrate |   1.0000
             |
             |
divorce_rate |   0.6933   1.0000
             |   0.0000
             |
      medage |  -0.4869  -0.2455   1.0000
             |   0.0003   0.0857
             |
```

The large difference in the results is caused by a single observation. Nevada's marriage rate is almost 10 times higher than the state with the next-highest marriage rate. An important feature of the Spearman rank correlation coefficient is its reduced sensitivity to extreme values compared to the Pearson correlation coefficient.

We can calculate Kendall's rank correlations by typing

```
. ktau mrgrate divorce_rate medage, stats(taua taub p)
(obs=50)

  +-----------------+
  | Key             |
  |-----------------|
  |   tau_a         |
  |   tau_b         |
  |   Sig. level    |
  +-----------------+

             |    mrgrate    divorc~e      medage
-------------+------------------------------------
     mrgrate |     0.9829
             |     1.0000
             |
             |
divorce_rate |     0.5110      0.9804
             |     0.5206      1.0000
             |     0.0000
             |
      medage |    -0.3486     -0.1698      0.9845
             |    -0.3544     -0.1728      1.0000
             |     0.0004      0.0828
             |
```

There are tied values for variables **mrgrate**, **divorce_rate**, and **medage**, so tied ranks are used. As a result, $\tau_a < 1$ on the diagonal (see *Methods and Formulas* for the definition of $\tau_a$).

◁

### ❑ Technical Note

According to Conover (1999, 323), "Spearman's $\rho$ tends to be larger than Kendall's $\tau$ in absolute value. However, as a test of significance, there is no strong reason to prefer one over the other because both will produce nearly identical results in most cases". ❑

### ▷ Example 2

We illustrate `spearman` and `ktau` using the auto data, which contains some missing values.

```
.use http://www.stata-press.com/data/r9/auto
(1978 Automobile Data)

. spearman mpg rep78

 Number of obs =      69
Spearman's rho =       0.3098

Test of Ho: mpg and rep78 are independent
    Prob > |t| =       0.0096
```

Because we specified two variables, `spearman` displayed the sample size, correlation, and $p$-value in tabular form. To obtain just the correlation coefficient displayed in matrix form, we type

```
. spearman mpg rep78, stats(rho) matrix
(obs=69)

             |      mpg    rep78
-------------+------------------
         mpg |   1.0000
       rep78 |   0.3098   1.0000
```

The `pw` option instructs `spearman` and `ktau` to use all nonmissing observations between a pair of variables when calculating their correlation coefficient. Notice in the output below that some correlations are based on 74 observations, while others are based on 69 because 5 observations contain a missing value for `rep78`.

```
. spearman mpg price rep78, pw stats(rho obs p)  star(0.01)

+-------------------+
| Key               |
|-------------------|
|  rho              |
|  Number of obs    |
|  Sig. level       |
+-------------------+

             |      mpg    price    rep78
-------------+---------------------------
         mpg |   1.0000
             |       74
             |
             |
       price |  -0.5419*  1.0000
             |       74       74
             |   0.0000
             |
       rep78 |   0.3143*  0.1025   1.0000
             |       69       69       69
             |   0.0085   0.4022
             |
```

Finally, the `bonferroni` and `sidak` options provide adjusted significance levels:

```
. ktau mpg price rep78, stats(taua taub score se p) bonferroni
(obs=69)

+-----------------+
| Key             |
|-----------------|
|   tau_a         |
|   tau_b         |
|   score         |
|   se of score   |
|   Sig. level    |
+-----------------+

             |       mpg     price     rep78
-------------+------------------------------
         mpg |    0.9471
             |    1.0000
             | 2222.0000
             |  191.8600
             |
             |
       price |   -0.3973    1.0000
             |   -0.4082    1.0000
             | -932.0000 2346.0000
             |  192.4561  193.0682
             |    0.0000
             |
       rep78 |    0.2076    0.0648    0.7136
             |    0.2525    0.0767    1.0000
             |  487.0000  152.0000 1674.0000
             |  181.7024  182.2233  172.2161
             |    0.0224    1.0000
             |
```

◁

Charles Edward Spearman (1863–1945) was a British psychologist who made contributions to correlation, factor analysis, test reliability, and psychometrics. After several years' military service, he obtained a Ph.D. in experimental psychology at Leipzig and became a professor at University College London, where he sustained a long program of work on the interpretation of intelligence tests. Ironically, the rank correlation version bearing his name is not the formula he advocated.

Maurice George Kendall (1907–1983) was a British statistician who contributed to rank correlation, time series, multivariate analysis, and other topics, and wrote many statistical texts. Most notably, perhaps, his advanced survey of the theory of statistics went through several editions, latterly with Alan Stuart; the baton has since passed to others. Kendall was employed in turn as a government and business statistician, as professor at the London School of Economics, as a consultant, and as director of the World Fertility Survey. He was knighted in 1974.

## Saved Results

`spearman` saves in `r()`:

Scalars

| | | | |
|---|---|---|---|
| `r(N)` | number of observations (last variable pair) | `r(p)` | two-sided $p$-value (last variable pair) |
| `r(rho)` | $\rho$ (last variable pair) | | |

Matrices

| | | | |
|---|---|---|---|
| `r(Nobs)` | number of observations | `r(P)` | two-sided $p$-value |
| `r(Rho)` | $\rho$ | | |

`ktau` saves in `r()`:

Scalars

| | | | |
|---|---|---|---|
| `r(N)` | number of observations (last variable pair) | `r(score)` | Kendall's score (last variable pair) |
| `r(tau_a)` | $\tau_{\rm a}$ (last variable pair) | `r(se_score)` | s.e. of score (last variable pair) |
| `r(tau_b)` | $\tau_{\rm b}$ (last variable pair) | `r(p)` | two-sided $p$-value (last variable pair) |

Matrices

| | | | |
|---|---|---|---|
| `r(Nobs)` | number of observations | `r(Score)` | Kendall's score |
| `r(Tau_a)` | $\tau_{\rm a}$ | `r(Se_Score)` | standard error of score |
| `r(Tau_b)` | $\tau_{\rm b}$ | `r(P)` | two-sided $p$-value |

## Methods and Formulas

`spearman` and `ktau` are implemented as ado-files.

Spearman's (1904) rank correlation is calculated as Pearson's correlation computed on the ranks and average ranks (Conover 1999, 314–315). Ranks are as calculated by `egen`; see [D] **egen**. The significance is calculated using the approximation

$$p = 2 \times \texttt{ttail}(n-2,\, |\widehat{\rho}|\,\sqrt{n-2}\,/\sqrt{1-\widehat{\rho}^2}\,)$$

For any two pairs of ranks $(x_i,\, y_i)$ and $(x_j,\, y_j)$ of one variable pair (*varname*$_1$, *varname*$_2$), $1 \le i,\, j \le n$, where $n$ is the number of observations, define them as concordant if

$$(x_i - x_j)(y_i - y_j) > 0$$

and discordant if this product is less than zero.

Kendall's (1938; also see Kendall and Gibbons 1990) score $S$ is defined as $C - D$, where $C$ ($D$) is the number of concordant (discordant) pairs. Let $N = n(n-1)/2$ be the total number of pairs, so $\tau_{\rm a}$ is given by

$$\tau_{\rm a} = S/N$$

and $\tau_{\rm b}$ is given by

$$\tau_{\rm b} = \frac{S}{\sqrt{N-U}\sqrt{N-V}},$$

where

$$U = \sum_{i=1}^{N_1} u_i(u_i - 1)/2$$

$$V = \sum_{j=1}^{N_2} v_j(v_j - 1)/2$$

and where $N_1$ is the number of sets of tied $x$ values, $u_i$ is the number of tied $x$ values in the $i$th set, $N_2$ is the number of sets of tied $y$ values, and $v_j$ is the number of tied $y$ values in the $j$th set. Under the null hypothesis of independence between *varname*$_1$ and *varname*$_2$, the variance of $S$ is exactly (Kendall and Gibbons 1990, 66)

$$\begin{aligned}\text{Var}(S) =& \frac{1}{18}\left\{n(n-1)(2n+5) - \sum_{i=1}^{N_1} u_i(u_i-1)(2u_i+5) - \sum_{j=1}^{N_2} v_j(v_j-1)(2v_j+5)\right\} \\ &+ \frac{1}{9n(n-1)(n-2)}\left\{\sum_{i=1}^{N_1} u_i(u_i-1)(u_i-2)\right\}\left\{\sum_{j=1}^{N_2} v_j(v_j-1)(v_j-2)\right\} \\ &+ \frac{1}{2n(n-1)}\left\{\sum_{i=1}^{N_1} u_i(u_i-1)\right\}\left\{\sum_{j=1}^{N_2} v_j(v_j-1)\right\}.\end{aligned}$$

Using a normal approximation with a continuity correction,

$$z = \frac{|S| - 1}{\sqrt{\text{Var}(S)}}$$

Note that for the hypothesis of independence, the statistics $S$, $\tau_\text{a}$, and $\tau_\text{b}$ produce equivalent tests and give the same significance.

In the case of Kendall's $\tau$, the normal approximation is surprisingly accurate for sample sizes as small as 8, at least for calculating $p$-values under the null hypothesis for continuous variables. (See Kendall and Gibbons, 1990, chapter 4, who also present some tables for calculating exact $p$-values for $n < 10$.) In the case of Spearman's $\rho$, the normal approximation requires larger samples to be valid.

Let $v$ be the number of variables specified so that $k = v(v-1)/2$ correlation coefficients are to be estimated. If `bonferroni` is specified, the adjusted significance level is $p' = \min(1, kp)$. If `sidak` is specified, $p' = \min\left\{1, 1 - (1-p)^n\right\}$. See *Methods and Formulas* in [R] **oneway** for a more complete description of the logic behind these adjustments.

Early work on rank correlation is surveyed by Kruskal (1958).

## Acknowledgment

The original version of `ktau` was written by Sean Becketti, a past editor of the *Stata Technical Bulletin*.

## References

Barnard, G. A. 1997. Kendall, Maurice George. In *Leading Personalities in Statistical Sciences from the Seventeenth Century to the Present*, ed. N. L. Johnson and S. Kotz, 130–132. New York: Wiley.

Conover, W. J. 1999. *Practical Nonparametric Statistics*. 3rd ed. New York: Wiley.

Jeffreys, H. 1961. *Theory of Probability*. Oxford: Oxford University Press.

Kendall, M. G. 1938. A new measure of rank correlation. *Biometrika* 30: 81–93.

Kendall, M. G. and J. D. Gibbons. 1990. *Rank Correlation Methods*. 5th ed. New York: Oxford University Press.

Kruskal, W. H. 1958. Ordinal measures of association. *Journal of the American Statistical Association* 53: 814–861.

Lovie, P. and A. D. Lovie. 1996. Charles Edward Spearman, F.R.S. (1863–1945). *Notes and Records of the Royal Society of London* 50: 1–14.

Newson, R. 2000a. snp15: somersd—Confidence intervals for nonparametric statistics and their differences. *Stata Technical Bulletin* 55: 47–55. Reprinted in *Stata Technical Bulletin Reprints*, vol. 10, pp. 312–322.

——. 2000b. snp15.1: Update to somersd. *Stata Technical Bulletin* 57: 35. Reprinted in *Stata Technical Bulletin Reprints*, vol. 10, pp. 322–323.

——. 2000c. snp15.2: Update to somersd. *Stata Technical Bulletin* 58: 30. Reprinted in *Stata Technical Bulletin Reprints*, vol. 10, p. 323.

——. 2001. snp15.3: Update to somersd. *Stata Technical Bulletin* 61: 22. Reprinted in *Stata Technical Bulletin Reprints*, vol. 10, p. 324.

Seed, P. T. 2001. sg159: Confidence intervals for correlations. *Stata Technical Bulletin* 59: 27–28. Reprinted in *Stata Technical Bulletin Reprints*, vol. 10, pp. 267–269.

Spearman, C. 1904. The proof and measurement of association between two things. *American Journal of Psychology* 15: 72–101.

Wolfe, F. 1997. sg64: pwcorrs: An enhanced correlation display. *Stata Technical Bulletin* 35: 22–25. Reprinted in *Stata Technical Bulletin Reprints*, vol. 6, pp. 163–167.

——. 1999. sg64.1: Update to pwcorrs. *Stata Technical Bulletin* 49: 17. Reprinted in *Stata Technical Bulletin Reprints*, vol. 9, p. 159.

## Also See

**Related:** [R] **correlate**, [R] **nptrend**

# Title

**spikeplot** — Spike plots and rootograms

# Syntax

`spikeplot` *varname* [*if*] [*in*] [*weight*] [, *options*]

| *options* | description |
|---|---|
| Main | |
| `round(#)` | round *varname* to nearest multiple of # (bin width) |
| `fraction` | make vertical scale the proportion of total values; default is frequencies |
| `root` | make vertical scale show square roots of frequencies |
| Plot | |
| *spike_options* | affect rendition of plotted spikes |
| Add plot | |
| `addplot(`*plot*`)` | add other plots to generated graph |
| Y-Axis, X-Axis, Title, Caption, Legend, Overall, By | |
| *twoway_options* | any options documented in [G] ***twoway_options*** |

`fweight`s, `aweight`s, and `iweight`s are allowed; see [U] **11.1.6 weight**

# Description

`spikeplot` produces a frequency plot for a variable in which the frequencies are depicted as vertical lines from zero. The frequency may be a count, a fraction, or the square root of the count (Tukey's rootogram, *circa* 1965). The vertical lines may also originate from a different baseline than zero at the user's option.

# Options

Main

`round(#)` rounds the values of *varname* to the nearest multiple of #. This effectively specifies the bin width.

`fraction` specifies that the vertical scale be the proportion of total values (percentage) rather than the count.

`root` specifies that the vertical scale show square roots. This option may not be specified if `frac` is specified.

Plot

*spike_options* affect the rendition of the plotted spikes; see [G] **graph twoway spike**.

Add plot

addplot(*plot*) provides a way to add other plots to the generated graph. See [G] ***addplot_option***.

Y-Axis, X-Axis, Title, Caption, Legend, Overall, By

*twoway_options* are any of the options documented in [G] ***twoway_options***. These include options for titling the graph (see [G] ***title_options***), options for saving the graph to disk (see [G] ***saving_option***), and the by() option (see [G] ***by_option***).

# Remarks

## ▷ Example 1

Cox and Brady (1997a) present an illustrative example using the age structure of the population of Ghana from the 1960 census (rounded to the nearest 1,000). The dataset has ages from 0 (less than one year) to 90. To view the distribution of ages, we would like to use each integer from 0 to 90 as the bins for the dataset.

```
. use http://www.stata-press.com/data/r9/ghanaage
. spikeplot age [fw=pop], ytitle("Population in 1000s") xlab(0(10)90) xmtick(5(10)85)
```

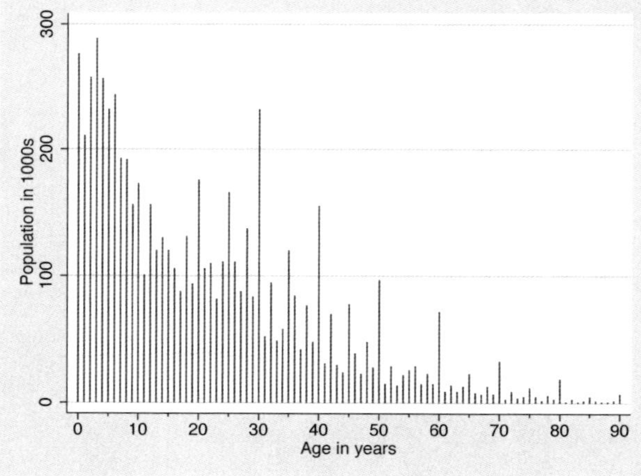


The resulting graph shows a "heaping" of ages at the multiples of 5. In addition, ages ending in even numbers are more frequent than ages ending in odd numbers (except for 5). This preference for reporting ages is well known in demography and other social sciences.

Note also that we used the ytitle() option to override the default title of "Frequency" and that we used the xlab() and xmtick() options with *numlists* to further customize the resulting graph. See [U] **11.1.8 numlist** for details on specifying *numlists*.

◁

## ▷ Example 2

The rootogram is a plot of the square-root transformation of the frequency counts. Note that the square root of a normal distribution is a multiple of another normal distribution:

```
. clear
. set seed 1234567
. set obs 5000
obs was 0, now 5000
. generate normal = invnormal(uniform())
. label variable normal "Gaussian(0,1) random numbers"
. spikeplot normal, round(.10) xlab(-4(1)4)
```

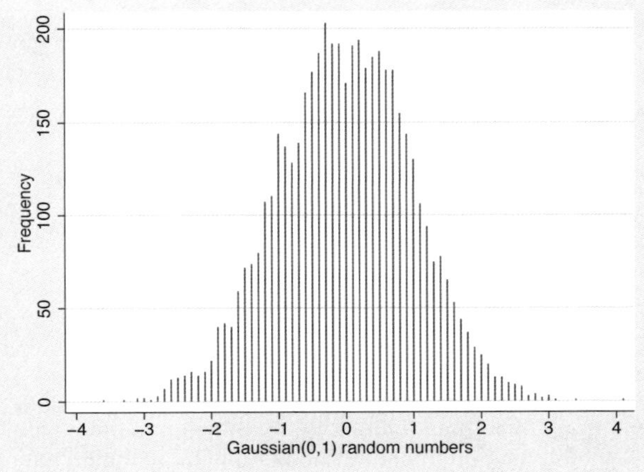


```
. spikeplot normal, round(.10) xlab(-4(1)4) root
```

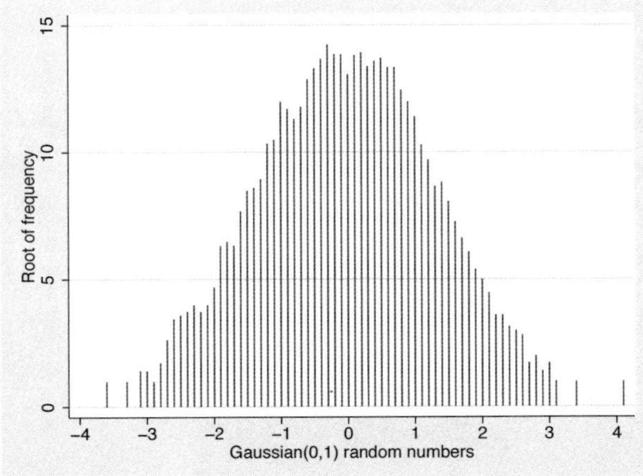


Interpreting a histogram in terms of normality is thus similar to interpreting the rootogram for normality.

This example also shows how the `round()` option is used to bin the values for a spike plot of a continuous variable. ◁

▷ Example 3

`spikeplot` can also be used to produce time-series plots. *varname* should be the time variable, and weights should be specified as the values for those times. To get a plot of daily rainfalls, we type

```
. spikeplot day [w=rain] if rain, ytitle("Daily rainfall in mm")
```

The `base()` option of `graph twoway spike` may be used to set a different baseline, such as when we want to show variations relative to an average or to some other measure of level. ◁

## Methods and Formulas

`spikeplot` is implemented as an ado-file.

## Acknowledgments

The original version of `spikeplot` was written by Nicholas J. Cox of the University of Durham and Anthony R. Brady of the Imperial College School of Medicine (1997).

## References

Cox, N. J. and A. R. Brady. 1997a. gr25: Spike plots for histograms, rootograms, and time-series plots. *Stata Technical Bulletin* 36: 8–11. Reprinted in *Stata Technical Bulletin Reprints*, vol. 6, pp. 50–54.

——. 1997b. gr25.1: Spike plots for histograms, rootograms, and time series plots: Update. *Stata Technical Bulletin* 40: 12. Reprinted in *Stata Technical Bulletin Reprints*, vol. 7, p. 58.

Tukey, J. W. 1965. The future of processes of data analysis. Reprinted in *The Collected Works of John W. Tukey, Volume IV: Philosophy and Principles of Data Analysis: 1965–1986*, ed. L. V. Jones, 517–547 (1986). Monterey, CA: Wadsworth & Brooks/Cole.

## Also See

**Related:** [R] **histogram**

**Background:** *Stata Graphics Reference Manual*

# Title

**ssc** — Install and uninstall packages from SSC

# Syntax

*Summary of packages most recently added or updated at SSC*

ssc whatsnew [, saving(*filename*[, replace]) type]

*Describe a specified package at SSC*

ssc describe {*pkgname* | *letter*} [, saving(*filename*[, replace])]

*Install a specified package from SSC*

ssc install *pkgname* [, all replace]

*Uninstall from your computer a previously installed package from SSC*

ssc uninstall *pkgname*

*Type a specific file stored at SSC*

ssc type *filename*

*Copy a specific file from SSC to your computer*

ssc copy *filename* [, plus personal replace public binary]

where *letter* in ssc describe is a–z or _.

# Description

ssc works with packages (and files) from the Statistical Software Components (SSC) archive, which is often called the Boston College Archive and is provided by *http://repec.org*.

The SSC has become the premier Stata download site for user-written software on the web. ssc provides a convenient command-line interface to the resources available there. For example, on Statalist (see *http://www.stata.com/statalist/*), users will often write

> The program can be found by typing ssc install newprogramname.

Typing that would load everything associated with newprogramname, including the help files, if a package of that name existed. Since there is no package named newprogramname, typing the above literally would just produce an error message, but no harm would be done.

If you are searching for what is available, see [R] **search**. search searches and provides a GUI interface from which programs can be loaded, including the programs at the SSC archive.

### Command overview

ssc whatsnew summarizes the packages made available or updated recently. Output is presented in the Stata Viewer, and from there you may click to find out more or install.

ssc describe *pkgname* describes, but does not install, the specified package. Use search to find packages; see [R] **search**. If you know the package name but do not know the exact spelling, type ssc describe followed by one letter, a–z or _ (underscore), to list all the packages starting with that letter.

ssc install *pkgname* installs the specified package. You do not have to describe a package before installing it. You may also install a package using net; see [R] **net**.

ssc uninstall *pkgname* removes the previously installed package from your computer. It does not matter whether the package was installed using ssc install or using some other method, such as net. (ssc uninstall is a synonym for ado uninstall, so it can uninstall any installed package, not just packages obtained from SSC.)

ssc type *filename* types a specific file stored at SSC. ssc cat is a synonym for ssc type, which may appeal to those familiar with Unix.

ssc copy *filename* copies a specific file stored at SSC to your computer. By default, the file is copied to the current directory, but you can use options to change this. ssc copy is a rarely used alternative to ssc install ..., all. ssc cp is a synonym for ssc copy.

## Options for use with ssc whatsnew

saving(*filename*[, replace]) specifies that the "what's new" summary be saved in *filename*. If *filename* is specified without a suffix, *filename*.smcl is assumed. If saving() is not specified, saving(ssc_result.smcl) is assumed.

type specifies that the "what's new" results be displayed in the Results window rather than in the Viewer.

## Option for use with ssc describe

saving(*filename*[, replace]) specifies that, in addition to the descriptions being displayed on your screen, it be saved in the specified file.

If *filename* is specified without an extension, .smcl will be assumed, and the file will be saved as a SMCL file.

If *filename* is specified with an extension, no default extension is added. If the extension is .log, the file will be stored as an ASCII text file.

If replace is specified, *filename* is replaced if it already exists.

## Options for use with ssc install

all specifies that any ancillary files associated with the package be downloaded to your current directory, in addition to the program and help files being installed. Ancillary files are files that do not end in .ado or .hlp and typically contain datasets or examples of the use of the new command.

You can find out which files are associated with the package by typing `ssc describe` *pkgname* before or after installing. If you install without using the `all` option and then want the ancillary files, you can `ssc install` again.

`replace` specifies that any files being downloaded that already exist on your computer be replaced by the downloaded files. If `replace` is not specified and any files already exist, none of the files from the package is downloaded or installed.

It is better not to specify the `replace` option and wait to see if there is a problem. If there is a problem, it is usually better to uninstall the old package using `ssc uninstall` or `ado uninstall` (which are, in fact, the same command).

## Options for use with ssc copy

`plus` specifies that the file be copied to the `PLUS` directory, the directory where user-written additions are installed. Typing `sysdir` will display the identity of the `PLUS` directory on your computer.

`personal` specifies that the file be copied to your `PERSONAL` directory as reported by `sysdir`.

If neither `plus` nor `personal` is specified, the default is to copy the file to the current directory.

`replace` specifies that, if the file already exists on your computer, the new file replace it.

`public` specifies that the new file be made readable by everyone; otherwise, the file will be created according to the default permission you have set with your operating system.

`binary` specifies that the file being copied is a binary file and that it is to be copied as is. The default is to assume that the file is a text file and change the end-of-line characters to those appropriate for your computer/operating system.

## Remarks

A new command to Stata is implemented as one or more ado-files, and together with their corresponding help files and any other associated files, they form a package.

Many packages are available at the SSC. Packages have names, such as `outreg`. When you type

```
. ssc install outreg
```

all the files associated with the package named `outreg` are downloaded and installed on your computer. Usually, names correspond to the name of the command being added to Stata, and so you would expect that installing the package `outreg` would add the new command `outreg` to Stata on your computer and that typing `help outreg` would provide the online help. That is indeed the situation in this case, but that is not always so. Before or after installing a package, it is a good idea to type `ssc describe` *pkgname* to find out the details.

### ▷ Example 1

`ssc whatsnew` summarizes the packages most recently made available or updated. Output is presented in the Viewer, from which you may click on a package name to find out more or install it. For example,

(*Continued on next page*)

```
. ssc whatsnew
(contacting http://repec.org)
  (output omitted)
IVREG2
 module for extended instrumental variables/2SLS and GMM estimation
 Authors: Christopher F Baum  Mark E Schaffer  Steven Stillman
 Revised: 2002-07-19
DECOMP
 module to compute decompositions of earnings gaps
 Authors: Ian Watson
 Revised: 2002-07-24
  (output omitted)
-------------------------------------------------------------------
 End of recent additions and updates
```

ssc describe *pkgname* describes, but does not install, the specified package. Note that you must already know the name of the package. See [R] **search** for assistance in searching for packages. Sometimes you know the package name, but you do not know the exact spelling. In that case, you can type ssc describe followed by one letter, a–z or _, to list all the packages starting with that letter; even so, using search is better.

```
. ssc describe decomp
-------------------------------------------------------------------
package decomp from http://fmwww.bc.edu/repec/bocode/d
-------------------------------------------------------------------
TITLE
      'DECOMP': module to compute decompositions of earnings gaps
DESCRIPTION/AUTHOR(S)
      decomp computes Blinder-Oaxaca wage decompositions. It compares
      the results from two regressions, using intermediate commands
      (himod and lomod), and produces a  table of output containing the
      decompositions. These decompositions show how much of  the wage
      gap is due to differing endowments between the two groups, and
      how much is due to discrimination (regarded as the portion of the
      wage gap due to the combined effect of  coefficients and slope
      intercepts for the two groups). decomp is designed for  Stata's
      regress command, but also works with other commands, such as
      ivreg, heckman and tobit.
      KW: wages
      KW: earnings
      KW: decomposition
      Distribution-Date: 20050128
      Author: Ian Watson , ACIRRT, University of Sydney
      Support: email  i.watson@econ.usyd.edu.au
INSTALLATION FILES                               (type net install decomp)
      decomp.ado
      decomp.hlp
      ../h/himod.ado
      ../l/lomod.ado
-------------------------------------------------------------------
(type -ssc install decomp- to install)
```

Note that the default setting for the saving() option is for the output to be saved with the .smcl extension. Alternatively you could save the file with a log extension, and in this case, the file would be stored as an ASCII text file.

```
. ssc describe d, saving(d.index)
  (output omitted)
. ssc describe decomp, saving(decomp.log)
  (output omitted)
```

ssc install *pkgname* installs the specified package. You do not have to describe a package before installing it. There are ways of installing packages other than ssc install, such as net. It does not matter how a package is installed. For instance, a package can be installed using net and still be uninstalled using ssc.

```
. ssc install decomp
checking decomp consistency and verifying not already installed...
installing into /home/lmg/ado/plus/...
installation complete.
```

ssc uninstall *pkgname* removes the specified, previously installed package from your computer. You can uninstall immediately after installation or at any time in the future. (Technical note: ssc uninstall is a synonym for ado uninstall, so it can uninstall any installed package, not just packages obtained from the SSC.)

```
. ssc uninstall decomp
package decomp from http://fmwww.bc.edu/repec/bocode/d
      'DECOMP': module to compute decompositions of earnings gaps
(package uninstalled)
```

ssc type *filename* types a specific file stored at the SSC. Although not shown in the syntax diagram, ssc cat is a synonym for ssc type, which may appeal to those familiar with Unix. To view only the help file for the decomp package, you would type

```
. ssc type decomp.hlp
-------------------------------------------------------------------------------
help for decomp
-------------------------------------------------------------------------------
Decomposition of wage gaps
--------------------------
  (output omitted)
```

ssc copy *filename* copies a specific file stored at the SSC to your computer. By default, the file is copied to the current directory, but you can use options to change this. ssc copy is a rarely used alternative to ssc install ..., all. ssc cp is a synonym for ssc copy.

```
. ssc copy decomp.ado
(file decomp.ado copied to current directory)
```

◁

# Acknowledgments

ssc is based on archutil by Nicholas J. Cox, University of Durham, and Christopher Baum, Boston College. The reworking of the original was done with their blessing and their participation.

# References

Baum, C. F. and N. J. Cox. 1999. ip29: Metadata for user-written contributions to the Stata programming language. *Stata Technical Bulletin* 52: 10–12. Reprinted in *Stata Technical Bulletin Reprints*, vol. 9, pp. 121–124.

Cox, N. J. and C. F. Baum. 2000. ip29.1: Metadata for user-written contributions to the Stata programming language. *Stata Technical Bulletin* 54: 21–22. Reprinted in *Stata Technical Bulletin Reprints*, vol. 9, pp. 124–126.

# Also See

**Complementary:** [R] **net**, [R] **search**, [R] **sj**,
[P] **sysdir**

# Title

**stem** — Stem-and-leaf displays

# Syntax

`stem` *varname* [*if*] [*in*] [, *options*]

| *options* | description |
|---|---|
| Main | |
| `prune` | do not print stems that have no leaves |
| `round(#)` | round data to this value; default is `round(1)` |
| `digits(#)` | digits per leaf; default is `digits(1)` |
| `lines(#)` | number of stems per interval of $10^{\texttt{digits}}$ |
| `width(#)` | stem width; equal to $10^{\texttt{digits}}$/`width` |

`by` may be used with `stem`; see [D] **by**.

# Description

`stem` displays stem-and-leaf plots.

# Options

Main

`prune` prevents printing any stems that have no leaves.

`round(#)` rounds the data to this value and displays the plot in these units. If `round()` is not specified, noninteger data will be rounded automatically.

`digits(#)` sets the number of digits per leaf. The default is 1.

`lines(#)` sets the number of stems per every data interval of $10^{\texttt{digits}}$. The value of `lines()` must divide $10^{\texttt{digits}}$; that is, if `digits(1)` is specified, then `lines()` must divide 10. If `digits(2)` is specified, then `lines()` must divide 100, etc. Only one of `lines()` or `width()` may be specified. If neither is specified, an appropriate value will be set automatically.

`width(#)` sets the width of a stem. `lines()` is equal to $10^{\texttt{digits}}$/`width`, and this option is merely an alternative way of setting `lines()`. The value of `width()` must divide $10^{\texttt{digits}}$. Only one of `width()` or `lines()` may be specified. If neither is specified, an appropriate value will be set automatically.

Note: If `lines()` or `width()` is not specified, `digits()` may be decreased in some circumstances to make a better-looking plot. If `lines()` or `width()` is set, the user-specified value of `digits()` will not be altered.

# Remarks

## ▷ Example 1

Stem-and-leaf displays are a compact way to present considerable information about a batch of data. For instance, using our automobile data (described in [U] **1.2.1 Sample datasets**):

```
. use http://www.stata-press.com/data/r9/auto
(1978 Automobile Data)
. stem mpg
Stem-and-leaf plot for mpg (Mileage (mpg))
  1t | 22
  1f | 44444455
  1s | 66667777
  1. | 88888888899999999
  2* | 00011111
  2t | 22222333
  2f | 444455555
  2s | 666
  2. | 8889
  3* | 001
  3t |
  3f | 455
  3s |
  3. |
  4* | 1
```

The stem-and-leaf display provides a way to list our data. The expression to the left of the vertical bar is called the stem; the digits to the right are called the leaves. All the stems that begin with the same digit and the corresponding leaves, written beside each other, reconstruct an observation of the data. Thus if we look at the four stems that begin with the digit 1 and their corresponding leaves, we see that we have two cars rated at 12 mpg, 6 cars at 14, 2 at 15, and so on. The car with the highest mileage rating in our data is rated at 41 mpg.

The above plot is a 5-line plot with `lines()` equal to 5 (5 lines per interval of 10) and `width()` equal to 2 (2 leaves per stem).

Instead, we could specify `lines(2)`:

```
. stem mpg, lines(2)
Stem-and-leaf plot for mpg (Mileage (mpg))
  1* | 22444444
  1. | 556666777788888888899999999
  2* | 00011111222223334444
  2. | 555556668889
  3* | 0014
  3. | 55
  4* | 1
```

`stem mpg, width(5)` would produce the same plot as above.

The stem-and-leaf display provides a crude histogram of our data, one not so pretty as that produced by `histogram` (see [R] **histogram**), but one that is nonetheless quite informative.

◁

### ▷ Example 2

The miles per gallon rating fits easily into a stem-and-leaf display because, in our data, it has two digits. This is not, however, required:

```
. stem price, lines(1) digits(3)

Stem-and-leaf plot for price (Price)

   3*** | 291,299,667,748,798,799,829,895,955,984,995
   4*** | 010,060,082,099,172,181,187,195,296,389,424,425,453,482,499, ... (26)
   5*** | 079,104,172,189,222,379,397,705,719,788,798,799,886,899
   6*** | 165,229,295,303,342,486,850
   7*** | 140,827
   8*** | 129,814
   9*** | 690,735
  10*** | 371,372
  11*** | 385,497,995
  12*** | 990
  13*** | 466,594
  14*** | 500
  15*** | 906
```

The (26) at the right of the second stem shows that there were 26 leaves on this stem—too many to display on one line.

We can make a more compact stem-and-leaf plot by rounding. To display `stem` in units of 100, we could type

```
. stem price, round(100)

Stem-and-leaf plot for price (Price)

price rounded to nearest multiple of 100
plot in units of 100

   3* | 33778889
   4* | 00001112222344455555667777899
   5* | 11222447788899
   6* | 2233359
   7* | 18
   8* | 18
   9* | 77
  10* | 44
  11* | 45
  12* | 0
  13* | 056
  14* | 5
  15* | 9
```

`price`, in our data, has four or five digits. `stem` presented the display in terms of units of 100, so a car that cost $3,291 was treated for display purposes as $3,300.

◁

### ❑ Technical Note

Stem-and-leaf diagrams have been used in Japanese railway timetables, as shown in Tufte (1990, 46–47).

❑

## Methods and Formulas

`stem` is implemented as an ado-file.

## References

Emerson, J. D. and D. C. Hoaglin. 1983. Stem-and-leaf displays. In *Understanding Robust and Exploratory Data Analysis*, ed. D. C. Hoaglin, F. Mosteller, and J. W. Tukey, 7–30. New York: Wiley.

Tufte, E. R. 1990. *Envisioning Information*. Cheshire, CT: Graphics Press.

Tukey, J. W. 1972. Some graphic and semigraphic displays. In *Statistical Papers in Honor of George W. Snedecor*, ed. T. A. Bancroft and S. A. Brown, 293–316. Ames, IA: Iowa State University Press.

——. 1977. *Exploratory Data Analysis*. Reading, MA: Addison–Wesley.

## Also See

**Related:** [R] **histogram**, [R] **lv**

# Title

**stepwise** — Stepwise estimation

# Syntax

`stepwise` [ , *options* ] : *command*

| *options* | description |
|---|---|
| Model | |
| *`pr(#)` | significance level for removal from the model |
| *`pe(#)` | significance level for addition to the model |
| Model2 | |
| `forward` | perform forward-stepwise selection |
| `hierarchical` | perform hierarchical selection |
| `lockterm1` | keep the first term |
| `lr` | perform likelihood-ratio test instead of Wald test |

* At least one of `pr(#)` or `pe(#)` must be specified.

`by` and `xi` may be used with `stepwise`; see [U] **11.1.10 Prefix commands**.

Weights are allowed if *command* allows them; see [U] **11.1.6 weight**.

All postestimation commands behave as they would after *command* without the `stepwise` prefix; see the postestimation manual entry for *command*.

See [U] **20 Estimation and postestimation commands** for additional capabilities of estimation commands.

# Description

`stepwise` performs stepwise estimation. Typing

```
. stepwise, pr(#): command
```

performs backward-selection estimation for *command*. The stepwise selection method is determined by the following option combinations:

| *options* | description |
|---|---|
| `pr(#)` | backward selection |
| `pr(#) hierarchical` | backward hierarchical selection |
| `pr(#) pe(#)` | backward stepwise |
| `pe(#)` | forward selection |
| `pe(#) hierarchical` | forward hierarchical selection |
| `pr(#) pe(#) forward` | forward stepwise |

*command* defines the estimation command to be executed. The following Stata commands are supported by `stepwise`:

`clogit cloglog cnreg glm intreg logistic logit nbreg ologit oprobit poisson probit qreg regress scobit stcox streg tobit` `stepwise` expects *command* to have the following form:

*command_name* [*depvar*] *term* [*term* ...] [*if*] [*in*] [*weight*] [, *command_options*]

where *term* is either *varname* or (*varlist*) (a *varlist* in parentheses indicates that this group of variables is to be included or excluded together). Note that *depvar* is not present when *command_name* is `streg` or `stcox`; otherwise, *depvar* is assumed to be present. For `intreg`, *depvar* is actually two dependent variable names ($depvar_1$ and $depvar_2$).

`sw` is a synonym for `stepwise`.

# Options

Model

`pr(`*#*`)` specifies the significance level for removal from the model; terms with $p \geq$ `pr()` are eligible for removal.

`pe(`*#*`)` specifies the significance level for addition to the model; terms with $p <$ `pe()` are eligible for addition.

Model 2

`forward` specifies the forward-stepwise method and may only be specified when both `pr()` and `pe()` are also specified. Specifying both `pr()` and `pe()` without `forward` results in backward-stepwise selection. Note that specifying only `pr()` results in backward selection, and specifying only `pe()` results in forward selection.

`hierarchical` specifies hierarchical selection.

`lockterm1` specifies that the first term be included in the model and not be subjected to the selection criteria.

`lr` specifies that the test of term significance be the likelihood-ratio test. The default is the less computationally expensive Wald test; that is, the test is based on the estimated variance–covariance matrix of the estimators.

# Remarks

Remarks are presented under the headings

*Introduction*
*Search logic for a step*
*Full search logic*
*Examples*
*Estimation sample considerations*
*Messages*
*Programming for stepwise*

## Introduction

Typing

```
. stepwise, pr(.10): regress y1 x1 x2 d1 d2 d3 x4 x5
```

performs a backward-selection search for the regression model y1 on x1, x2, d1, d2, d3, x4, and x5. In this search, each explanatory variable is said to be a term. Typing

```
. stepwise, pr(.10): regress y1 x1 x2 (d1 d2 d3) (x4 x5)
```

performs a similar backward-selection search, but the variables d1, d2, and d3 are treated as a single term, as are x4 and x5. That is, d1, d2, and d3 may or may not appear in the final model, but they appear or do not appear together.

### ▷ Example 1

Using the automobile dataset, we fit a backward-selection model of mpg:

```
. use http://www.stata-press.com/data/r9/auto
. gen weight2 = weight*weight
. stepwise, pr(.2): regress mpg weight weight2 displ gear turn headroom
> foreign price
                      begin with full model
p = 0.7116 >= 0.2000  removing headroom
p = 0.6138 >= 0.2000  removing displacement
p = 0.3278 >= 0.2000  removing price

      Source |       SS       df       MS              Number of obs =      74
-------------+------------------------------           F(  5,    68) =   33.39
       Model |  1736.31455     5  347.262911           Prob > F      =  0.0000
    Residual |  707.144906    68  10.3991898           R-squared     =  0.7106
-------------+------------------------------           Adj R-squared =  0.6893
       Total |  2443.45946    73  33.4720474           Root MSE      =  3.2248

------------------------------------------------------------------------------
         mpg |      Coef.   Std. Err.      t    P>|t|     [95% Conf. Interval]
-------------+----------------------------------------------------------------
      weight |  -.0158002   .0039169    -4.03   0.000    -.0236162   -.0079842
     weight2 |   1.77e-06   6.20e-07     2.86   0.006     5.37e-07    3.01e-06
     foreign |  -3.615107   1.260844    -2.87   0.006    -6.131082   -1.099131
  gear_ratio |   2.011674   1.468831     1.37   0.175    -.9193321     4.94268
        turn |  -.3087038   .1763099    -1.75   0.084    -.6605248    .0431172
       _cons |   59.02133     9.3903     6.29   0.000     40.28327    77.75938
------------------------------------------------------------------------------
```

This estimation treated each variable as its own term and thus considered each one separately. The engine displacement and gear ratio should really be considered together:

*(Continued on next page)*

```
. stepwise, pr(.2): regress mpg weight weight2 (displ gear) turn headroom
> foreign price
                      begin with full model
p = 0.7116 >= 0.2000  removing headroom
p = 0.3944 >= 0.2000  removing displacement gear_ratio
p = 0.2798 >= 0.2000  removing price

      Source |       SS       df       MS              Number of obs =      74
-------------+------------------------------           F(  4,    69) =   40.76
       Model |  1716.80842     4  429.202105           Prob > F      =  0.0000
    Residual |  726.651041    69  10.5311745           R-squared     =  0.7026
-------------+------------------------------           Adj R-squared =  0.6854
       Total |  2443.45946    73  33.4720474           Root MSE      =  3.2452

------------------------------------------------------------------------------
         mpg |      Coef.   Std. Err.      t    P>|t|     [95% Conf. Interval]
-------------+----------------------------------------------------------------
      weight |  -.0160341    .0039379    -4.07   0.000    -.0238901   -.0081782
     weight2 |   1.70e-06    6.21e-07     2.73   0.008     4.58e-07    2.94e-06
     foreign |  -2.758668    1.101772    -2.50   0.015    -4.956643   -.5606925
        turn |  -.2862724     .176658    -1.62   0.110    -.6386955    .0661508
       _cons |   65.39216    8.208778     7.97   0.000      49.0161    81.76823
------------------------------------------------------------------------------
```

◁

## Search logic for a step

Before discussing the complete search logic, consider the logic for a step—the first step—in detail. The other steps follow the same logic. If you type

```
. stepwise, pr(.20): regress y1 x1 x2 (d1 d2 d3) (x4 x5)
```

the logic is

1. Fit the model y on `x1 x2 d1 d2 d3 x4 x5`.
2. Consider dropping `x1`.
3. Consider dropping `x2`.
4. Consider dropping `d1 d2 d3`.
5. Consider dropping `x4 x5`.
6. Find the term above that is least significant. If its significance level is $\geq .20$, remove that term.

If you type

```
. stepwise, pr(.20) hierarchical: regress y1 x1 x2 (d1 d2 d3) (x4 x5)
```

the logic would be different because the `hierarchical` option states that the terms are ordered. The initial logic would become

1. Fit the model y on `x1 x2 d1 d2 d3 x4 x5`.
2. Consider dropping `x4 x5`—the last term.
3. If the significance of this last term is $\geq .20$, remove the term.

The process would then stop or continue. It would stop if `x4 x5` were not dropped, and otherwise, `stepwise` would continue to consider the significance of the next-to-last term, `d1 d2 d3`.

Specifying `pe()` rather than `pr()` switches to forward estimation. If you type

```
. stepwise, pe(.20): regress y1 x1 x2 (d1 d2 d3) (x4 x5)
```

`stepwise` performs forward-selection search. The logic for the first step is

1. Fit a model of `y` on nothing (meaning a constant).
2. Consider adding `x1`.
3. Consider adding `x2`.
4. Consider adding `d1 d2 d3`.
5. Consider adding `x4 x5`.
6. Find the term above that is most significant. If its significance level is $< .20$, add that term.

As with backward estimation, if you specify `hierarchical`,

```
. stepwise, pe(.20) hierarchical: regress y1 x1 x2 (d1 d2 d3) (x4 x5)
```

the search for the most significant term is restricted to the next term:

1. Fit a model of `y` on nothing (meaning a constant).
2. Consider adding `x1`—the first term.
3. If the significance is $< .20$, add the term.

If `x1` were added, `stepwise` would next consider `x2`; otherwise, the search process would stop.

`stepwise` can also employ a stepwise selection logic that alternates between adding and removing terms. The full logic for all the possibilities is given below.

(*Continued on next page*)

## Full search logic

<table>
<tr><th>Option</th><th>Logic</th></tr>
<tr><td><code>pr()</code><br>(backward selection)</td><td>Fit the full model on all explanatory variables.<br>While the least-significant term is “insignificant”, remove it and re-estimate.</td></tr>
<tr><td><code>pr() hierarchical</code><br>(backward hierarchical selection)</td><td>Fit full model on all explanatory variables.<br>While the last term is “insignificant”, remove it and re-estimate.</td></tr>
<tr><td><code>pr() pe()</code><br>(backward stepwise)</td><td>Fit full model on all explanatory variables.<br>If the least-significant term is “insignificant”, remove it and re-estimate; otherwise, stop.<br>Do that again: If the least-significant term is “insignificant”, remove it and re-estimate; otherwise, stop.<br>Repeatedly,<br>if the most-significant excluded term is “significant”, add it and re-estimate;<br>if the least-significant included term is “insignificant”, remove it and re-estimate;<br>until neither is possible.</td></tr>
<tr><td><code>pe()</code><br>(forward selection)</td><td>Fit “empty” model.<br>While the most-significant excluded term is “significant”, add it and re-estimate.</td></tr>
<tr><td><code>pe() hierarchical</code><br>(forward hierarchical selection)</td><td>Fit “empty” model.<br>While the next term is “significant”, add it and re-estimate.</td></tr>
<tr><td><code>pr() pe() forward</code><br>(forward stepwise)</td><td>Fit “empty” model.<br>If the most-significant excluded term is “significant”, add it and re-estimate; otherwise, stop.<br>If the most-significant excluded term is “significant”, add it and re-estimate; otherwise, stop.<br>Repeatedly,<br>if the least-significant included term is “insignificant”, remove it and re-estimate;<br>if the most-significant excluded term is “significant”, add it and re-estimate;<br>until neither is possible.</td></tr>
</table>

## Examples

The following two statements are equivalent; both include solely single-variable terms:

```
. stepwise, pr(.2): reg price mpg weight displ
. stepwise, pr(.2): reg price (mpg) (weight) (displ)
```

The following two statements are equivalent; the last term in each is `r1`, ..., `r4`:

```
. stepwise, pr(.2) hierarchical: reg price mpg weight displ (r1-r4)
. stepwise, pr(.2) hierarchical: reg price (mpg) (weight) (displ) (r1-r4)
```

To group variables `weight` and `displ` into a single term, type

```
. stepwise, pr(.2) hierarchical: reg price mpg (weight displ) (r1-r4)
```

`stepwise` can be used with commands other than `regress`; for instance,

```
. stepwise, pr(.2): logit outcome (sex weight) treated1 treated2
. stepwise, pr(.2): logistic outcome (sex weight) treated1 treated2
```

Either statement would fit the same model because `logistic` and `logit` both perform logistic regression; they differ only in how they report results; see [R] **logit** and [R] **logistic**.

We use the `lockterm1` option to force the first term to be included in the model. To keep `treated1` and `treated2` in the model no matter what, we type

```
. stepwise, pr(.2) lockterm1: logistic outcome (treated1 treated2) ...
```

After `stepwise` estimation, we can type `stepwise` without arguments to redisplay results,

```
. stepwise
  (output from logistic appears)
```

or type the underlying estimation command:

```
. logistic
  (output from logistic appears)
```

At estimation time, we can specify options unique to the command being stepped:

```
. stepwise, pr(.2): logit outcome (sex weight) treated1 treated2, or
```

`or` is `logit`'s option to report odds ratios rather than coefficients; see [R] **logit**.

## Estimation sample considerations

Whether you use backward or forward estimation, `stepwise` forms an estimation sample by taking observations with nonmissing values of all the variables specified (except for $depvar_1$ and $depvar_2$ for `intreg`). The estimation sample is held constant throughout the stepping. Thus if you type

```
. stepwise, pr(.2) hierarchical: regress amount sk edul sval
```

and variable `sval` is missing in half the data, that half of the data will not be used in the reported model, even if `sval` is not included in the final model.

The function `e(sample)` identifies the sample that was used. `e(sample)` contains 1 for observations used and 0 otherwise. For instance, if you type

```
. stepwise, pr(.2) pe(.10): logistic outcome x1 x2 (x3 x4) (x5 x6 x7)
```

and the final model is outcome on x1, x5, x6, and x7, you could recreate the final regression by typing

```
. logistic outcome x1 x5 x6 x7 if e(sample)
```

You could obtain summary statistics within the estimation sample of the independent variables by typing

```
. summarize x1 x5 x6 x7 if e(sample)
```

If you fit another model, e(sample) will automatically be redefined. Typing

```
. stepwise, lock pr(.2): logistic outcome (x1 x2) (x3 x4) (x5 x6 x7)
```

would automatically drop e(sample) and recreate it.

## Messages

**note: ________ dropped due to collinearity**

Each term is checked for collinearity, and variables within the term are dropped if collinearity is found. For instance, say that you type

```
. stepwise, pr(.2): regress y x1 x2 (r1-r4) (x3 x4)
```

and assume that variables r1 through r4 are mutually exclusive and exhaustive dummy variables—perhaps r1, ..., r4 indicate in which of four regions the subject resides. One of the r1, ..., r4 variables will be automatically dropped to identify the model.

This message should cause you no concern.

**Error message: between-term collinearity, variable ________**

After removing any within-term collinearity, if stepwise still finds collinearity between terms, it refuses to continue. For instance, assume that you type

```
. stepwise, pr(.2): regress y1 x1 x2 (d1-d8) (r1-r4)
```

Assume that r1, ..., r4 identify in which of four regions the subject resides, and that d1, ..., d8 identify the same sort of information, but more finely. r1, say, amounts to d1 and d2; r2 to d3, d4, and d5; r3 to d6 and d7; and r4 to d8. You can estimate the d* variables or the r* variables, but not both.

It is your responsibility to specify noncollinear terms.

**note: ________ dropped due to estimability**
**note: ________ obs. dropped due to estimability**

You probably received this message in fitting a logistic or probit model. Regardless of estimation strategy, stepwise checks that the full model can be fitted. The indicated variable had a 0 or infinite standard error.

In the case of logistic, logit, and probit, this is typically caused by one-way causation. Assume that you type

```
. stepwise, pr(.2): logistic outcome (x1 x2 x3) d1
```

and assume that variable d1 is an indicator (dummy) variable. Further assume that whenever d1 = 1, outcome = 1 in the data. Then the coefficient on d1 is infinite. One (conservative) solution to this problem is to drop the d1 variable and the d1==1 observations. The underlying estimation commands probit, logit, and logistic report the details of the difficulty and solution; stepwise simply accumulates such problems and reports the above summary messages. Thus if you see this message, you could type

```
. logistic outcome x1 x2 x3 d1
```

to see the details. While you should think carefully about such situations, Stata's solution of dropping the offending variables and observations is, in general, appropriate.

### Programming for stepwise

stepwise requires that *command_name* follow standard Stata syntax and allow the if qualifier; see [U] **11 Language syntax**. Furthermore, *command_name* must have sw or swml as a program property; see [P] **program**. If *command_name* has swml as a property, *command_name* must save the log-likelihood value in e(ll) and model degrees of freedom in e(df_m).

## Saved Results

stepwise saves whatever is saved by the underlying estimation command.

In addition, stepwise saves the "stepwise" in e(stepwise).

## Methods and Formulas

stepwise is implemented as an ado-file.

Some statisticians do not recommend stepwise procedures; see Sribney (1998) for a summary.

## References

Beale, E. M. L. 1970. Note on procedures for variable selection in multiple regression. *Technometrics* 12: 909–914.

Bendel, R. B. and A. A. Afifi. 1977. Comparison of stopping rules in forward "stepwise" regression. *Journal of the American Statistical Association* 72: 46–53.

Berk, K. N. 1978. Comparing subset regression procedures. *Technometrics* 20: 1–6.

Draper, N. and H. Smith. 1998. *Applied Regression Analysis*. 3rd ed. New York: Wiley.

Efroymson, M. A. 1960. Multiple regression analysis. In *Mathematical Methods for Digital Computers*, ed. A. Ralston and H. S. Wilf, 191–203. New York: Wiley.

Gorman, J. W. and R. J. Toman. 1966. Selection of variables for fitting equations to data. *Technometrics* 8: 27–51.

Hocking, R. R. 1976. The analysis and selection of variables in linear regression. *Biometrics* 32: 1–50.

Hosmer, D. W., Jr., and S. Lemeshow. 2000. *Applied Logistic Regression*. 2nd ed. New York: Wiley.

Kennedy, W. J., Jr., and T. A. Bancroft. 1971. Model-building for prediction in regression based on repeated significance tests. *Annals of Mathematical Statistics* 42: 1273–1284.

Mantel, N. 1970. Why stepdown in variable selection. *Technometrics* 12: 621–625.

——. 1971. More on variable selection and an alternative approach (letter to the editor). *Technometrics* 13: 455–457.

Sribney, W. M. 1998. FAQ: What are some problems with stepwise regression? *http://www.stata.com/support/faqs/stat.*

Wang, Z. 2000. sg134: Model selection using the Akaike information criterion. *Stata Technical Bulletin* 54: 47–49. Reprinted in *Stata Technical Bulletin Reprints*, vol. 9, pp. 335–337.

# Also See

**Complementary:** [R] **clogit**, [R] **cloglog**, [R] **cnreg**, [R] **glm**, [R] **intreg**, [R] **logistic**, [R] **logit**, [R] **nbreg**, [R] **ologit**, [R] **oprobit**, [R] **poisson**, [R] **probit**, [R] **qreg**, [R] **regress**, [R] **scobit**, [R] **tobit**, [ST] **stcox**, [ST] **streg**

**Background:** [U] **11.1.10 Prefix commands**, [U] **20 Estimation and postestimation commands**, [U] **26 Overview of Stata estimation commands**

# Title

**suest** — Seemingly unrelated estimation

# Syntax

`suest` *namelist* [ , *options* ]

| *options* | description |
|---|---|
| Model | |
| `cluster(`*varname*`)` | cluster variable |
| Reporting | |
| `level(`*#*`)` | set confidence level; default is `level(95)` |
| `dir` | display a table describing the models |
| `eform(`*string*`)` | report exponentiated coefficients and label as *string* |

where *namelist* is a list of one or more names under which estimation results were saved via `estimates store`. Wildcards may be used. `*` and `_all` refer to all stored results. A single period (`.`) may be used to refer to the last estimation results, even if they have not (yet) been stored.

# Description

`suest` is a postestimation command; see [U] **20 Estimation and postestimation commands**.

`suest` combines the estimation results—parameter estimates and associated (co)variance matrices—stored under *namelist* into a single parameter vector and *simultaneous* (co)variance matrix of the sandwich/robust type. This (co)variance matrix is appropriate even if the estimates were obtained on the same or on overlapping data.

Typical applications of `suest` are tests for intramodel and cross-model hypotheses using `test` or `testnl`, for example, a generalized Hausman specification test. `lincom` and `nlcom` may be used after `suest` to estimate linear combinations and nonlinear functions of coefficients. `suest` may also be used to adjust a standard VCE for clustering or survey design effects.

Different estimators are allowed, for example, a `regress` model and a `probit` model. The only requirement is that `predict` produce equation-level scores with the `score` option after an estimation command. The models may be estimated on different samples, either due to explicit `if` or `in` selection or to missing values. If weights are applied, the same weights (type and values) should be applied to all models in *namelist*. The estimators should be estimated without `robust` or `cluster()` options. `suest` returns the robust VCE, has a `cluster()` option, and automatically works with results from the `svy` prefix command (only for `vce(linearized)`).

Since `suest` posts its results like a proper estimation command, its results can be stored via `estimates store`. Moreover, like other estimation commands, `suest` typed without arguments replays the results.

# Options

Model

`cluster(`*varname*`)` specifies that the observations are independent across groups (clusters), but not necessarily independent within groups. *varname* specifies to which group each observation belongs, for example, `cluster(personid)` in data with repeated observations on individuals.

`cluster()` may not be combined with estimation results from the `svy` prefix command.

Reporting

`level(`*#*`)` specifies the confidence level, as a percentage, for confidence intervals of the coefficients; see [R] **level**.

`dir` displays a table describing the models in *namelist* just like `estimates dir` *namelist*.

`eform(`*str*`)` displays the coefficient table in exponentiated form: for each coefficient, $\exp(b)$ rather than $b$ is displayed, and standard errors and confidence intervals are transformed. Display of the intercept, if any, is suppressed. *str* is the table header that will be displayed above the transformed coefficients and must be 11 characters or fewer, for example, `eform("Odds ratio")`.

# Remarks

Remarks are presented under the headings



## Using suest

If you plan to use `suest`, you must take precautions when fitting the original models. These restrictions are relaxed when using survey data; see *Using suest with survey data* below.

1. `suest` works with estimation commands that allow `predict` to generate equation-level score variables when supplied with the `score` (or `scores`) option. For example, equation-level score variables are generated after running `mlogit` by typing

   ```
   . predict sc*, scores
   ```

2. Estimation should take place *without* the options `robust` or `cluster()`. `suest` always computes the robust estimator of the (co)variance, and `suest` has a `cluster()` option.

   The within-model covariance matrices computed by `suest` are identical to those obtained by specifying a `robust` or `cluster()` option during estimation. `suest`, however, also estimates the between-model covariances of parameter estimates.

3. Finally, the estimation results to be combined should be stored by `estimates store`; see [R] **estimates**.

After estimating and storing a series of estimation results, you are ready to combine the estimation results with `suest`,

. `suest` *name1* [ *name2* ... ] [ , `cluster(` *varname* `)` ]

and you can subsequently use postestimation commands, such as `test`, to test hypotheses. Here an important issue is how `suest` assigns names to the equations. If you specify a single model *name*, the original equation names are left unchanged; otherwise, `suest` constructs new equation names. The coefficients of a single-equation model (such as `logit` and `poisson`) that was `estimate stored` under name *X* are collected under equation *X*. With a multiequation model stored under name *X*, `suest` prefixes *X_* to an original equation name *eq*, forming equation name, *X_eq*.

❑ Technical Note

In very rare circumstances, `suest` may have to truncate equation names to 32 characters. When equation names are not unique because of truncation, `suest` numbers the equations within models, using equations named *X_#*.

❑

## Remarks on regress

`regress` does not include its ancillary parameter, the residual variance, in its coefficient vector and (co)variance matrix. Moreover, while the `score` option is allowed with `predict` after `regress`, a score variable is generated for the mean but not for the variance parameter. `suest` contains special code that assigns the equation name `mean` to the coefficients for the mean, adds the equation `lnvar` for the log variance, and computes the appropriate two score variables itself.

## Testing the assumption of the independence of irrelevant alternatives (IIA)

The multinomial logit model and the closely related conditional logit model satisfy a probabilistic version of the assumption of the independence of irrelevant alternatives, implying that the ratio of the probabilities for two alternatives does not depend on what other alternatives are available. Hausman and McFadden (1984) proposed a test for this assumption that is implemented in the `hausman` command. The standard Hausman test has a number of limitations. First, the test statistic may be undefined because the estimated VCE does not satisfy the required asymptotic properties of the test. Second, the classic Hausman test only applies to the test of the equality of two estimators. Third, it requires access to a fully efficient estimator; such an estimator may not be available, for example, if you are analyzing complex survey data. Using `suest` can overcome these three limitations.

▷ Example 1

In our first example, we follow the analysis of the type of health insurance reported in [R] **mlogit** and demonstrate the `hausman` command with the `suest/test` combination. We fit the full multinomial logit model for all three alternatives, and two restricted multinomial models in which one alternative is excluded. After fitting each of these models, we store the results using the `store` subcommand of `estimates`. `title()` simply documents the models.

(*Continued on next page*)

```
. use http://www.stata-press.com/data/r9/sysdsn4
(Health insurance data)

. mlogit insure age male
Iteration 0:   log likelihood = -555.85446
Iteration 1:   log likelihood = -551.32973
Iteration 2:   log likelihood = -551.32802
Multinomial logistic regression                   Number of obs   =        615
                                                  LR chi2(4)      =       9.05
                                                  Prob > chi2     =     0.0598
Log likelihood = -551.32802                       Pseudo R2       =     0.0081

------------------------------------------------------------------------------
      insure |      Coef.   Std. Err.      z    P>|z|     [95% Conf. Interval]
-------------+----------------------------------------------------------------
Prepaid      |
         age |  -.0100251   .0060181    -1.67   0.096    -.0218204    .0017702
        male |   .5095747   .1977893     2.58   0.010     .1219148    .8972345
       _cons |   .2633838   .2787574     0.94   0.345    -.2829708    .8097383
-------------+----------------------------------------------------------------
Uninsure     |
         age |  -.0051925   .0113821    -0.46   0.648     -.027501     .017116
        male |   .4748547   .3618446     1.31   0.189    -.2343477    1.184057
       _cons |  -1.756843   .5309591    -3.31   0.001    -2.797504   -.7161824
------------------------------------------------------------------------------
(insure==Indemnity is the base outcome)
. estimates store m1, title(all three insurance forms)
. quietly mlogit insure age male if insure != "Uninsure":insure
. estimates store m2, title(insure != "Uninsure":insure)
. quietly mlogit insure age male if insure != "Prepaid":insure
. estimates store m3, title(insure != "Prepaid":insure)
```

Having performed the three estimations, we inspect the results. `estimates dir` provides short descriptions of the models that were stored using `estimates store`. Typing `estimates table` lists the coefficients, displaying blanks for a coefficient not contained in a model.

```
. est dir

-------------------------------------------------------------------------
model        | command      depvar         npar  title
-------------+-----------------------------------------------------------
          m1 | mlogit       insure            6  all three insurance forms
          m2 | mlogit       insure            3  insure != Uninsure :insure
          m3 | mlogit       insure            3  insure != Prepaid :insure
-------------------------------------------------------------------------
```

```
. est table m1 m2 m3, star stats(N ll)

    Variable         m1              m2              m3

Prepaid
         age  -.01002511      -.01015205
        male   .50957468**     .51440033**
       _cons   .26338378       .26780432

Uninsure
         age  -.00519249                      -.00410547
        male   .47485472                       .45910724
       _cons  -1.7568431***                   -1.8017743***

Statistics
           N         615             570             338
          ll  -551.32802      -390.48643      -131.76807

                         legend: * p<0.05; ** p<0.01; *** p<0.001
```

Comparing the coefficients between models does not suggest substantial differences. We can formally test that coefficients are the same for the full model `m1` and the restricted models `m2` and `m3` using the **hausman** command. Note that **hausman** expects the models to be specified in the order "always consistent" first and "efficient under $H_0$" second.

```
. hausman m2 m1, alleqs constant

                 ---- Coefficients ----
                  (b)          (B)            (b-B)     sqrt(diag(V_b-V_B))
                   m2           m1         Difference          S.E.

         age   -.0101521    -.0100251       -.0001269              .
        male    .5144003     .5095747        .0048256          .012334
       _cons    .2678043     .2633838        .0044205              .

                        b = consistent under Ho and Ha; obtained from mlogit
         B = inconsistent under Ha, efficient under Ho; obtained from mlogit

    Test:  Ho:  difference in coefficients not systematic

                  chi2(3) = (b-B)'[(V_b-V_B)^(-1)](b-B)
                          =        0.08
                Prob>chi2 =      0.9944
                (V_b-V_B is not positive definite)

. hausman m3 m1, alleqs constant

                 ---- Coefficients ----
                  (b)          (B)            (b-B)     sqrt(diag(V_b-V_B))
                   m3           m1         Difference          S.E.

         age   -.0041055    -.0051925         .001087         .0021357
        male    .4591072     .4748547       -.0157475              .
       _cons   -1.801774    -1.756843       -.0449311         .1333464

                        b = consistent under Ho and Ha; obtained from mlogit
         B = inconsistent under Ha, efficient under Ho; obtained from mlogit

    Test:  Ho:  difference in coefficients not systematic

                  chi2(3) = (b-B)'[(V_b-V_B)^(-1)](b-B)
                          =     -0.18    chi2<0 ==> model fitted on these
                                         data fails to meet the asymptotic
                                         assumptions of the Hausman test;
                                         see suest for a generalized test
```

According to the test of m1 against m2, we cannot reject the hypothesis that the coefficients of m1 and m2 are the same. The second Hausman test is not well defined—something that happens fairly frequently. The problem is due to the estimator of the variance V(b-B) as V(b)-V(B), which is a feasible estimator only asymptotically. In this case, it simply is not a proper variance matrix, and the Hausman test becomes undefined.

**suest m1 m2** estimates the simultaneous (co)variance of the coefficients of models m1 and m2. The variance estimates of m1 and m2 are identical to what you would have obtained by specifying the **robust** option with **mlogit**. Although **suest** is technically a postestimation command, it acts like an estimation command in that it stores the simultaneous coefficients in e(b) and the full (co)variance matrix in e(V). We use the command **estat vce** to display this full (co)variance matrix to show that the cross-model covariances have indeed been estimated. Typically, we do not expect to have a direct interest in e(V).

```
. suest m1 m2
Simultaneous results for m1, m2
                                              Number of obs   =        615
```

| | Coef. | Robust Std. Err. | z | P>\|z\| | [95% Conf. | Interval] |
|---|---|---|---|---|---|---|
| m1_Prepaid | | | | | | |
| age | -.0100251 | .0059403 | -1.69 | 0.091 | -.0216679 | .0016176 |
| male | .5095747 | .1988158 | 2.56 | 0.010 | .1199029 | .8992465 |
| _cons | .2633838 | .277307 | 0.95 | 0.342 | -.280128 | .8068955 |
| m1_Uninsure | | | | | | |
| age | -.0051925 | .0109004 | -0.48 | 0.634 | -.0265569 | .0161719 |
| male | .4748547 | .3677294 | 1.29 | 0.197 | -.2458817 | 1.195591 |
| _cons | -1.756843 | .4971363 | -3.53 | 0.000 | -2.731212 | -.782474 |
| m2_Prepaid | | | | | | |
| age | -.0101521 | .0058988 | -1.72 | 0.085 | -.0217135 | .0014094 |
| male | .5144003 | .1996132 | 2.58 | 0.010 | .1231656 | .9056351 |
| _cons | .2678043 | .2744018 | 0.98 | 0.329 | -.2700133 | .805622 |

(*Continued on next page*)

```
. estat vce
Covariance matrix of coefficients of suest model

             | m1_Prepaid                             | m1_Unins~e
        e(V) |        age         male        _cons   |        age         male
-------------+----------------------------------------+-------------------------
m1_Prepaid   |
         age |  .00003529
        male | -.00011951    .03952773
       _cons | -.00154346   -.00415209    .07689918
-------------+----------------------------------------+-------------------------
m1_Uninsure  |
         age |  .00001529   -.00010823   -.00067479  |  .00011882
        male | -.00008815     .0216828   -.00037879  | -.00102264    .13522494
       _cons | -.00067516    .00041156    .03409656  | -.00501568    .00833699
-------------+----------------------------------------+-------------------------
m2_Prepaid   |
         age |  .00003501   -.00013472   -.00152737  |  .00001604   -.00011163
        male | -.00013269    .03967268   -.00359027  | -.00011897    .02204034
       _cons | -.00152755    -.0035127    .07603858  | -.00070546    .00058334

             | m1_Unins~e | m2_Prepaid
        e(V) |      _cons |        age         male        _cons
-------------+------------+--------------------------------------
m1_Uninsure  |
       _cons |  .24714447
-------------+------------+--------------------------------------
m2_Prepaid   |
         age | -.00070509 |   .0000348
        male |  .00091027 | -.00014889    .03984544
       _cons |   .0353047 | -.00151433   -.00291391    .07529635
```

**suest** created equation names by combining the name under which we stored the results using **estimates store** with the original equation names. Thus in the simultaneous estimation result, equation **Prepaid** originating in model **m1** is named **m1_Prepaid**. According to the McFadden–Hausman specification of a test for IIA, the coefficients of the equations **m1_PrePaid** and **m2_PrePaid** should be equal. This can be tested easily with the **test** command. The option **cons** specifies that the intercept **_cons** be included in the test.

```
. test [m1_Prepaid = m2_Prepaid], cons
 ( 1)  [m1_Prepaid]age - [m2_Prepaid]age = 0
 ( 2)  [m1_Prepaid]male - [m2_Prepaid]male = 0
 ( 3)  [m1_Prepaid]_cons - [m2_Prepaid]_cons = 0

           chi2(  3) =    0.89
         Prob > chi2 =    0.8267
```

The Hausman test via **suest** is comparable to that computed by **hausman**, but they use different estimators of the variance of the difference of the estimates. The **hausman** command estimates $V(b-B)$ by $V(b) - V(B)$, whereas **suest** estimates $V(b - B)$ by $V(b) - \text{cov}(b, B) - \text{cov}(B, b) + V(B)$. One advantage of the second estimator is that it is always admissible, so the resulting test is always well defined. This is illustrated in the Hausman-type test of IIA comparing models **m1** and **m3**.

(*Continued on next page*)

```
. suest m1 m3
Simultaneous results for m1, m3
                                                  Number of obs   =        615

------------------------------------------------------------------------------
             |               Robust
             |      Coef.   Std. Err.      z    P>|z|     [95% Conf. Interval]
-------------+----------------------------------------------------------------
m1_Prepaid   |
         age |  -.0100251   .0059403    -1.69   0.091    -.0216679    .0016176
        male |   .5095747   .1988158     2.56   0.010     .1199029    .8992465
       _cons |   .2633838    .277307     0.95   0.342     -.280128    .8068955
-------------+----------------------------------------------------------------
m1_Uninsure  |
         age |  -.0051925   .0109004    -0.48   0.634    -.0265569    .0161719
        male |   .4748547   .3677294     1.29   0.197    -.2458817    1.195591
       _cons |  -1.756843   .4971363    -3.53   0.000    -2.731212    -.782474
-------------+----------------------------------------------------------------
m3_Uninsure  |
         age |  -.0041055   .0111185    -0.37   0.712    -.0258974    .0176865
        male |   .4591072   .3601307     1.27   0.202    -.2467359     1.16495
       _cons |  -1.801774   .5226351    -3.45   0.001     -2.82612   -.7774283
------------------------------------------------------------------------------

. test [m1_Uninsure = m3_Uninsure], cons
 ( 1)  [m1_Uninsure]age - [m3_Uninsure]age = 0
 ( 2)  [m1_Uninsure]male - [m3_Uninsure]male = 0
 ( 3)  [m1_Uninsure]_cons - [m3_Uninsure]_cons = 0

           chi2(  3) =    1.49
         Prob > chi2 =    0.6843
```

While the classic Hausman test computed by **hausman** is not defined in this case, the suest-based test is just fine. We cannot reject the equality of the common coefficients across m1 and m3.

A second advantage of the suest approach is that we can estimate the (co)variance matrix of the multivariate normal distribution of the estimators of the three models m1, m2, and m3 and test that the common coefficients are equal.

(*Continued on next page*)

```
. suest m*
Simultaneous results for m1, m2, m3
                                              Number of obs    =        615

------------------------------------------------------------------------------
             |               Robust
             |      Coef.   Std. Err.      z    P>|z|     [95% Conf. Interval]
-------------+----------------------------------------------------------------
m1_Prepaid   |
         age |  -.0100251   .0059403    -1.69   0.091    -.0216679    .0016176
        male |   .5095747   .1988158     2.56   0.010     .1199029    .8992465
       _cons |   .2633838    .277307     0.95   0.342     -.280128    .8068955
-------------+----------------------------------------------------------------
m1_Uninsure  |
         age |  -.0051925   .0109004    -0.48   0.634    -.0265569    .0161719
        male |   .4748547   .3677294     1.29   0.197    -.2458817    1.195591
       _cons |  -1.756843   .4971363    -3.53   0.000    -2.731212    -.782474
-------------+----------------------------------------------------------------
m2_Prepaid   |
         age |  -.0101521   .0058988    -1.72   0.085    -.0217135    .0014094
        male |   .5144003   .1996132     2.58   0.010     .1231656    .9056351
       _cons |   .2678043   .2744018     0.98   0.329    -.2700133     .805622
-------------+----------------------------------------------------------------
m3_Uninsure  |
         age |  -.0041055   .0111185    -0.37   0.712    -.0258974    .0176865
        male |   .4591072   .3601307     1.27   0.202    -.2467359     1.16495
       _cons |  -1.801774   .5226351    -3.45   0.001     -2.82612   -.7774283
------------------------------------------------------------------------------
. test [m1_Prepaid = m2_Prepaid] , cons notest
 ( 1)  [m1_Prepaid]age - [m2_Prepaid]age = 0
 ( 2)  [m1_Prepaid]male - [m2_Prepaid]male = 0
 ( 3)  [m1_Prepaid]_cons - [m2_Prepaid]_cons = 0
. test [m1_Uninsure = m3_Uninsure], cons acc
 ( 1)  [m1_Prepaid]age - [m2_Prepaid]age = 0
 ( 2)  [m1_Prepaid]male - [m2_Prepaid]male = 0
 ( 3)  [m1_Prepaid]_cons - [m2_Prepaid]_cons = 0
 ( 4)  [m1_Uninsure]age - [m3_Uninsure]age = 0
 ( 5)  [m1_Uninsure]male - [m3_Uninsure]male = 0
 ( 6)  [m1_Uninsure]_cons - [m3_Uninsure]_cons = 0
           chi2(  6) =    1.95
         Prob > chi2 =    0.9239
```

Again we do not find evidence against the correct specification of the multinomial logit for type of insurance. The classic Hausman test assumes that one of the estimators (named `B` in `hausman`) is efficient, that is, it has minimal (asymptotic) variance. This assumption ensures that $V(b) - V(B)$ is an admissible, viable estimator for $V(b - B)$. The assumption that we have an efficient estimator is a restrictive one. It is violated, for instance, if our data are clustered. We want to adjust for clustering via a `cluster(`*varname*`)` option by requesting the cluster-adjusted sandwich estimator of variance. Consequently, in such a case, `hausman` cannot be used. This problem does not exist with the `suest` version of the Hausman test. To illustrate this feature, we suppose that the data are clustered by city—we constructed an imaginary variable `cityid` for this illustration. If we plan to apply `suest`, we would not specify the `cluster()` option at the time of estimation. `suest` has a `cluster()` option. Thus we do not need to refit the models; we can call `suest` and `test` right away.

```
. suest m1 m2, cluster(cityid)
Simultaneous results for m1, m2
                                                    Number of obs   =        615
                                      (Std. Err. adjusted for 260 clusters in cityid)

------------------------------------------------------------------------------
             |               Robust
             |      Coef.   Std. Err.      z    P>|z|     [95% Conf. Interval]
-------------+----------------------------------------------------------------
m1_Prepaid   |
         age |  -.0100251    .005729    -1.75   0.080    -.0212538    .0012035
        male |   .5095747   .1910495     2.67   0.008     .1351246    .8840248
       _cons |   .2633838   .2698797     0.98   0.329    -.2655708    .7923383
-------------+----------------------------------------------------------------
m1_Uninsure  |
         age |  -.0051925   .0104374    -0.50   0.619    -.0256494    .0152644
        male |   .4748547   .3773989     1.26   0.208    -.2648334    1.214543
       _cons |  -1.756843   .4916593    -3.57   0.000    -2.720478   -.7932087
-------------+----------------------------------------------------------------
m2_Prepaid   |
         age |  -.0101521   .0057164    -1.78   0.076    -.0213559    .0010518
        male |   .5144003   .1921384     2.68   0.007      .137816    .8909847
       _cons |   .2678043   .2682193     1.00   0.318    -.2578958    .7935044
------------------------------------------------------------------------------

. test [m1_Prepaid = m2_Prepaid], cons
 ( 1)  [m1_Prepaid]age - [m2_Prepaid]age = 0
 ( 2)  [m1_Prepaid]male - [m2_Prepaid]male = 0
 ( 3)  [m1_Prepaid]_cons - [m2_Prepaid]_cons = 0

           chi2(  3) =    0.79
         Prob > chi2 =    0.8529
```

Note that suest provides some descriptive information about the clustering on cityid. Like any other estimation command, suest informs us that the standard errors are adjusted for clustering. The standard errors reported by suest are exactly what we would have obtained from mlogit if we had specified cluster(cityid) at estimation time. The Hausman-type test obtained from the test command uses a simultaneous (co)variance of m1 and m2 appropriately adjusted for clustering. In this example, we still do not have reason to conclude that the multinomial logit model in this application is misspecified, that is, that IIA is violated.

◁

The multinomial logistic regression model is a special case of the conditional logistic regression model, see [R] **clogit**. Like the multinomial logistic regression model, the conditional logistic regression model also makes the IIA assumption. [R] **clogit** discusses an example in which the demand for American, Japanese, and European cars is modeled in terms of the number of local dealers of the respective brands and of some individual attributes incorporated in interaction with the nationality of cars. We want to perform a Hausman-type test for IIA comparing the decision between all nationalities with the decision between non-American cars. The following code fragment demonstrates how to conduct a Hausman test for IIA via suest in this case.

```
. clogit choice japan europe maleJap maleEur incJap incEur dealer, group(id)
. estimates store allcars
. clogit choice japan maleJap incJap dealer if car!=1 , group(id)
. estimates store foreign
. suest allcars foreign
. test [allcars=foreign], common
```

## Testing proportionality

The applications of suest that we have discussed so far concern Hausman-type tests for misspecification. To test such a hypothesis, we compared two estimators that have the same probability limit if the hypothesis holds true, but otherwise have different limits. We may also want to compare the coefficients of models (estimators) for other substantive reasons. While we most frequently want to test whether coefficients differ between models or estimators, occasionally we may want to test other constraints (see Ruud and Hausman 1987).

### ▷ Example 2

In this example, using simulated labor market data for siblings, we consider two dependent variables, income (inc) and whether or not a person was promoted in the last year (promo). We apply familiar economic arguments regarding human capital, according to which employees have a higher income and a higher probability of being promoted, by having more human capital. Human capital is acquired through formal education (edu) and on-the-job training experience (exp). We study whether income and promotion are "two sides of the same coin", that is, whether they reflect a common latent variable, "human capital". Accordingly, we want to compare the effects of different aspects of human capital on different outcome variables.

We estimate fairly simple labor market equations. The income model is estimated with regress, and the estimation results are stored under the name Inc.

```
. use http://www.stata-press.com/data/r9/income
. regress inc edu exp male
      Source |       SS       df       MS              Number of obs =     277
-------------+------------------------------           F(  3,   273) =   42.34
       Model |  2058.44672     3  686.148908           Prob > F      =  0.0000
    Residual |  4424.05183   273  16.2053181           R-squared     =  0.3175
-------------+------------------------------           Adj R-squared =  0.3100
       Total |  6482.49855   276  23.4873136           Root MSE      =  4.0256

------------------------------------------------------------------------------
         inc |      Coef.   Std. Err.      t    P>|t|     [95% Conf. Interval]
-------------+----------------------------------------------------------------
         edu |   2.213707    .243247     9.10   0.000     1.734828    2.692585
         exp |    1.47293    .231044     6.38   0.000     1.018076    1.927785
        male |   .5381153   .4949466     1.09   0.278     -.436282    1.512513
       _cons |   1.255497   .3115808     4.03   0.000      .642091    1.868904
------------------------------------------------------------------------------
. est store Inc
```

Note that, being sibling data, the observations are clustered on family of origin, famid. In the estimation of the regression parameters, we did not specify a cluster(famid) option in order to adjust standard errors for clustering on family (famid). Thus the standard errors reported by regress are potentially flawed. This will, however, be corrected by specifying a cluster() option with suest.

Next we estimate the promotion equation with probit and again store the results under an appropriate name.

*(Continued on next page)*

```
. probit promo edu exp male, nolog
Probit regression                                 Number of obs   =        277
                                                  LR chi2(3)      =      49.76
                                                  Prob > chi2     =     0.0000
Log likelihood = -158.43888                       Pseudo R2       =     0.1357

------------------------------------------------------------------------------
       promo |      Coef.   Std. Err.      z    P>|z|     [95% Conf. Interval]
-------------+----------------------------------------------------------------
         edu |   .4593002   .0898535     5.11   0.000     .2831906    .6354097
         exp |   .3593023   .0805773     4.46   0.000     .2013737    .5172309
        male |   .2079983   .1656412     1.26   0.209    -.1166524     .532649
       _cons |   -.464622   .1088164    -4.27   0.000    -.6778982   -.2513458
------------------------------------------------------------------------------

. est store Promo
```

The coefficients in the income and promotion equations definitely seem to be different. However, since the scales of the two variables are different, we would not expect the coefficients to be equal. The correct hypothesis in this case is that the proportionality of the coefficients of the two models, apart from the constant, are equal. This formulation would still reflect that the relative effects of the different aspects of human capital do not differ between the dependent variables. We can obtain a nonlinear Wald test for the hypothesis of proportionality using the `testnl` command on the combined estimation results of the two estimators. Thus we first have to form the combined estimation results. Note that at this point we specify the `cluster(famid)` option to adjust for the clustering of observations on `famid`.

```
. suest Inc Promo, cluster(famid)
Simultaneous results for Inc, Promo

                                                  Number of obs   =        277
                                 (Std. Err. adjusted for 135 clusters in famid)

------------------------------------------------------------------------------
             |               Robust
             |      Coef.   Std. Err.      z    P>|z|     [95% Conf. Interval]
-------------+----------------------------------------------------------------
Inc_mean     |
         edu |   2.213707   .2483907     8.91   0.000      1.72687    2.700543
         exp |    1.47293   .1890583     7.79   0.000     1.102383    1.843478
        male |   .5381153   .4979227     1.08   0.280    -.4377952    1.514026
       _cons |   1.255497   .3374977     3.72   0.000      .594014    1.916981
-------------+----------------------------------------------------------------
Inc_lnvar    |
       _cons |   2.785339    .079597    34.99   0.000     2.629332    2.941347
-------------+----------------------------------------------------------------
Promo        |
         edu |   .4593002   .0886977     5.18   0.000     .2854558    .6331445
         exp |   .3593023   .0797717     4.50   0.000     .2029526    .5156521
        male |   .2079983   .1691049     1.23   0.219    -.1234412    .5394379
       _cons |   -.464622   .1042165    -4.46   0.000    -.6688826   -.2603614
------------------------------------------------------------------------------
```

The standard errors reported by suest are identical to those reported by the respective estimation commands when invoked with the option cluster(famid). We are now ready to test for proportionality:

$$H_0: \frac{\beta^{\text{Income}}_{\text{edu}}}{\beta^{\text{Promotion}}_{\text{edu}}} = \frac{\beta^{\text{Income}}_{\text{exp}}}{\beta^{\text{Promotion}}_{\text{exp}}} = \frac{\beta^{\text{Income}}_{\text{male}}}{\beta^{\text{Promotion}}_{\text{male}}}$$

It is straightforward to translate this into syntax suitable for testnl, recalling that the coefficient of variable *v* in equation *eq* is denoted by [*eq*]*v*.

```
. testnl [Inc_mean]edu/[Promo]edu = ///
>        [Inc_mean]exp/[Promo]exp = ///
>        [Inc_mean]male/[Promo]male

  (1)  [Inc_mean]edu/[Promo]edu = [Inc_mean]exp/[Promo]exp
  (2)  [Inc_mean]edu/[Promo]edu = [Inc_mean]male/[Promo]male

               chi2(2) =        0.61
           Prob > chi2 =        0.7385
```

Based on the evidence, we fail to reject the hypotheses that the coefficients of the income and promotion equations are proportional. Thus it is not unreasonable to assume that income and promotion can be explained by the same latent variable, "labor market success".

A disadvantage of the nonlinear Wald test is that it is not invariant with respect to representation: A Wald test for a mathematically equivalent formulation of the nonlinear constraint usually leads to a different test result. An equivalent formulation of the proportionality hypothesis is

$$H_0:\ \beta^{\text{Income}}_{\text{edu}}\beta^{\text{Promotion}}_{\text{exp}} = \beta^{\text{Promotion}}_{\text{edu}}\beta^{\text{Income}}_{\text{exp}} \quad \text{and}$$
$$\beta^{\text{Income}}_{\text{edu}}\beta^{\text{Promotion}}_{\text{male}} = \beta^{\text{Promotion}}_{\text{edu}}\beta^{\text{Income}}_{\text{male}}$$

This formulation is "more linear" in the coefficients. The asymptotic $\chi^2$ distribution of the nonlinear Wald statistic can be expected to be more accurate for this representation.

```
. testnl ([Inc_mean]edu*[Promo]exp = [Inc_mean]exp*[Promo]edu) ///
>        ([Inc_mean]edu*[Promo]male = [Inc_mean]male*[Promo]edu)

  (1)  [Inc_mean]edu*[Promo]exp = [Inc_mean]exp*[Promo]edu
  (2)  [Inc_mean]edu*[Promo]male = [Inc_mean]male*[Promo]edu

               chi2(2) =        0.46
           Prob > chi2 =        0.7936
```

In this case, the two representations lead to very similar test statistics and $p$-values. As before, we fail to reject the hypothesis of proportionality of the coefficients of the two models.

◁

## Testing cross-model hypotheses

### ▷ Example 3

In this example, we demonstrate how some cross-model hypotheses can be tested using the facilities already available in most estimation commands. This demonstration will explain the intricate relationship between the cluster adjustment of the robust estimator of variance and the suest command. It will also be made clear that a new facility is required to perform more general cross-model testing.

We want to test whether the effect of $x_1$ on the binary variable $y_1$ is the same as the effect of $x_2$ on the binary $y_2$; see Clogg et al. 1995. In this setting, $x_1$ may equal $x_2$, and $y_1$ may equal $y_2$. We assume that logistic regression models can be used to model the responses, and for simplicity, we ignore further predictor variables in these models. If the two logit models are fitted on independent samples so that the estimators are (stochastically) independent, a Wald test for `_b[x1] = _b[x2]` rejects the null hypothesis if

$$\frac{\widehat{b}(x_1) - \widehat{b}(x_2)}{\left[\widehat{\sigma}^2\left\{\widehat{b}(x_1)\right\} + \widehat{\sigma}^2\left\{\widehat{b}(x_2)\right\}\right]^{1/2}}$$

is larger than the appropriate $\chi^2_1$ threshold. If the models are fitted on the *same* sample (or on dependent samples), so that the estimators are stochastically dependent, the above test that ignores the covariance between the estimators is not appropriate.

It is instructive to see how this problem can be tackled by "stacking" data. In the stacked format, we doubled the number of observations. The dependent variable is $y_1$ in the first half of the data and is $y_2$ in the second half of the data. The predictor variable $z_1$ is set to $x_1$ in the first half of the expanded data and to 0 in the rest. Similarly, $z_2$ is 0 in the first half and $x_2$ in the second half. The following diagram illustrates the transformation, in the terminology of the `reshape` command, from "wide" to "long" format.

| id | $y_1$ | $y_2$ | $x_1$ | $x_2$ |
|---|---|---|---|---|
| 1 | $y_{11}$ | $y_{21}$ | $x_{11}$ | $x_{21}$ |
| 2 | $y_{12}$ | $y_{22}$ | $x_{12}$ | $x_{22}$ |
| 3 | $y_{13}$ | $y_{23}$ | $x_{13}$ | $x_{23}$ |

$\Longrightarrow$

| id | $y$ | $z_1$ | $z_2$ | model |
|---|---|---|---|---|
| 1 | $y_{11}$ | $x_{11}$ | 0 | 1 |
| 2 | $y_{12}$ | $x_{12}$ | 0 | 1 |
| 3 | $y_{13}$ | $x_{13}$ | 0 | 1 |
| 1 | $y_{21}$ | 0 | $x_{21}$ | 2 |
| 2 | $y_{22}$ | 0 | $x_{22}$ | 2 |
| 3 | $y_{23}$ | 0 | $x_{23}$ | 2 |

The observations in the "long" data organization are *clustered* on the original subjects and are identified with the identifier `id`. The clustering on `id` has to be accounted for when fitting a simultaneous model. The simplest way to deal with clustering is to use the cluster adjustment of the robust or sandwich estimator; see [P] **_robust**. The data manipulation can be accomplished easily with the `stack` command; see [D] **stack**. Subsequently, we fit a simultaneous logit model and perform a Wald test for the hypothesis that the coefficients of `z1` and `z2` are the same. A full setup to obtain the cross-model Wald test could then be as follows:

```
. generate zero = 0            // a variable that is always 0
. generate one  = 1            // a variable that is always 1
. generate two  = 2            // a variable that is always 2
. stack  id y1 x1 zero one   id y2 zero x2 two, into(id y z1 z2 model)
. generate model2 = (model==2)
. logit y model2 z1 z2, cluster(id)
. test _b[z1] = _b[z2]
```

The coefficient of `z1` represents the effect of `x1` on `y1`, and similarly, `z2` for the effect of `x2` on `y2`. The variable `model2` is a dummy for the "second model", which is included to allow the intercept in the second model to differ from that in the first model. The estimates of the coefficient of `z1` and its standard error in the combined model are the same as the estimates of the coefficient of `z1` and its standard error if we fit the model on the unstacked data.

```
. logit y1 x1, robust
```

The cluster() option specified with the logit command for the stacked data ensures that the covariances of _b[z1] and _b[z2] are indeed estimated. This ensures that the Wald test for the equality of the coefficients is correct. If we had not specified the cluster() option, the (co)variance matrix of the coefficients would have been block-diagonal; i.e., the covariances of _b[z1] and _b[z2] would have been 0. In that case, test would have effectively used the invalid formula for the Wald test for two independent samples.

In this example, the two logit models were fitted on the same data. The same setup would apply, without modification, when the two logit models were fitted on overlapping data that resulted, for instance, if the $y$ or $x$ variables were missing in some observations.

The suest command allows us to obtain the above Wald test more efficiently by avoiding the data manipulation, obviating the need to fit a model with twice the number of coefficients. The test statistic produced by the above code fragment is *identical* to that obtained via suest on the original (unstacked) data:

```
. logit y1 x1
. estimates store M1
. logit y2 x2
. estimates store M2
. suest M1 M2
. test [M1]x1=[M2]x2
```

The stacking method can be applied not only to the testing of cross-model hypotheses for logit models, but also to other estimation commands that support the cluster option, for example, regress, poisson, oprobit, heckman, and streg. The stacking approach clearly generalizes to stacking more than two logit or other models, testing more general linear hypotheses, and testing nonlinear cross-model hypotheses (see [R] **testnl**). In all these cases, suest would yield identical statistical results but at smaller costs in terms of data management, computer storage, and computer time.

Is suest nothing but a convenience command? No, there are two disadvantages to the stacking method, both of which are resolved via suest. First, if the models include ancillary parameters (in a regression model, the residual variance; in an ordinal response model, the cutpoints; and in lognormal survival-time regression, the time scale parameter), these parameters are constrained to be equal between the stacked models. In suest, this constraint is relaxed. Second, the stacking method does not generalize to compare different statistical models, such as a probit model and a regression model. As demonstrated in the previous section, suest can deal with this situation.

◁

## Using suest with survey data

suest can be used to obtain the variance estimates for a series of estimators that used the svy prefix. To use suest for this purpose, perform the following steps:

1. Be sure to set the survey design characteristics correctly using svyset. Do not use the vce() option to change the default variance estimator from the linearized variance estimator. vce(brr) and vce(jackknife) are not supported by suest.
2. Fit the model or models using the svy prefix command, optionally including subpopulation estimation with the subpop() option.
3. Store the estimation results with estimates store *name*.

## ▷ Example 4

As an example, we start with the analysis of self-reported health status in the *Nhanes II* data using the svy variant of the ordinal logistic regression model. This analysis is also reported in [SVY] **svy: ologit**:

```
. use http://www.stata-press.com/data/r9/nhanes2f
. svyset psuid [pw=finalwgt], strata(stratid)
      pweight: finalwgt
          VCE: linearized
     Strata 1: stratid
         SU 1: psuid
        FPC 1: <zero>
. svy: ologit health female black age age2
(running ologit on estimation sample)
Survey: Ordered logistic regression
Number of strata   =        31              Number of obs      =     10335
Number of PSUs     =        62              Population size    = 1.170e+08
                                            Design df          =        31
                                            F(   4,      28)   =    223.27
                                            Prob > F           =    0.0000
```

| health | Coef. | Linearized Std. Err. | t | P>\|t\| | [95% Conf. | Interval] |
|---|---|---|---|---|---|---|
| female | -.1615219 | .0523678 | -3.08 | 0.004 | -.2683266 | -.0547171 |
| black | -.986568 | .0790276 | -12.48 | 0.000 | -1.147746 | -.8253901 |
| age | -.0119491 | .0082974 | -1.44 | 0.160 | -.0288717 | .0049736 |
| age2 | -.0003234 | .000091 | -3.55 | 0.001 | -.000509 | -.0001377 |
| /cut1 | -4.566229 | .1632559 | -27.97 | 0.000 | -4.899192 | -4.233266 |
| /cut2 | -3.057415 | .1699943 | -17.99 | 0.000 | -3.404121 | -2.710709 |
| /cut3 | -1.520596 | .1714341 | -8.87 | 0.000 | -1.870238 | -1.170954 |
| /cut4 | -.242785 | .1703964 | -1.42 | 0.164 | -.5903107 | .1047407 |

The self-reported health variable takes 5 categories. Categories 1 and 2 denote negative categories, whereas categories 4 and 5 denote positive categories. We wonder whether the distinctions between the two positive categories and between the two negative categories are produced in accordance with a single latent dimension, which is an assumption of the ordinal logistic model. To test one-dimensionality, we will collapse the 5-point health measure into a 3-point measure, refit the ordinal logistic model, and compare the regression coefficients and cutpoints between the two analyses. If the single latent variable assumption is valid, the coefficients and cutpoints should match. This can be seen as a Hausman-style specification test. Estimation of the ordinal logistic model parameters for survey data is by maximum pseudolikelihood. Neither estimator is fully efficient, and thus the assumptions for the classic Hausman test and for the hausman command are not satisfied. With suest, we can obtain an appropriate Hausman test for survey data.

To perform the Hausman test, we are already almost halfway there by following steps 1 and 2 for one of the models. We just need to store the current estimation results before moving on to the next model. Here we store the results with estimates store under the name H5, indicating that in this analysis, the dependent variable health has 5 categories.

```
. estimates store H5
```

We proceed by generating a new dependent variable `health3`, which maps values 1 and 2 into 2, 3 into 3, and 4 and 5 into 4. This transformation is conveniently accomplished with the function `clip()`. We then fit an `ologit` model with this new dependent variable and store the estimation results under the name `H3`.

```
. gen health3 = clip(health,2,4)
(2 missing values generated)

. svy: ologit health3 female black age age2
(running ologit on estimation sample)

Survey: Ordered logistic regression

Number of strata   =         31                Number of obs      =      10335
Number of PSUs     =         62                Population size    = 1.170e+08
                                               Design df          =         31
                                               F(   4,      28)   =     197.08
                                               Prob > F           =     0.0000

------------------------------------------------------------------------------
             |             Linearized
     health3 |      Coef.   Std. Err.      t    P>|t|     [95% Conf. Interval]
-------------+----------------------------------------------------------------
      female |  -.1551238   .0563808    -2.75   0.010    -.2701132   -.0401343
       black |  -1.046316   .0728273   -14.37   0.000    -1.194848   -.8977838
         age |  -.0365408   .0073653    -4.96   0.000    -.0515624   -.0215192
        age2 |    -.00009   .0000791    -1.14   0.264    -.0002512    .0000713
-------------+----------------------------------------------------------------
       /cut1 |  -3.655498   .1610209   -22.70   0.000    -3.983902   -3.327094
       /cut2 |  -2.109584   .1597055   -13.21   0.000    -2.435305   -1.783862
------------------------------------------------------------------------------

. estimates store H3
```

We can now obtain the combined estimation results of the two models stored under `H5` and `H3` with design-based standard errors.

*(Continued on next page)*

```
. suest H5 H3

Simultaneous survey results for H5, H3

Number of strata   =        31                 Number of obs      =      10335
Number of PSUs     =        62                 Population size    = 1.170e+08
                                               Design df          =         31

------------------------------------------------------------------------------
             |             Linearized
             |      Coef.   Std. Err.      t    P>|t|     [95% Conf. Interval]
-------------+----------------------------------------------------------------
H5_health    |
      female |  -.1615219   .0523678    -3.08   0.004    -.2683266   -.0547171
       black |   -.986568   .0790276   -12.48   0.000    -1.147746   -.8253901
         age |  -.0119491   .0082974    -1.44   0.160    -.0288717    .0049736
        age2 |  -.0003234    .000091    -3.55   0.001     -.000509   -.0001377
-------------+----------------------------------------------------------------
H5_cut1      |
       _cons |  -4.566229   .1632559   -27.97   0.000    -4.899192   -4.233266
-------------+----------------------------------------------------------------
H5_cut2      |
       _cons |  -3.057415   .1699943   -17.99   0.000    -3.404121   -2.710709
-------------+----------------------------------------------------------------
H5_cut3      |
       _cons |  -1.520596   .1714341    -8.87   0.000    -1.870238   -1.170954
-------------+----------------------------------------------------------------
H5_cut4      |
       _cons |   -.242785   .1703964    -1.42   0.164    -.5903107    .1047407
-------------+----------------------------------------------------------------
H3_health3   |
      female |  -.1551238   .0563808    -2.75   0.010    -.2701132   -.0401343
       black |  -1.046316   .0728273   -14.37   0.000    -1.194848   -.8977838
         age |  -.0365408   .0073653    -4.96   0.000    -.0515624   -.0215192
        age2 |   -.00009    .0000791    -1.14   0.264    -.0002512    .0000713
-------------+----------------------------------------------------------------
H3_cut1      |
       _cons |  -3.655498   .1610209   -22.70   0.000    -3.983902   -3.327094
-------------+----------------------------------------------------------------
H3_cut2      |
       _cons |  -2.109584   .1597055   -13.21   0.000    -2.435305   -1.783862
------------------------------------------------------------------------------
```

The coefficients of `H3` and `H5` look rather similar. We now use `test` to perform a formal Hausman-type test for the hypothesis that the regression coefficients are indeed the same, as we would expect if there is indeed a one-dimensional latent dimension for health. Thus we test that the coefficients in the equation `H5_mean` are equal to those in `H3_mean`.

```
. test [H5_health=H3_health3]

Adjusted Wald test

 ( 1)  [H5_health]female - [H3_health3]female = 0
 ( 2)  [H5_health]black - [H3_health3]black = 0
 ( 3)  [H5_health]age - [H3_health3]age = 0
 ( 4)  [H5_health]age2 - [H3_health3]age2 = 0

       F(  4,    28) =   17.13
            Prob > F =    0.0000
```

We can reject the null hypothesis, which indicates that the ordinal logistic regression model is indeed misspecified. An additional specification test can be conducted with respect to the cutpoints. Variable `health3` was constructed from `health` by collapsing the two top categories into value 4 and the two bottom categories into value 2. This effectively has removed two cutpoints, but if the model fits the data, it should not affect the other two cutpoints. The comparison is hampered by a difference in the names of the cutpoints between the models, as illustrated in the figure below:

```
                       cut1     cut2     cut3     cut4
H5     latent     -------x--------x--------x--------x-------
       observed      1       2        3        4        5

                                cut1     cut2
H3     latent     ----------------x--------x----------------
       observed                2       3        4
```

cutpoint _cut2 of model H5 should be compared with cutpoint _cut1 of H3, and similarly, _cut3 of H5 with _cut2 of H3.

```
. test ([H5_cut2]_cons=[H3_cut1]_cons) ([H5_cut3]_cons=[H3_cut2]_cons)
Adjusted Wald test
 ( 1)  [H5_cut2]_cons - [H3_cut1]_cons = 0
 ( 2)  [H5_cut3]_cons - [H3_cut2]_cons = 0
       F(  2,    30) =   33.49
            Prob > F =    0.0000
```

We conclude that the invariance of the cutpoints under the collapse of categories is not supported by the data, again providing evidence against the correct specification of the ordinal logistic model in this case.

◁

# Saved Results

suest saves in e():

| | |
|---|---|
| Scalars | |
| e(N) | number of observations |
| e(N_clust) | number of clusters |
| e(rank) | rank of e(V) |
| Macros | |
| e(cmd) | suest |
| e(eqnames#) | original names of equations of model # |
| e(names) | list of model names |
| e(wtype) | weight type |
| e(wexp) | weight expression |
| e(clustvar) | name of cluster variable |
| e(vcetype) | title used to label Std. Err. |
| Matrices | |
| e(b) | stacked coefficient vector of the models |
| e(V) | (co)variance matrix |
| Functions | |
| e(sample) | marks estimation sample |

# Methods and Formulas

`suest` is implemented as an ado-file.

The estimation of the simultaneous (co)variance of a series of $k$ estimators is a nonstandard application of the sandwich estimator, as implemented by the command [P] **_robust**. You may want to read this manual entry before reading further.

The starting point is that we have fitted $k$ different models on the *same* data—partially overlapping or nonoverlapping data are special cases. We want to derive the *simultaneous* distribution of these $k$ estimators, for instance, in order to test a cross-estimator hypothesis $H_0$. As in the framework of Hausman testing, $H_0$ will often be of the form that different estimators have the same probability limit under some hypothesis, while the estimators have different limits if the hypothesis is violated.

We consider (vector) estimators $\widehat{\boldsymbol{\beta}}_i$ to be defined as "the" solution of the estimation equations $\mathbf{G}_i$,

$$\mathbf{G}_i(\mathbf{b}_i) = \sum_j w_{ij}\mathbf{u}_{ij}(\mathbf{b}_i) = \mathbf{0}, \qquad i = 1, \dots, k$$

We refer to the $\mathbf{u}_{ij}$ as the "scores". Note that specifying some weights $w_{ij} = 0$ trivially accommodates for partially overlapping or even disjointed data. Under "suitable regularity conditions" (see White 1982 and 1994 for details), the $\widehat{\boldsymbol{\beta}}_i$ are asymptotically normally distributed, with the variance estimated consistently by the sandwich estimator

$$V_i = \mathrm{Var}(\widehat{\boldsymbol{\beta}}_i) = \mathbf{D}_i^{-1} \sum_j w_{ij}\mathbf{u}_{ij}\mathbf{u}'_{ij}\, \mathbf{D}_i^{-1}$$

where $\mathbf{D}_i$ is the Jacobian of $\mathbf{G}_i$ evaluated at $\widehat{\boldsymbol{\beta}}_i$. In the context of maximum likelihood estimation, $\mathbf{D}_i$ can be estimated consistently by (minus) the Hessian of the log likelihood or by the Fisher information matrix. If the model is also well specified, the sandwiched term ($\sum_j w_{ij}\mathbf{u}_{ij}\mathbf{u}'_{ij}$) converges in probability to $\mathbf{D}_i$, so $V_i$ may be consistently estimated by $\mathbf{D}_i^{-1}$.

To derive the simultaneous distribution of the estimators, we consider the "stacked" estimation equation,

$$\mathbf{G}(\widehat{\boldsymbol{\beta}}) = \left[\mathbf{G}_1(\widehat{\boldsymbol{\beta}}_1)' \;\; \mathbf{G}_1(\widehat{\boldsymbol{\beta}}_2)' \;\; \dots \;\; \mathbf{G}_k(\widehat{\boldsymbol{\beta}}_k)'\right]' = \mathbf{0}$$

Under "suitable regularity conditions" (see White 1994 for details), $\widehat{\boldsymbol{\beta}}$ is asymptotically *jointly* normally distributed. The Jacobian and scores of the simultaneous equation are easily expressed in the Jacobian and scores of the separate equations. The Jacobian of $\mathbf{G}$,

$$\mathbf{D}(\widehat{\boldsymbol{\beta}}) = \left.\frac{d\mathbf{G}(\boldsymbol{\beta})}{d\boldsymbol{\beta}}\right|_{\boldsymbol{\beta}=\widehat{\boldsymbol{\beta}}}$$

is block diagonal with blocks $\mathbf{D}_1, \dots, \mathbf{D}_k$. The inverse of $\mathbf{D}(\widehat{\boldsymbol{\beta}})$ is again block diagonal, with the inverses of $\mathbf{D}_i$ on the diagonal. The scores $\mathbf{u}$ of $\mathbf{G}$ are simply obtained as the *concatenated* scores of the separate equations:

$$\mathbf{u}_j = (\mathbf{u}'_{1j} \;\; \mathbf{u}'_{2j} \;\; \dots \;\; \mathbf{u}'_{kj})'$$

Out-of-sample (i.e., where $w_{ij} = 0$) values of the score variables are defined as 0 (thus we drop the $i$ subscript from the common weight variable). The sandwich estimator for the asymptotic variance of $\widehat{\boldsymbol{\beta}}$ reads

$$V = \mathrm{Var}(\widehat{\boldsymbol{\beta}}) = \mathbf{D}(\widehat{\boldsymbol{\beta}})^{-1} \left(\sum_j w_j\mathbf{u}_j\mathbf{u}'_j\right) \mathbf{D}(\widehat{\boldsymbol{\beta}})^{-1}$$

Taking a "partitioned" look at this expression, we see that $V(\widehat{\boldsymbol{\beta}}_i)$ is estimated by

$$\mathbf{D}_i^{-1}\left(\sum_j w_j \mathbf{u}_{ij}\mathbf{u}_{ij}'\right)\mathbf{D}_i^{-1}$$

which is, yet again, the familiar sandwich estimator for $\widehat{\boldsymbol{\beta}}_i$ based on the separate estimation equation $\mathbf{G}_i$. Thus considering several estimators simultaneously in this way does not affect the estimators of the asymptotic variances of these estimators. However, as a bonus of stacking, we obtained a sandwich-type estimate of the *covariance* $V_{jh}$ of estimators $\widehat{\boldsymbol{\beta}}_i$ and $\widehat{\boldsymbol{\beta}}_h$,

$$V_{jh} = \mathrm{Cov}(\widehat{\boldsymbol{\beta}}_i, \widehat{\boldsymbol{\beta}}_h) = \mathbf{D}_i^{-1}\left(\sum_j w_j \mathbf{u}_{ij}\mathbf{u}_{jh}'\right)\mathbf{D}_h^{-1}$$

which is also obtained by White (1982).

Note that this estimator for the covariance of estimators is an application of the cluster modification of the sandwich estimator proposed by Rogers (1993). Consider the stacked data format as discussed in the logit example, and assume that Stata would be able to estimate a "stacked model" in which different models apply to different observations, for example, a probit model for the first half, a regression model for the second half, and a one-to-one cluster relation between the first and second half. If there are no common parameters to both models, the score statistics of parameters for the stacked models are zero in the half of the data in which they do not occur. In Rogers' method, we have to sum the score statistics over the observations within a cluster. This boils down to concatenating the score statistics at the level of the cluster.

We compare the sandwich estimator of the (co)variance $V_{12}$ of two estimators with the estimator of variance $\widetilde{V}_{12}$ applied in the classic Hausman test. Hausman (1978) showed that if $\widehat{\boldsymbol{\beta}}_1$ is consistent under $H_0$ and $\widehat{\boldsymbol{\beta}}_2$ is efficient under $H_0$, then asymptotically

$$\mathrm{Cov}(\widehat{\boldsymbol{\beta}}_1, \widehat{\boldsymbol{\beta}}_2) = \mathrm{Var}(\widehat{\boldsymbol{\beta}}_2)$$

and so $\mathrm{var}(\widehat{\boldsymbol{\beta}}_1 - \widehat{\boldsymbol{\beta}}_2)$ is consistently estimated by $V_1 - V_2$.

## Acknowledgment

`suest` was written by Jeroen Weesie, Department of Sociology, Utrecht University, the Netherlands. This research is supported by grant PGS 50-370 by the Netherlands Organization for Scientific Research.

An earlier version of `suest` was published in the *Stata Technical Bulletin* (1999). The current version of `suest` is not backward compatible with the STB version due to the introduction of new ways to manage estimation results via the `estimates` command.

## References

Arminger, G. 1990. Testing against misspecification in parametric rate models. In *Event History Analysis in Life Course Research*, ed. K. U. Mayer and N. B. Tuma. University of Wisconsin Press.

Clogg, C. C., E. Petkova, and A. Haritou. 1995. Statistical methods for comparing regression coefficients between models. *American Journal of Sociology* 100(5): 1271–1293. With comments by P. D. Allison and a reply by the authors.

Gourieroux, C. and A. Monfort. 1989. *Statistics and Econometric Models*, vol. 2. New York: Springer.

Hausman, J. 1978. Specification tests in econometrics. *Econometrica* 46: 1251–1271.

Hausman, J., and D. McFadden. 1984. Specification tests in econometrics. *Econometrica* 52: 1219–1240.

Huber, P. J. 1967. The behavior of maximum likelihood estimates under nonstandard conditions. *Proceedings of the Fifth Symposium on Mathematical Statistics and Probability*, vol. 1, pp. 221–233. Berkeley: University of California Press.

Rogers, W. H. 1993. sg17: Regression standard errors in clustered samples. *Stata Technical Bulletin* 13: 19–23. Reprinted in *Stata Technical Bulletin Reprints*, vol. 3, pp. 88–94.

Ruud, B., and J. Hausman. 1987. Specifying and testing econometric models for rank-ordered data. *Journal of Econometrics* 34: 83–104.

Weesie, J. 1999. sg121: Seemingly unrelated estimation and the cluster-adjusted sandwich estimator. *Stata Technical Bulletin* 52: 34–47. Reprinted in *Stata Technical Bulletin Reprints*, vol. 9, pp. 231–248.

White, H. 1982. Maximum-likelihood estimation of misspecified models. *Econometrica* 50: 1–25.

——. 1994. *Estimation, Inference and Specification Analysis*. Cambridge: Cambridge University Press.

## Also See

**Complementary:** [R] **estimates**, [R] **hausman**, [R] **lincom**, [R] **nlcom**, [R] **test**, [R] **testnl**, [P] **_robust**

**Background:** [R] **estimation options**

# Title

**summarize** — Summary statistics

# Syntax

`summarize` [*varlist*] [*if*] [*in*] [*weight*] [, *options*]

| *options* | description |
|---|---|
| Main | |
| `detail` | display additional statistics |
| `meanonly` | suppress the display; calculate only the mean; programmer's option |
| `format` | use variable's display format |
| `separator(#)` | draw separator line after every # variables; default is `separator(5)` |

*varlist* may contain time-series operators; see [U] **11.4.3 Time-series varlists**.

`by` may be used with `summarize`; see [D] **by**.

`aweights`, `fweights`, and `iweights` are allowed. However, `iweights` may not be used with the `detail` option; see [U] **11.1.6 weight**.

# Description

`summarize` calculates and displays a variety of univariate summary statistics. If no *varlist* is specified, summary statistics are calculated for all the variables in the dataset.

Also see [R] **ci** for calculating the standard error and confidence intervals of the mean.

# Options

Main

`detail` produces additional statistics, including skewness, kurtosis, the four smallest and largest values, and various percentiles.

`meanonly`, which is allowed only when `detail` is not specified, suppresses the display of results and calculation of the variance. Ado-file writers will find this useful for fast calls.

`format` requests that the summary statistics be displayed using the display formats associated with the variables, rather than the default `g` display format; see [U] **12.5 Formats: controlling how data are displayed**.

`separator(#)` specifies how often to insert separation lines into the output. The default is `separator(5)`, meaning that a line is drawn after every 5 variables. `separator(10)` would draw a line after every 10 variables. `separator(0)` suppresses the separation line.

# Remarks

`summarize` can produce two different sets of summary statistics. Without the `detail` option, the number of nonmissing observations, the mean and standard deviation, and the minimum and maximum values are presented. With `detail`, the same information is presented along with the variance, skewness, and kurtosis; the four smallest and four largest values; and the 1st, 5th, 10th, 25th, 50th (median), 75th, 90th, 95th, and 99th percentiles.

### ▷ Example 1

We have data containing information on various automobiles, among which is the variable mpg, the mileage rating. We can obtain a quick summary of the mpg variable by typing

```
. use http://www.stata-press.com/data/r9/auto
(1978 Automobile Data)
. summarize mpg
    Variable |       Obs        Mean    Std. Dev.       Min        Max
-------------+--------------------------------------------------------
         mpg |        74     21.2973    5.785503         12         41
```

We see that we have 74 observations. The mean of mpg is 21.3 miles per gallon, and the standard deviation is 5.79. The minimum is 12, and the maximum is 41.

If we had not specified the variable (or variables) we wanted to summarize, we would have obtained summary statistics on all the variables in the dataset:

```
. summarize, separator(4)
    Variable |       Obs        Mean    Std. Dev.       Min        Max
-------------+--------------------------------------------------------
        make |         0
       price |        74    6165.257    2949.496       3291      15906
         mpg |        74     21.2973    5.785503         12         41
       rep78 |        69    3.405797    .9899323          1          5
-------------+--------------------------------------------------------
    headroom |        74    2.993243    .8459948        1.5          5
       trunk |        74    13.75676    4.277404          5         23
      weight |        74    3019.459    777.1936       1760       4840
      length |        74    187.9324    22.26634        142        233
-------------+--------------------------------------------------------
        turn |        74    39.64865    4.399354         31         51
displacement |        74    197.2973    91.83722         79        425
  gear_ratio |        74    3.014865    .4562871       2.19       3.89
     foreign |        74    .2972973    .4601885          0          1
```

Notice that there are only 69 observations on rep78, so some of the observations are missing. There are no observations on make since it is a string variable.

◁

### ▷ Example 2

The detail option provides all the information of a normal summarize and more. The format of the output also differs:

```
. summarize mpg, detail
                        Mileage (mpg)
-------------------------------------------------------------
      Percentiles      Smallest
 1%           12             12
 5%           14             12
10%           14             14       Obs                  74
25%           18             14       Sum of Wgt.          74

50%           20                      Mean            21.2973
                        Largest       Std. Dev.      5.785503
75%           25             34
90%           29             35       Variance       33.47205
95%           34             35       Skewness       .9487176
99%           41             41       Kurtosis       3.975005
```

As in the previous example, we see that the mean of `mpg` is 21.3 miles per gallon and that the standard deviation is 5.79. We also see the various percentiles. The median of `mpg` (the 50th percentile) is 20 miles per gallon. The 25th percentile is 18, and the 75th percentile is 25.

When we performed `summarize`, we learned that the minimum and maximum were 12 and 41, respectively. We now see that the four smallest values in our dataset are 12, 12, 14, and 14. The four largest values are 34, 35, 35, and 41. The skewness of the distribution is 0.95, and the kurtosis is 3.98. (A normal distribution would have a skewness of 0 and a kurtosis of 3.)

*Skewness* is a measure of the lack of symmetry of a distribution. If the coefficient of skewness is 0, the distribution is symmetric. If the coefficient is negative, the median is usually greater than the mean and the distribution is said to be skewed left. If the coefficient is positive, the median is usually less than the mean and the distribution is said to be skewed right. *Kurtosis* (from the Greek kyrtosis meaning curvature) is a measure of peakedness of a distribution. The smaller the coefficient of kurtosis, the flatter the distribution. The normal distribution has a coefficient of kurtosis of 3 and provides a convenient benchmark.

On a historical note, see Plackett (1958) for a history of the concept of the mean. ◁

The idea of the mean is quite old (Plackett 1958), but its extension to a scheme of moment-based measures was not done until the end of the 19th century. Between 1893 and 1905, Pearson discussed and named the standard deviation, skewness, and kurtosis, but he was not the first to use any of these. Thiele (1889), in contrast, had earlier grasped firmly the notion that the $m_r$ provide a systematic basis for discussing distributions. However, even earlier anticipations can also be found. For example, Euler in 1778 used $m_2$ and $m_3$ in passing in a treatment of estimation (Hald 1998, 87), but seemingly did not build on that.

Similarly, the idea of the median is quite old. The history of the interquartile range is tangled up with that of the probable error, a long-popular measure. Extending this in various ways to a more general approach based on quantiles (to use a later term) occurred to several people in the nineteenth century. Galton (1875) is a nice example, particularly because he seems so near to the key idea of the quantiles as a function, which took another century to re-emerge strongly.

Thorvald Nicolai Thiele (1838–1910) was a Danish scientist who worked in astronomy, mathematics, actuarial science and statistics. He made many pioneering contributions to statistics, several of which were overlooked until very recently. Thiele advocated graphical analysis of residuals checking for trends, symmetry of distributions, and changes of sign, and even warned against over-interpreting such graphs.

## ▷ Example 3

`summarize` can usefully be combined with the `by` *varlist*: prefix. In our dataset, we have a variable, `foreign`, that distinguishes foreign and domestic cars. We can obtain summaries of `mpg` and `weight` within each subgroup by typing

(*Continued on next page*)

```
. by foreign: summarize mpg weight

-> foreign = Domestic

    Variable |       Obs        Mean    Std. Dev.       Min        Max
-------------+--------------------------------------------------------
         mpg |        52    19.82692    4.743297         12         34
      weight |        52    3317.115    695.3637       1800       4840

-> foreign = Foreign

    Variable |       Obs        Mean    Std. Dev.       Min        Max
-------------+--------------------------------------------------------
         mpg |        22    24.77273    6.611187         14         41
      weight |        22    2315.909    433.0035       1760       3420
```

Domestic cars in our dataset average 19.8 miles per gallon, whereas foreign cars average 24.8.

Since `by` *varlist*`:` can be combined with `summarize`, it can also be combined with `summarize, detail`:

```
. by foreign: summarize mpg, detail

-> foreign = Domestic
                        Mileage (mpg)
-------------------------------------------------------------
      Percentiles      Smallest
 1%           12             12
 5%           14             12
10%           14             14       Obs                  52
25%         16.5             14       Sum of Wgt.          52

50%           19                      Mean           19.82692
                        Largest       Std. Dev.      4.743297
75%           22             28
90%           26             29       Variance       22.49887
95%           29             30       Skewness       .7712432
99%           34             34       Kurtosis       3.441459

-> foreign = Foreign
                        Mileage (mpg)
-------------------------------------------------------------
      Percentiles      Smallest
 1%           14             14
 5%           17             17
10%           17             17       Obs                  22
25%           21             18       Sum of Wgt.          22

50%         24.5                      Mean           24.77273
                        Largest       Std. Dev.      6.611187
75%           28             31
90%           35             35       Variance       43.70779
95%           35             35       Skewness        .657329
99%           41             41       Kurtosis        3.10734
```

◁

## ❑ Technical Note

`summarize` respects display formats if we specify the `format` option. When we type `summarize price weight`, we obtain

```
. summarize price weight

    Variable |       Obs        Mean    Std. Dev.       Min        Max
-------------+--------------------------------------------------------
       price |        74    6165.257    2949.496       3291      15906
      weight |        74    3019.459    777.1936       1760       4840
```

The display is accurate but is not as aesthetically pleasing as we may wish, particularly if we plan to use the output directly in published work. By placing formats on the variables, we can control how the table appears:

```
. format price weight %9.2fc

. summarize price weight, format

    Variable |       Obs        Mean    Std. Dev.       Min        Max
-------------+--------------------------------------------------------
       price |        74    6,165.26    2,949.50   3,291.00  15,906.00
      weight |        74    3,019.46      777.19   1,760.00   4,840.00
```

❑

If you specify a weight (see [U] **11.1.6 weight**), each observation is multiplied by the value of the weighting expression before the summary statistics are calculated so that the weighting expression is interpreted as the discrete density of each observation.

## ▷ Example 4

We have 1980 census data on each of the 50 states. Included in our variables is `medage`, the median age of the population of each state. If we type `summarize medage`, we obtain unweighted statistics:

```
. use http://www.stata-press.com/data/r9/census
(1980 Census data by state)

. summarize medage

    Variable |       Obs        Mean    Std. Dev.       Min        Max
-------------+--------------------------------------------------------
      medage |        50       29.54    1.693445       24.2       34.7
```

Also among our variables is `pop`, the population in each state. Typing `summarize medage [w=pop]` produces population-weighted statistics:

```
. summarize medage [w=pop]
(analytic weights assumed)

    Variable |     Obs      Weight        Mean   Std. Dev.       Min        Max
-------------+-----------------------------------------------------------------
      medage |      50   225907472    30.11047    1.66933       24.2       34.7
```

The number listed under `Weight` is the sum of the weighting variable, `pop`, indicating that there are roughly 226 million people in the U.S. The `pop`-weighted mean of `medage` is 30.11 (as compared with 29.54 for the unweighted statistic), and the weighted standard deviation is 1.67 (as compared with 1.69).

◁

### ▷ Example 5

We can obtain detailed summaries of weighted data as well. When we do this, *all* the statistics are weighted, including the percentiles.

```
. summarize medage [w=pop], detail
(analytic weights assumed)
                            Median age
-------------------------------------------------------------
      Percentiles      Smallest
 1%         27.1           24.2
 5%         27.7           26.1
10%         28.2           27.1       Obs                  50
25%         29.2           27.4       Sum of Wgt.   225907472

50%         29.9                      Mean           30.11047
                        Largest       Std. Dev.       1.66933
75%         30.9             32
90%         32.1           32.1       Variance       2.786661
95%         32.2           32.2       Skewness       .5281972
99%         34.7           34.7       Kurtosis       4.494223
```

◁

### ❑ Technical Note

If you are writing a program and need to access the mean of a variable, the `meanonly` option provides for fast calls. For example, suppose your program reads as follows:

```
program mean
        summarize `1', meanonly
        display "  mean = " r(mean)
end
```

The result of executing this is

```
. mean price
  mean = 6165.2568
```

❑

## Saved Results

`summarize` saves in `r()`:

Scalars

| | | | |
|---|---|---|---|
| `r(N)` | number of observations | `r(p50)` | 50th percentile (`detail` only) |
| `r(mean)` | mean | `r(p75)` | 75th percentile (`detail` only) |
| `r(skewness)` | skewness (`detail` only) | `r(p90)` | 90th percentile (`detail` only) |
| `r(min)` | minimum | `r(p95)` | 95th percentile (`detail` only) |
| `r(max)` | maximum | `r(p99)` | 99th percentile (`detail` only) |
| `r(sum_w)` | sum of the weights | `r(Var)` | variance |
| `r(p1)` | 1st percentile (`detail` only) | `r(kurtosis)` | kurtosis (`detail` only) |
| `r(p5)` | 5th percentile (`detail` only) | `r(sum)` | sum of variable |
| `r(p10)` | 10th percentile (`detail` only) | `r(sd)` | standard deviation |
| `r(p25)` | 25th percentile (`detail` only) | | |

## Methods and Formulas

Let $x$ denote the variable on which we want to calculate summary statistics, and let $x_i$, $i = 1, \dots, n$, denote an individual observation on $x$. Let $v_i$ be the weight, and if no weight is specified, define $v_i = 1$ for all $i$.

Define $V$ as the *sum of the weight*:

$$V = \sum_{i=1}^{n} v_i$$

Define $w_i$ to be $v_i$ normalized to sum to $n$, $w_i = v_i(n/V)$.

The *mean*, $\overline{x}$, is defined as

$$\overline{x} = \frac{1}{n}\sum_{i=1}^{n} w_i x_i$$

The *variance*, $s^2$, is defined as

$$s^2 = \frac{1}{n-1}\sum_{i=1}^{n} w_i (x_i - \overline{x})^2$$

The *standard deviation*, $s$, is defined as $\sqrt{s^2}$.

Define $m_r$ as the $r$th moment about the mean $\overline{x}$:

$$m_r = \frac{1}{n}\sum_{i=1}^{n} w_i (x_i - \overline{x})^r$$

The *coefficient of skewness* is then defined as $m_3 m_2^{-3/2}$. The *coefficient of kurtosis* is defined as $m_4 m_2^{-2}$.

Let $x_{(i)}$ refer to the $x$ in ascending order, and let $w_{(i)}$ refer to the corresponding weights of $x_{(i)}$. The four smallest values are $x_{(1)}$, $x_{(2)}$, $x_{(3)}$, and $x_{(4)}$. The four largest values are $x_{(n)}$, $x_{(n-1)}$, $x_{(n-2)}$, and $x_{(n-3)}$.

To obtain the $p$th *percentile*, which we will denote as $x_{[p]}$, let $P = np/100$. Let

$$W_{(i)} = \sum_{j=1}^{i} w_{(j)}$$

Find the first index $i$ such that $W_{(i)} > P$. The $p$th *percentile* is then

$$x_{[p]} = \begin{cases} \dfrac{x_{(i-1)} + x_{(i)}}{2} & \text{if } W_{(i-1)} = P \\ x_{(i)} & \text{otherwise} \end{cases}$$

## References

David, H. A. 2001. First (?) occurrence of common terms in statistics and probability. In *Annotated Readings in the History of Statistics*, ed. H. David and A. Edwards, 209–246. New York: Springer.

Galton, F. 1875. Statistics by intercomparison, with remarks on the law of frequency of error. *Philosophical Magazine* 49: 33–46.

Gleason, J. R. 1997. sg67: Univariate summaries with boxplots. *Stata Technical Bulletin* 36: 23–25. Reprinted in *Stata Technical Bulletin Reprints*, vol. 6, pp. 179–183.

——. 1999. sg67.1: Update to univar. *Stata Technical Bulletin* 51: 27–28. Reprinted in *Stata Technical Bulletin Reprints*, vol. 9, pp. 159–161.

Hald, A. 1998. *A History of Mathematical Statistics from 1750 to 1930.* New York: Wiley.

Hamilton, L. C. 1996. *Data Analysis for Social Scientists.* Pacific Grove, CA: Brooks/Cole.

——. 2004. *Statistics with Stata.* Belmont, CA: Brooks/Cole.

Kirkwood, B. R. and J. A. C. Sterne. 2003. *Essential Medical Statistics.* 2nd ed. Malden, MA: Blackwell.

Lauritzen, S. L. 2002. *Thiele: Pioneer in Statistics.* Oxford: Oxford University Press.

Plackett, R. L. 1958. The principle of the arithmetic mean. *Biometrika* 45: 130–135.

Stuart, A. and J. K. Ord. 1994. *Kendall's Advanced Theory of Statistics, Vol. I.* 6th ed. London: Arnold.

Thiele, T. N. 1889. *Almindelig Iagttagelseslære: Sandsynlighedsregning og mindste Kvadraters Methode.* Kjøbenhavn: C.A. Reitzel. (English translation included in Lauritzen 2002.)

Weisberg, H. F. 1992. *Central Tendency and Variability.* Newbury Park, CA: Sage.

## Also See

**Related:** [R] **ameans**, [R] **centile**, [R] **ci**, [R] **lv**, [R] **mean**, [R] **table**, [R] **tabstat**,
[R] **tabulate, summarize()**,
[D] **cf**, [D] **codebook**, [D] **compare**, [D] **describe**, [D] **egen**, [D] **inspect**, [D] **pctile**,
[ST] **stsum**, [SVY] **svy: mean**, [XT] **xtsum**

# Title

**sunflower** — Density-distribution sunflower plots

# Syntax

`sunflower` *yvar xvar* [*if*] [*in*] [*weight*] [ , *sunflower_options*]

| *sunflower_options* | description |
|---|---|
| Main | |
| `binwidth(#)` | width of the hexagonal bins |
| `binar(#)` | aspect ratio of the hexagonal bins |
| `light(#)` | minimum observations for a light sunflower; default is `light(3)` |
| `dark(#)` | minimum observations for a dark sunflower; default is `dark(13)` |
| `xcenter(#)` | $x$-coordinate of the reference bin |
| `ycenter(#)` | $y$-coordinate of the reference bin |
| `petalweight(#)` | observations in a dark sunflower petal |
| `petallength(#)` | length of sunflower petal as a percentage |
| `flowersonly` | show petals only; do not render bins |
| `nosinglepetal` | suppress single petals |
| *sunflower_style_options* | (see the next table) |
| *marker_options* | affect rendition of markers drawn at the plotted points |
| `notable` | do not show summary table; implied when `by()` is specified |
| `nograph` | do not show graph |
| Add plot | |
| `addplot(`*plot*`)` | add other plots to generated graph |
| Y-Axis, X-Axis, Title, Caption, Legend, Overall, By | |
| *twoway_options* | any options documented in [G] ***twoway_options*** |

| *sunflower_style_options* | description |
|---|---|
| [l\|d]`bstyle(`*areastyle*`)` | overall look of hexagonal bins |
| [l\|d]`bcolor(`*colorstyle*`)` | outline and fill color |
| [l\|d]`bfcolor(`*colorstyle*`)` | fill color |
| [l\|d]`blstyle(`*linestyle*`)` | overall look of outline |
| [l\|d]`blcolor(`*colorstyle*`)` | outline color |
| [l\|d]`blwidth(`*linewidthstyle*`)` | thickness of outline |
| [l\|d]`flstyle(`*linestyle*`)` | overall style of sunflower petals |
| [l\|d]`flcolor(`*colorstyle*`)` | color of sunflower petals |
| [l\|d]`flwidth(`*linewidthstyle*`)` | thickness of sunflower petals |

All options are *rightmost*; see [G] **concept: repeated options**.

`fweight`s are allowed; see [U] **11.1.6 weight**.

## Description

`sunflower` draws density-distribution sunflower plots (Dupont and Plummer 2002). These plots are useful for displaying bivariate data whose density is too great for conventional scatterplots to be effective.

A sunflower is a number of line segments of equal length, called petals, that radiate from a central point. There are two varieties of sunflowers: light and dark. Each petal of a light sunflower represents one observation. Each petal of a dark sunflower represents a number of observations. Dark and light sunflowers represent high- and medium-density regions of the data, and marker symbols represent individual observations in low-density regions.

The plane defined by the variables *yvar* and *xvar* is divided into contiguous hexagonal bins. The number of observations contained within a bin determines how the bin will be represented.

1. When there are fewer than `light(#)` observations in a bin, each point is plotted using the usual marker symbols in a scatter plot.
2. Bins with at least `light(#)` but fewer than `dark(#)` observations are represented by a light sunflower.
3. Bins with at least `dark(#)` observations are represented by a dark sunflower.

## Options

Main

`binwidth(#)` specifies the horizontal width of the hexagonal bins in the same units as *xvar*. By default,

$$binwidth = \texttt{max}(\text{rbw}, \text{nbw})$$

where

$$\text{rbw} = \text{range of } xvar/40$$

$$\text{nbw} = \text{range of } xvar/\texttt{max(1,nb)}$$

and

$$\text{nb} = \texttt{int(min(sqrt(}n\texttt{),10 * log10(}n\texttt{)))}$$

where

$$n = \text{the number of observations in the dataset}$$

`binar(#)` specifies the aspect ratio for the hexagonal bins. The height of the bins is given by

$$binheight = binwidth \times \# \times \sqrt{3}$$

The default is `binar(`*r*`)`, where *r* results in the rendering of regular hexagons.

`light(#)` specifies the minimum number of observations needed for a bin to be represented by a light sunflower. The default is `light(3)`.

`dark(#)` specifies the minimum number of observations needed for a bin to be represented by a dark sunflower. The default is `dark(13)`.

`xcenter(`*#*`)` and `ycenter(`*#*`)` specify the center of the reference bin. The default values are the median values of *xvar* and *yvar*, respectively. The centers of the other bins are implicitly defined by the location of the reference bin together with the common bin width and height.

`petalweight(`*#*`)` specifies the number of observations represented by each petal of a dark sunflower. The default value is chosen so that the maximum number of petals on a dark sunflower is 14.

`petallength(`*#*`)` specifies the length of petals in the sunflowers. The value specified is interpreted as a percentage of half the bin width. The default is 100 percent.

`flowersonly` suppresses rendering of the bins. This option is equivalent to specifying `lbcolor(none)` and `dbcolor(none)`.

`nosinglepetal` suppresses flowers from being drawn in light bins that contain only one observation and dark bins that contain as many observations as the petal weight (see the `petalweight()` option).

*sunflower_style_options* affect how the hexagonal bins and sunflower petals are rendered.

`lbstyle(`*areastyle*`)` and `dbstyle(`*areastyle*`)` specify the look of the light and dark hexagonal bins, respectively. The options listed below allow you to change each attribute, but `lbstyle()` and `dbstyle()` provide the starting points. See [G] ***areastyle*** for a list of available area styles.

`lbcolor(`*colorstyle*`)` and `dbcolor(`*colorstyle*`)` specify a single color to be used both to outline the shape and to fill the interior of the light and dark hexagonal bins, respectively. See [G] ***colorstyle*** for a list of color choices.

`lbfcolor(`*colorstyle*`)` and `dbfcolor(`*colorstyle*`)` specify the color to be used to fill the interior of the light and dark hexagonal bins, respectively. See [G] ***colorstyle*** for a list of color choices.

`lblstyle(`*linestyle*`)` and `dblstyle(`*linestyle*`)` specify the overall style of the line used to outline the area, which includes its pattern (solid, dashed, etc.), thickness, and color. The other options listed below allow you to change the line's attributes, but `lblstyle()` and `lblstyle()` are the starting points. See [G] ***linestyle*** for a list of choices.

`lblcolor(`*colorstyle*`)` and `dblcolor(`*colorstyle*`)` specify the color to be used to outline the light and dark hexagonal bins, respectively. See [G] ***colorstyle*** for a list of color choices.

`lblwidth(`*linewidthstyle*`)` and `dblwidth(`*linewidthstyle*`)` specify the thickness of the line to be used to outline the light and dark hexagonal bins, respectively. See [G] ***linewidthstyle*** for a list of choices.

`lflstyle(`*linestyle*`)` and `dflstyle(`*linestyle*`)` specify the overall style of the light and dark sunflower petals, respectively.

`lflcolor(`*colorstyle*`)` and `dflcolor(`*colorstyle*`)` specify the color of the light and dark sunflower petals, respectively.

`lflwidth(`*linewidthstyle*`)` and `dflwidth(`*linewidthstyle*`)` specify the width of the light and dark sunflower petals, respectively.

*marker_options* affect the rendition of markers drawn at the plotted points, including their shape, size, color, and outline; see [G] ***marker_options***.

`notable` prevents the summary table from being displayed. This option is implied when the `by()` option is specified.

`nograph` prevents the graph from being generated.

### Add plot

`addplot(`*plot*`)` provides a way to add other plots to the generated graph; see [G] ***addplot_option***.

Y-Axis, X-Axis, Title, Caption, Legend, Overall, By

*twoway_options* are any of the options documented in [G] ***twoway_options***. These include options for titling the graph (see [G] ***title_options***), options for saving the graph to disk (see [G] ***saving_option***), and the `by()` option (see [G] ***by_option***).

# Remarks

## ▷ Example 1

Using the auto dataset, we want to examine the relationship between `weight` and `mpg`. To do that, we type

```
. use http://www.stata-press.com/data/r9/auto
(1978 Automobile Data)
. sunflower mpg weight, binwid(500) petalw(2) dark(8)
Bin width           =       500
Bin height          =   8.38703
Bin aspect ratio    =  .0145268
Max obs in a bin    =        15
Light               =         3
Dark                =         8
X-center            =      3190
Y-center            =        20
Petal weight        =         2
```

| flower type | petal weight | No. of petals | No. of flowers | estimated obs. | actual obs. |
|---|---|---|---|---|---|
| none | | | | 10 | 10 |
| light | 1 | 3 | 1 | 3 | 3 |
| light | 1 | 4 | 2 | 8 | 8 |
| light | 1 | 7 | 3 | 21 | 21 |
| dark | 2 | 4 | 1 | 8 | 8 |
| dark | 2 | 5 | 1 | 10 | 9 |
| dark | 2 | 8 | 1 | 16 | 15 |
| | | | | 76 | 74 |

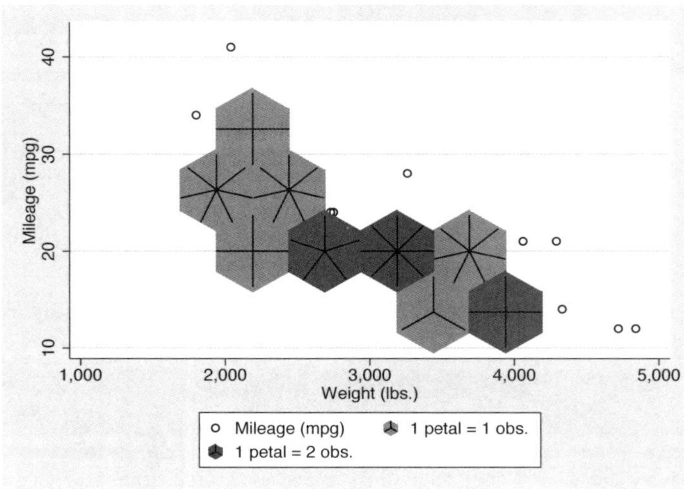

The three darkly shaded sunflowers immediately catch our eyes, indicating a group of 8 cars that are heavy (nearly 4000 pounds) and fuel inefficient and two groups of cars that get about 20 miles per gallon and weight in the neighborhood of 3000 pounds, one with 10 cars and one with 8 cars. The lighter sunflowers with 7 petals each indicate groups of 7 cars that share similar weight and fuel economy characteristics. To obtain the number of cars in each group, we counted the number of petals in each flower and consulted the graph legend to see how many observations each petal represents.

◁

# Methods and Formulas

`sunflower` is implemented as an ado-file.

# Acknowledgments

We would like to thank William D. Dupont and W. Dale Plummer (Vanderbilt University), authors of the original `sunflower` command, for their assistance in producing this version.

# References

Carr, D. B., R. J. Littlefield, W. L. Nicholson, and J. S. Littlefield. 1987. Scatterplot matrix techniques for large N. *Journal of the American Statistical Association* 82: 424–436.

Cleveland, W. S. and R. McGill. 1984. The many faces of a scatterplot. *Journal of the American Statistical Association* 79: 807–822.

Dupont, W. D. and W. D. Plummer, Jr. 2003. Density distribution sunflower plots. *Journal of Statistical Software* 8(3): 1–11.

Huang, C., J. A. McDonald, and W. Stuetzle. 1997. Variable resolution bivariate plots. *Journal of Computational and Graphical Statistics* 6: 383–396.

Levy, D. 1999. *50 years of discovery: Medical milestones from the National Heart, Lung, and Blood Institute's Framingham Heart Study.* Hackensack, NJ: Center for Bio-Medical Communication Inc.

Steichen, T. J. and N. J. Cox. 1999. flower: Stata module to draw sunflower plots. Stata program and help file downloadable from *http://ideas.repec.org/c/boc/bocode/s393001.html.* Accessed December 6, 2002.

# Also See

**Complementary:** [G] **graph**

# Title

**sureg** — Zellner's seemingly unrelated regression

# Syntax

*Basic syntax*

`sureg` (*depvar*$_1$ *varlist*$_1$) (*depvar*$_2$ *varlist*$_2$) ... (*depvar*$_N$ *varlist*$_N$)

[*if*] [*in*] [*weight*]

*Full syntax*

`sureg` ([*eqname*$_1$`:`]*depvar*$_{1a}$ [*depvar*$_{1b}$ ... =]*varlist*$_1$ [`, noconstant`])

([*eqname*$_2$`:`]*depvar*$_{2a}$ [*depvar*$_{2b}$ ... =]*varlist*$_2$ [`, noconstant`])

...

([*eqname*$_N$`:`]*depvar*$_{Na}$ [*depvar*$_{Nb}$ ... =]*varlist*$_N$ [`, noconstant`])

[*if*] [*in*] [*weight*] [`,` *options*]

Explicit equation naming (*eqname*`:`) cannot be combined with multiple dependent variables in an equation specification.

| *options* | description |
|---|---|
| Model | |
| `isure` | iterate until estimates converge |
| `constraints(`*constraints*`)` | apply specified linear constraints |
| df adj. | |
| `small` | report small-sample statistics |
| `dfk` | use small-sample adjustment |
| `dfk2` | use alternate adjustment |
| Reporting | |
| `level(`*#*`)` | set confidence level; default is `level(95)` |
| `corr` | perform Breusch–Pagan test |
| Opt options | |
| *optimization_options* | control the optimization process; seldom used |
| †`noheader` | suppress header table from above coefficient table |
| †`notable` | suppress coefficient table |

†`noheader` and `notable` do not appear in the dialog box.

The *depvars* and the *varlists* may contain time-series operators; see [U] **11.4.3 Time-series varlists**.

`bootstrap`, `by`, `jackknife`, `rolling`, `statsby`, and `xi` are allowed; see [U] **11.1.10 Prefix commands**.

`aweight`s and `fweight`s are allowed; see [U] **11.1.6 weight**.

See [U] **20 Estimation and postestimation commands** for additional capabilities of estimation commands.

# Description

sureg fits seemingly unrelated regression models (Zellner 1962, Zellner and Huang 1962, Zellner 1963). The acronyms SURE and SUR are often used for the estimator.

# Options

Model

isure specifies that sureg iterate over the estimated disturbance covariance matrix and parameter estimates until the parameter estimates converge. Under seemingly unrelated regression, this iteration converges to the maximum likelihood results. If this option is not specified, sureg produces two-step estimates.

constraints(*constraints*); see [R] **estimation options**.

df adj.

small specifies that small-sample statistics be computed. It shifts the test statistics from chi-squared and $Z$ statistics to $F$ statistics and $t$ statistics. While the standard errors from each equation are computed using the degrees of freedom for the equation, the degrees of freedom for the $t$ statistics are all taken to be those for the first equation.

dfk specifies the use of an alternate divisor in computing the covariance matrix for the equation residuals. As an asymptotically justified estimator, sureg by default uses the number of sample observations ($n$) as a divisor. When the dfk option is set, a small-sample adjustment is made and the divisor is taken to be $\sqrt{(n-k_i)(n-k_j)}$, where $k_i$ and $k_j$ are the numbers of parameters in equations $i$ and $j$, respectively.

dfk2 specifies the use of an alternate divisor in computing the covariance matrix for the equation residuals. When the dfk2 option is set, the divisor is taken to be the mean of the residual degrees of freedom from the individual equations.

Reporting

level(#); see [R] **estimation options**.

corr displays the correlation matrix of the residuals between equations and performs a Breusch–Pagan test for independent equations; that is, the disturbance covariance matrix is diagonal.

Opt options

*optimization_options* control the iterative process that minimizes the sum of squared errors when isure is specified. These options are seldom used.

iterate(#) specifies the maximum number of iterations. When the number of iterations equals #, the optimizer stops and presents the current results, even if the convergence tolerance has not been reached. The default value of iterate() is the current value of set maxiter, which is iterate(16000) if maxiter has not been changed.

trace adds to the iteration log a display of the current parameter vector

nolog suppresses the display of the iteration log.

tolerance(#) specifies the tolerance for the coefficient vector. When the relative change in the coefficient vector from one iteration to the next is less than or equal to #, the optimization process is stopped. tolerance(1e-6) is the default.

The following options are available with `sureg` but are not shown in the dialog box:

`noheader` suppresses display of the table reporting $F$ statistics, $R$-squared, and root mean squared error above the coefficient table.

`notable` suppresses display of the coefficient table.

# Remarks

Seemingly unrelated regression models are so called because they appear to be joint estimates from several regression models, each with its own error term. The regressions are related because the (contemporaneous) errors associated with the dependent variables may be correlated.

## ▷ Example 1

When we fit models with the same set of right-hand-side variables, the seemingly unrelated regression results (in terms of coefficients and standard errors) are the same as fitting the models separately (using, say, `regress`). The same is true when the models are nested. Even in such cases, `sureg` is useful when we want to perform joint tests. For instance, let us assume that we think

$$\texttt{price} = \beta_0 + \beta_1\texttt{foreign} + \beta_2\texttt{length} + u_1$$

$$\texttt{weight} = \gamma_0 + \gamma_1\texttt{foreign} + \gamma_2\texttt{length} + u_2$$

Since the models have the same set of explanatory variables, we could estimate the two equations separately. Yet, we might still choose to estimate them with `sureg` because we want to perform the joint test $\beta_1 = \gamma_1 = 0$.

We use the `small` and `dfk` options to obtain small-sample statistics comparable with `regress` or `mvreg`.

```
. use http://www.stata-press.com/data/r9/auto
(1978 Automobile Data)
. sureg (price foreign length) (weight foreign length), small dfk
Seemingly unrelated regression
------------------------------------------------------------------------------
Equation          Obs  Parms        RMSE    "R-sq"       F-Stat        P
------------------------------------------------------------------------------
price              74      2    2474.593    0.3154        16.35   0.0000
weight             74      2    250.2515    0.8992       316.54   0.0000
------------------------------------------------------------------------------

------------------------------------------------------------------------------
             |      Coef.   Std. Err.      t    P>|t|     [95% Conf. Interval]
-------------+----------------------------------------------------------------
price        |
     foreign |   2801.143    766.117     3.66   0.000     1286.674    4315.611
      length |   90.21239   15.83368     5.70   0.000     58.91219    121.5126
       _cons |  -11621.35   3124.436    -3.72   0.000    -17797.77    -5444.93
-------------+----------------------------------------------------------------
weight       |
     foreign |  -133.6775   77.47615    -1.73   0.087    -286.8332     19.4782
      length |   31.44455   1.601234    19.64   0.000     28.27921    34.60989
       _cons |   -2850.25   315.9691    -9.02   0.000    -3474.861   -2225.639
------------------------------------------------------------------------------
```

These two equations have a common set of regressors, and we could have used a shorthand syntax to specify the equations:

```
. sureg (price weight = foreign length), small dfk
```

In this case, the results presented by `sureg` are the same as if we had estimated the equations separately:

```
. regress price foreign length
  (output omitted)
. regress weight foreign length
  (output omitted)
```

There is, however, a difference. We have allowed $u_1$ and $u_2$ to be correlated and have estimated the full variance–covariance matrix of the coefficients. `sureg` has estimated the correlations, but it does not report them unless we specify the `corr` option. We did not remember to specify `corr` when we fitted the model, but we can redisplay the results:

```
. sureg, notable noheader corr

Correlation matrix of residuals:

          price  weight
 price   1.0000
weight   0.5840  1.0000

Breusch-Pagan test of independence: chi2(1) =    25.237, Pr = 0.0000
```

The `notable` and `noheader` options prevented `sureg` from redisplaying the header and coefficient tables. We find that, for the same cars, the correlation of the residuals in the `price` and `weight` equations is .5840 and that we can reject the hypothesis that this correlation is zero.

We can test that the coefficients on `foreign` are jointly zero in both equations—as we set out to do—by typing `test foreign`; see [R] **test**. When we type a variable without specifying the equation, that variable is tested for zero in all equations in which it appears:

```
. test foreign

 ( 1)  [price]foreign = 0
 ( 2)  [weight]foreign = 0

       F(  2,   142) =   17.99
            Prob > F =    0.0000
```

◁

## ▷ Example 2

When the models do not have the same set of explanatory variables and are not nested, `sureg` may lead to more efficient estimates than running the models separately as well as allowing joint tests. This time, let us assume that we believe

$$\texttt{price} = \beta_0 + \beta_1\texttt{foreign} + \beta_2\texttt{mpg} + \beta_3\texttt{displ} + u_1$$
$$\texttt{weight} = \gamma_0 + \gamma_1\texttt{foreign} + \gamma_2\texttt{length} + u_2$$

To fit this model, we type

Arnold Zellner (1927– ) was born in New York. He studied physics at Harvard and economics at Berkeley, and then taught economics at the Universities of Washington and Wisconsin before settling in Chicago in 1966. Among his many major contributions to econometrics and statistics are his work on seemingly unrelated regression, three-stage least squares, and Bayesian econometrics.

```
. sureg (price foreign mpg displ) (weight foreign length), corr
Seemingly unrelated regression
------------------------------------------------------------------------
Equation          Obs  Parms        RMSE    "R-sq"       chi2        P
------------------------------------------------------------------------
price              74      3    2165.321    0.4537      49.64   0.0000
weight             74      2    245.2916    0.8990     661.84   0.0000
------------------------------------------------------------------------

------------------------------------------------------------------------------
             |      Coef.   Std. Err.      z    P>|z|     [95% Conf. Interval]
-------------+----------------------------------------------------------------
price        |
     foreign |    3058.25   685.7357     4.46   0.000     1714.233    4402.267
         mpg |  -104.9591   58.47209    -1.80   0.073    -219.5623    9.644042
displacement |   18.18098   4.286372     4.24   0.000     9.779842    26.58211
       _cons |   3904.336   1966.521     1.99   0.047      50.0263    7758.645
-------------+----------------------------------------------------------------
weight       |
     foreign |  -147.3481   75.44314    -1.95   0.051    -295.2139     .517755
      length |   30.94905   1.539895    20.10   0.000     27.93091    33.96718
       _cons |  -2753.064   303.9336    -9.06   0.000    -3348.763   -2157.365
------------------------------------------------------------------------------

Correlation matrix of residuals:

         price  weight
 price  1.0000
weight  0.3285  1.0000

Breusch-Pagan test of independence: chi2(1) =     7.984, Pr = 0.0047
```

In comparison, if we had fitted the `price` model separately,

```
. regress price foreign mpg displ
      Source |       SS       df       MS              Number of obs =      74
-------------+------------------------------           F(  3,    70) =   20.13
       Model |   294104790     3  98034929.9           Prob > F      =  0.0000
    Residual |   340960606    70  4870865.81           R-squared     =  0.4631
-------------+------------------------------           Adj R-squared =  0.4401
       Total |   635065396    73  8699525.97           Root MSE      =    2207

------------------------------------------------------------------------------
       price |      Coef.   Std. Err.      t    P>|t|     [95% Conf. Interval]
-------------+----------------------------------------------------------------
     foreign |   3545.484   712.7763     4.97   0.000     2123.897    4967.072
         mpg |  -98.88559   63.17063    -1.57   0.122    -224.8754    27.10426
displacement |   22.40416   4.634239     4.83   0.000     13.16146    31.64686
       _cons |    2796.91   2137.873     1.31   0.195    -1466.943    7060.763
------------------------------------------------------------------------------
```

The coefficients are slightly different, but the standard errors are uniformly larger. This would still be true if we specified the **dfk** option to make a small-sample adjustment to the estimated covariance of the disturbances.

◁

## ❑ Technical Note

Constraints can be applied to SURE models using Stata's standard syntax for constraints. For a general discussion of constraints, see [R] **constraint**; for examples similar to seemingly unrelated regression models, see [R] **reg3**.

❑

## Saved Results

sureg saves in e():

Scalars

| | |
|---|---|
| e(N) | number of observations |
| e(k) | number of parameters in system |
| e(k_eq) | number of equations |
| e(mss_#) | model sum of squares for equation # |
| e(df_m#) | model degrees of freedom for equation # |
| e(rss_#) | residual sum of squares for equation # |
| e(df_r) | residual degrees of freedom |
| e(r2_#) | $R$-squared for equation # |
| e(F_#) | $F$ statistic for equation # (small only) |
| e(rmse_#) | root mean squared error for equation # |
| e(ll) | log likelihood |
| e(chi2_#) | $\chi^2$ for equation # |
| e(p_#) | significance for equation # |
| e(chi2_bp) | Breusch–Pagan $\chi^2$ |
| e(df_bp) | degrees of freedom for Breusch–Pagan $\chi^2$ test |
| e(cons_#) | 1 when equation # has a constant; 0, otherwise |
| e(ic) | number of iterations |

Macros

| | |
|---|---|
| e(cmd) | sureg |
| e(depvar) | names of dependent variables |
| e(exog) | names of exogenous variables |
| e(eqnames) | names of equations |
| e(wtype) | weight type |
| e(wexp) | weight expression |
| e(method) | sure or isure |
| e(corr) | correlation structure |
| e(small) | small |
| e(dfk) | alternate divisor (dfk or dfk2 only) |
| e(predict) | program used to implement predict |
| e(properties) | b V |

Matrices

| | |
|---|---|
| e(b) | coefficient vector |
| e(V) | variance–covariance matrix of the estimators |
| e(Sigma) | $\widehat{\Sigma}$ matrix |

Functions

| | |
|---|---|
| e(sample) | marks estimation sample |

## Methods and Formulas

sureg is implemented as an ado-file.

sureg uses the asymptotically efficient, feasible, generalized least-squares algorithm described in Greene (2003, 340–362). The computing formulas are given on page 342.

The $R$-squared reported is the percent of variance explained by the predictors. It may be used for descriptive purposes, but $R$-squared is not a well-defined concept when GLS is used.

`sureg` will refuse to compute the estimators if the same equation is named more than once or the covariance matrix of the residuals is singular.

The Breusch and Pagan (1980) $\chi^2$ statistic—a Lagrange multiplier statistic—is given by

$$\lambda = T \sum_{m=1}^{M} \sum_{n=1}^{m-1} r_{mn}^2$$

where $r_{mn}$ is the estimated correlation between the residuals of the $M$ equations and $T$ is the number of observations. It is distributed as $\chi^2$ with $M(M-1)/2$ degrees of freedom.

## References

Breusch, T. and A. Pagan. 1980. The LM test and its applications to model specification in econometrics. *Review of Economic Studies* 47: 239–254.

Greene, W. H. 2003. *Econometric Analysis*. 5th ed. Upper Saddle River, NJ: Prentice Hall.

McDowell, A. W. 2004. From the help desk: Seemingly unrelated regression with unbalanced equations. *Stata Journal* 4: 442–448.

Weesie, J. 1999. sg121: Seemingly unrelated estimation and the cluster-adjusted sandwich estimator. *Stata Technical Bulletin* 52: 34–47. Reprinted in *Stata Technical Bulletin Reprints*, vol. 9, pp. 231–248.

Zellner, A. 1962. An efficient method of estimating seeming unrelated regressions and tests for aggregation bias. *Journal of the American Statistical Association* 57: 348–368.

——. 1963. Estimators for seemingly unrelated regression equations: Some exact finite sample results. *Journal of the American Statistical Association* 58: 977–992.

Zellner, A. and D. S. Huang. 1962. Further properties of efficient estimators for seemingly unrelated regression equations. *International Economic Review* 3: 300–313.

## Also See

**Complementary:** [R] **sureg postestimation**; [R] **constraint**

**Related:** [R] **ivreg**, [R] **mvreg**, [R] **reg3**, [R] **regress**

**Background:** [U] **11.1.10 Prefix commands**,
[U] **20 Estimation and postestimation commands**,
[R] **estimation options**, [R] **maximize**

# Title

**sureg postestimation** — Postestimation tools for sureg

# Description

The following postestimation commands are available for `sureg`:

| command | description |
|---|---|
| `estat` | AIC, BIC, VCE, and estimation sample summary |
| `estimates` | cataloging estimation results |
| `lincom` | point estimates, standard errors, testing, and inference for linear combinations of coefficients |
| `mfx` | marginal effects or elasticities |
| `nlcom` | point estimates, standard errors, testing, and inference for nonlinear combinations of coefficients |
| `predict` | predictions, residuals, influence statistics, and other diagnostic measures |
| `predictnl` | point estimates, standard errors, testing, and inference for generalized predictions |
| `test` | Wald tests for simple and composite linear hypotheses |
| `testnl` | Wald tests of nonlinear hypotheses |

See the corresponding entries in the *Stata Base Reference Manual* for details.

# Syntax for predict

`predict` [*type*] *newvar* [*if*] [*in*] [, `equation(`*eqno*[`,`*eqno*]`)` *statistic*]

| *statistic* | description |
|---|---|
| `xb` | $\mathbf{x}_j\mathbf{b}$, fitted values; the default |
| `stdp` | standard error of the prediction |
| `difference` | difference between the linear predictions of two equations |
| `stddp` | standard error of the difference in linear predictions |
| `residuals` | residuals |

These statistics are available both in and out of sample; type `predict ... if e(sample) ...` if wanted only for the estimation sample.

(*Continued on next page*)

## Options for predict

`equation(`*eqno*[`,`*eqno*]`)` specifies to which equation(s) you are referring.

`equation()` is filled in with one *eqno* for options `xb`, `stdp`, and `residuals`. `equation(#1)` would mean that the calculation is to be made for the first equation, `equation(#2)` would mean the second, and so on. Alternatively, you could refer to the equations by their names. `equation(income)` would refer to the equation named income and `equation(hours)` to the equation named hours.

If you do not specify `equation()`, the results are the same as if you specified `equation(#1)`.

`difference` and `stddp` refer to between-equation concepts. To use these options, you must specify two equations, for example, `equation(#1,#2)` or `equation(income,hours)`. When two equations must be specified, `equation()` is required.

`xb`, the default, calculates the fitted values—the prediction of $\mathbf{x}_j\mathbf{b}$ for the specified equation.

`stdp` calculates the standard error of the prediction for the specified equation. It can be thought of as the standard error of the predicted expected value or mean for the observation's covariate pattern. This is also referred to as the standard error of the fitted value.

`difference` calculates the difference between the linear predictions of two equations in the system. With `equation(#1,#2)`, `difference` computes the prediction of `equation(#1)` minus the prediction of `equation(#2)`.

`stddp` is allowed only after you have previously fitted a multiple-equation model. The standard error of the difference in linear predictions $(\mathbf{x}_{1j}\mathbf{b} - \mathbf{x}_{2j}\mathbf{b})$ between equations 1 and 2 is calculated.

`residuals` calculates the residuals.

For more information on using `predict` after multiple-equation estimation commands, see [R] **predict**.

## Methods and Formulas

All postestimation commands listed above are implemented as ado-files.

## Also See

**Complementary:** [R] **sureg**; [R] **estimates**, [R] **lincom**, [R] **mfx**, [R] **nlcom**, [R] **predictnl**, [R] **test**, [R] **testnl**

**Background:** [U] **13.5 Accessing coefficients and standard errors**, [U] **20 Estimation and postestimation commands**, [R] **estat**, [R] **predict**

# Title

**swilk** — Shapiro–Wilk and Shapiro–Francia tests for normality

# Syntax

*Shapiro–Wilk normality test*

`swilk` *varlist* [*if*] [*in*] [, *options*]

*Shapiro–Francia normality test*

`sfrancia` *varlist* [*if*] [*in*]

| *options* | description |
|---|---|
| Main | |
| `generate(`*newvar*`)` | create *newvar* containing $W$ test coefficients |
| `lnnormal` | test for three-parameter log normality |
| `noties` | do not use average ranks for tied values |

`by` may be used with `swilk` and `sfrancia`; see [D] **by**.

# Description

`swilk` performs the Shapiro–Wilk $W$ test for normality, and `sfrancia` performs the Shapiro–Francia $W'$ test for normality. `swilk` can be used with $4 \leq n \leq 2000$ observations, and `sfrancia` can be used with $5 \leq n \leq 5000$ observations; see [R] **sktest** for a test allowing a larger number of observations.

# Options

Main

`generate(`*newvar*`)` creates new variable *newvar* containing the $W$ test coefficients.

`lnnormal` specifies that the test be for three-parameter log normality, meaning that $\ln(X-k)$ is tested for normality, where $k$ is calculated from the data as the value that makes the skewness coefficient zero. When simply testing $\ln(X)$ for normality, do not specify this option. See [R] **lnskew0** for estimation of $k$.

`noties` suppresses use of averaged ranks for tied values when calculating the $W$ test coefficients.

(*Continued on next page*)

# Remarks

## ▷ Example 1

Using our automobile dataset, we will test whether the variables `mpg` and `trunk` are normally distributed:

```
. use http://www.stata-press.com/data/r9/auto
(1978 Automobile Data)

. swilk mpg trunk

                   Shapiro-Wilk W test for normal data
    Variable |    Obs       W           V         z       Prob>z
-------------+-----------------------------------------------------
         mpg |     74    0.94821      3.335     2.627    0.00430
       trunk |     74    0.97921      1.339     0.637    0.26215

. sfrancia mpg trunk

                  Shapiro-Francia W' test for normal data
    Variable |    Obs       W'          V'        z       Prob>z
-------------+-----------------------------------------------------
         mpg |     74    0.94872      3.629     2.490    0.00639
       trunk |     74    0.98446      1.100     0.190    0.42477
```

We can reject the hypothesis that `mpg` is normally distributed, but we cannot reject that `trunk` is normally distributed.

The values reported under $W$ and $W'$ are the Shapiro–Wilk and Shapiro–Francia test statistics. The tests also report $V$ and $V'$, which are more appealing indexes for departure from normality. The median values of $V$ and $V'$ are 1 for samples from normal populations. Large values indicate non-normality. The 95% critical values of $V$ ($V'$), which depend on the sample size, are between 1.2 and 2.4 (2.0 and 2.8); see Royston (1991a). There is no more information in $V$ ($V'$) than in $W$ ($W'$)—one is just the transform of the other.

◁

## ▷ Example 2

We have data on a variable called `chol`, which we suspect is distributed lognormally:

```
. generate lchol = ln(chol)

. swilk lchol

                   Shapiro-Wilk W test for normal data
    Variable |    Obs       W           V         z       Prob>z
-------------+-----------------------------------------------------
       lchol |     78    0.84530     10.401     5.124    0.00000
```

We can reject the lognormal assumption. Note that we do *not* specify the `lnnormal` option when testing for log normality. The `lnnormal` option is for 3-parameter log normality.

◁

## ▷ Example 3

Having discovered that $\ln($`chol`$)$ is not distributed normally, we now test that $\ln($`chol` $- k)$ is normally distributed, where $k$ is chosen so that the resulting skewness is zero. We obtain the estimate for $k$ from `lnskew0`; see [R] **lnskew0**:

```
. lnskew0 lcholk = chol, level(95)
      Transform |       k      [95% Conf. Interval]        Skewness
----------------+--------------------------------------------------
      ln(cholk) |   4.984322    4.566498   5.069748        .0000399
. swilk lcholk, lnnormal
          Shapiro-Wilk W test for 3-parameter lognormal data
   Variable |    Obs        W          V          z     Prob>z
------------+------------------------------------------------------
     lcholk |     78    0.98316      1.132      0.770  0.22053
```

We cannot reject the hypothesis that $\ln(\texttt{chol} - 4.98)$ is distributed normally. Note that we do specify the `lnnormal` option when using an estimated value of $k$.

◁

## Saved Results

`swilk` and `sfrancia` save in `r()`:

Scalars

| | | | |
|---|---|---|---|
| `r(N)` | number of observations | `r(W)` | $W$ or $W'$ |
| `r(p)` | significance | `r(V)` | $V$ or $V'$ |
| `r(z)` | $z$ statistic | | |

## Methods and Formulas

`swilk` and `sfrancia` are implemented as ado-files.

The Shapiro–Wilk test is based on Shapiro and Wilk (1965) with a new approximation accurate for $4 \leq n \leq 2000$ (Royston 1992). The calculations made by `swilk` are based on Royston (1982, 1992, 1993).

The Shapiro–Francia test (Shapiro and Francia, 1972; Royston 1983) is an approximate test that is similar to the Shapiro–Wilk test for very large samples.

Samuel S. Shapiro (1930– ) earned degrees in statistics and engineering from City College of New York, Columbia, and Rutgers. After employment in the U.S. army and industry, he joined the faculty at Florida International University in 1972. Shapiro has co-authored various texts in statistics and published several papers on distributional testing and other statistical topics.

## Acknowledgment

`swilk` and `sfrancia` were written by Patrick Royston of the MRC Clinical Trials Unit, London.

# References

Gould, W. W. 1992. sg3.7: Final summary of tests of normality. *Stata Technical Bulletin* 5: 10–11. Reprinted in *Stata Technical Bulletin Reprints*, vol. 1, pp. 114–115.

Royston, P. 1982. An extension of Shapiro and Wilk's $W$ test for normality to large samples. *Applied Statistics* 31: 115–124.

——. 1983. A simple method for evaluating the Shapiro–Francia $W'$ test of non-normality. *Applied Statistics* 32: 297–300.

——. 1991a. Estimating departure from normality. *Statistics in Medicine* 10: 1283–1293.

——. 1991b. sg3.2: Shapiro–Wilk and Shapiro–Francia tests. *Stata Technical Bulletin* 3: 19. Reprinted in *Stata Technical Bulletin Reprints*, vol. 1, p. 105.

——. 1992. Approximating the Shapiro–Wilk $W$-test for non-normality. *Statistics and Computing* 2: 117–119.

——. 1993. A toolkit for testing for non-normality in complete and censored samples. *Statistician* 42: 37–43.

Shapiro, S. S. and R. S. Francia. 1972. An approximate analysis of variance test for normality. *Journal of the American Statistical Association* 67: 215–216.

Shapiro, S. S. and M. B. Wilk. 1965. An analysis of variance test for normality (complete samples). *Biometrika* 52: 591–611.

# Also See

**Related:** [R] **lnskew0**, [R] **lv**, [R] **sktest**

# Title

**symmetry** — Symmetry and marginal homogeneity tests

# Syntax

*Symmetry and marginal homogeneity tests*

`symmetry` *casevar controlvar* [*if*] [*in*] [*weight*] [, *options*]

*Immediate form of symmetry and marginal homogeneity tests*

`symmi` $\#_{11}$ $\#_{12}$ [...] \ $\#_{21}$ $\#_{22}$ [...] [\...] [*if*] [*in*] [, *options*]

| *options* | description |
|---|---|
| Main | |
| `notable` | suppress output of contingency table |
| `contrib` | report contribution of each off-diagonal cell pair |
| `exact` | perform exact test of table symmetry |
| `mh` | perform two marginal homogeneity tests |
| `trend` | perform a test for linear trend in the (log) relative risk (RR) |
| `cc` | use continuity correction when calculating test for linear trend |

`fweight`s are allowed; see [U] **11.1.6 weight**.

# Description

`symmetry` performs asymptotic symmetry and marginal homogeneity tests, and an exact symmetry test on $K \times K$ tables where there is a 1-to-1 matching of cases and controls (nonindependence). This is used to analyze matched-pair case–control data with multiple discrete levels of the exposure (outcome) variable. In genetics, the test is known as the Transmission/Disequilibrium test (TDT) and is used to test the association between transmitted and nontransmitted parental marker alleles to an affected child (Spieldman, McGinnis, and Ewens 1993). In the case of $2 \times 2$ tables, the asymptotic test statistics reduce to the McNemar test statistic, and the exact symmetry test produces an exact McNemar test; see [ST] **epitab**. For numeric exposure variables, `symmetry` can optionally perform a test for linear trend in the log relative risk.

`symmetry` expects the data to be in the wide format; that is, each observation contains the matched case and control values in variables *casevar* and *controlvar*. Variables can be numeric or string.

`symmi` is the immediate form of `symmetry`. The `symmi` command uses the values specified on the command line; rows are separated by '\', and options are the same as for `symmetry`. See [U] **19 Immediate commands** for a general introduction to immediate commands.

*(Continued on next page)*

# Options

Main

`notable` suppresses the output of the contingency table. By default, `symmetry` displays the $n \times n$ contingency table at the top of the output.

`contrib` reports the contribution of each off-diagonal cell pair to the overall symmetry $\chi^2$.

`exact` performs an exact test of table symmetry. This option is recommended for sparse tables. CAUTION: the exact test requires substantial amounts of time and computer memory for large tables.

`mh` performs two marginal homogeneity tests that do not require the inversion of the variance–covariance matrix.

By default, `symmetry` produces the Stuart–Maxwell test statistic, which requires the inversion of the nondiagonal variance–covariance matrix, $\mathbf{V}$. When the table is sparse, the matrix may not be of full rank, and, in that case, the command substitutes a generalized inverse $\mathbf{V}^*$ for $\mathbf{V}^{-1}$. `mh` calculates optional marginal homogeneity statistics that do not require the inversion of the variance–covariance matrix. These tests may be preferred in certain situations. See *Methods and Formulas* and Bickenböller and Clerget-Darpoux (1995) for details on these test statistics.

`trend` performs a test for linear trend in the (log) relative risk (RR). This option is only allowed for numeric exposure (outcome) variables, and its use should be restricted to measurements on the ordinal or the interval scales.

`cc` specifies that the continuity correction be used when calculating the test for linear trend. This correction should only be specified when the levels of the exposure variable are equally spaced.

# Remarks

`symmetry` and `symmi` may be used to analyze 1-to-1 matched case–control data with multiple discrete levels of the exposure (outcome) variable.

## ▷ Example 1

Consider a survey of 344 individuals (BMDP 1990, 267–270) who were asked in October 1986 whether they agreed with President Reagan's handling of foreign affairs. In January 1987, after the "Iran-Contra" affair became public, these same individuals were surveyed again and asked the same question. We would like to know if public opinion changed over this time period.

We first describe the dataset and list a few observations.

```
. use http://www.stata-press.com/data/r9/iran
. describe
Contains data from http://www.stata-press.com/data/r9/iran.dta
  obs:           344
 vars:             2                          29 Jan 2005 02:37
 size:         2,064 (99.9% of memory free)
-------------------------------------------------------------------------------
              storage  display     value
variable name   type   format      label      variable label
-------------------------------------------------------------------------------
before          byte   %8.0g       vlab       Public Opinion before IC
after           byte   %8.0g       vlab       Public Opinion after IC
-------------------------------------------------------------------------------
Sorted by:
```

```
. list in 1/5

     +---------------------+
     |   before      after |
     |---------------------|
  1. |    agree      agree |
  2. |    agree   disagree |
  3. |    agree     unsure |
  4. | disagree      agree |
  5. | disagree   disagree |
     +---------------------+
```

Each observation corresponds to one of the 344 individuals. The data are in wide form so that each observation has a before and an after measurement. We now perform the test without options.

```
. symmetry before after

Public
Opinion          Public Opinion after IC
before IC    agree   disagree    unsure      Total

    agree       47         56        38        141
 disagree       28         61        31        120
   unsure       26         47        10         83

    Total      101        164        79        344

                                            chi2        df       Prob>chi2

Symmetry (asymptotic)                      14.87         3          0.0019
Marginal homogeneity (Stuart-Maxwell)      14.78         2          0.0006
```

The test first tabulates the data in a $K \times K$ table and then performs Bowker's (1948) test for table symmetry and the Stuart–Maxwell (Stuart 1955, Maxwell 1970) test for marginal homogeneity.

Both the symmetry test and the marginal homogeneity test are highly significant, thus indicating a shift in public opinion.

An exact test of symmetry is provided for use on sparse tables. This test is computationally intensive, so it should not be used on large tables. Since we are working on a fast computer, we will run the symmetry test again and this time include the `exact` option. We will suppress the output of the contingency table by specifying `notable` and also include the `contrib` option so that we may further examine the cells responsible for the significant result.

```
. symmetry before after, contrib exact mh notable
                      Contribution
                       to symmetry
     Cells             chi-squared

  n1_2 & n2_1             9.3333
  n1_3 & n3_1             2.2500
  n2_3 & n3_2             3.2821
                                            chi2        df       Prob>chi2

Symmetry (asymptotic)                      14.87         3          0.0019
Marginal homogeneity (Stuart-Maxwell)      14.78         2          0.0006
Marginal homogeneity (Bickenboller)        13.53         2          0.0012
Marginal homogeneity (no diagonals)        15.25         2          0.0005

Symmetry (exact significance probability)                           0.0018
```

The largest contribution to the symmetry $\chi^2$ is due to cells $n_{12}$ and $n_{21}$. These correspond to changes between the agree and disagree categories. Of the 344 individuals, 56 (16.3%) changed from the agree to the disagree response, whereas only 28 (8.1%) changed in the opposite direction.

For these data, the results from the exact test are similar to those from the asymptotic test.

◁

## ▷ Example 2

Breslow and Day (1980, 163) reprinted data from Mack et al. (1976) from a case–control study of the effect of exogenous estrogen on the risk of endometrial cancer. The data consist of 59 elderly women diagnosed with endometrial cancer and 59 disease-free controls living in the same community as the cases. Cases and controls were matched on age, marital status, and time living in the community. The data collected included information on the daily dose of conjugated estrogen therapy. Breslow and Day analyzed these data by creating four levels of the dose variable. Here are the data as entered into a Stata dataset:

```
. use http://www.stata-press.com/data/r9/bd163
. list, noobs divider
```

| case | control | count |
|---|---|---|
| 0 | 0 | 6 |
| 0 | 0.1-0.299 | 2 |
| 0 | 0.3-0.625 | 3 |
| 0 | 0.626+ | 1 |
| 0.1-0.299 | 0 | 9 |
| 0.1-0.299 | 0.1-0.299 | 4 |
| 0.1-0.299 | 0.3-0.625 | 2 |
| 0.1-0.299 | 0.626+ | 1 |
| 0.3-0.625 | 0 | 9 |
| 0.3-0.625 | 0.1-0.299 | 2 |
| 0.3-0.625 | 0.3-0.625 | 3 |
| 0.3-0.625 | 0.626+ | 1 |
| 0.626+ | 0 | 12 |
| 0.626+ | 0.1-0.299 | 1 |
| 0.626+ | 0.3-0.625 | 2 |
| 0.626+ | 0.626+ | 1 |

This dataset is in a different format than the previous example. Instead of each observation representing a single matched pair, each observation represents possibly multiple pairs indicated by the `count` variable. For instance, the first observation corresponds to 6 matched pairs where neither the case nor the control was on estrogen; the second observation corresponds to 2 matched pairs where the case was not on estrogen and the control was on 0.1 to 0.299 mg/day; etc.

To use `symmetry` to analyze this dataset, we must specify `fweight` to indicate that in our data there are observations corresponding to more than one matched pair.

```
. symmetry case control [fweight=count]
```

| case | control 0 | 0.1-0.299 | 0.3-0.625 | 0.626+ | Total |
|---|---|---|---|---|---|
| 0 | 6 | 2 | 3 | 1 | 12 |
| 0.1-0.299 | 9 | 4 | 2 | 1 | 16 |
| 0.3-0.625 | 9 | 2 | 3 | 1 | 15 |
| 0.626+ | 12 | 1 | 2 | 1 | 16 |
| Total | 36 | 9 | 10 | 4 | 59 |

| | chi2 | df | Prob>chi2 |
|---|---|---|---|
| Symmetry (asymptotic) | 17.10 | 6 | 0.0089 |
| Marginal homogeneity (Stuart-Maxwell) | 16.96 | 3 | 0.0007 |

Both the test of symmetry and the test of marginal homogeneity are highly significant, thus leading us to reject the null hypothesis that there is no effect of exposure to estrogen on the risk of endometrial cancer.

Breslow and Day perform a test for trend assuming that the estrogen exposure levels were equally spaced by recoding the exposure levels as 1, 2, 3, and 4.

We can easily reproduce their results by recoding our data in this way and by specifying the `trend` option. Two new numeric variables were created, `ca` and `co`, corresponding to the variables `case` and `control`, respectively. Below we list some of the data and our results from `symmetry`:

```
. encode case, gen(ca)
. encode control, gen(co)
. label values ca
. label values co
. list in 1/4
```

| | case | control | count | ca | co |
|---|---|---|---|---|---|
| 1. | 0 | 0 | 6 | 1 | 1 |
| 2. | 0 | 0.1-0.299 | 2 | 1 | 2 |
| 3. | 0 | 0.3-0.625 | 3 | 1 | 3 |
| 4. | 0 | 0.626+ | 1 | 1 | 4 |

```
. symmetry ca co [fw=count], notable trend cc
```

| | chi2 | df | Prob>chi2 |
|---|---|---|---|
| Symmetry (asymptotic) | 17.10 | 6 | 0.0089 |
| Marginal homogeneity (Stuart-Maxwell) | 16.96 | 3 | 0.0007 |
| Linear trend in the (log) RR | 14.43 | 1 | 0.0001 |

Note that we requested the continuity correction by specifying `cc`. This is appropriate because our coded exposure levels are equally spaced.

The test for trend was highly significant, indicating an increased risk of endometrial cancer with increased dosage of conjugated estrogen.

You must be cautious: the way in which you code the exposure variable affects the linear trend statistic. If instead of coding the levels as 1, 2, 3, and 4, we had instead used 0, .2, .46, and .7

(roughly the midpoint in the range of each level), we would have obtained a $\chi^2$ statistic of 11.19 for these data.

◁

## Saved Results

`symmetry` saves in `r()`:

Scalars

| | |
|---|---|
| `r(N_pair)` | number of matched pairs |
| `r(chi2)` | asymptotic symmetry $\chi^2$ |
| `r(df)` | asymptotic symmetry degrees of freedom |
| `r(p)` | asymptotic symmetry $p$-value |
| `r(chi2_sm)` | MH (Stuart–Maxwell) $\chi^2$ |
| `r(df_sm)` | MH (Stuart–Maxwell) degrees of freedom |
| `r(p_sm)` | MH (Stuart–Maxwell) $p$-value |
| `r(chi2_b)` | MH (Bickenböller) $\chi^2$ |
| `r(df_b)` | MH (Bickenböller) degrees of freedom |
| `r(p_b)` | MH (Bickenböller) $p$-value |
| `r(chi2_nd)` | MH (no diagonals) $\chi^2$ |
| `r(df_nd)` | MH (no diagonals) degrees of freedom |
| `r(p_nd)` | MH (no diagonals) $p$-value |
| `r(chi2_t)` | $\chi^2$ for linear trend |
| `r(p_trend)` | $p$-value for linear trend |
| `r(p_exact)` | exact symmetry $p$-value |

## Methods and Formulas

`symmetry` and `symmi` are implemented as ado-files.

### Asymptotic tests

Consider a square table with $K$ exposure categories, that is, $K$ rows and $K$ columns. Let $n_{ij}$ be the count corresponding to row $i$ and column $j$ of the table, $N_{ij} = n_{ij} + n_{ji}$, for $i, j = 1, 2, \ldots, K$, and $n_{i.}$, and let $n_{.j}$ be the marginal totals for row $i$ and column $j$, respectively. Asymptotic tests for symmetry and marginal homogeneity for this $K \times K$ table are calculated as follows:

The null hypothesis of complete symmetry $p_{ij} = p_{ji}$, $i \neq j$, is tested by calculating the test statistic (Bowker 1948),

$$T_{\text{cs}} = \sum_{i<j} \frac{(n_{ij} - n_{ji})^2}{n_{ij} + n_{ji}}$$

which is asymptotically distributed as $\chi^2$ with $K(K-1)/2 - R$ degrees of freedom, where $R$ is the number of off-diagonal cells with $N_{ij} = 0$.

The null hypothesis of marginal homogeneity, $p_{i.} = p_{.i}$, is tested by calculating the Stuart–Maxwell test statistic (Stuart 1955, Maxwell 1970),

$$T_{\text{sm}} = \mathbf{d}'\mathbf{V}^{-1}\mathbf{d}$$

where $\mathbf{d}$ is a column vector with elements equal to the differences $d_i = n_{i.} - n_{.i}$ for $i = 1, 2, \dots, K$, and $\mathbf{V}$ is the variance–covariance matrix with elements

$$v_{ii} = n_{i.} + n_{.i} - 2n_{ii}$$
$$v_{ij} = -(n_{ij} + n_{ji}), \quad i \neq j$$

$T_{\text{sm}}$ is asymptotically $\chi^2$ with $K - 1$ degrees of freedom.

This test statistic properly accounts for the dependence between the table's rows and columns. When the matrix $\mathbf{V}$ is not of full rank, a generalized inverse $\mathbf{V}^*$ is substituted for $\mathbf{V}^{-1}$.

The Bickenböller and Clerget-Darpoux (1995) marginal homogeneity test statistic is calculated by

$$T_{\text{mh}} = \sum_i \frac{(n_{i.} - n_{.i})^2}{n_{i.} + n_{.i}}$$

This statistic is asymptotically distributed, under the assumption of marginal independence, as $\chi^2$ with $K - 1$ degrees of freedom.

The marginal homogeneity (no diagonals) test statistic $T^0_{\text{mh}}$ is calculated in the same way as $T_{\text{mh}}$, except that the diagonal elements do not enter into the calculation of the marginal totals. Unlike the previous test statistic, $T^0_{\text{mh}}$ reduces to a McNemar test statistic for $2 \times 2$ tables. The test statistic $\{(K-1)/2\}T^0_{\text{mh}}$ is asymptotically distributed as $\chi^2$ with $K - 1$ degrees of freedom (Cleves, Olson, and Jacobs 1997; Spieldman and Ewens 1996).

Breslow and Day's test statistic for linear trend in the (log) of relative risk is

$$\frac{\left\{\sum_{i<j}(n_{ij} - n_{ji})(X_j - X_i) - cc\right\}^2}{\sum_{i<j}(n_{ij} + n_{ji})(j - i)^2}$$

where the $X_j$ are the "doses" associated with the various levels of exposure and $cc$ is the continuity correction; it is asymptotically distributed as $\chi^2$ with 1 degree of freedom.

The continuity correction option is only applicable when the levels of the exposure variable are equally spaced.

### Exact symmetry test

The exact test is based on a permutation algorithm applied to the null distribution. The distribution of the off-diagonal elements $n_{ij}$, $i \neq j$, conditional on the sum of the complementary off-diagonal cells, $N_{ij} = n_{ij} + n_{ji}$, can be written as the product of $K(K-1)/2$ binomial random variables,

$$P(\mathbf{n}) = \prod_{i<j} \binom{N_{ij}}{n_{ij}} {\pi_{ij}}^{n_{ij}} (1 - \pi_{ij})^{n_{ij}}$$

where $\mathbf{n}$ is a vector with elements $n_{ij}$ and $\pi_{ij} = E(n_{ij}/N_{ij} | N_{ij})$. Under the null hypothesis of complete symmetry, $\pi_{ij} = \pi_{ji} = 1/2$, and thus the permutation distribution is given by

$$P_0(\mathbf{n}) = \prod_{i<j} \binom{N_{ij}}{n_{ij}} \left(\frac{1}{2}\right)^{N_{ij}}$$

The exact significance test is performed by evaluating

$$P_{\text{cs}} = \sum_{n \in p} P_0(\mathbf{n})$$

where $p = \{n : P_0(\mathbf{n}) < P_0(\mathbf{n}^*)\}$ and $\mathbf{n}^*$ is the observed contingency table data vector. The algorithm evaluates $p_{\text{cs}}$ exactly. For information about permutation tests, see Good (2001 and 2000).

# References


Bickenböller, H. and F. Clerget-Darpoux. 1995. Statistical properties of the allelic and genotypic transmission/disequilibrium test for multiallelic markers. *Genetic Epidemiology* 12: 865–870.

BMDP. 1990. *BMDP statistical software manual.* Example 4F2.9. Los Angeles: BMDP Statistical Software, Inc.

Bowker, A. H. 1948. A test for symmetry in contingency tables. *Journal of the American Statistical Association* 43: 572–574.

Breslow, N. E. and N. E. Day. 1980. *Statistical Methods in Cancer Research*, vol. 1, 182–198. Lyon: IARC.

Cleves, M. A. 1997. sg74: Symmetry and marginal homogeneity test/TDT. *Stata Technical Bulletin* 40: 23–27. Reprinted in *Stata Technical Bulletin Reprints*, vol. 7, pp. 193–197.

——. 1999. sg110: Hardy–Weinberg equilibrium test and allele frequency estimation. *Stata Technical Bulletin* 48: 34–37. Reprinted in *Stata Technical Bulletin Reprints*, vol. 8, pp. 280–284.

Cleves, M. A., J. M. Olson, and K. B. Jacobs. 1997. Exact transmission–disequilibrium tests with multiallelic markers. *Genetic Epidemiology* 14: 337–347.

Cui, J. 2000. sg150: Hardy–Weinberg equilibrium test in case–control studies. *Stata Technical Bulletin* 57: 17–19. Reprinted in *Stata Technical Bulletin Reprints*, vol. 10, pp. 218–220.

Good, P. 2000. *Permutation Tests: A Practical Guide to Resampling Methods for Testing Hypotheses.* 2nd ed. New York: Springer.

——. 2001. *Resampling Methods: A Practical Guide to Data Analysis.* 2nd ed. Boston: Birkhäuser.

Mack, T. M., M. C. Pike, B. E. Henderson, R. I. Pfeffer, V. R. Gerkins, B. S. Arthur, and S. E. Brown. 1976. Estrogens and endometrial cancer in a retirement community. *New England Journal of Medicine* 294: 1262–1267.

Mander, A. 2000. sbe38: Haplotype frequency estimation using an EM algorithm and log-linear modeling. *Stata Technical Bulletin* 57: 5–7.

Maxwell, A. E. 1970. Comparing the classification of subjects by two independent judges. *British Journal of Psychiatry* 116: 651–655.

Spieldman, R. S. and W. J. Ewens. 1996. The TDT and other family-based tests for linkage disequilibrium and association. *American Journal of Human Genetics* 59: 983–989.

Spieldman, R. S., R. E. McGinnis, and W. J. Ewens. 1993. Transmission test for linkage disequilibrium: The insulin gene region and insulin-dependence diabetes mellitus. *American Journal of Human Genetics* 52: 506–516.

Stuart, A. 1955. A test for homogeneity of the marginal distributions in a two-way classification. *Biometrika* 42: 412–416.


# Also See

**Related:** [ST] **epitab**

# Title

**table** — Tables of summary statistics

# Syntax

`table` *rowvar* [*colvar* [*supercolvar*]] [*if*] [*in*] [*weight*] [, *options*]

| *options* | description |
|---|---|
| Main | |
| `contents(`*clist*`)` | contents of table cells; select up to 5 statistics; default is `contents(freq)` |
| `by(`*superrowvarlist*`)` | superrow variables |
| Options | |
| `cw` | perform casewise deletion |
| `cellwidth(`*#*`)` | cell width |
| `csepwidth(`*#*`)` | column separation width |
| `stubwidth(`*#*`)` | stub width |
| `scsepwidth(`*#*`)` | supercolumn separation width |
| `center` | center-align table cells; default is right-align |
| `left` | left-align table cells; default is right-align |
| `row` | add row totals |
| `col` | add column totals |
| `scol` | add supercolumn totals |
| `format(%`*fmt*`)` | display format for numbers in cells; default is `format(%9.0g)` |
| `concise` | suppress rows with all missing entries |
| `missing` | show missing statistics with period |
| `replace` | replace current data with table statistics |
| `name(`*string*`)` | name new variables with prefix *string* |

`by` may be used with `table`; see [D] **by**.

`fweight`s, `aweight`s, `iweight`s, and `pweight`s are allowed; see [U] **11.1.6 weight**.

`sd` (standard deviation) is not allowed with `pweight`s.

where the elements of *clist* may be

| | | | |
|---|---|---|---|
| `freq` | frequency | `median` *varname* | median |
| `mean` *varname* | mean of *varname* | `p1` *varname* | 1st percentile |
| `sd` *varname* | standard deviation | `p2` *varname* | 2nd percentile |
| `sum` *varname* | sum | ... | 3rd–49th percentiles |
| `rawsum` *varname* | sums ignoring optionally specified weight | `p50` *varname* | 50th percentile (`median`) |
| `count` *varname* | count of nonmissing observations | ... | 51st–97th percentiles |
| `n` *varname* | same as `count` | `p98` *varname* | 98th percentile |
| `max` *varname* | maximum | `p99` *varname* | 99th percentile |
| `min` *varname* | minimum | `iqr` *varname* | interquartile range |

Rows, columns, supercolumns, and superrows are thus defined as

| | |
|---|---|
| row 1 | . |
| row 2 | . |

| | supercol 1 | | supercol 2 | |
|---|---|---|---|---|
| | col 1 | col 2 | col 1 | col 2 |
| row 1 | . | . | . | . |
| row 2 | . | . | . | . |

| | col 1 | col 2 |
|---|---|---|
| row 1 | . | . |
| row 2 | . | . |

| | supercol 1 | | supercol 2 | |
|---|---|---|---|---|
| | col 1 | col 2 | col 1 | col 2 |
| superrow 1: | | | | |
| row 1 | . | . | . | . |
| row 2 | . | . | . | . |
| superrow 2: | | | | |
| row 1 | . | . | . | . |
| row 2 | . | . | . | . |

# Description

`table` calculates and displays tables of statistics.

# Options

Main

`contents(`*clist*`)` specifies the contents of the table's cells; if not specified, `contents(freq)` is used by default. `contents(freq)` produces a table of frequencies. `contents(mean mpg)` produces a table of the means of variable `mpg`. `contents(freq mean mpg sd mpg)` produces a table of frequencies together with the mean and standard deviation of variable `mpg`. Up to five statistics may be specified.

`by(`*superrowvarlist*`)` specifies that numeric or string variables be treated as superrows. Up to four variables may be specified in *superrowvarlist*. Note that the `by()` option may be specified with the `by` prefix.

Options

`cw` specifies casewise deletion. If `cw` is not specified, all observations possible are used to calculate each of the specified statistics. `cw` is relevant only when you request a table containing statistics on multiple variables. For instance, `contents(mean mpg mean weight)` would produce a table reporting the means of variables `mpg` and `weight`. Consider an observation in which `mpg` is known but `weight` is missing. By default, that observation will be used in the calculation of the mean of `mpg`. If you specify `cw`, the observation will be excluded in the calculation of the means of both `mpg` and `weight`.

`cellwidth(#)` specifies the width of the cell in units of digit widths; 10 means the space occupied by 10 digits, which is `1234567890`. The default `cellwidth()` is not a fixed number, but a number chosen by `table` to spread the table out while presenting a reasonable number of columns across the page.

`csepwidth(#)` specifies the separation between columns in units of digit widths. The default is not a fixed number, but a number chosen by `table` according to what it thinks looks best.

`stubwidth(#)` specifies the width, in units of digit widths, to be allocated to the left stub of the table. The default is not a fixed number, but a number chosen by `table` according to what it thinks looks best.

`scsepwidth(#)` specifies the separation between supercolumns in units of digit widths. The default is not a fixed number, but a number chosen by `table` to present the results best.

`center` specifies that results be centered in the table's cells. The default is to right-align results. For centering to work well, you typically need to specify a display format as well. `center format(%9.2f)` is popular.

`left` specifies that column labels be left-aligned. The default is to right-align column labels to distinguish them from supercolumn labels, which are left-aligned. If you specify `left`, both column and supercolumn labels are left-aligned.

`row` specifies that a row be added to the table reflecting the total across the rows.

`col` specifies that a column be added to the table reflecting the total across columns.

`scol` specifies that a supercolumn be added to the table reflecting the total across supercolumns.

`format(%`*fmt*`)` specifies the display format for presenting numbers in the table's cells. `format(%9.0g)` is the default; `format(%9.2f)` and `format(%9.2fc)` are popular alternatives. The width of the format you specify does not matter, except that `%`*fmt* must be valid. The width of the cells is chosen by `table` to present the results best. Option `cellwidth()` allows you to override `table`'s choice.

`concise` specifies that rows with all missing entries not be displayed.

`missing` specifies that missing statistics be shown in the table as periods (Stata's missing-value indicator). The default is that missing entries be left blank.

`replace` specifies that the data in memory be replaced with data containing one observation per cell (row, column, supercolumn, and superrow) and with variables containing the statistics designated in `contents()`.

This option is rarely specified. If you do not specify this option, the data in memory remain unchanged.

If you do specify this option, the first statistic will be named `table1`, the second `table2`, and so on. For instance, if `contents(mean mpg sd mpg)` was specified, the means of `mpg` would be in variable `table1` and the standard deviations in `table2`.

`name(`*string*`)` is relevant only if you specify `replace`. `name()` allows changing the default stub name that `replace` uses to name the new variables associated with the statistics. If you specify `name(stat)`, the first statistic will be placed in variable `stat1`, the second in `stat2`, and so on.

## Limits

Up to 4 variables may be specified in the `by()`, so with the three row, column, and supercolumn variables, seven-way tables may be displayed.

Up to 5 statistics may be displayed in each cell of the table.

The sum of the number of rows, columns, supercolumns, and superrows is called the number of margins. A table may contain up to 3,000 margins. Thus a one-way table may contain 3,000 rows. A two-way table could contain 2,998 rows and 2 columns, 2,997 rows and 3 columns, ..., 1,500 rows and 1,500 columns, ..., 2 rows and 2,998 columns. A three-way table is similarly limited by the sum of the number of rows, columns, and supercolumns. A $r \times c \times d$ table is feasible if $r+c+d \leq 3{,}000$. Note that the limit is set in terms of the sum of the rows, columns, supercolumns, and superrows, and not, as you might expect, in terms of their product.

# Remarks

Remarks are presented under the headings

*One-way tables*
*Two-way tables*
*Three-way tables*
*Four-way and higher-dimensional tables*

## One-way tables

### ▷ Example 1

Using the automobile dataset, here is a simple one-way table:

```
. use http://www.stata-press.com/data/r9/auto
(1978 Automobile Data)
. table rep78, contents(mean mpg)
```

| Repair Record 1978 | mean(mpg) |
|---:|---:|
| 1 | 21 |
| 2 | 19.125 |
| 3 | 19.4333 |
| 4 | 21.6667 |
| 5 | 27.3636 |

We are not limited to including only a single statistic:

```
. table rep78, c(n mpg  mean mpg  sd mpg  median mpg)
```

| Repair Record 1978 | N(mpg) | mean(mpg) | sd(mpg) | med(mpg) |
|---:|---:|---:|---:|---:|
| 1 | 2 | 21 | 4.24264 | 21 |
| 2 | 8 | 19.125 | 3.758324 | 18 |
| 3 | 30 | 19.4333 | 4.141325 | 19 |
| 4 | 18 | 21.6667 | 4.93487 | 22.5 |
| 5 | 11 | 27.3636 | 8.732385 | 30 |

Note that we abbreviated `contents()` as `c()`. The `format()` option will allow us to better format the numbers in the table:

```
. table rep78, c(n mpg  mean mpg  sd mpg  median mpg) format(%9.2f)
```

| Repair Record 1978 | N(mpg) | mean(mpg) | sd(mpg) | med(mpg) |
|---|---|---|---|---|
| 1 | 2 | 21.00 | 4.24 | 21.00 |
| 2 | 8 | 19.12 | 3.76 | 18.00 |
| 3 | 30 | 19.43 | 4.14 | 19.00 |
| 4 | 18 | 21.67 | 4.93 | 22.50 |
| 5 | 11 | 27.36 | 8.73 | 30.00 |

The `center` option will center the results under the headings:

```
. table rep78, c(n mpg  mean mpg  sd mpg  median mpg) format(%9.2f) center
```

| Repair Record 1978 | N(mpg) | mean(mpg) | sd(mpg) | med(mpg) |
|---|---|---|---|---|
| 1 | 2 | 21.00 | 4.24 | 21.00 |
| 2 | 8 | 19.12 | 3.76 | 18.00 |
| 3 | 30 | 19.43 | 4.14 | 19.00 |
| 4 | 18 | 21.67 | 4.93 | 22.50 |
| 5 | 11 | 27.36 | 8.73 | 30.00 |

◁

## Two-way tables

### ▷ Example 2

In example 1, when we typed '`table rep78, ...`', we obtained a one-way table. If we were to type '`table rep78 foreign, ...`', we would obtain a two-way table:

```
. table rep78 foreign, c(mean mpg)
```

| Repair Record 1978 | Car type | |
|---|---|---|
| | Domestic | Foreign |
| 1 | 21 | |
| 2 | 19.125 | |
| 3 | 19 | 23.3333 |
| 4 | 18.4444 | 24.8889 |
| 5 | 32 | 26.3333 |

Note the missing cells. Certain combinations of repair record and car type do not exist in our dataset.

As with one-way tables, we can specify a display format for the cells and center the numbers within the cells if we wish.

```
. table rep78 foreign, c(mean mpg) format(%9.2f) center
```

| Repair Record 1978 | Car type Domestic | Foreign |
|---|---|---|
| 1 | 21.00 | |
| 2 | 19.12 | |
| 3 | 19.00 | 23.33 |
| 4 | 18.44 | 24.89 |
| 5 | 32.00 | 26.33 |

We can obtain row totals by specifying the `row` option and obtain column totals by specifying the `col` option. We specify both below:

```
. table rep78 foreign, c(mean mpg) format(%9.2f) center row col
```

| Repair Record 1978 | Domestic | Car type Foreign | Total |
|---|---|---|---|
| 1 | 21.00 | | 21.00 |
| 2 | 19.12 | | 19.12 |
| 3 | 19.00 | 23.33 | 19.43 |
| 4 | 18.44 | 24.89 | 21.67 |
| 5 | 32.00 | 26.33 | 27.36 |
| Total | 19.54 | 25.29 | 21.29 |

`table` can display multiple statistics within cells, but once we move beyond one-way tables, the table becomes busy:

```
. table foreign rep78, c(mean mpg  n mpg) format(%9.2f) center
```

| Car type | Repair Record 1978 1 | 2 | 3 | 4 | 5 |
|---|---|---|---|---|---|
| Domestic | 21.00<br>2 | 19.12<br>8 | 19.00<br>27 | 18.44<br>9 | 32.00<br>2 |
| Foreign | | | 23.33<br>3 | 24.89<br>9 | 26.33<br>9 |

This two-way table with two statistics per cell works well in this case. That was, in part, helped along by our interchanging the rows and columns. We turned the table around by typing `table foreign rep78` rather than `table rep78 foreign`.

Another way to display two-way tables is to specify a row and superrow rather than a row and column. We do that below and display three statistics per cell:

```
. table foreign, by(rep78) c(mean mpg  sd mpg  n mpg) format(%9.2f) center
```

| Repair Record 1978 and Car type | mean(mpg) | sd(mpg) | N(mpg) |
|---|---|---|---|
| 1 | | | |
| Domestic | 21.00 | 4.24 | 2 |
| Foreign | | | |
| 2 | | | |
| Domestic | 19.12 | 3.76 | 8 |
| Foreign | | | |
| 3 | | | |
| Domestic | 19.00 | 4.09 | 27 |
| Foreign | 23.33 | 2.52 | 3 |
| 4 | | | |
| Domestic | 18.44 | 4.59 | 9 |
| Foreign | 24.89 | 2.71 | 9 |
| 5 | | | |
| Domestic | 32.00 | 2.83 | 2 |
| Foreign | 26.33 | 9.37 | 9 |

◁

## Three-way tables

### ▷ Example 3

We have data on the prevalence of byssinosis, a form of pneumoconiosis to which workers exposed to cotton dust are susceptible. The dataset is on 5,419 workers in a large cotton mill. We know whether each worker smokes, his or her race, and the dustiness of the work area. The categorical variables are

| | |
|---|---|
| `smokes` | Smoker or nonsmoker in the last five years. |
| `race` | White or other. |
| `workplace` | 1 (most dusty), 2 (less dusty), 3 (least dusty). |

Moreover, this dataset includes a frequency-weight variable pop. Here is a three-way table showing the fraction of workers with byssinosis:

```
. use http://www.stata-press.com/data/r9/byssin
(Byssinosis incidence)
. table workplace smokes race [fw=pop], c(mean prob)
```

| Dustiness of workplace | Race and Smokes — other — no | yes | — white — no | yes |
|---|---|---|---|---|
| least | .0107527 | .0101523 | .0081549 | .0162774 |
| less | .02 | .0081633 | .0136612 | .0143149 |
| most | .0820896 | .1679105 | .0833333 | .2295082 |

This table would look better if we showed the fraction to four digits:

```
. table workplace smokes race [fw=pop], c(mean prob) format(%9.4f)

----------------------------------------------
Dustiness |           Race and Smokes
of        |  --- other ---      --- white ---
workplace |      no     yes         no     yes
----------+-----------------------------------
    least |  0.0108  0.0102     0.0082  0.0163
     less |  0.0200  0.0082     0.0137  0.0143
     most |  0.0821  0.1679     0.0833  0.2295
----------------------------------------------
```

In this table, the rows are the dustiness of the workplace, the columns are whether the worker smokes, and the supercolumns are the worker's race.

Now we request that the table include the supercolumn totals by specifying the option `sctotal`, which we can abbreviate as `sc`:

```
. table workplace smokes race [fw=pop], c(mean prob) format(%9.4f) sc

--------------------------------------------------------------
Dustiness |                  Race and Smokes
of        |  --- other ---      --- white ---      --- Total ---
workplace |      no     yes         no     yes         no     yes
----------+---------------------------------------------------
    least |  0.0108  0.0102     0.0082  0.0163     0.0090  0.0145
     less |  0.0200  0.0082     0.0137  0.0143     0.0159  0.0123
     most |  0.0821  0.1679     0.0833  0.2295     0.0826  0.1929
--------------------------------------------------------------
```

The supercolumn total is the total over `race` and is divided into its columns based on `smokes`. Here is the table with the column rather than the supercolumn totals:

```
. table workplace smokes race [fw=pop], c(mean prob) format(%9.4f) col

------------------------------------------------------------
Dustiness |                  Race and Smokes
of        |  ------- other -------   ------- white -------
workplace |      no     yes   Total       no     yes   Total
----------+-------------------------------------------------
    least |  0.0108  0.0102  0.0104   0.0082  0.0163  0.0129
     less |  0.0200  0.0082  0.0135   0.0137  0.0143  0.0140
     most |  0.0821  0.1679  0.1393   0.0833  0.2295  0.1835
------------------------------------------------------------
```

Here is the table with both column and supercolumn totals:

```
. table workplace smokes race [fw=pop], c(mean prob) format(%9.4f) sc col

---------------------------------------------------------------------------------
Dustin  |
ess of  |                             Race and Smokes
workpl  |  ------- other -------  ------- white -------  ------- Total -------
ace     |      no     yes   Total      no     yes   Total      no     yes   Total
--------+------------------------------------------------------------------------
  least |  0.0108  0.0102  0.0104  0.0082  0.0163  0.0129  0.0090  0.0145  0.0122
   less |  0.0200  0.0082  0.0135  0.0137  0.0143  0.0140  0.0159  0.0123  0.0138
   most |  0.0821  0.1679  0.1393  0.0833  0.2295  0.1835  0.0826  0.1929  0.1570
---------------------------------------------------------------------------------
```

Note that `table` is struggling to keep this table from becoming too wide—notice how it divided the words in the title in the top-left stub. In this case, if the table had more columns, or, if we demanded more digits, `table` would be forced to segment the table and present it in pieces, which it would do:

```
. table workplace smokes race [fw=pop], c(mean prob) format(%9.6f) sc col
```

| Dustiness of workplace | Race and Smokes: other no | other yes | other Total | white no | white yes | white Total |
|---|---|---|---|---|---|---|
| least | 0.010753 | 0.010152 | 0.010417 | 0.008155 | 0.016277 | 0.012949 |
| less | 0.020000 | 0.008163 | 0.013483 | 0.013661 | 0.014315 | 0.014035 |
| most | 0.082090 | 0.167910 | 0.139303 | 0.083333 | 0.229508 | 0.183521 |

| Dustiness of workplace | Race and Smokes: Total no | Total yes | Total Total |
|---|---|---|---|
| least | 0.008990 | 0.014471 | 0.012174 |
| less | 0.015901 | 0.012262 | 0.013846 |
| most | 0.082569 | 0.192905 | 0.156951 |

In this case, three digits is probably enough, so here is the table including all the row, column, and supercolumn totals:

```
. table workplace smokes race [fw=pop], c(mean prob) format(%9.3f)
> sc col row
```

| Dustiness of workplace | Race and Smokes: other no | other yes | other Total | white no | white yes | white Total | Total no | Total yes | Total Total |
|---|---|---|---|---|---|---|---|---|---|
| least | 0.011 | 0.010 | 0.010 | 0.008 | 0.016 | 0.013 | 0.009 | 0.014 | 0.012 |
| less | 0.020 | 0.008 | 0.013 | 0.014 | 0.014 | 0.014 | 0.016 | 0.012 | 0.014 |
| most | 0.082 | 0.168 | 0.139 | 0.083 | 0.230 | 0.184 | 0.083 | 0.193 | 0.157 |
| Total | 0.025 | 0.048 | 0.038 | 0.014 | 0.035 | 0.026 | 0.018 | 0.039 | 0.030 |

We can show multiple statistics:

```
. table workplace smokes race [fw=pop], c(mean prob  n prob) format(%9.3f)
> sc col row
```

| Dustiness of workplace | Race and Smokes: other no | other yes | other Total | white no | white yes | white Total | Total no | Total yes | Total Total |
|---|---|---|---|---|---|---|---|---|---|
| least | 0.011 | 0.010 | 0.010 | 0.008 | 0.016 | 0.013 | 0.009 | 0.014 | 0.012 |
| | 465 | 591 | 1,056 | 981 | 1,413 | 2,394 | 1,446 | 2,004 | 3,450 |
| less | 0.020 | 0.008 | 0.013 | 0.014 | 0.014 | 0.014 | 0.016 | 0.012 | 0.014 |
| | 200 | 245 | 445 | 366 | 489 | 855 | 566 | 734 | 1,300 |
| most | 0.082 | 0.168 | 0.139 | 0.083 | 0.230 | 0.184 | 0.083 | 0.193 | 0.157 |
| | 134 | 268 | 402 | 84 | 183 | 267 | 218 | 451 | 669 |
| Total | 0.025 | 0.048 | 0.038 | 0.014 | 0.035 | 0.026 | 0.018 | 0.039 | 0.030 |
| | 799 | 1,104 | 1,903 | 1,431 | 2,085 | 3,516 | 2,230 | 3,189 | 5,419 |

◁

## Four-way and higher-dimensional tables

### ▷ Example 4

Let's pretend that our byssinosis dataset also recorded each worker's sex (it does not, and we have made up this extra information). We obtain a four-way table just as we would a three-way table, but we specify the fourth variable as a superrow by including it in the `by()` option:

```
. use http://www.stata-press.com/data/r9/byssin1
(Byssinosis incidence)
. table workplace smokes race [fw=pop], by(sex) c(mean prob) format(%9.3f)
> sc col row
```

| Sex and Dustiness of workplace | Race and Smokes: other no | other yes | other Total | white no | white yes | white Total | Total no | Total yes | Total Total |
|---|---|---|---|---|---|---|---|---|---|
| Female | | | | | | | | | |
| least | 0.006 | 0.009 | 0.008 | 0.009 | 0.021 | 0.016 | 0.009 | 0.018 | 0.014 |
| less | 0.020 | 0.008 | 0.010 | 0.015 | 0.015 | 0.015 | 0.016 | 0.012 | 0.014 |
| most | 0.057 | 0.154 | 0.141 | | | | 0.057 | 0.154 | 0.141 |
| Total | 0.017 | 0.051 | 0.043 | 0.011 | 0.020 | 0.016 | 0.012 | 0.032 | 0.024 |
| Male | | | | | | | | | |
| least | 0.013 | 0.011 | 0.012 | 0.006 | 0.007 | 0.006 | 0.009 | 0.008 | 0.009 |
| less | 0.020 | 0.000 | 0.019 | 0.000 | 0.013 | 0.011 | 0.016 | 0.013 | 0.014 |
| most | 0.091 | 0.244 | 0.136 | 0.083 | 0.230 | 0.184 | 0.087 | 0.232 | 0.167 |
| Total | 0.029 | 0.041 | 0.033 | 0.020 | 0.056 | 0.043 | 0.025 | 0.052 | 0.039 |

If our dataset also included work group and we wanted a five-way table, we could include both the sex and work-group variables in the `by()` option. You may include up to four variables in `by()`, and so produce up to 7-way tables.

◁

## Methods and Formulas

`table` is implemented as an ado-file. The contents of cells are calculated by `collapse` and are displayed by `tabdisp`; see [D] **collapse** and [P] **tabdisp**.

## Also See

**Related:** [R] **adjust**, [R] **tabstat**, [R] **tabulate oneway**, [R] **tabulate twoway**, [D] **collapse**, [P] **tabdisp**

**Background:** [U] **12.6 Dataset, variable, and value labels**, [U] **25 Dealing with categorical variables**

# Title

**tabstat** — Display table of summary statistics

# Syntax

`tabstat` *varlist* [*if*] [*in*] [*weight*] [, *options*]

| *options* | description |
|---|---|
| Main | |
| `by(`*varname*`)` | group statistics by variable |
| `statistics(`*statname*[...]`)` | report specified statistics |
| Options | |
| `casewise` | perform casewise deletion of observations |
| `nototal` | do not report overall statistics; use with `by()` |
| `missing` | report statistics for missing values of `by()` variable |
| `noseparator` | do not use separator line between `by()` categories |
| `labelwidth(`*#*`)` | width for `by()` variable labels; default is `labelwidth(16)` |
| `varwidth(`*#*`)` | variable width; default is `varwidth(12)` |
| `columns(`*variables*`)` | display variables in table columns; the default |
| `columns(`*statistics*`)` | display statistics in table columns |
| `longstub` | make left table stub wider |
| `format`[`(`*%fmt*`)`] | display format for statistics; default format is `%9.0g` |
| `save` | save summary statistics in `r()` |

`by` may be used with `tabstat`; see [D] **by**.
`aweight`s and `fweight`s are allowed; see [U] **11.1.6 weight**.

# Description

`tabstat` displays summary statistics for a series of numeric variables in a single table, possibly broken down on (conditioned by) another variable.

Without the `by()` option, `tabstat` is a useful alternative to `summarize` because it allows you to specify the list of statistics to be displayed.

With the `by()` option, `tabstat` resembles `tabulate` used with its `summarize()` option in that both report statistics of *varlist* for the different values of *varname*. `tabstat` allows more flexibility in terms of the statistics presented and the format of the table.

`tabstat` is sensitive to the `linesize`; it widens the table, if possible, and wraps, if necessary.

(*Continued on next page*)

# Options

Main

`by(`*varname*`)` specifies that the statistics be displayed separately for each unique value of *varname*; *varname* may be numeric or string. For instance, `tabstat height` would present the overall mean of height. `tabstat height, by(sex)` would present the mean height of males, and of females, and the overall mean height. Do not confuse the `by()` option with the `by` prefix; both may be specified.

`statistics(`*statname* [...]`)` specifies the statistics to be displayed; the default is equivalent to specifying `statistics(mean)`. (`stats()` is a synonym for `statistics()`.) Multiple statistics may be specified and are separated by white space, such as `statistics(mean sd)`. Available statistics are

| *statname* | definition | *statname* | definition |
|---|---|---|---|
| `mean` | mean | `p1` | 1st percentile |
| `count` | count of nonmissing observations | `p5` | 5th percentile |
| `n` | same as `count` | `p10` | 10th percentile |
| `sum` | sum | `p25` | 25th percentile |
| `max` | maximum | `median` | median (same as `p50`) |
| `min` | minimum | `p50` | 50th percentile (same as `median`) |
| `sd` | standard deviation | `p75` | 75th percentile |
| `variance` | variance | `p90` | 90th percentile |
| `cv` | coefficient of variation (sd/mean) | `p95` | 95th percentile |
| `semean` | standard error of mean ($sd/\sqrt{n}$) | `p99` | 99th percentile |
| `skewness` | skewness | `range` | range = `max` − `min` |
| `kurtosis` | kurtosis | `iqr` | interquartile range = `p75` − `p25` |
| | | `q` | equivalent to specifying `p25 p50 p75` |

Options

`casewise` specifies casewise deletion of observations. Statistics are to be computed for the sample that is not missing for any of the variables in *varlist*. The default is to use all the nonmissing values for each variable.

`nototal` is for use with `by()`; it specifies that the overall statistics not be reported.

`missing` specifies that missing values of the `by()` variable be treated just like any other value and that statistics should be displayed for them. The default is not to report the statistics for the `by()==` *missing* group. If the `by()` variable is a string variable, `by()==""` is considered to mean missing.

`noseparator` specifies that a separator line between the `by()` categories not be displayed.

`labelwidth(`*#*`)` specifies the maximum width to be used within the stub to display the labels of the `by()` variable. The default is `labelwidth(16)`. $8 \leq \# \leq 32$.

`varwidth(`*#*`)` specifies the maximum width to be used within the stub to display the names of the variables. The default is `varwidth(12)`. `varwidth()` is effective only with `columns(statistics)`. Setting `varwidth()` implies `longstub`. $8 \leq \# \leq 16$.

`columns(variables | statistics)` specifies whether to display variables or statistics in the columns of the table. `columns(variables)` is the default when more than one variable is specified.

`longstub` specifies that the left stub of the table be made wider so that it can include names of the statistics or variables in addition to the categories of `by(`*varname*`)`. The default is to describe the statistics or variables in a header. `longstub` is ignored if `by(`*varname*`)` is not specified.

`format` and `format(%fmt)` specify how the statistics are to be formatted. The default is to use a `%9.0g` format.

`format` specifies that each variable's statistics be formatted with the variable's display format; see [D] **format**.

`format(%fmt)` specifies the format to be used for all statistics. The maximum width of the specified format should not exceed 9 characters.

`save` specifies that the summary statistics be returned in `r()`. The overall (unconditional) statistics are returned in matrix `r(StatTotal)` (rows are statistics, columns are variables). The conditional statistics are returned in the matrices `r(Stat1)`, `r(Stat2)`, ..., and the names of the corresponding variables are returned in the macros `r(name1)`, `r(name2)`, ....

## Remarks

This command is probably most easily understood by going through a series of examples.

### ▷ Example 1

We have data on the price, weight, mileage rating, and repair record of 22 foreign and 52 domestic 1978 automobiles. We want to summarize these variables for the different origins of the automobiles.

```
. use http://www.stata-press.com/data/r9/auto
(1978 Automobile Data)

. tabstat price weight mpg rep78, by(foreign)

Summary statistics: mean
  by categories of: foreign (Car type)

 foreign |     price    weight       mpg     rep78
---------+----------------------------------------
Domestic |  6072.423  3317.115  19.82692  3.020833
 Foreign |  6384.682  2315.909  24.77273  4.285714
---------+----------------------------------------
   Total |  6165.257  3019.459   21.2973  3.405797
--------------------------------------------------
```

Additional summary statistics can be requested via the option `statistics()`. The group totals can be suppressed with the `nototal` option.

```
. tabstat price weight mpg rep78, by(foreign) stat(mean sd min max) nototal

Summary statistics: mean, sd, min, max
  by categories of: foreign (Car type)

 foreign |     price    weight       mpg     rep78
---------+----------------------------------------
Domestic |  6072.423  3317.115  19.82692  3.020833
         |  3097.104  695.3637  4.743297   .837666
         |      3291      1800        12         1
         |     15906      4840        34         5
---------+----------------------------------------
 Foreign |  6384.682  2315.909  24.77273  4.285714
         |  2621.915  433.0035  6.611187  .7171372
         |      3748      1760        14         3
         |     12990      3420        41         5
--------------------------------------------------
```

While the header of the table describes the statistics running vertically in the "cells", the table may become hard to read, especially with a large number of variables or statistics. The option `longstub` specifies that a column be added describing the contents of the cells. The option `format` can be issued to specify that `tabstat` display the statistics using the display format of the variables rather than the overall default `%9.0g`.

```
. tabstat price weight mpg rep78, by(foreign) stat(mean sd min max) long format
foreign       stats |    price    weight       mpg     rep78
--------------------+----------------------------------------
Domestic       mean |  6,072.4   3,317.1   19.8269   3.02083
                 sd |  3,097.1   695.364    4.7433   .837666
                min |    3,291     1,800        12         1
                max |   15,906     4,840        34         5
--------------------+----------------------------------------
Foreign        mean |  6,384.7   2,315.9   24.7727   4.28571
                 sd |  2,621.9   433.003   6.61119   .717137
                min |    3,748     1,760        14         3
                max |   12,990     3,420        41         5
--------------------+----------------------------------------
Total          mean |  6,165.3   3,019.5   21.2973    3.4058
                 sd |  2,949.5   777.194    5.7855   .989932
                min |    3,291     1,760        12         1
                max |   15,906     4,840        41         5
-------------------------------------------------------------
```

We can specify a layout of the table in which the statistics run horizontally and the variables run vertically by specifying the `col(statistics)` option.

```
. tabstat price weight mpg rep78, by(foreign) stat(min mean max) col(stat) long
foreign       variable |       min      mean       max
-----------------------+------------------------------
Domestic         price |      3291  6072.423     15906
                weight |      1800  3317.115      4840
                   mpg |        12  19.82692        34
                 rep78 |         1  3.020833         5
-----------------------+------------------------------
Foreign          price |      3748  6384.682     12990
                weight |      1760  2315.909      3420
                   mpg |        14  24.77273        41
                 rep78 |         3  4.285714         5
-----------------------+------------------------------
Total            price |      3291  6165.257     15906
                weight |      1760  3019.459      4840
                   mpg |        12   21.2973        41
                 rep78 |         1  3.405797         5
------------------------------------------------------
```

Finally, `tabstat` can also be used to enhance `summarize` so we can specify the statistics to be displayed. For instance, we can display the number of observations, the mean, the coefficient of variation, and the 25%, 50%, and 75% quantiles for a list of variables.

```
. tabstat price weight mpg rep78, stat(n mean cv q) col(stat)
    variable |         N      mean        cv       p25       p50       p75
-------------+------------------------------------------------------------
       price |        74  6165.257   .478406      4195    5006.5      6342
      weight |        74  3019.459  .2573949      2240      3190      3600
         mpg |        74   21.2973  .2716543        18        20        25
       rep78 |        69  3.405797   .290661         3         3         4
--------------------------------------------------------------------------
```

Since we did not specify the `by()` option, these statistics were not displayed for the subgroups of the data formed by the categories of the `by()` variable. ◁

## Methods and Formulas

`tabstat` is implemented as an ado-file.

## Acknowledgments

The `tabstat` command was written by Jeroen Weesie and Vincent Buskens of Utrecht University in the Netherlands.

## Also See

**Related:** [R] **summarize**, [R] **table**, [R] **tabulate, summarize()**,
[D] **collapse**

# Title

**tabulate oneway** — One-way tables of frequencies

# Syntax

*One-way tables of frequencies*

`tabulate` *varname* [*if*] [*in*] [*weight*] [, *tabulate1_options*]

*One-way table for each variable—a convenience tool*

`tab1` *varlist* [*if*] [*in*] [*weight*] [, *tab1_options*]

| *tabulate1_options* | description |
|---|---|
| Main | |
| `subpop(`*varname*`)` | exclude observations for which *varname* = 0 |
| `missing` | treat missing values like other values |
| `nofreq` | do not display frequencies |
| `nolabel` | display numeric codes rather than value labels |
| `plot` | produce a bar chart of the relative frequencies |
| `sort` | display the table in descending order of frequency |
| Advanced | |
| `generate(`*varname*`)` | create indicator variables for *varname* |
| `matcell(`*matname*`)` | save frequencies in *matname*; programmer's option |
| `matrow(`*matname*`)` | save unique values of *varname* in *matname*; programmer's option |

| *tab1_options* | description |
|---|---|
| Main | |
| `missing` | treat missing values like other values |
| `nolabel` | display numeric codes rather than value labels |
| `plot` | produce a bar chart of the relative frequencies |
| `sort` | display the table in descending order of frequency |

`by` may be used with `tabulate` and `tab1`; see [D] **by**.
`fweights`, `aweights`, and `iweights` are allowed by `tabulate`. `fweights` are allowed by `tab1`. See [U] **11.1.6 weight**.

# Description

`tabulate` produces one-way tables of frequency counts.

For information about two-way tables of frequency counts along with various measures of association, including the common Pearson $\chi^2$, the likelihood-ratio $\chi^2$, Cramér's $V$, Fisher's exact test, Goodman and Kruskal's gamma, and Kendall's $\tau_b$, see [R] **tabulate twoway**.

`tab1` produces a one-way tabulation for each variable specified in *varlist*.

Also see [R] **table** and [R] **tabstat** if you want one-, two-, or $n$-way tables of frequencies and a wide variety of summary statistics. See [R] **tabulate, summarize()** for a description of `tabulate` with the `summarize()` option; it produces tables (breakdowns) of means and standard deviations. `table` is better than `tabulate, summarize()`, but `tabulate, summarize()` is faster. See [ST] **epitab** for $2 \times 2$ tables with statistics of interest to epidemiologists.

# Options

Main

`subpop(`*varname*`)` excludes observations for which *varname* = 0 in tabulating frequencies. The mathematical results of `tabulate ..., subpop(myvar)` are the same as `tabulate ... if myvar !=0`, but the table may be presented differently. The identities of the rows and columns will be determined from all the data, including the `myvar` = 0 group, so there may be entries in the table with frequency 0.

Consider tabulating `answer`, a variable that takes on values 1, 2, and 3, but consider tabulating it just for the `male==1` subpopulation. Assume that `answer` is never 2 in this group. `tabulate answer if male==1` produces a table with two rows: one for answer 1 and one for answer 3. There will be no row for answer 2 because answer 2 was never observed. `tabulate answer, subpop(male)` produces a table with three rows. The row for answer 2 will be shown as having 0 frequency.

`missing` requests that missing values be treated like other values in calculations of counts, percentages, and other statistics.

`nofreq` suppresses the printing of the frequencies.

`nolabel` causes the numeric codes to be displayed rather than the value labels.

`plot` produces a bar chart of the relative frequencies in a one-way table. (Also see [R] **histogram**.)

`sort` puts the table in descending order of frequency (and ascending order of the variable within equal values of frequency).

Advanced

`generate(`*varname*`)` creates a set of indicator variables reflecting the observed values of the tabulated variable. The `generate()` option may not be used with the `by` prefix.

`matcell(`*matname*`)` saves the reported frequencies in *matname*. This option is for use by programmers.

`matrow(`*matname*`)` saves the numeric values of the $r \times 1$ row stub in *matname*. This option is for use by programmers. `matrow()` may not be specified if the row variable is a string.

## Limits

One-way tables may have a maximum of 12,000 rows (Stata/SE), 3,000 rows (Intercooled Stata), or 500 rows (Small Stata).

# Remarks

Remarks are presented under the headings

*tabulate*
*tab1*

For each value of a specified variable, `tabulate` reports the number of observations with that value. The number of times a value occurs is called its *frequency*.

## tabulate

### ▷ Example 1

We have data summarizing the speed limit, the number of access points (on-ramps and off-ramps) per mile, and the accident rate per million vehicle miles along various Minnesota highways in 1973. The variable containing the speed limit is called `spdlimit`. If we `summarize` the variable, we obtain its mean and standard deviation:

```
. summarize spdlimit

    Variable |       Obs        Mean    Std. Dev.       Min        Max
-------------+--------------------------------------------------------
    spdlimit |        39          55    5.848977         40         70
```

The average speed limit is 55 miles per hour. We can learn more about this variable by tabulating it:

```
. tabulate spdlimit

Speed limit |      Freq.     Percent        Cum.
------------+-----------------------------------
         40 |          1        2.56        2.56
         45 |          3        7.69       10.26
         50 |          7       17.95       28.21
         55 |         15       38.46       66.67
         60 |         11       28.21       94.87
         65 |          1        2.56       97.44
         70 |          1        2.56      100.00
------------+-----------------------------------
      Total |         39      100.00
```

We see that one highway has a speed limit of 40 miles per hour, three have speed limits of 45, 7 of 50, and so on. The column labeled `Percent` shows the percentage of highways in the dataset that have the indicated speed limit. For instance, 38.46% of highways in our dataset have a speed limit of 55 miles per hour. The final column shows the cumulative percentage. We see that 66.67% of highways in our dataset have a speed limit of 55 miles per hour or less.

◁

## ▷ Example 2

The `plot` option places a sideways histogram alongside the table:

```
. tabulate spdlimit, plot
Speed limit |      Freq.
------------+------------+-----------------------------------------------------
         40 |          1 |*
         45 |          3 |***
         50 |          7 |*******
         55 |         15 |***************
         60 |         11 |***********
         65 |          1 |*
         70 |          1 |*
------------+------------+-----------------------------------------------------
      Total |         39
```

Of course, `graph` can produce better-looking histograms; see [R] **histogram**.

◁

## ▷ Example 3

`tabulate` labels tables using *variable* and *value labels* if they exist. To demonstrate how this works, let's add a new variable to our dataset that categorizes `spdlimit` into three categories. We will call this new variable `spdcat`:

```
. generate spdcat=recode(spdlimit,50,60,70)
```

The `recode()` function divides `spdlimit` into 50 miles per hour or below, 51–60, and above 60; see [D] **functions**. We specified the break points in the arguments `(spdlimit,50,60,70)`. The first argument is the variable to be recoded. The second argument is the first break point, the third argument is the second break point, and so on. We can specify as many break points as we wish.

`recode()` used our arguments not only as the break points, but to label the results as well. If `spdlimit` is less than or equal to 50, `spdcat` is set to 50; if `spdlimit` is between 51 and 60, `spdcat` is 60; otherwise, `spdcat` is arbitrarily set to 70. (See [U] **25 Dealing with categorical variables**.)

Since we just created the variable `spdcat`, it is not yet labeled. When we make a table using this variable, `tabulate` uses the variable's name to label it:

```
. tabulate spdcat
     spdcat |      Freq.     Percent        Cum.
------------+-----------------------------------
         50 |         11       28.21       28.21
         60 |         26       66.67       94.87
         70 |          2        5.13      100.00
------------+-----------------------------------
      Total |         39      100.00
```

Even through the table is not well labeled, `recode()`'s coding scheme provides us with clues as to the table's meaning. The first line of the table corresponds to 50 miles per hour and below, the next to 51 through 60 miles per hour, and the last to above 60 miles per hour.

We can improve this table by labeling the values and variables:

```
. label define scat 50 "40 to 50" 60 "55 to 60" 70 "Above 60"
. label values spdcat scat
. label variable spdcat "Speed Limit Category"
```

We define a *value label* called `scat` that attaches labels to the numbers 50, 60, and 70 using the `label define` command; see [U] **12.6.3 Value labels**. We label the value 50 as '`40 to 50`', since we looked back at our original tabulation in the first example and saw that the speed limit was never less than 40. Similarly, we could have labeled the last category '`65 to 70`' since the speed limit is never greater than 70 miles per hour.

Next, we requested that Stata label the values of the new variable `spdcat` using the value label `scat`. Finally, we labeled our variable `Speed Limit Category`. We are now ready to `tabulate` the result:

```
. tabulate spdcat

Speed Limit |
   Category |      Freq.     Percent        Cum.
------------+-----------------------------------
   40 to 50 |         11       28.21       28.21
   55 to 60 |         26       66.67       94.87
   Above 60 |          2        5.13      100.00
------------+-----------------------------------
      Total |         39      100.00
```

◁

## ▷ Example 4

If we have missing values in our dataset, `tabulate` ignores them unless we explicitly indicate otherwise. We have no missing data in our example, so let's add some:

```
. replace spdcat=. in 1
(1 real change made, 1 to missing)
```

We changed the first observation on `spdcat` to *missing*. Let's now `tabulate` the result:

```
. tabulate spdcat

Speed Limit |
   Category |      Freq.     Percent        Cum.
------------+-----------------------------------
   40 to 50 |         11       28.95       28.95
   55 to 60 |         26       68.42       97.37
   Above 60 |          1        2.63      100.00
------------+-----------------------------------
      Total |         38      100.00
```

Comparing this output with that in the previous example, we see that the total frequency count is now one less than it was—38 rather than 39. Also the '`Above 60`' category now has only one observation where it used to have two, so we evidently changed a road with a high speed limit.

We want `tabulate` to treat missing values just as it treats numbers, so we specify the `missing` option:

```
. tabulate spdcat, missing

Speed Limit |
   Category |      Freq.     Percent        Cum.
------------+-----------------------------------
   40 to 50 |         11       28.21       28.21
   55 to 60 |         26       66.67       94.87
   Above 60 |          1        2.56       97.44
          . |          1        2.56      100.00
------------+-----------------------------------
      Total |         39      100.00
```

We now see our missing value—the last category, labeled '.', shows a frequency count of 1. The table sum is once again 39.

Let's put our dataset back as it was originally:

```
. replace spdcat=70 in 1
(1 real change made)
```

◁

## ❑ Technical Note

`tabulate` also has the ability to automatically create indicator variables from categorical variables. We will briefly review that capability here, but see [U] **25 Dealing with categorical variables** for a complete description. Let's begin by describing our highway dataset:

```
. describe

Contains data from hiway.dta
  obs:            39                          Minnesota Highway Data, 1973
 vars:             5                          18 Jan 2005 11:42
 size:           936 (99.5% of memory free)
-------------------------------------------------------------------------------
              storage  display     value
variable name   type   format      label      variable label
-------------------------------------------------------------------------------
acc_rate        float  %9.0g                  Accident rate
spdlimit        float  %9.0g                  Speed limit
acc_pts         float  %9.0g                  Access points per mile
rate            float  %9.0g       rcat       Accident rate per million
                                                vehicle miles
spdcat          float  %9.0g       scat       Speed Limit Category
-------------------------------------------------------------------------------
Sorted by:
     Note:  dataset has changed since last saved
```

Our dataset contains five variables. We will type `tabulate spdcat, generate(spd)`, `describe` our data, and then explain what happened.

```
. tabulate spdcat, generate(spd)

Speed Limit |
   Category |      Freq.     Percent        Cum.
------------+-----------------------------------
   40 to 50 |         11       28.21       28.21
   55 to 60 |         26       66.67       94.87
   Above 60 |          2        5.13      100.00
------------+-----------------------------------
      Total |         39      100.00
```

(*Continued on next page*)

```
. describe

Contains data from hiway.dta
  obs:            39                          Minnesota Highway Data, 1973
 vars:             8                          18 Jan 2005 11:42
 size:         1,053 (99.5% of memory free)

              storage  display     value
variable name   type   format      label      variable label

acc_rate        float  %9.0g                  Accident rate
spdlimit        float  %9.0g                  Speed limit
acc_pts         float  %9.0g                  Access points per mile
rate            float  %9.0g       rcat       Accident rate per million
                                                vehicle miles
spdcat          float  %9.0g       scat       Speed Limit Category
spd1            byte   %8.0g                  spdcat==40 to 50
spd2            byte   %8.0g                  spdcat==55 to 60
spd3            byte   %8.0g                  spdcat==Above 60

Sorted by:
     Note:  dataset has changed since last saved
```

When we typed `tabulate` with the `generate()` option, Stata responded by producing a one-way frequency table, so it appeared that the option did nothing. Yet when we `describe` our dataset, we find that we now have *eight* variables instead of the original five. The new variables are named `spd1`, `spd2`, and `spd3`.

When we specify the `generate()` option, we are telling Stata not only to produce the table, but also to create a set of indicator variables that correspond to that table. Stata adds a numeric suffix to the name we specify in the parentheses. `spd1` refers to the first line of the table, `spd2` to the second line, and so on. In addition, Stata labels the variables so that we know what they mean. `spd1` is an indicator variable that is *true* (takes on the value 1) when `spdcat` is between 40 and 50; otherwise, it is zero. (There is an exception: If `spdcat` is missing, so are the `spd1`, `spd2`, and `spd3` variables. This did not happen in our dataset.)

We want to prove our claim. Since we have not yet introduced two-way tabulations, we will use the `summarize` statement:

```
. summarize spdlimit if spd1==1

    Variable |     Obs        Mean    Std. Dev.       Min        Max
-------------+--------------------------------------------------------
    spdlimit |      11    47.72727    3.437758         40         50

. summarize spdlimit if spd2==1

    Variable |     Obs        Mean    Std. Dev.       Min        Max
-------------+--------------------------------------------------------
    spdlimit |      26    57.11538    2.519157         55         60

. summarize spdlimit if spd3==1

    Variable |     Obs        Mean    Std. Dev.       Min        Max
-------------+--------------------------------------------------------
    spdlimit |       2        67.5    3.535534         65         70
```

Notice the indicated minimum and maximum in each of the tables above. When we restrict the sample to `spd1`, `spdlimit` is between 40 and 50; when we restrict the sample to `spd2`, `spdlimit` is between 55 and 60; when we restrict the sample to `spd3`, `spdlimit` is between 65 and 70.

Thus `tabulate` provides an easy way to create indicator (sometimes called dummy) variables. We could now use these variables in, for instance, regression analysis. See [U] **25 Dealing with categorical variables** for an example.

❑

### tab1

`tab1` is a convenience tool. Typing

```
. tab1 myvar thisvar thatvar, plot
```

is equivalent to typing

```
. tabulate myvar, plot
. tabulate thisvar, plot
. tabulate thatvar, plot
```

## Saved Results

`tabulate` and `tab1` save in `r()`:

Scalars

| | | | |
|---|---|---|---|
| `r(N)` | number of observations | `r(r)` | number of rows |

## Methods and Formulas

`tab1` is implemented as an ado-file.

## Also See

**Complementary:** [D] **encode**

**Related:** [R] **table**, [R] **tabstat**, [R] **tabulate twoway**, [R] **tabulate, summarize()**, [D] **collapse**, [ST] **epitab**, [SVY] **svy: tabulate oneway**, [SVY] **svy: tabulate twoway**, [XT] **xtdes**, [XT] **xttab**

**Background:** [U] **12.6.3 Value labels**, [U] **25 Dealing with categorical variables**

## Title

**tabulate twoway** — Two-way tables of frequencies

## Syntax

*Two-way tables*

`tabulate` *varname*$_1$ *varname*$_2$ [*if*] [*in*] [*weight*] [, *options*]

*Two-way tables for all possible combinations—a convenience tool*

`tab2` *varlist* [*if*] [*in*] [*weight*] [, *options*]

*Immediate form of two-way tabulations*

`tabi` $\#_{11}$ $\#_{12}$ [...] \ $\#_{21}$ $\#_{22}$ [...] [\ ...] [, `replace` *options*]

| *options* | description |
|---|---|
| Main | |
| `chi2` | report Pearson's $\chi^2$ |
| `exact`[(#)] | report Fisher's exact test |
| `gamma` | report Goodman and Kruskal's gamma |
| `lrchi2` | report likelihood-ratio $\chi^2$ |
| `taub` | report Kendall's $\tau_b$ |
| `V` | report Cramér's $V$ |
| `cchi2` | report Pearson's $\chi^2$ in each cell |
| `column` | report relative frequency within its column of each cell |
| `row` | report relative frequency within its row of each cell |
| `clrchi2` | report likelihood-ratio $\chi^2$ in each cell |
| `cell` | report the relative frequency of each cell |
| `expected` | report expected frequency in each cell |
| `nofreq` | do not display frequencies |
| `missing` | treat missing values like other values |
| `wrap` | do not wrap wide tables |
| [`no`]`key` | report/suppress cell contents key |
| `nolabel` | display numeric codes rather than value labels |
| Advanced | |
| `matcell(`*matname*`)` | save frequencies in *matname*; programmer's option |
| `matrow(`*matname*`)` | save unique values of *varname*$_1$ in *matname*; programmer's option |
| `matcol(`*matname*`)` | save unique values of *varname*$_2$ in *matname*; programmer's option |
| †`all` | equivalent to specifying `chi2 lrchi2 V gamma taub` |

†`all` does not appear in the dialog box.

`by` may be used with `tabulate` and `tab2` but not with `tabi`; see [D] **by**.

`fweight`s, `aweight`s, and `iweight`s are allowed by `tabulate`. `fweight`s are allowed by `tab2`. See [U] **11.1.6 weight**.

# Description

`tabulate` produces two-way tables of frequency counts, along with various measures of association, including the common Pearson $\chi^2$, the likelihood-ratio $\chi^2$, Cramér's $V$, Fisher's exact test, Goodman and Kruskal's gamma, and Kendall's $\tau_b$.

Line size is respected. That is, if you resize the Results window (in windowed versions of Stata) before running `tabulate`, the resulting two-way tabulation will take advantage of the available horizontal space. Stata for Unix(console) users can instead use the `set linesize` command to take advantage of this feature.

`tab2` produces all possible two-way tabulations of the variables specified in *varlist*.

`tabi` displays the $r \times c$ table using the values specified; rows are separated by '\'. If no options are specified, it is as if `exact` were specified for $2 \times 2$ tables and `chi2` were specified otherwise. See [U] **19 Immediate commands** for a general description of immediate commands. Specifics for `tabi` can be found toward the end of [R] **tabulate twoway**.

See [R] **tabulate oneway** if you want one-way tables of frequencies. See [R] **table** and [R] **tabstat** if you want one-, two-, or $n$-way tables of frequencies and a wide variety of summary statistics. See [R] **tabulate, summarize()** for a description of `tabulate` with the `summarize()` option; it produces tables (breakdowns) of means and standard deviations. `table` is better than `tabulate, summarize()`, but `tabulate, summarize()` is faster. See [ST] **epitab** for $2 \times 2$ tables with statistics of interest to epidemiologists.

# Options

Main

`chi2` calculates and displays Pearson's $\chi^2$ for the hypothesis that the rows and columns in a two-way table are independent. `chi2` may not be specified if `aweights` or `iweights` are specified.

`exact`[`(`*#*`)`] displays the significance calculated by Fisher's exact test and may be applied to $r \times c$ as well as to $2 \times 2$ tables. In the case of $2 \times 2$ tables, both one- and two-sided probabilities are displayed. `exact` may not be specified if `aweights` or `iweights` are specified. The optional positive integer # is a multiplier on the amount of memory that the command is permitted to consume. The default is 1. This option should not be necessary for reasonable $r \times c$ tables. If the command terminates with error 911, try `exact(2)`.

`gamma` displays Goodman and Kruskal's gamma, along with its asymptotic standard error. `gamma` is appropriate only when both variables are ordinal. `gamma` may not be specified if `aweights` or `iweights` are specified.

`lrchi2` displays the likelihood-ratio $\chi^2$ statistic. The request is ignored if any cell of the table contains no observations. `lrchi2` may not be specified if `aweights` or `iweights` are specified.

`taub` displays Kendall's $\tau_b$ along with its asymptotic standard error. `taub` is appropriate only when both variables are ordinal. `taub` may not be specified if `aweights` or `iweights` are specified.

`V` (note capitalization) displays Cramér's $V$. `V` may not be specified if `aweights` or `iweights` are specified.

`cchi2` displays the contribution to Pearson's chi-squared in each cell of a two-way table.

`column` displays in each cell of a two-way table the relative frequency of that cell within its column.

`row` displays in each cell of a two-way table the relative frequency of that cell within its row.

`clrchi2` displays the contribution to the likelihood-ratio chi-squared in each cell of a two-way table.

`cell` displays the relative frequency of each cell in a two-way table.

`expected` displays the expected frequency in each cell of a two-way table.

`nofreq` suppresses the printing of the frequencies.

`missing` requests that missing values be treated like other values in calculations of counts, percentages, and other statistics.

`wrap` requests that Stata take no action on wide, two-way tables to make them readable. Unless `wrap` is specified, wide tables are broken into pieces to enhance readability.

[`no`]`key` suppresses or forces the display of a key above two-way tables. The default is to display the key if more than one cell statistic is requested, and otherwise to omit it. `key` forces the display of the key. `nokey` suppresses its display.

`nolabel` causes the numeric codes to be displayed rather than the value labels.

Advanced

`matcell(`*matname*`)` saves the reported frequencies in *matname*. This option is for use by programmers.

`matrow(`*matname*`)` saves the numeric values of the $r \times 1$ row stub in *matname*. This option is for use by programmers. `matrow()` may not be specified if the row variable is a string.

`matcol(`*matname*`)` saves the numeric values of the $1 \times c$ column stub in *matname*. This option is for use by programmers. `matcol()` may not be specified if the column variable is a string.

`replace` indicates that the immediate data specified as arguments to the `tabi` command be left as the current data in place of whatever data were there.

The following option is available with `tabulate` but is not shown in the dialog box:

`all` is equivalent to specifying `chi2 lrchi2 V gamma taub`. Note the omission of `exact`. When `all` is specified, `no` may be placed in front of the other options. `all noV` requests all measures of association but Cramér's $V$ (and Fisher's exact). `all exact` requests all association measures including Fisher's exact test. `all` may not be specified if `aweights` or `iweights` are specified.

## Limits

Two-way tables may have a maximum of 12,000 rows and 80 columns (Stata/SE), 300 rows and 20 columns (Intercooled Stata), or 160 rows and 20 columns (Small Stata). If larger tables are needed, see [R] **table**.

# Remarks

Remarks are presented under the headings

*tabulate*
*Measures of association*
*N-way tables*
*Weighted data*
*Tables with immediate data*
*tab2*

For each value of a specified variable (or a set of values for a pair of variables), `tabulate` reports the number of observations with that value. The number of times a value occurs is called its *frequency*.

## tabulate

### ▷ Example 1

`tabulate` will make two-way tables if we specify two variables following the word `tabulate`. In our highway dataset, we have a variable called `rate` that divides the accident rate into three categories: below 4, 4–7, and above 7 per million vehicle miles. Let's make a table of the speed limit category and the accident-rate category:

```
. tabulate spdcat rate
     Speed |     Accident rate per million
     Limit |           vehicle miles
  Category |   Below 4        4-7    Above 7 |     Total
-----------+---------------------------------+----------
  40 to 50 |         3          5          3 |        11
  55 to 60 |        19          6          1 |        26
  Above 60 |         2          0          0 |         2
-----------+---------------------------------+----------
     Total |        24         11          4 |        39
```

The table indicates that 3 stretches of highway have an accident rate below 4 and a speed limit of 40 to 50 miles per hour. The table also shows the row and column sums (called the *marginals*). The number of highways with a speed limit of 40 to 50 miles per hour is 11, which is the same result we obtained in our previous one-way tabulations.

Stata can present this basic table in a number of ways—16, to be precise—and we will show just a few below. It might be easier to read the table if we included the row percentages. For instance, out of 11 highways in the lowest speed limit category, 3 are also in the lowest accident-rate category. Three-elevenths amounts to some 27.3%. We can ask Stata to fill in this information for us by using the `row` option:

```
. tabulate spdcat rate, row

+----------------+
| Key            |
|----------------|
|   frequency    |
| row percentage |
+----------------+

     Speed |     Accident rate per million
     Limit |           vehicle miles
  Category |   Below 4        4-7    Above 7 |     Total
-----------+---------------------------------+----------
  40 to 50 |         3          5          3 |        11
           |     27.27      45.45      27.27 |    100.00
-----------+---------------------------------+----------
  55 to 60 |        19          6          1 |        26
           |     73.08      23.08       3.85 |    100.00
-----------+---------------------------------+----------
  Above 60 |         2          0          0 |         2
           |    100.00       0.00       0.00 |    100.00
-----------+---------------------------------+----------
     Total |        24         11          4 |        39
           |     61.54      28.21      10.26 |    100.00
```

The number listed below each frequency is the percentage of cases that each cell represents out of its row. That is easy to remember because we see 100% listed in the "`Total`" column. The bottom row is also informative. We see that 61.54% of all the highways in our dataset fall into the lowest accident-rate category, that 28.21% are in the middle category, and that 10.26% are in the highest.

`tabulate` can calculate column percentages and cell percentages, as well. It does so when we specify the `column` or `cell` options, respectively. We can even specify them together. Below is a table that includes everything:

```
. tabulate spdcat rate, row column cell
```

| Key |
|---|
| *frequency*<br>*row percentage*<br>*column percentage*<br>*cell percentage* |

| Speed Limit Category | Accident rate per million vehicle miles | | | |
|---|---|---|---|---|
| | Below 4 | 4-7 | Above 7 | Total |
| 40 to 50 | 3<br>27.27<br>12.50<br>7.69 | 5<br>45.45<br>45.45<br>12.82 | 3<br>27.27<br>75.00<br>7.69 | 11<br>100.00<br>28.21<br>28.21 |
| 55 to 60 | 19<br>73.08<br>79.17<br>48.72 | 6<br>23.08<br>54.55<br>15.38 | 1<br>3.85<br>25.00<br>2.56 | 26<br>100.00<br>66.67<br>66.67 |
| Above 60 | 2<br>100.00<br>8.33<br>5.13 | 0<br>0.00<br>0.00<br>0.00 | 0<br>0.00<br>0.00<br>0.00 | 2<br>100.00<br>5.13<br>5.13 |
| Total | 24<br>61.54<br>100.00<br>61.54 | 11<br>28.21<br>100.00<br>28.21 | 4<br>10.26<br>100.00<br>10.26 | 39<br>100.00<br>100.00<br>100.00 |

The number at the top of each cell is the frequency count. The second number is the row percentage—they sum to 100% going across the table. The third number is the column percentage—they sum to 100% going down the table. The bottom number is the cell percentage—they sum to 100% going down all the columns and across all the rows. For instance, highways with a speed limit above 60 miles per hour and in the lowest accident rate category account for 100% of highways with a speed limit above 60 miles per hour; 8.33% of highways in the lowest accident-rate category; and 5.13% of all our data.

A fourth option, `nofreq`, tells Stata not to print the frequency counts. To construct a table consisting of only row percentages, we type

```
. tabulate spdcat rate, row nofreq
```

| Speed Limit Category | Accident rate per million vehicle miles | | | |
|---|---|---|---|---|
| | Below 4 | 4-7 | Above 7 | Total |
| 40 to 50 | 27.27 | 45.45 | 27.27 | 100.00 |
| 55 to 60 | 73.08 | 23.08 | 3.85 | 100.00 |
| Above 60 | 100.00 | 0.00 | 0.00 | 100.00 |
| Total | 61.54 | 28.21 | 10.26 | 100.00 |

◁

## Measures of association

### ▷ Example 2

`tabulate` will calculate the Pearson $\chi^2$ test for the independence of the rows and columns if we specify the `chi2` option. Suppose that we have 1980 census data on 956 cities in the U.S. and wish to compare the age distribution across regions of the country. Assume that `agecat` is the median age in each city and that `region` denotes the region of the country in which the city is located.

```
. use http://www.stata-press.com/data/r9/citytemp2
(City Temperature Data)

. tabulate region agecat, chi2

    Census |               agecat
    Region |     19-29      30-34        35+ |     Total
-----------+---------------------------------+----------
        NE |        46         83         37 |       166 
   N Cntrl |       162         92         30 |       284 
     South |       139         68         43 |       250 
      West |       160         73         23 |       256 
-----------+---------------------------------+----------
     Total |       507        316        133 |       956 

          Pearson chi2(6) =  61.2877   Pr = 0.000
```

We obtain the standard two-way table and, at the bottom, a summary of the $\chi^2$ test. Stata informs us that the $\chi^2$ associated with this table has 6 degrees of freedom and is 61.29. The observed differences are quite significant.

The table is, perhaps, easier to understand if we suppress the frequencies and print just the row percentages:

```
. tabulate region agecat, row nofreq chi2

    Census |               agecat
    Region |     19-29      30-34        35+ |     Total
-----------+---------------------------------+----------
        NE |     27.71      50.00      22.29 |    100.00 
   N Cntrl |     57.04      32.39      10.56 |    100.00 
     South |     55.60      27.20      17.20 |    100.00 
      West |     62.50      28.52       8.98 |    100.00 
-----------+---------------------------------+----------
     Total |     53.03      33.05      13.91 |    100.00 

          Pearson chi2(6) =  61.2877   Pr = 0.000
```

◁

### ▷ Example 3

We have data on dose level and outcome for a set of patients and wish to evaluate the association between the two variables. We can obtain all the association measures by specifying the `all` and `exact` options:

(*Continued on next page*)

```
. use http://www.stata-press.com/data/r9/dose

. tabulate dose function, all exact

           |              Function
    Dosage |    < 1 hr     1 to 4         4+ |     Total
-----------+---------------------------------+----------
     1/day |        20         10          2 |        32 
     2/day |        16         12          4 |        32 
     3/day |        10         16          6 |        32 
-----------+---------------------------------+----------
     Total |        46         38         12 |        96 
          Pearson chi2(4) =   6.7780   Pr = 0.148
 likelihood-ratio chi2(4) =   6.9844   Pr = 0.137
               Cramer's V =   0.1879
                    gamma =   0.3689  ASE = 0.129
          Kendall's tau-b =   0.2378  ASE = 0.086
           Fisher's exact =                 0.145
```

We find evidence of association but not enough to be truly convincing.

If we had not also specified the **exact** option, we would not have obtained Fisher's exact test. Stata can calculate this statistic both for $2 \times 2$ tables and for $r \times c$. For $2 \times 2$ tables, the calculation is almost instant. On more general tables, however, the calculation can take longer.

Note that we carefully constructed our example so that **all** would be meaningful. Kendall's $\tau_b$ and Goodman and Kruskal's gamma are relevant only when both dimensions of the table can be ordered, say, from low to high or from worst to best. The other statistics, however, are applicable in all cases.

◁

### ❑ Technical Note

Be careful when attempting to compute the $p$-value for Fisher's exact test since the number of tables that contribute to the $p$-value can be extremely large and a solution may not be feasible. The errors that are indicative of this situation are errors 911, exceeded memory limitations, and 1401, integer overflow due to large row margin frequencies. If execution terminates because of memory limitations, use `exact(`*2*`)` to permit the algorithm to consume twice the memory, `exact(`*3*`)` for three times the memory, etcetera. The default memory usage should be sufficient for reasonable tables.

❑

## N-way tables

If you need more than two-way tables, your best alternative to is use `table`, not `tabulate`; see [R] **table**.

The technical note below shows you how to use `tabulate` to create a sequence of two-way tables that together form, in effect, a three-way table, but using `table` is easy and produces prettier results:

```
. table birthcat region agecat, c(freq)
```

| birthcat | 19-29 NE | 19-29 N Cntrl | 19-29 South | 19-29 West | 30-34 NE | 30-34 N Cntrl | 30-34 South | 30-34 West |
|---|---|---|---|---|---|---|---|---|
| 29-136 | 11 | 23 | 11 | 11 | 34 | 27 | 10 | 8 |
| 137-195 | 31 | 97 | 65 | 46 | 48 | 58 | 45 | 42 |
| 196-529 | 4 | 38 | 59 | 91 | 1 | 3 | 12 | 21 |

(agecat and Census Region)

| birthcat | 35+ NE | 35+ N Cntrl | 35+ South | 35+ West |
|---|---|---|---|---|
| 29-136 | 34 | 26 | 27 | 18 |
| 137-195 | 3 | 4 | 7 | 4 |
| 196-529 | | | 4 | |

(agecat and Census Region)

## ❑ Technical Note

We can make $n$-way tables by combining the `by` *varlist*: prefix with `tabulate`. Continuing with the dataset of 956 cities, say that we want to make a table of age category by birth-rate category by region of the country. The birth-rate category variable is named `birthcat` in our dataset. To make separate tables for each age category, we would type

```
. by agecat, sort: tabulate birthcat region
```

-> agecat = 19-29

| birthcat | NE | N Cntrl | South | West | Total |
|---|---|---|---|---|---|
| | Census Region | | | | |
| 29-136 | 11 | 23 | 11 | 11 | 56 |
| 137-195 | 31 | 97 | 65 | 46 | 239 |
| 196-529 | 4 | 38 | 59 | 91 | 192 |
| Total | 46 | 158 | 135 | 148 | 487 |

-> agecat = 30-34

| birthcat | NE | N Cntrl | South | West | Total |
|---|---|---|---|---|---|
| | Census Region | | | | |
| 29-136 | 34 | 27 | 10 | 8 | 79 |
| 137-195 | 48 | 58 | 45 | 42 | 193 |
| 196-529 | 1 | 3 | 12 | 21 | 37 |
| Total | 83 | 88 | 67 | 71 | 309 |

-> agecat = 35+

| birthcat | NE | N Cntrl | South | West | Total |
|---|---|---|---|---|---|
| | Census Region | | | | |
| 29-136 | 34 | 26 | 27 | 18 | 105 |
| 137-195 | 3 | 4 | 7 | 4 | 18 |
| 196-529 | 0 | 0 | 4 | 0 | 4 |
| Total | 37 | 30 | 38 | 22 | 127 |

❑

## Weighted data

### ▷ Example 4

`tabulate` can process weighted as well as unweighted data. As with all Stata commands, we indicate the weight by specifying the [*weight*] modifier; see [U] **11.1.6 weight**.

Continuing with our dataset of 956 cities, we also have a variable called `pop`, the population of each city. We can make a table of region by age category, weighted by population, by typing

```
. tabulate region agecat [freq=pop]

    Census |             Age Category
    region |     19-29      30-34        35+ |     Total
-----------+---------------------------------+----------
        NE |   4257167   17290828    5015443 |  26563438
   N Cntrl |  17161373    5548927    1348988 |  24059288
     South |  17607696    4809089    2612535 |  25029320
      West |  12862832    9089231    1856258 |  23808321
-----------+---------------------------------+----------
     Total |  51889068   36738075   10833224 |  99460367
```

If we specify the `cell`, `column`, or `row` options, they will also be appropriately weighted. Below we repeat the table, suppressing the counts and substituting row percentages:

```
. tabulate region agecat [freq=pop], nofreq row

    Census |             Age Category
    region |     19-29      30-34        35+ |     Total
-----------+---------------------------------+----------
        NE |     16.03      65.09      18.88 |    100.00
   N Cntrl |     71.33      23.06       5.61 |    100.00
     South |     70.35      19.21      10.44 |    100.00
      West |     54.03      38.18       7.80 |    100.00
-----------+---------------------------------+----------
     Total |     52.17      36.94      10.89 |    100.00
```

◁

## Tables with immediate data

### ▷ Example 5

`tabi` ignores the dataset in memory and uses as the table the values that we specify on the command line:

```
. tabi 30 18 \ 38 14

           |          col
       row |         1          2 |     Total
-----------+----------------------+----------
         1 |        30         18 |        48
         2 |        38         14 |        52
-----------+----------------------+----------
     Total |        68         32 |       100

           Fisher's exact =                 0.289
   1-sided Fisher's exact =                 0.179
```

We may specify any of the options of `tabulate` and are not limited to 2 × 2 tables:

```
. tabi 30 18 38 \ 13 7 22, chi2 exact

           |                 col
       row |         1          2          3 |     Total
-----------+---------------------------------+----------
         1 |        30         18         38 |        86 
         2 |        13          7         22 |        42 
-----------+---------------------------------+----------
     Total |        43         25         60 |       128 

          Pearson chi2(2) =   0.7967   Pr = 0.671
           Fisher's exact =                 0.707

. tabi 30 13 \ 18 7 \ 38 22, all exact col

+-------------------+
| Key               |
|-------------------|
|     frequency     |
| column percentage |
+-------------------+

           |          col
       row |         1          2 |     Total
-----------+----------------------+----------
         1 |        30         13 |        43 
           |     34.88      30.95 |     33.59 
-----------+----------------------+----------
         2 |        18          7 |        25 
           |     20.93      16.67 |     19.53 
-----------+----------------------+----------
         3 |        38         22 |        60 
           |     44.19      52.38 |     46.88 
-----------+----------------------+----------
     Total |        86         42 |       128 
           |    100.00     100.00 |    100.00 

          Pearson chi2(2) =   0.7967   Pr = 0.671
 likelihood-ratio chi2(2) =   0.7985   Pr = 0.671
               Cramer's V =   0.0789
                    gamma =   0.1204  ASE = 0.160
          Kendall's tau-b =   0.0630  ASE = 0.084
           Fisher's exact =                 0.707
```

Note that, for 2 × 2 tables, both one- and two-sided Fisher's exact probabilities are displayed; this is true of both `tabulate` and `tabi`. See *Cumulative incidence data* and *Case–control data* in [ST] **epitab** for more discussion on the relationship between one- and two-sided probabilities.

◁

## ❑ Technical Note

`tabi`, as with all immediate commands, leaves any data in memory undisturbed. With the `replace` option, however, the data in memory are replaced by the data from the table:

*(Continued on next page)*

```
. tabi 30 18 \ 38 14, replace

           |          col
       row |         1          2 |     Total
-----------+----------------------+----------
         1 |        30         18 |        48 
         2 |        38         14 |        52 
-----------+----------------------+----------
     Total |        68         32 |       100 

           Fisher's exact =                 0.289
   1-sided Fisher's exact =                 0.179
. list

     +-----------------+
     | row   col   pop |
     |-----------------|
  1. |   1     1    30 |
  2. |   1     2    18 |
  3. |   2     1    38 |
  4. |   2     2    14 |
     +-----------------+
```

With this dataset, you could recreate the above table by typing

```
. tabulate row col [freq=pop], exact

           |          col
       row |         1          2 |     Total
-----------+----------------------+----------
         1 |        30         18 |        48 
         2 |        38         14 |        52 
-----------+----------------------+----------
     Total |        68         32 |       100 

           Fisher's exact =                 0.289
   1-sided Fisher's exact =                 0.179
```

❑

## tab2

tab2 is a convenience tool. Typing

```
. tab2 myvar thisvar thatvar, chi2
```

is equivalent to typing

```
. tabulate myvar thisvar, chi2
. tabulate myvar thatvar, chi2
. tabulate thisvar thatvar, chi2
```

## Saved Results

tabulate, tab2, and tabi save in r():

Scalars

| | | | |
|---|---|---|---|
| r(N) | number of observations | r(p_exact) | Fisher's exact $p$ |
| r(r) | number of rows | r(chi2_lr) | likelihood-ratio $\chi^2$ |
| r(c) | number of columns | r(p_lr) | significance of likelihood-ratio $\chi^2$ |
| r(chi2) | Pearson's $\chi^2$ | r(CramersV) | Cramér's V |
| r(p) | significance of Pearson's $\chi^2$ | r(ase_gam) | ASE of gamma |
| r(gamma) | gamma | r(ase_taub) | ASE of $\tau_b$ |
| r(p1_exact) | one-sided Fisher's exact $p$ | r(taub) | $\tau_b$ |

r(p1_exact) is defined only for 2×2 tables. In addition, the matrow(), matcol(), and matcell() options allow you to obtain the row values, column values, and frequencies, respectively.

## Methods and Formulas

tab2 and tabi are implemented as ado-files.

Let $n_{ij}$, $i = 1, \dots, I$, $j = 1, \dots, J$, be the number of observations in the $i$th row and $j$th column. If the data are not weighted, $n_{ij}$ is just a count. If the data are weighted, $n_{ij}$ is the sum of the weights of all data corresponding to the $(i, j)$ cell.

Define the row and column marginals as

$$n_{i\cdot} = \sum_{j=1}^{J} n_{ij} \qquad\qquad n_{\cdot j} = \sum_{i=1}^{I} n_{ij}$$

and let $n = \sum_i \sum_j n_{ij}$ be the overall sum. Also define the concordance and discordance as

$$A_{ij} = \sum_{k>i}\sum_{l>j} n_{kl} + \sum_{k<i}\sum_{l<j} n_{kl} \qquad\qquad D_{ij} = \sum_{k>i}\sum_{l<j} n_{kl} + \sum_{k<i}\sum_{l>j} n_{kl}$$

along with twice the number of concordances $P = \sum_i \sum_j n_{ij} A_{ij}$ and twice the number of discordances $Q = \sum_i \sum_j n_{ij} D_{ij}$.

The Pearson $\chi^2$ statistic with $(I-1)(J-1)$ degrees of freedom (so called because it is based on Pearson 1900; see Conover 1999, 240 and Fienberg 1980, 9) is defined as

$$X^2 = \sum_i \sum_j \frac{(n_{ij} - m_{ij})^2}{m_{ij}}$$

where $m_{ij} = n_{i\cdot} n_{\cdot j}/n$.

The likelihood-ratio $\chi^2$ statistic with $(I-1)(J-1)$ degrees of freedom (Fienberg 1980, 40) is defined as

$$G^2 = 2 \sum_i \sum_j n_{ij} \ln(n_{ij}/m_{ij})$$

Cramér's $V$ (Cramér 1946; also see Agresti 1984, 23–24) is a measure of association designed so that the attainable upper bound is 1. For $2 \times 2$ tables, $-1 \leq V \leq 1$, and otherwise, $0 \leq V \leq 1$.

$$V = \begin{cases} (n_{11}n_{22} - n_{12}n_{21})/(n_{1\cdot}n_{2\cdot}n_{\cdot 1}n_{\cdot 2})^{1/2} & \text{for } 2 \times 2 \\ \{(X^2/n)/\min(I-1, J-1)\}^{1/2} & \text{otherwise} \end{cases}$$

Gamma (Goodman and Kruskal 1954, 1959, 1963, 1972; also see Agresti 1984, 159–161) ignores tied pairs and is based only on the number of concordant and discordant pairs of observations, $-1 \leq \gamma \leq 1$,

$$\gamma = (P - Q)/(P + Q)$$

with asymptotic variance

$$16 \sum_i \sum_j n_{ij}(QA_{ij} - PD_{ij})^2/(P+Q)^4$$

Kendall's $\tau_b$ (Kendall 1945; also see Agresti 1984, 161–163), $-1 \leq \tau_b \leq 1$, is similar to gamma, except that it uses a correction for ties,

$$\tau_b = (P - Q)/(w_r w_c)^{1/2}$$

with asymptotic variance

$$\frac{\sum_i \sum_j n_{ij}(2w_r w_c d_{ij} + \tau_b v_{ij})^2 - n^3 \tau_b^2 (w_r + w_c)^2}{(w_r w_c)^4}$$

where

$$\begin{aligned} w_r &= n^2 - \sum_i n_{i\cdot}^2 \\ w_c &= n^2 - \sum_j n_{\cdot j}^2 \\ d_{ij} &= A_{ij} - D_{ij} \\ v_{ij} &= n_{i\cdot} w_c + n_{\cdot j} w_r \end{aligned}$$

Fisher's exact test (Fisher 1935 and Finney 1948; see Zelterman and Louis 1992, 293–301, for the $2 \times 2$ case) yields the probability of observing a table that gives at least as much evidence of association as the one actually observed under the assumption of no association. Holding row and column marginals fixed, the hypergeometric probability $P$ of every possible table $A$ is computed, and the

$$P = \sum_{T \in A} \Pr(T)$$

where $A$ is the set of all tables with the same marginals as the observed table, $T^\star$, such that $\Pr(T) \leq \Pr(T^\star)$. In the case of $2 \times 2$ tables, the one-sided probability is calculated by further restricting $A$ to tables in the same tail as $T^\star$. The first algorithm extending this calculation to $r \times c$ tables was Pagano and Halvorsen (1981); the one implemented here is the FEXACT algorithm by Mehta and Patel (1986). This is a search-tree clipping method originally published by Mehta and Patel (1983) with further refinements by Joe (1988) and Clarkson, Fan, and Joe (1993). Fisher's exact test is a permutation test. For more information on permutation tests, see Good (2001 and 2000).

# References


Agresti, A. 1984. *Analysis of Ordinal Categorical Data*. New York: Wiley.

Campbell, M. J. and D. Machin. 1999. *Medical Statistics: A Commonsense Approach*. 3rd ed. New York: Wiley.

Clarkson, D. B., Y. Fan, and H. Joe. 1993. A remark on Algorithm 643: FEXACT: An algorithm for performing Fisher's exact test in $r \times c$ contingency tables. *ACM Transactions on Mathematical Software* 19: 484–488.

Conover, W. J. 1999. *Practical Nonparametric Statistics*. 3rd ed. New York: Wiley.

Cox, N. J. 1996. sg57: An immediate command for two-way tables. *Stata Technical Bulletin* 33: 7–9. Reprinted in *Stata Technical Bulletin Reprints*, vol. 6, pp. 140–143.

——. 1999. sg113: Tabulation of modes. *Stata Technical Bulletin* 50: 26–27. Reprinted in *Stata Technical Bulletin Reprints*, vol. 9, pp. 180–181.

Cramér, H. 1946. *Mathematical Methods of Statistics*. Princeton: Princeton University Press.

Fienberg, S. E. 1980. *The Analysis of Cross-Classified Categorical Data*. 2nd ed. Cambridge, MA: MIT Press.

Finney, D. J. 1948. The Fisher–Yates test of significance in $2 \times 2$ contingency tables. *Biometrika* 35: 145–156.

Fisher, R. A. 1935. The logic of inductive inference. *Journal of the Royal Statistical Society, Series A* 98: 39–54.

Good, P. 2000. *Permutation Tests: A Practical Guide to Resampling Methods for Testing Hypotheses*. 2nd ed. New York: Springer.

——. 2001. *Resampling Methods: A Practical Guide to Data Analysis*. 2nd ed. Boston: Birkhäuser.

Goodman, L. A. and W. H. Kruskal. 1954. Measures of association for cross classifications. *Journal of the American Statistical Association* 49: 732–764.

——. 1959. Measures of association for cross classifications II: Further discussion and references. *Journal of the American Statistical Association* 54: 123–163.

——. 1963. Measures of association for cross classifications III: Approximate sampling theory. *Journal of the American Statistical Association* 58: 310–364.

——. 1972. Measures of association for cross classifications IV: Simplification of asymptotic variances. *Journal of the American Statistical Association* 67: 415–421.

Joe, H. 1988. Extreme probabilities for contingency tables under row and column independence with application to Fisher's exact test. *Communications in Statistics—Theory and Methods* 17: 3677–3685.

Judson, D. H. 1992. sg12: Extended tabulate utilities. *Stata Technical Bulletin* 10: 22–23. Reprinted in *Stata Technical Bulletin Reprints*, vol. 2, pp. 140–141.

Kendall, M. G. 1945. The treatment of ties in rank problems. *Biometrika* 33: 239–251.

Mehta, C. R. and N. R. Patel. 1983. A network algorithm for performing Fisher's exact test in $r \times c$ contingency tables. *Journal of the American Statistical Association* 78: 427–434.

——. 1986. Algorithm 643 FEXACT: A FORTRAN subroutine for Fisher's exact test on unordered $r \times c$ contingency tables. *ACM Transactions on Mathematical Software* 12: 154–161.

Newson, R. 2002. Parameters behind "nonparametric" statistics: Kendall's tau, Somers' D, and median differences. *Stata Journal* 2: 45–64.

Pagano, M. and K. Halvorsen. 1981. An algorithm for finding the exact significance levels of $r \times c$ tables. *Journal of the American Statistical Association* 76: 931–934.

Pearson, K. 1900. On the criterion that a given system of deviations from the probable in the case of a correlated system of variables is such that it can be reasonably supposed to have arisen from random sampling. *Philosophical Magazine*, Series 5, 50: 157–175.

Weesie, J. 2001. dm91: Patterns of missing values. *Stata Technical Bulletin* 61: 5–7. Reprinted in *Stata Technical Bulletin Reprints*, vol. 10, pp. 49–51.

Wolfe, R. 1999. sg118: Partitions of Pearson's $\chi^2$ for analyzing two-way tables that have ordered columns. *Stata Technical Bulletin* 51: 37–40. Reprinted in *Stata Technical Bulletin Reprints*, vol. 9, pp. 203–207.

Zelterman, D. and T. A. Louis. 1992. Contingency tables in medical studies. In *Medical Uses of Statistics*, ed. J. C. Bailar III and F. Mosteller, 293–310. 2nd ed. Boston: New England Journal of Medicine Books.

## Also See

**Complementary:** [D] **encode**

**Related:** [R] **table**, [R] **tabstat**, [R] **tabulate oneway**, [R] **tabulate, summarize()**, [D] **collapse**, [ST] **epitab**, [SVY] **svy: tabulate oneway**, [SVY] **svy: tabulate twoway**, [XT] **xtdes**, [XT] **xttab**

**Background:** [U] **12.6.3 Value labels**, [U] **19 Immediate commands**, [U] **25 Dealing with categorical variables**

## Title

**tabulate, summarize()** — One- and two-way tables of summary statistics

## Syntax

`tabulate` *varname*$_1$ [*varname*$_2$] [*if*] [*in*] [*weight*] [, *options*]

| *options* | description |
|---|---|
| Main | |
| `summarize(`*varname*$_3$`)` | report summary statistics for *varname*$_3$ |
| [`no`]`means` | include or suppress means |
| [`no`]`standard` | include or suppress standard deviations |
| [`no`]`freq` | include or suppress frequencies |
| [`no`]`obs` | include or suppress number of observations |
| `nolabel` | show numeric codes, not labels |
| `wrap` | do not break wide tables |
| `missing` | treat missing values of *varname*$_1$ and *varname*$_2$ as categories |

`by` may be used with `tabulate, summarize()`; see [D] **by**.
`aweight`s and `fweight`s are allowed; see [U] **11.1.6 weight**.

## Description

`tabulate, summarize()` produces one- and two-way tables (breakdowns) of means and standard deviations. See [R] **tabulate oneway** and [R] **tabulate twoway** for one- and two-way frequency tables. See [R] **table** for a more flexible command that produces one-, two-, and $n$-way tables of frequencies and a wide variety of summary statistics. `table` is better, but `tabulate, summarize()` is faster. Also see [R] **tabstat** for yet another alternative.

## Options

Main

`summarize(`*varname*$_3$`)` identifies the name of the variable for which summary statistics are to be reported. If you do not specify this option, a table of frequencies is produced; see [R] **tabulate oneway** and [R] **tabulate twoway**. The description here concerns `tabulate` when this option is specified.

(*Continued on next page*)

[no]means includes or suppresses only the means from the table.

The summarize() table normally includes the mean, standard deviation, frequency, and, if the data are weighted, number of observations. Individual elements of the table may be included or suppressed by the [no]means, [no]standard, [no]freq, and [no]obs options. For example, typing

```
. tabulate category, summarize(myvar) means standard
```

produces a summary table by category containing only the means and standard deviations of myvar. You could also achieve the same result by typing

```
. tabulate category, summarize(myvar) nofreq
```

[no]standard includes or suppresses only the standard deviations from the table; see [no]means option above.

[no]freq includes or suppresses only the frequencies from the table; see [no]means option above.

[no]obs includes or suppresses only the reported number of observations from the table. If the data are not weighted, the number of observations is identical to the frequency, and by default only the frequency is reported. If the data are weighted, the frequency refers to the sum of the weights. See [no]means option above.

nolabel causes the numeric codes to be displayed rather than the label values.

wrap requests that no action be taken on wide tables to make them readable. Unless wrap is specified, wide tables are broken into pieces to enhance readability.

missing requests that missing values of *varname*$_1$ and *varname*$_2$ be treated as categories rather than as observations to be omitted from the analysis.

# Remarks

tabulate with the summarize() option produces one- and two-way tables of summary statistics. When combined with the by prefix, it can produce $n$-way tables as well.

Remarks are presented under the headings

*One-way tables*
*Two-way tables*

## One-way tables

### ▷ Example 1

We have data on 74 automobiles. Included in our dataset are the variables foreign, which marks domestic and foreign cars, and mpg, the car's mileage rating. Typing tabulate foreign displays a breakdown of the number of observations we have by the values of the foreign variable.

```
. use http://www.stata-press.com/data/r9/auto
(1978 Automobile Data)

. tabulate foreign

   Car type |      Freq.     Percent        Cum.
------------+-----------------------------------
   Domestic |         52       70.27       70.27
    Foreign |         22       29.73      100.00
------------+-----------------------------------
      Total |         74      100.00
```

We discover that we have 52 domestic cars and 22 foreign cars in our dataset. If we add the `summarize(`*varname*`)` option, however, `tabulate` produces a table of summary statistics for *varname*:

```
. tabulate foreign, summarize(mpg)

            |     Summary of Mileage (mpg)
   Car type |        Mean   Std. Dev.       Freq.
------------+------------------------------------
   Domestic |   19.826923   4.7432972          52
    Foreign |   24.772727   6.6111869          22
------------+------------------------------------
      Total |   21.297297   5.7855032          74
```

We also discover that the average gas mileage for domestic cars is about 20 mpg and the average foreign is almost 25 mpg. Overall, the average is 21 mpg in our dataset.

◁

## ❑ Technical Note

We might now wonder if the difference in gas mileage between foreign and domestic cars is statistically significant. We can use the `oneway` command to find out; see [R] **oneway**. To obtain an analysis-of-variance table of `mpg` on `foreign`, we type

```
. oneway mpg foreign
                        Analysis of Variance
    Source              SS         df      MS            F     Prob > F
------------------------------------------------------------------------
Between groups      378.153515      1   378.153515     13.18     0.0005
 Within groups      2065.30594     72   28.6848048
------------------------------------------------------------------------
    Total           2443.45946     73   33.4720474
Bartlett's test for equal variances:  chi2(1) =   3.4818  Prob>chi2 = 0.062
```

The $F$ statistic is 13.18, and the difference between foreign and domestic cars' mileage ratings is significant at the 0.05% level.

There are a number of ways that we could have statistically compared mileage ratings—see, for instance, [R] **anova**, [R] **oneway**, [R] **regress**, and [R] **ttest**—but `oneway` seemed the most convenient.

❑

## Two-way tables

### ▷ Example 2

`tabulate, summarize` can be used to obtain two-way as well as one-way breakdowns. For instance, we obtained summary statistics on `mpg` decomposed by `foreign` by typing `tabulate foreign, summarize(mpg)`. We can specify up to two variables before the comma:

*(Continued on next page)*

```
. generate wgtcat = autocode(weight,4,1760,4840)
. tabulate wgtcat foreign, summarize(mpg)
         Means, Standard Deviations and Frequencies of Mileage (mpg)

           |       Car type
    wgtcat |  Domestic    Foreign |     Total
-----------+----------------------+----------
      2530 | 28.285714    27.0625 | 27.434783
           | 3.0937725  5.9829619 | 5.2295149
           |         7         16 |        23
-----------+----------------------+----------
      3300 |     21.75       19.6 | 21.238095
           | 2.4083189  3.4351128 | 2.7550819
           |        16          5 |        21
-----------+----------------------+----------
      4070 |  17.26087         14 |    17.125
           | 1.8639497          0 | 1.9406969
           |        23          1 |        24
-----------+----------------------+----------
      4840 | 14.666667          . | 14.666667
           |   3.32666          . |   3.32666
           |         6          0 |         6
-----------+----------------------+----------
     Total | 19.826923  24.772727 | 21.297297
           | 4.7432972  6.6111869 | 5.7855032
           |        52         22 |        74
```

In addition to the means, standard deviations, and frequencies for each weight–mileage cell, also reported are the summary statistics by weight, by mileage, and overall. For instance, the last row of the table reveals that the average mileage of domestic cars is 19.83 and that of foreign cars is 24.77—domestic cars yield poorer mileage than foreign cars. But we now see that domestic cars yield better gas mileage within weight class—the reason domestic cars yield poorer gas mileage is because they are, on average, heavier.

◁

## ▷ Example 3

If we do not specify the statistics to be included in a table, `tabulate` reports the mean, standard deviation, and frequency. We can specify the statistics that we want to see using the `means`, `standard`, and `freq` options:

```
. tabulate wgtcat foreign, summarize(mpg) means
                     Means of Mileage (mpg)

           |       Car type
    wgtcat |  Domestic    Foreign |     Total
-----------+----------------------+----------
      2530 | 28.285714    27.0625 | 27.434783
      3300 |     21.75       19.6 | 21.238095
      4070 |  17.26087         14 |    17.125
      4840 | 14.666667          . | 14.666667
-----------+----------------------+----------
     Total | 19.826923  24.772727 | 21.297297
```

When we specify one or more of the `means`, `standard`, and `freq` options, only those statistics are displayed. Thus we could obtain a table containing just the means and standard deviations by typing `means standard` after the `summarize(mpg)` option. We can also suppress selected statistics by placing `no` in front of the option name. Another way of obtaining only the means and standard deviations is to add the `nofreq` option:

```
. tabulate wgtcat foreign, summarize(mpg) nofreq
              Means and Standard Deviations of Mileage (mpg)

           |       Car type       |
    wgtcat |  Domestic    Foreign |     Total
-----------+----------------------+-----------
      2530 | 28.285714    27.0625 | 27.434783
           | 3.0937725  5.9829619 | 5.2295149
-----------+----------------------+-----------
      3300 |     21.75       19.6 | 21.238095
           | 2.4083189  3.4351128 | 2.7550819
-----------+----------------------+-----------
      4070 |  17.26087         14 |    17.125
           | 1.8639497          0 | 1.9406969
-----------+----------------------+-----------
      4840 | 14.666667          . | 14.666667
           |   3.32666          . |   3.32666
-----------+----------------------+-----------
     Total | 19.826923  24.772727 | 21.297297
           | 4.7432972  6.6111869 | 5.7855032
```

◁

# Also See

**Related:** [R] **adjust**, [R] **oneway**, [R] **table**, [R] **tabstat**, [R] **tabulate oneway**, [R] **tabulate twoway**, [D] **collapse**, [SVY] **svy: tabulate oneway**, [SVY] **svy: tabulate twoway**

**Background:** [U] **12.6 Dataset, variable, and value labels**, [U] **25 Dealing with categorical variables**

# Title

**test** — Test linear hypotheses after estimation

# Syntax

*Basic syntax* (see [R] **anova postestimation** for `test` after `anova`;
see [MV] **manova postestimation** for `test` after `manova`)

`test` *coeflist* (*Syntax 1*)

`test` *exp*=*exp*[`=`...] (*Syntax 2*)

`test [`*eqno*`]` [: *varlist*] (*Syntax 3*)

`test [`*eqno*=*eqno*[`=`...]`]` [: *varlist*] (*Syntax 4*)

`testparm` *varlist* [, `equal` `equation(`*eqno*`)`]

*Full syntax*

`test (`*spec*`)` [`(`*spec*`)` ...] [, *test_options*]

| *test_options* | description |
|---|---|
| Options | |
| `mtest`[`(`*opt*`)`] | test each condition separately |
| `coef` | report estimated constrained coefficients |
| `accumulate` | test hypothesis jointly with previously tested hypotheses |
| `notest` | suppress the output |
| `common` | test only variables common to all the equations |
| `constant` | include the constant in coefficients to be tested |
| `nosvyadjust` | carry out the Wald test as $W/k \sim F(k,d)$; for use with `svy` estimation commands |
| `minimum` | perform test with the constant, drop terms until the test becomes nonsingular, and test without the constant on the remaining terms; highly technical |
| †`matvlc(`*matname*`)` | save the variance–covariance matrix; programmer's option |

†`matvlc(`*matname*`)` does not appear in the dialog box.
*varlist* and *varname* may contain time-series operators; see [U] **11.4.3 Time-series varlists**.

Syntax 1 tests that coefficients are 0.

Syntax 2 tests that linear expressions are equal.

Syntax 3 tests that coefficients in *eqno* are 0.

Syntax 4 tests equality of coefficients between equations.

*spec* is one of

*coeflist*
*exp*=*exp* [ =*exp* ]
[*eqno*] [ : *varlist* ]
[$eqno_1$=$eqno_2$ [=... ]] [ : *varlist* ]

*coeflist* is

*varlist*
[*eqno*]*varname* [ [*eqno*]*varname*... ]
[*eqno*]_b[*varname*] [ [*eqno*]_b[*varname*]... ]

*exp* is a linear expression containing

*varname*
_b[*varname*]
[*eqno*]*varname*
[*eqno*]_b[*varname*]

*eqno* is

*##*
*name*

Distinguish between [], which are to be typed, and [ ], which indicate optional arguments.

While not shown in the syntax diagram, parentheses around *spec* are only required with multiple specifications. Also, the diagram does not show that `test` may be called without arguments to redisplay the results from the last `test`.

## Description

`test` performs Wald tests for simple and composite linear hypotheses about the parameters of the most recently fitted model.

`test` supports `svy` estimators, carrying out an adjusted Wald test by default in such cases.

`testparm` provides a useful alternative to `test` that permits *varlist* rather than a list of coefficients (which is often nothing more than a list of variables), allowing the use of standard Stata notation, including '-' and '*', which are given the *exp*ression interpretation by `test`.

`test` and `testparm` perform Wald tests. For likelihood-ratio tests, see [R] **lrtest**. For Wald-type tests of nonlinear hypotheses, see [R] **testnl**. To display estimates for one-dimensional linear or nonlinear expressions of coefficients, see [R] **lincom** and [R] **nlcom**.

See [R] **anova postestimation** for `test` after `anova`.

See [MV] **manova postestimation** for `test` after `manova`.

## Options for testparm

`equal` tests that the variables appearing in *varlist*, which also appear in the previously fitted model, are equal to each other rather than jointly equal to zero.

`equation(`*eqno*`)` is only relevant for multiple-equation models, such as `mvreg`, `mlogit`, and `heckman`. It specifies the equation for which the all-zero or all-equal hypothesis is tested. `equation(#1)` specifies that the test be conducted regarding the first equation `#1`. `equation(price)` specifies that the test concern the equation named `price`.

## Options for test

Options

`mtest`[`(`*opt*`)`] specifies that tests be performed for each condition separately. *opt* specifies the method for adjusting $p$-values for multiple testing. Valid values for *opt* are

| | |
|---|---|
| `bonferroni` | Bonferroni's method |
| `holm` | Holm's method |
| `sidak` | Šidák's method |
| `noadjust` | no adjustment is to be made |

Specifying `mtest` without an argument is equivalent to `mtest(noadjust)`. `mtest` is not allowed after `svy: mean`, `svy: total`, or `svy: ratio`.

`coef` specifies that the constrained coefficients be displayed.

`accumulate` allows a hypothesis to be tested jointly with the previously tested hypotheses.

`notest` suppresses the output. This option is useful when you are interested only in the joint test of a number of hypotheses, specified in a subsequent call of `test, accumulate`.

`common` specifies that when you use the `[`*eqno1*`=`*eqno2*[`=...`]`]` form of *spec*, the variables common to the equations *eqno1*, *eqno2*, etc., be tested. The default action is to complain if the equations have variables not in common.

`constant` specifies that `_cons` be included in the list of coefficients to be tested when using the `[`*eqno1*`=`*eqno2*[`=...`]`]` or `[`*eqno*`]` forms of *spec*. The default is not to include `_cons`.

`nosvyadjust` is for use with `svy` estimation commands. It specifies that the Wald test be carried out as $W/k \sim F(k,d)$ rather than as $(d-k+1)W/(kd) \sim F(k,d-k+1)$, where $k$ = the dimension of the test and $d$ = the total number of sampled PSUs minus the total number of strata.

`minimum` is a highly technical option. It first performs the test with the constant added. If this test is singular, coefficients are dropped until the test becomes nonsingular. Then the test without the constant is performed with the remaining terms.

The following option is available with `test` but is not shown in the dialog box:

`matvlc(`*matname*`)`, a programmer's option, saves the variance–covariance matrix of the linear combinations involved in the suite of tests. For the test of the linear constraints $Lb = c$, *matname* contains $L\mathbf{V}L'$, where $\mathbf{V}$ is the estimated variance–covariance matrix of $b$.

## Remarks

Remarks are presented under the headings

*Introductory examples*
*Special syntaxes after multiple-equation estimation*
*Constrained coefficients*
*Multiple testing*

## Introductory examples

test performs $F$ or $\chi^2$ tests of linear restrictions applied to the most recently fitted model (e.g., regress or svy: regress in the linear regression case; logit, stcox, svy: logit ... in the single-equation maximum-likelihood case; and mlogit, mvreg, streg ... in the multiple-equation maximum-likelihood case). test may be used after *any* estimation command, although in the case of maximum likelihood techniques, test produces a Wald test that depends only on the estimate of the covariance matrix—you may prefer to use the more computationally expensive likelihood-ratio test; see [U] **20 Estimation and postestimation commands** and [R] **lrtest**.

There are several variations on the syntax for test. The second syntax,

test *exp*=*exp*[=...]

is allowed after any form of estimation, although it is not that useful after anova unless you wish to concoct your own test. The anova case is discussed in [R] **anova**, so we will ignore it. Putting aside anova, after fitting a model of *depvar* on x1, x2, and x3, typing test x1+x2=x3 tests the restriction that the coefficients on x1 and x2 sum to the coefficient on x3. The expressions can be arbitrarily complicated; for instance, typing test x1+2*(x2+x3)=x2+3*x3 is the same as typing test x1+x2=x3.

As a convenient shorthand, test also allows you to specify equality for multiple expressions; for example, test x1+x2 = x3+x4 = x5+x6 tests that the three specified pairwise sums of coefficients are equal.

Note that test understands that when you type x1, you are referring to the coefficient on x1. You could also more explicitly type test _b[x1]+_b[x2]=_b[x3]; or you could test _coef[x1]+_coef[x2]=_coef[x3], or test [#1]x1+[#1]x2= [#1]x3, or many other things since there is more than one way to refer to an estimated coefficient; see [U] **13.5 Accessing coefficients and standard errors**. The shorthand involves less typing. On the other hand, you must be more explicit after estimation of multiple-equation models since there may be more than one coefficient associated with an independent variable. You might type, for instance, test [#2]x1+[#2]x2=[#2]x3 to test the constraint in equation 2 or, more readably, test [ford]x1+[ford]x2=[ford]x3, meaning that Stata test the constraint on the equation corresponding to ford, which might be equation 2. ford would be an equation name after, say, sureg, or, after mlogit, ford would be one of the outcomes. In the case of mlogit, you could also type test [2]x1+[2]x2=[2]x3—note the lack of the #—meaning not equation 2, but the equation corresponding to the numeric outcome 2. You can even test constraints across equations: test [ford]x1+[ford]x2=[buick]x3.

The syntax,

test *coeflist*

is available after all estimation commands except anova and is a convenient way to test that multiple coefficients are zero following estimation. A *coeflist* can simply be a list of variable names,

test *varname* [*varname* ...]

and it is most often specified that way. After you have fitted a model of depvar on x1, x2, and x3, typing test x1 x3 tests that the coefficients on x1 and x3 are jointly zero. After multiple-equation estimation, this would test that the coefficients on x1 and x3 are zero in all equations that contain them. Alternatively, you can be more explicit and type, for instance, test [ford]x1 [ford]x3 to test that the coefficients on x1 and x3 are zero in the equation for ford.

In the multiple-equation case, there are more alternatives. You could also test that the coefficients on x1 and x3 are zero in the equation for ford by typing test [ford]: x1 x3. You could test that all coefficients except the coefficient on the constant are zero in the equation for ford by typing test [ford]. You could test that the coefficients on x1 and x3 in the equation for ford are equal to the corresponding coefficients in the equation corresponding to buick by typing test[ford=buick]: x1 x3. You could test that all the corresponding coefficients except the constant in three equations are equal by typing test [ford=buick=volvo].

testparm is much like the first syntax of test and cannot be used after anova. Its usefulness will be demonstrated below.

The examples below use regress, but what is said applies equally after any single-equation estimation command (such as logistic, etc.). It also applies after multiple-equation estimation commands as long as references to coefficients are qualified with an equation name or number in square brackets placed before them. The convenient syntaxes for dealing with tests of many coefficients in multiple-equation models are demonstrated in *Special syntaxes after multiple-equation estimation* below.

## ▷ Example 1

We have 1980 census data on the 50 states recording the birth rate in each state (brate), the median age (medage) and its square (medagesq), and the region of the country in which each state is located.

The variable reg1 is 1 if the state is located in the Northeast and zero otherwise, whereas reg2 marks the North Central, reg3 the South, and reg4 the West. We estimate the following regression:

```
. use http://www.stata-press.com/data/r9/census4
(birth rate, median age)

. regress brate medage medagesq reg2-reg4

      Source |       SS       df       MS              Number of obs =      50
-------------+------------------------------           F(  5,    44) =  100.63
       Model |  38803.419     5  7760.68381           Prob > F      =  0.0000
    Residual | 3393.40095    44  77.1227489           R-squared     =  0.9196
-------------+------------------------------           Adj R-squared =  0.9104
       Total |   42196.82    49  861.159592           Root MSE      =   8.782

------------------------------------------------------------------------------
       brate |      Coef.   Std. Err.      t    P>|t|     [95% Conf. Interval]
-------------+----------------------------------------------------------------
      medage |  -109.0957   13.52452    -8.07   0.000    -136.3526   -81.83886
    medagesq |   1.635208   .2290536     7.14   0.000     1.173581    2.096835
        reg2 |   15.00284   4.252068     3.53   0.001     6.433365    23.57233
        reg3 |   7.366435   3.953336     1.86   0.069    -.6009898    15.33386
        reg4 |   21.39679   4.650602     4.60   0.000     12.02412    30.76946
       _cons |    1947.61   199.8405     9.75   0.000     1544.858    2350.362
------------------------------------------------------------------------------
```

test can now be used to perform a variety of statistical tests. We can test the hypothesis that the coefficient on reg3 is zero by typing

```
. test reg3=0

 ( 1)  reg3 = 0

       F(  1,    44) =    3.47
            Prob > F =    0.0691
```

The $F$ statistic with 1 numerator and 44 denominator degrees of freedom is 3.47. The significance level of the test is 6.91%—we can reject the hypothesis at the 10% level but not at the 5% level.

This result from `test` is identical to one presented in the output from `regress`, which indicates that the $t$ statistic on the `reg3` coefficient is 1.863 and that its significance level is 0.069. The $t$ statistic presented in the output can be used to test the hypothesis that the corresponding coefficient is zero, although it states the test in slightly different terms. The $F$ distribution with 1 numerator degree of freedom is, however, identical to the $t^2$ distribution. We note that $1.863^2 \approx 3.47$ and that the significance levels in each test agree, although one extra digit is presented by the `test` command.

◁

## ❑ Technical Note

After all estimation commands, including those that use the maximum likelihood method, the test that a single variable is zero is identical to that reported by the command's output. The tests are performed in the same way—using the estimated covariance matrix—and are known as Wald tests. If the estimation command reports significance levels and confidence intervals using $z$ rather than $t$ statistics, `test` reports results using the $\chi^2$ rather than the $F$ statistic.

❑

## ▷ Example 2

If that were all `test` could do, it would be useless. We can use `test`, however, to perform other tests. For instance, we can `test` the hypothesis that the coefficient on `reg2` is 21 by typing

```
. test reg2=21
 ( 1)  reg2 = 21
       F(  1,    44) =    1.99
            Prob > F =    0.1654
```

We find that we cannot reject that hypothesis, or at least we cannot reject it at any significance level below 16.5%.

◁

## ▷ Example 3

The previous test is useful, but we could almost as easily perform it by hand using the results presented in the regression output if we were well read on our statistics. We could type

```
. display Ftail(1,44,((_coef[reg2]-21)/4.252068)^2)
.16544972
```

So, now let's `test` something a bit more difficult; whether the coefficient on `reg2` is the same as the coefficient on `reg4`:

```
. test reg2=reg4
 ( 1)  reg2 - reg4 = 0
       F(  1,    44) =    2.84
            Prob > F =    0.0989
```

We find that we cannot reject the equality hypothesis at the 5% level, but we can at the 10% level.

◁

## ▷ Example 4

When we tested the equality of the `reg2` and `reg4` coefficients, Stata rearranged our algebra. When Stata displayed its interpretation of the specified test, it indicated that we were testing whether `reg2` *minus* `reg4` is zero. The rearrangement is innocuous and, in fact, allows Stata to perform much more complicated algebra, for instance,

```
. test 2*(reg2-3*(reg3-reg4))=reg3+reg2+6*(reg4-reg3)
 ( 1)  reg2 - reg3 = 0
       F(  1,    44) =    5.06
            Prob > F =    0.0295
```

Although we requested what appeared to be a lengthy hypothesis, once Stata simplified the algebra, it realized that all we wanted to do was test whether the coefficient on `reg2` is the same as the coefficient on `reg3`.

◁

## ❑ Technical Note

Stata's ability to simplify and test complex hypotheses is limited to *linear* hypotheses. If you attempt to `test` a nonlinear hypothesis, you will be told that it is not possible:

```
. test reg2/reg3=reg2+reg3
not possible with test
r(131);
```

To test a nonlinear hypothesis, see [R] **testnl**.

❑

## ▷ Example 5

The real power of `test` is demonstrated when we test *joint* hypotheses. Perhaps we wish to test whether the region variables, taken as a whole, are significant by testing whether the coefficients on `reg2`, `reg3`, and `reg4` are simultaneously zero. `test` allows us to specify multiple conditions to be tested, each embedded within parentheses.

```
. test (reg2=0) (reg3=0) (reg4=0)
 ( 1)  reg2 = 0
 ( 2)  reg3 = 0
 ( 3)  reg4 = 0
       F(  3,    44) =    8.85
            Prob > F =    0.0001
```

`test` displays the set of conditions and reports an $F$ statistic of 8.85. `test` also reports the degrees of freedom of the test to be 3, the "dimension" of the hypothesis, and the residual degrees of freedom, 44. The significance level of the test is very close to 0, so we can strongly reject the hypothesis of no difference between the regions.

An alternative method to specify simultaneous hypotheses uses the convenient shorthand of conditions with multiple equality operators.

```
. test reg2=reg3=reg4=0
 ( 1)  reg2 - reg3 = 0
 ( 2)  reg2 - reg4 = 0
 ( 3)  reg2 = 0
       F(  3,    44) =    8.85
            Prob > F =    0.0001
```

◁

## ❑ Technical Note

Another method to test simultaneous hypotheses is to specify a `test` for each constraint and `accumulate` it with the previous constraints:

```
. test reg2=0
 ( 1)  reg2 = 0
       F(  1,    44) =   12.45
            Prob > F =    0.0010
. test reg3=0, accumulate
 ( 1)  reg2 = 0
 ( 2)  reg3 = 0
       F(  2,    44) =    6.42
            Prob > F =    0.0036
. test reg4=0, accumulate
 ( 1)  reg2 = 0
 ( 2)  reg3 = 0
 ( 3)  reg4 = 0
       F(  3,    44) =    8.85
            Prob > F =    0.0001
```

We tested the hypothesis that the coefficient on `reg2` was zero by typing `test reg2=0`. We then tested whether the coefficient on `reg3` was also zero by typing `test reg3=0, accumulate`. The `accumulate` option told Stata that this was not the start of a new test but a continuation of a previous one. Stata responded by showing us the two equations and reporting an $F$ statistic of 6.42. The significance level associated with those two coefficients being zero is 0.36%.

When we added the last constraint `test reg4=0, accumulate`, we discovered that the three region variables are quite significant. If all we wanted was the overall significance and we did not want to bother seeing the interim results, we could have used the `notest` option:

```
. test reg2=0, notest
 ( 1)  reg2 = 0
. test reg3=0, accumulate notest
 ( 1)  reg2 = 0
 ( 2)  reg3 = 0
. test reg4=0, accumulate
 ( 1)  reg2 = 0
 ( 2)  reg3 = 0
 ( 3)  reg4 = 0
       F(  3,    44) =    8.85
            Prob > F =    0.0001
```

❑

## ▷ Example 6

Since tests that coefficients are zero are so common in applied statistics, the `test` command has a more convenient syntax to accommodate this case.

```
. test reg2 reg3 reg4
 ( 1)  reg2 = 0
 ( 2)  reg3 = 0
 ( 3)  reg4 = 0
       F(  3,    44) =    8.85
            Prob > F =    0.0001
```

◁

## ▷ Example 7

We will now show how to use `testparm`. In its first syntax, `test` accepts a list of variable names but not a *varlist*.

```
. test reg2-reg4
- not found
r(111);
```

In a *varlist*, `reg2-reg3` means variables `reg2` and `reg3` and all the variables between, yet we received an error. `test` is confused because the `-` has two meanings: it means subtraction in an expression and "through" in a *varlist*. Similarly, '`*`' means "any set of characters" in a *varlist* and multiplication in an expression. `testparm` avoids this confusion—it allows only a *varlist*.

```
. testparm reg2-reg4
 ( 1)  reg2 = 0
 ( 2)  reg3 = 0
 ( 3)  reg4 = 0

       F(  3,    44) =    8.85
            Prob > F =    0.0001
```

`testparm` has another advantage. We have five variables in our dataset that start with the characters `reg`: `region`, `reg1`, `reg2`, `reg3`, and `reg4`. `reg*` thus means those five variables:

```
. describe reg*
              storage  display     value
variable name   type   format      label      variable label
---------------------------------------------------------------------
region          int    %8.0g       region     Census Region
reg1            byte   %9.0g                  region==NE
reg2            byte   %9.0g                  region==N Cntrl
reg3            byte   %9.0g                  region==South
reg4            byte   %9.0g                  region==West
```

We cannot type `test reg*` because, in an expression, '`*`' means multiplication, but here is what would happen if we attempted to test all the variables that begin with `reg`:

```
. test region reg1 reg2 reg3 reg4
region not found
r(111);
```

The variable `region` was not included in our model, so it was not found. However, with `testparm`,

```
. testparm reg*
 ( 1)  reg2 = 0
 ( 2)  reg3 = 0
 ( 3)  reg4 = 0

       F(  3,    44) =    8.85
            Prob > F =    0.0001
```

That is, `testparm` took `reg*` to mean all variables that start with `reg` that were in our model.

◁

## ❑ Technical Note

Actually, `reg*` means what it always does—all variables in our dataset that begin with `reg`—in this case, `region reg1 reg2 reg3 reg4`. `testparm` just ignores any variables you specify that are not in the model.

❑

## ▷ Example 8

We just used test (testparm, actually, but it does not matter) to test the hypothesis that reg2, reg3, and reg4 are jointly zero. We can review the results of our last test by typing test without arguments:

```
. test

 ( 1)  reg2 = 0
 ( 2)  reg3 = 0
 ( 3)  reg4 = 0

       F(  3,    44) =    8.85
            Prob > F =    0.0001
```

◁

## ❑ Technical Note

test does not care how we build joint hypotheses; we may freely mix different forms of syntax. (We can even start with testparm, but we cannot use it thereafter because it does not have an accumulate option.)

Say that we type test reg2 reg3 reg4 to test that the coefficients on our region dummies are jointly zero. We could then add a fourth constraint, say, that medage = 100, by typing test medage=100, accumulate. Or, if we had introduced the medage constraint first (our first test command had been test medage=100), we could then add the region dummy test by typing test reg2 reg3 reg4, accumulate or test (reg2=0) (reg3=0) (reg4=0), accumulate.

Remember that all previous tests are cleared when we do not specify the accumulate option. No matter what tests we performed in the past, if we type test medage medagesq, omitting the accumulate option, we would test that medage and medagesq are jointly zero.

❑

## ▷ Example 9

Let's test the hypothesis that all the included regions have the *same* coefficient—that the Northeast is significantly different from the rest of the nation:

```
. test reg2=reg3=reg4

 ( 1)  reg2 - reg3 = 0
 ( 2)  reg2 - reg4 = 0

       F(  2,    44) =    8.23
            Prob > F =    0.0009
```

We find that they are not all the same. The syntax reg2=reg3=reg4 with multiple = operators is just a convenient shorthand for typing that the first expression equals the second expression and that the first expression equals the third expression,

```
. test (reg2=reg3) (reg2=reg4)
```

We performed the test for equality of the three regions by imposing two constraints: region 2 has the same coefficient as region 3, and region 2 has the same coefficient as region 4. Alternatively we could have tested that the coefficients on regions 2 and 3 are the same and that the coefficients on regions 3 and 4 are the same. We would obtain the same results in either case.

To test for equality of the three regions, we might, likely by mistake, type equality constraints for *all* pairs of regions:

```
. test (reg2=reg3) (reg2=reg4) (reg3=reg4)
 ( 1)  reg2 - reg3 = 0
 ( 2)  reg2 - reg4 = 0
 ( 3)  reg3 - reg4 = 0
       Constraint 3 dropped
       F(  2,    44) =    8.23
            Prob > F =    0.0009
```

Equality of regions 2 and 3 and of regions 2 and 4, however, implies equality of regions 3 and 4. `test` recognized that the last constraint is implied by the other constraints and hence dropped it.

◁

## ❑ Technical Note

Generally, Stata uses = for assignment, as in `gen` *newvar* = *exp*, and == as the operator for testing equality in expressions. For your convenience, `test` allows both = and == to be used.

❑

## ▷ Example 10

The test for the equality of the regions is also possible with the `testparm` command. When we include the `equal` option, `testparm` tests that the coefficients of all the variables specified are equal:

```
. testparm reg*, equal
 ( 1) - reg2 + reg3 = 0
 ( 2) - reg2 + reg4 = 0
       F(  2,    44) =    8.23
            Prob > F =    0.0009
```

We can also obtain the equality test by accumulating single equality tests.

```
. test reg2=reg4, notest
 ( 1)  reg2 - reg4 = 0
. test reg3=reg4, accum
 ( 1)  reg2 - reg4 = 0
 ( 2)  reg3 - reg4 = 0
       F(  2,    44) =    8.23
            Prob > F =    0.0009
```

◁

## ❑ Technical Note

If we specify a set of inconsistent constraints, `test` will tell us by dropping the constraint or constraints that led to the inconsistency. For instance, let's `test` that the coefficients on region 2 and region 4 are the same, add the test that the coefficient on region 2 is 20, and finally add the test that the coefficient on region 4 is 21:

```
. test (reg2==reg4) (reg2=20) (reg4=21)
 ( 1)  reg2 - reg4 = 0
 ( 2)  reg2 = 20
 ( 3)  reg4 = 21
       Constraint 2 dropped
       F(  2,    44) =    1.82
            Prob > F =    0.1737
```

Note that `test` informed us that it was dropping constraint 2. All three equations cannot be simultaneously true, so `test` drops whatever it takes to get back to something that makes sense.

❑

## Special syntaxes after multiple-equation estimation

Everything said above about tests after single-equation estimation applies to tests after multiple-equation estimation, as long as you remember to specify the equation name. To demonstrate, let's estimate a seemingly unrelated regression using `sureg`; see [R] **sureg**.

```
. use http://www.stata-press.com/data/r9/auto
(1978 Automobile Data)

. sureg (price foreign mpg displ) (weight foreign length)

Seemingly unrelated regression
----------------------------------------------------------------------
Equation          Obs  Parms        RMSE    "R-sq"       chi2        P
----------------------------------------------------------------------
price              74      3    2165.321    0.4537      49.64   0.0000
weight             74      2    245.2916    0.8990     661.84   0.0000
----------------------------------------------------------------------

------------------------------------------------------------------------------
             |      Coef.   Std. Err.      z    P>|z|     [95% Conf. Interval]
-------------+----------------------------------------------------------------
price        |
     foreign |    3058.25   685.7357     4.46   0.000     1714.233    4402.267
         mpg |  -104.9591   58.47209    -1.80   0.073    -219.5623    9.644042
displacement |   18.18098   4.286372     4.24   0.000     9.779842    26.58211
       _cons |   3904.336   1966.521     1.99   0.047      50.0263    7758.645
-------------+----------------------------------------------------------------
weight       |
     foreign |  -147.3481   75.44314    -1.95   0.051    -295.2139     .517755
      length |   30.94905   1.539895    20.10   0.000     27.93091    33.96718
       _cons |  -2753.064   303.9336    -9.06   0.000    -3348.763   -2157.365
------------------------------------------------------------------------------
```

To test the significance of `foreign` in the `price` equation, we could type

```
. test [price]foreign

 ( 1)  [price]foreign = 0

           chi2(  1) =   19.89
         Prob > chi2 =    0.0000
```

which is the same result reported by `sureg`: $4.460^2 \approx 19.89$. To test `foreign` in both equations, we could type

```
. test [price]foreign [weight]foreign

 ( 1)  [price]foreign = 0
 ( 2)  [weight]foreign = 0

           chi2(  2) =   31.61
         Prob > chi2 =    0.0000
```

or

```
. test foreign

 ( 1)  [price]foreign = 0
 ( 2)  [weight]foreign = 0

           chi2(  2) =   31.61
         Prob > chi2 =    0.0000
```

This last syntax—typing the variable name by itself—tests the coefficients in all equations in which they appear. The variable `length` appears in only the `weight` equation, so typing

```
. test length

 ( 1)  [weight]length = 0

           chi2(  1) =  403.94
         Prob > chi2 =    0.0000
```

yields the same result as typing `test [weight]length`. We may also specify a linear expression rather than a list of coefficients:

```
. test mpg=displ

 ( 1)  [price]mpg - [price]displ = 0

           chi2(  1) =    4.85
         Prob > chi2 =    0.0277
```

or

```
. test [price]mpg = [price]displ

 ( 1)  [price]mpg - [price]displ = 0

           chi2(  1) =    4.85
         Prob > chi2 =    0.0277
```

A variation on this syntax can be used to test cross-equation constraints:

```
. test [price]foreign = [weight]foreign

 ( 1)  [price]foreign - [weight]foreign = 0

           chi2(  1) =   23.07
         Prob > chi2 =    0.0000
```

Typing an equation name in square brackets by itself tests all the coefficients except the intercept in that equation:

```
. test [price]

 ( 1)  [price]foreign = 0
 ( 2)  [price]mpg = 0
 ( 3)  [price]displacement = 0

           chi2(  3) =   49.64
         Prob > chi2 =    0.0000
```

Typing an equation name in square brackets, a colon, and a list of variable names tests those variables in the specified equation:

```
. test [price]: foreign displ

 ( 1)  [price]foreign = 0
 ( 2)  [price]displacement = 0

           chi2(  2) =   25.19
         Prob > chi2 =    0.0000
```

`test [`*eqname*$_1$`=`*eqname*$_2$`]` tests that all the coefficients in the two equations are equal. We cannot use that syntax here because there are different variables in the model:

```
. test [price=weight]
variables differ between equations
(to test equality of coefficients in common, specify option -common-)
r(111);
```

The option common specifies a test of the equality coefficients common to the equations price and weight,

```
. test [price=weight], common
 ( 1)  [price]foreign - [weight]foreign = 0
           chi2(  1) =   23.07
         Prob > chi2 =    0.0000
```

By default, test does not include the constant, the coefficient of the constant variable _cons, in the test. The option cons specifies that the constant be included.

```
. test [price=weight], common cons
 ( 1)  [price]foreign - [weight]foreign = 0
 ( 2)  [price]_cons - [weight]_cons = 0
           chi2(  2) =   51.23
         Prob > chi2 =  0.0000
```

We can also use a modification of this syntax with the model if we also type a colon and the names of the variables we want to test:

```
. test [price=weight]: foreign
 ( 1)  [price]foreign - [weight]foreign = 0
           chi2(  1) =   23.07
         Prob > chi2 =    0.0000
```

We have only one variable in common between the two equations, but if there had been more, we could have listed them.

Finally, a simultaneous test of multiple constraints may be specified just as after single-equation estimation.

```
. test ([price]: foreign) ([weight]: foreign)
 ( 1)  [price]foreign = 0
 ( 2)  [weight]foreign = 0
           chi2(  2) =   31.61
         Prob > chi2 =  0.0000
```

test can also test for equality of coefficients across more than two equations. For instance, test [eq1=eq2=eq3] specifies a test that the coefficients in the three equations eq1, eq2, and eq3 are equal. This requires that the same variables be included in the three equations. If some variables are only entered in some of the equation, you can type test [eq1=eq2=eq3], common to test that the coefficients of the variables common to all three equations are equal. Alternatively, you can explicitly list the variables for which equality of coefficients across the equations is to be tested. For instance, test [eq1=eq2=eq3]: time money tests that the coefficients of the variables time and money do not differ between the equations.

## ❑ Technical Note

test [eq1=eq2=eq3], common tests the equality of the coefficients common to all equations, but it does *not* test the equality of all common coefficients. Consider the case where

| | |
|---|---|
| eq1 | contains the variables var1 var2 var3 |
| eq2 | contains the variables var1 var2 var4 |
| eq3 | contains the variables var1 var3 var4 |

Obviously, only `var1` is common to all three equations. Thus `test [eq1=eq2=eq3], common` tests that the coefficients of `var1` do not vary across the equations, so it is equivalent to `test [eq1=eq2=eq3]: var1`. To perform a test of the coefficients of variables common to two equations, you could explicitly list the constraints to be tested,

```
. test ([eq1=eq2=eq3]:var1) ([eq1=eq2]:var2) ([eq1=eq3]:var3) ([eq2=eq3]:var4)
```

or use `test` with the `accumulate` option, and maybe also with the `notest` option, to form the appropriate joint hypothesis:

```
. test [eq1=eq2], common notest
. test [eq1=eq3], common accumulate notest
. test [eq2=eq3], common accumulate
```

❑

## Constrained coefficients

If the test indicates that the data do not allow you to conclude that the constraints are not satisfied, you may want to inspect the constrained coefficients. The option `coef` specified that the constrained results, estimated by GLS, are shown.

```
. test [price=weight], common coef
 ( 1)  [price]foreign - [weight]foreign = 0
           chi2(  1) =   23.07
         Prob > chi2 =    0.0000
Constrained coefficients
```

| | Coef. | Std. Err. | z | P>\|z\| | [95% Conf. | Interval] |
|---|---|---|---|---|---|---|
| price | | | | | | |
| foreign | -216.4015 | 74.06083 | -2.92 | 0.003 | -361.558 | -71.2449 |
| mpg | -121.5717 | 58.36972 | -2.08 | 0.037 | -235.9742 | -7.169116 |
| displacement | 7.632566 | 3.681114 | 2.07 | 0.038 | .4177148 | 14.84742 |
| _cons | 7312.856 | 1834.034 | 3.99 | 0.000 | 3718.215 | 10907.5 |
| weight | | | | | | |
| foreign | -216.4015 | 74.06083 | -2.92 | 0.003 | -361.558 | -71.2449 |
| length | 30.34875 | 1.534815 | 19.77 | 0.000 | 27.34057 | 33.35693 |
| _cons | -2619.719 | 302.6632 | -8.66 | 0.000 | -3212.928 | -2026.51 |

The constrained coefficient of `foreign` is −216.40 with standard error 74.06 in equations `price` and `weight`. Note that the other coefficients and their standard errors are affected by imposing the equality constraint of the two coefficients of `foreign` because the unconstrained estimates of these two coefficients were correlated with the estimates of the other coefficients.

### ❑ Technical Note

The two-step constrained coefficients $b_c$ displayed by `test, coef` are asymptotically equivalent to the one-stage constrained estimates that are computed by specifying the constraints during estimation using the `constraint()` option of estimation commands (Gourieroux and Monfort 1998, chapter 10). Generally, one-step constrained estimates have better small-sample properties. For inspection and interpretation, however, two-step constrained estimates are a convenient alternative. Moreover, some estimation commands (e.g., `stcox`, `logit`, many `xt`-estimators, ...) do not have a `constraint()` option.

❑

## Multiple testing

When performing the test of a joint hypothesis, you might want to inspect the underlying one-degree-of-freedom hypotheses. Which constraint "is to blame"? `test` displays the 'univariate' as well as the simultaneous test if the option `mtest` is specified. For example,

```
. test [price=weight], common cons mtest
 ( 1)  [price]foreign - [weight]foreign = 0
 ( 2)  [price]_cons - [weight]_cons = 0
```

| | chi2 | df | p |
|---|---|---|---|
| (1) | 23.07 | 1 | 0.0000 # |
| (2) | 11.17 | 1 | 0.0008 # |
| all | 51.23 | 2 | 0.0000 |

*# unadjusted p-values*

Both coefficients seem to contribute to the highly significant result. The one-degree-of-freedom test shown here is identical to those if `test` had been invoked to test just this simple hypotheses. There is, of course, a real risk in inspecting these simple hypotheses. Especially in high dimensional hypotheses, you may easily find one hypothesis that happens to be significant. Multiple testing procedures are designed to provide some safeguard against this risk. $p$-values of the univariate hypotheses are modified so that the probability of falsely rejecting one of the null hypotheses is bounded. `test` provides the methods due to Bonferroni, Šidák, and Holm.

```
. test [price=weight], common cons mtest(b)
 ( 1)  [price]foreign - [weight]foreign = 0
 ( 2)  [price]_cons - [weight]_cons = 0
```

| | chi2 | df | p |
|---|---|---|---|
| (1) | 23.07 | 1 | 0.0000 # |
| (2) | 11.17 | 1 | 0.0017 # |
| all | 51.23 | 2 | 0.0000 |

*# Bonferroni adjusted p-values*

# Saved Results

`test` and `testparm` save in `r()`:

Scalars

| | |
|---|---|
| `r(p)` | two-sided $p$-value |
| `r(F)` | $F$ statistic |
| `r(df)` | test constraints degrees of freedom |
| `r(df_r)` | residual degrees of freedom |
| `r(dropped_i)` | index of $i$th constraint dropped |
| `r(chi2)` | $\chi^2$ |
| `r(ss)` | model sum of squares |
| `r(rss)` | residual sum of squares |
| `r(drop)` | 1 if constraints were dropped, 0 otherwise |

Macros

| | |
|---|---|
| `r(mtmethod)` | method of adjustment for multiple testing |

Matrix

| | |
|---|---|
| `r(mtest)` | multiple test results |

`r(ss)` and `r(rss)` are defined only when `test` is used for testing effects after `anova`.

## Methods and Formulas

`test` and `testparm` are implemented as ado-files.

`test` and `testparm` perform Wald tests. Let the estimated coefficient vector be $\mathbf{b}$ and the estimated variance–covariance matrix be $\mathbf{V}$. Let $\mathbf{Rb} = \mathbf{r}$ denote the set of $q$ linear hypotheses to be tested jointly.

The Wald test statistic is (Judge et al. 1985, 20–28)

$$W = (\mathbf{Rb} - \mathbf{r})'(\mathbf{RVR}')^{-1}(\mathbf{Rb} - \mathbf{r})$$

If the estimation command reports its significance levels using $Z$ statistics, a chi-squared distribution with $q$ degrees of freedom,

$$W \sim \chi^2_q$$

is used for computation of the significance level of the hypothesis test.

If the estimation command reports its significance levels using $t$ statistics with $d$ degrees of freedom, an $F$ statistic,

$$F = \frac{1}{q}W$$

is computed, and an $F$ distribution with $q$ numerator degrees of freedom and $d$ denominator degrees of freedom computes the significance level of the hypothesis test.

The two-step constrained estimates $b_c$ displayed by `test` with the option `coef` are the GLS estimates of the unconstrained estimates $b$ subject to the specified constraints $Rb = c$ (Gourieroux and Monfort 1998: chapter 10),

$$\mathbf{b_c} = \mathbf{b} - \mathbf{R}'(\mathbf{RVR}')^{-1}\mathbf{R}(\mathbf{Rb} - \mathbf{r})$$

with variance–covariance matrix

$$\mathbf{V_c} = \mathbf{V} - \mathbf{VR}'(\mathbf{RVR}')^{-1}\mathbf{RV}$$

If `test` displays a Wald test for joint (simultaneous) hypotheses, it can also display all one-degree-of-freedom tests, with $p$-values adjusted for multiple testing. Let $p_1, p_2, \ldots, p_k$ be the unadjusted $p$-values of these one-degree-of-freedom tests. The Bonferroni-adjusted $p$-values are defined as $p_i^b = \min(1, kp_i)$. The Šidák-adjusted $p$-values are $p_i^s = 1 - (1 - p_i)^k$. Holm's method for adjusting $p$-values is defined as $p_i^h = \min(1, k_i p_i)$, where $k_i$ is the number of $p$-values at least as large as $p_i$. Note that $p_i^h < p_i^b$, reflecting that Holm's method is strictly less conservative than the widely used Bonferroni method.

If `test` is used after a `svy` command, it carries out an adjusted Wald test—this adjustment should not be confused with the adjustment for multiple testing. Both adjustments may actually be combined. Specifically, the survey adjustment uses an approximate $F$ statistic $(d-k+1)W/(kd)$, where $W$ is the Wald test statistic, $k$ is the dimension of the hypothesis test, and $d =$ the total number of sampled PSUs minus the total number of strata. Under the null hypothesis, $(d-k+1)F/(kd) \sim F(k, d-k+1)$, where $F(k, d-k+1)$ is an $F$ distribution with $k$ numerator degrees of freedom and $d-k+1$ denominator degrees of freedom. If `svynoadjust` is specified, the $p$-value is computed using $W/k \sim F(k, d)$.

See Korn and Graubard (1990) for a detailed description of the Bonferroni adjustment technique and for a discussion of the relative merits of it and of the adjusted and unadjusted Wald tests.

# Acknowledgment

The svy adjustment code was adopted from another command developed in collaboration with John L. Eltinge, Bureau of Labor Statistics.

# References

Beale, E. M. L. 1960. Confidence regions in nonlinear estimation. *Journal of the Royal Statistical Society, Series B* 22: 41–88.

Eltinge, J. L. and W. M. Sribney. 1996. svy5: Estimates of linear combinations and hypothesis tests for survey data. *Stata Technical Bulletin* 31: 31–42. Reprinted in *Stata Technical Bulletin Reprints*, vol. 6, pp. 246–259.

Gourieroux, C. and A. Monfort. 1995/1998. *Statistics and Econometric Models*. 2 Volumes. Cambridge: Cambridge University Press.

Holm, C. 1979. A simple sequentially rejective multiple test procedure. *Scandinavian Journal of Statistics* 6: 65–70.

Judge, G. G., W. E. Griffiths, R. C. Hill, H. Lütkepohl, and T.-C. Lee. 1985. *The Theory and Practice of Econometrics*. 2nd ed. New York: Wiley.

Korn, E. L. and B. I. Graubard. 1990. Simultaneous testing of regression coefficients with complex survey data: Use of Bonferroni $t$ statistics. *The American Statistician* 44: 270–276.

Weesie, J. 1999. sg100: Two-stage linear constrained estimation. *Stata Technical Bulletin* 47: 24–30. Reprinted in *Stata Technical Bulletin Reprints*, vol. 8, pp. 217–225.

# Also See

**Related:** [R] **anova**, [R] **lincom**, [R] **linktest**, [R] **lrtest**, [R] **nlcom**, [R] **testnl**, [SVY] **svy postestimation**

**Background:** [U] **13.5 Accessing coefficients and standard errors**, [U] **20 Estimation and postestimation commands**

## Title

**testnl** — Test nonlinear hypotheses after estimation

## Syntax

`testnl` *exp*=*exp* [=*exp*... ] [, *options*]

`testnl` (*exp*=*exp* [=*exp*... ]) [ (*exp*=*exp* [=*exp*... ]) ... ] [, *options*]

| *options* | description |
|---|---|
| `mtest` [(*opt*)] | test each condition separately |
| `nosvyadjust` | carry out the Wald test as $W/k \sim F(k, d)$; for use with `svy` estimation commands |
| `iterate(#)` | use maximum # of iterations to find the optimal step size |

The second syntax means that if more than one constraint is specified, each must be surrounded by parentheses.

## Description

`testnl` tests (linear or nonlinear) hypotheses about the estimated parameters from the most recently fitted model.

`testnl` produces Wald-type tests of smooth nonlinear (or linear) hypotheses about the estimated parameters from the most recently fitted model. The $p$-values are based on the "delta method", an approximation appropriate in large samples.

`testnl` supports `svy` regression-type commands (`svy: regress`, `svy: logit`, etc.). See the *Survey Data Reference Manual*.

The format ($exp_1$=$exp_2$=$exp_3$ ...) for a simultaneous-equality hypothesis is just a convenient shorthand for a list ($exp_1$=$exp_2$) ($exp_1$=$exp_3$) etc.

`testnl` may also be used to test linear hypotheses. `test` is faster if you want to test only linear hypotheses. `testnl` is the only option for testing linear and nonlinear hypotheses simultaneously.

## Options

`mtest` [(*opt*)] specifies that tests be performed for each condition separately. *opt* specifies the method for adjusting $p$-values for multiple testing. Valid values for *opt* are

| | |
|---|---|
| `bonferroni` | Bonferroni's method |
| `holm` | Holm's method |
| `sidak` | Šidák's method |
| `noadjust` | no adjustment is to be made |

Specifying `mtest` without an argument is equivalent to specifying `mtest(noadjust)`.

`nosvyadjust` is for use with `svy` estimation commands. It specifies that the Wald test be carried out as $W/k \sim F(k,d)$ rather than as $(d-k+1)W/(kd) \sim F(k,d-k+1)$, where $k =$ the dimension of the test and $d =$ the total number of sampled PSUs minus the total number of strata.

`iterate(#)` specifies the maximum number of iterations used to find the optimal step size in the calculation of numerical derivatives of the test expressions. By default, the maximum number of iterations is 100, but convergence is usually achieved after only a few iterations. You should rarely have to use this option.

# Remarks

Remarks are presented under the headings



## Introduction

### ▷ Example 1

We have just estimated the parameters of an earnings model on cross-sectional time-series data using one of Stata's more sophisticated estimators:

```
. use http://www.stata-press.com/data/r9/earnings
(NLS Women 14-24 in 1968)
. xtgee ln_w grade age age2, i(id) t(t) corr(unstr) nolog
GEE population-averaged model                 Number of obs      =      1326
Group and time vars:              idcode t    Number of groups   =       269
Link:                             identity    Obs per group: min =         1
Family:                           Gaussian                   avg =       4.9
Correlation:                  unstructured                   max =         9
                                              Wald chi2(3)       =    283.21
Scale parameter:                  .0977882    Prob > chi2        =    0.0000

------------------------------------------------------------------------------
     ln_wage |      Coef.   Std. Err.      z    P>|z|     [95% Conf. Interval]
-------------+----------------------------------------------------------------
       grade |   .0742647   .0066998    11.08   0.000     .0611333    .0873961
         age |   .1191284   .0275896     4.32   0.000     .0650539     .173203
        age2 |  -.0018869   .0005578    -3.38   0.001    -.0029803   -.0007935
       _cons |  -.9840205   .3251321    -3.03   0.002    -1.621268   -.3467733
------------------------------------------------------------------------------
```

An implication of this model is that peak earnings occur at age `-_b[age]/(2*_b[age2])`, which here is equal to 31.6. Say that we have a theory that peak earnings should occur at age `16 + 1/_b[grade]`.

```
. testnl -_b[age]/(2*_b[age2]) = 16 + 1/_b[grade]
  (1)  -_b[age]/(2*_b[age2]) = 16 + 1/_b[grade]
               chi2(1) =        0.62
           Prob > chi2 =        0.4305
```

These data do not reject our theory. ◁

## Using testnl to perform linear tests

testnl may be used to test linear constraints, but test is faster; see [R] **test**. You could type

```
. testnl _b[x4] = _b[x1]
```

but it would take less computer time if you typed

```
. test _b[x4] = _b[x1]
```

## Specifying constraints

The constraints to be tested can be formulated in many different ways. You could type

```
. testnl _b[mpg]*_b[weight] = 1
```

or

```
. testnl _b[mpg] = 1/_b[weight]
```

or you could express the constraint any other way you wished. (To say that testnl allows constraints to be specified in different ways does not mean that the test itself does not depend on the formulation. This point is briefly discussed later.) In formulating the constraints, you must, however, exercise one caution: Users of test often refer to the coefficient on a variable by specifying the variable name. For example,

```
. test mpg = 0
```

More formally, they should type

```
. test _b[mpg] = 0
```

but test allows the _b[] surrounding the variable name to be omitted. testnl does not allow this shorthand. Typing

```
. testnl mpg=0
```

specifies the constraint that the value of variable mpg in the first observation is zero. If you make this mistake, in some cases testnl will catch it:

```
. testnl mpg=0
equation (1) contains reference to X rather than _b[X]
r(198);
```

In other cases, testnl may not catch the mistake; in that case, the constraint will be dropped because it does not make sense:

```
. testnl mpg=0
  Constraint (1) dropped
```

(There are reasons other than this for constraints being dropped.) The worst case, however, is

```
. testnl _b[weight]*mpg = 1
```

when what you mean is not that _b[weight] equals the reciprocal of the value of mpg in the first observation, but rather that

```
. testnl _b[weight]*_b[mpg] = 1
```

Sometimes this mistake will be caught by the "contains reference to X rather than _b[X]" error, and sometimes it will not. Be careful.

testnl, like test, can be used after any Stata estimation command, including the survey estimators. When you use it after a multiple-equation command, such as nlogit or heckman, you refer to coefficients using Stata's standard syntax: [*eqname*]_b[*varname*].

Stata's single-equation estimation output looks like this:

```
             |    Coef   ...
-------------+------------------------------------------------
      weight |   12.27   ...    <-  coefficient is   _b[weight]
         mpg |    3.21   ...
--------------------------------------------------------------
```

Stata's multiple-equation output looks like this:

```
             |    Coef   ...
-------------+------------------------------------------------
cat1         |           ...
      weight |   12.27   ...    <-  coefficient is   [cat1]_b[weight]
         mpg |    3.21   ...
-------------+------------------------------------------------
8            |           ...
      weight |    5.83   ...    <-  coefficient is   [8]_b[weight]
         mpg |    7.43   ...
--------------------------------------------------------------
```

## Dropped constraints

testnl automatically drops constraints when

1. They are nonbinding, e.g., _b[mpg]=_b[mpg]. More subtle cases include

```
_b[mpg]*_b[weight] = 4
_b[weight] = 2
_b[mpg] = 2
```

   In this example, the third constraint is nonbinding since it is implied by the first two.

2. They are contradictory, e.g., _b[mpg]=2 and _b[mpg]=3. More subtle cases include

```
_b[mpg]*_b[weight] = 4
_b[weight] = 2
_b[mpg] = 3
```

   The third constraint contradicts the first two.

## Output

testnl reports the constraints being tested followed by a $F$ or $\chi^2$ test:

```
. use http://www.stata-press.com/data/r9/auto2
(1978 Automobile Data)

. regress price mpg weight weightsq foreign
  (output omitted)
```

```
. testnl (39*_b[mpg]^2 = _b[foreign]) (_b[mpg]/_b[weight] = 4)
  (1)  39*_b[mpg]^2 = _b[foreign]
  (2)  _b[mpg]/_b[weight] = 4
               F(2, 69) =        0.08
               Prob > F =        0.9195
. logit foreign price weight mpg
  (output omitted)
. testnl (45*_b[mpg]^2 = _b[price]) (_b[mpg]/_b[weight] = 4)
  (1)  45*_b[mpg]^2 = _b[price]
  (2)  _b[mpg]/_b[weight] = 4
               chi2(2) =        2.45
           Prob > chi2 =        0.2945
```

## Multiple constraints

### ▷ Example 2

We illustrate the simultaneous test of a series of constraints using simulated data on labor-market promotion in a given year. We fit a probit model with separate effects for education, experience, and experience-squared for men and women.

```
. use http://www.stata-press.com/data/r9/promotion
. generate yedu_m  = yedu  * (male==1)
. generate yexp_m  = yexp  * (male==1)
. generate yexp2_m = yexp2 * (male==1)
. generate yedu_f  = yedu  * (male==0)
. generate yexp_f  = yexp  * (male==0)
. generate yexp2_f = yexp2 * (male==0)
. probit promo male yedu_m yexp_m yexp2_m  yedu_f yexp_f yexp2_f, nolog
Probit regression                                 Number of obs   =        775
                                                  LR chi2(7)      =     424.42
                                                  Prob > chi2     =     0.0000
Log likelihood = -245.42768                       Pseudo R2       =     0.4637
```

| promo | Coef. | Std. Err. | z | P>\|z\| | [95% Conf. | Interval] |
|---|---|---|---|---|---|---|
| male | .6489974 | .203739 | 3.19 | 0.001 | .2496763 | 1.048318 |
| yedu_m | 1.390517 | .1527288 | 9.10 | 0.000 | 1.091174 | 1.68986 |
| yexp_m | 1.422539 | .1544255 | 9.21 | 0.000 | 1.11987 | 1.725207 |
| yexp2_m | -.3749457 | .1160113 | -3.23 | 0.001 | -.6023236 | -.1475677 |
| yedu_f | .9730237 | .1056136 | 9.21 | 0.000 | .7660248 | 1.180023 |
| yexp_f | .4559544 | .0901169 | 5.06 | 0.000 | .2793285 | .6325803 |
| yexp2_f | -.1027149 | .0573059 | -1.79 | 0.073 | -.2150325 | .0096026 |
| _cons | .9872018 | .1148215 | 8.60 | 0.000 | .7621559 | 1.212248 |

```
note: 1 failure and 2 successes completely determined.
```

The effects of human capital seem to differ between men and women. A formal test confirms this.

```
. test (_b[yedu_m]=_b[yedu_f]) (_b[yexp_m]=_b[yexp_f]) (_b[yexp2_m]=_b[yexp2_f])
 ( 1)  yedu_m - yedu_f = 0
 ( 2)  yexp_m - yexp_f = 0
 ( 3)  yexp2_m - yexp2_f = 0
           chi2(  3) =   35.43
         Prob > chi2 =    0.0000
```

How do we interpret this gender difference? It has repeatedly been stressed (see, for example, Long 1997, 47–50; Allison 1999) that comparison of groups in binary-response models, and similarly in other latent-variable models, is hampered by an identification problem: with $\beta$ the regression coefficients for the latent variable and $\sigma$ the standard deviation of the latent residual, only the $\beta/\sigma$ are identified. In fact, in terms of the latent regression, the probit coefficients should be interpreted as $\beta/\sigma$, not as the $\beta$. If we cannot claim convincingly that the residual standard deviation $\sigma$ does not vary between the sexes, equality of the regression coefficients $\beta$ implies that the coefficients of the probit model for men and women are *proportional* but not necessarily equal. This is a nonlinear hypothesis in terms of the probit coefficients, not a linear one.

```
. testnl _b[yedu_m]/_b[yedu_f] = _b[yexp_m]/_b[yexp_f] = _b[yexp2_m]/_b[yexp2_f]
  (1)  _b[yedu_m]/_b[yedu_f] = _b[yexp_m]/_b[yexp_f]
  (2)  _b[yedu_m]/_b[yedu_f] = _b[yexp2_m]/_b[yexp2_f]
               chi2(2) =        9.21
           Prob > chi2 =        0.0100
```

We conclude that we find fairly strong evidence against the proportionality of the coefficients, and hence we have to conclude that success in the labor market is produced in different ways by men and women. (But remember, these were simulated data.)

◁

## ▷ Example 3

The syntax for specifying the equality of multiple expressions is just a convenient shorthand for specifying a series of constraints, namely, that the first expression equals the second expression, the first expression also equals the third expression, etc. The Wald test performed and the output of `testnl` are exactly the same whether we use the shorthand or we specify the series of constraints. The lengthy specification as a series of constraints can be simplified using the continuation symbols `///`.

```
. testnl (_b[yedu_m]/_b[yedu_f] = _b[yexp_m]/_b[yexp_f]) ///
         (_b[yedu_m]/_b[yedu_f] = _b[yexp2_m]/_b[yexp2_f])
  (1)  _b[yedu_m]/_b[yedu_f] = _b[yexp_m]/_b[yexp_f]
  (2)  _b[yedu_m]/_b[yedu_f] = _b[yexp2_m]/_b[yexp2_f]
               chi2(2) =        9.21
           Prob > chi2 =        0.0100
```

Having established differences between men and women, we would like to do multiple testing between the ratios. Since we did not specify hypotheses in advance, we prefer to adjust the $p$-values of tests using, in this case, Bonferroni's method.

```
. testnl _b[yedu_m]/_b[yedu_f] = _b[yexp_m]/_b[yexp_f] = ///
                                 _b[yexp2_m]/_b[yexp2_f], mtest(b)
  (1)  _b[yedu_m]/_b[yedu_f] = _b[yexp_m]/_b[yexp_f]
  (2)  _b[yedu_m]/_b[yedu_f] = _b[yexp2_m]/_b[yexp2_f]
```

| | chi2 | df | p |
|---|---|---|---|
| (1) | 6.89 | 1 | 0.0173 # |
| (2) | 0.93 | 1 | 0.6713 # |
| all | 9.21 | 2 | 0.0100 |

*# Bonferroni adjusted p-values*

◁

## Manipulability

While testnl allows you to specify constraints in different ways that are mathematically equivalent, as noted above, this does not mean that the tests are the same. This is known as the manipulability of the Wald test for nonlinear hypotheses; also see [R] **boxcox**. The test might even be significant for one formulation, but not significant for another formulation that is mathematically equivalent. It goes without saying that "trying out" different specifications to find a formulation with the desired $p$-value is totally inappropriate, though it may actually be fun to try. There is no variance under representation because the nonlinear Wald test is actually a standard Wald test for a linearization of the constraint, which depends on the particular specification. We note that the likelihood-ratio test is not manipulable in this sense.

From a statistical point of view, it is best to choose a specification of the constraints that is "as linear is possible". This usually improves the accuracy of the approximation of the null-distribution of the test by a $\chi2$ or $F$ distribution. The example above used the nonlinear Wald test to test whether the coefficients of human capital variables for men were proportional to those of women. A specification of proportionality of coefficients in terms of ratios of coefficients is "fairly nonlinear" if the coefficients in the denominator are close to 0. A "more linear" version of the test results from a "bilinear" formulation. Thus instead of

```
. testnl _b[yedu_m]/_b[yedu_f] = _b[yexp_m]/_b[yexp_f]
  (1)  _b[yedu_m]/_b[yedu_f] = _b[yexp_m]/_b[yexp_f]
               chi2(1) =       6.89
           Prob > chi2 =       0.0087
```

perhaps

```
. testnl _b[yedu_m]*_b[yexp_f] =  _b[yedu_f]*_b[yexp_m]
  (1)  _b[yedu_m]*_b[yexp_f] = _b[yedu_f]*_b[yexp_m]
               chi2(1) =      13.95
           Prob > chi2 =       0.0002
```

is better, and in fact it has been suggested that the latter version of the test is more reliable. This is confirmed by performing simulations and is in line with theoretical results of Philips and Park (1988). There is strong evidence against the proportionality of human capital effects between men and women, implying for this example that differences in the residual variances between the sexes can be ruled out as the explanation of the sex differences in the analysis of labor market participation.

# Saved Results

testnl saves in r():

Scalars

| | |
|---|---|
| r(df) | degrees of freedom |
| r(df_r) | residual degrees of freedom |
| r(chi2) | $\chi^2$ |
| r(p) | significance |
| r(F) | $F$ statistic |

Matrices

| | |
|---|---|
| r(G) | derivatives of $R(\mathbf{b})$ with respect to $\mathbf{b}$; see *Methods and Formulas* below. |
| r(R) | $R(\mathbf{b})-\mathbf{q}$; see *Methods and Formulas* below. |

## Methods and Formulas

`testnl` is implemented as an ado-file.

After fitting a model, define $\mathbf{b}$ as the resulting $1 \times k$ parameter vector and $\mathbf{V}$ as the $k \times k$ covariance matrix. The (linear or nonlinear) hypothesis is given by $R(\mathbf{b}) = \mathbf{q}$, where $R$ is a function returning a $j \times 1$ vector. The Wald test formula is (Greene 2003, 487)

$$W = \Big\{R(\mathbf{b}) - \mathbf{q}\Big\}' \Big(\mathbf{GVG}'\Big)^{-1} \Big\{R(\mathbf{b}) - \mathbf{q}\Big\}$$

where $\mathbf{G}$ is the derivative matrix of $R(\mathbf{b})$ with respect to $\mathbf{b}$. $W$ is distributed as $\chi^2$ if $\mathbf{V}$ is an asymptotic covariance matrix. $F = W/j$ is distributed as $F$ in the case of linear regression.

The adjustment methods for multiple testing are described in [R] **test**. The adjustment for survey design effects is described in [SVY] **svy postestimation**.

## References

Allison, P. D. 1999. Comparing logit and probit coefficients across groups. *Sociological Methods and Research* 28: 186–208.

Gould, W. W. 1996. crc43: Wald test of nonlinear hypotheses after model estimation. *Stata Technical Bulletin* 29: 2–4. Reprinted in *Stata Technical Bulletin Reprints*, vol. 5, pp. 15–18.

Greene, W. H. 2003. *Econometric Analysis*. 5th ed. Upper Saddle River, NJ: Prentice Hall.

Long, J. S. 1997. *Regression Models for Categorical and Limited Dependent Variables*. Thousand Oaks CA: Sage.

Phillips, P. C. and J. Y. Park. 1988. On the formulation of Wald tests of nonlinear restrictions. *Econometrica* 56: 1065–1083.

## Also See

**Related:** [R] **lincom**, [R] **lrtest**, [R] **test**

**Background:** [U] **20 Estimation and postestimation commands**

# Title

**tetrachoric** — Tetrachoric correlations for binary variables

# Syntax

`tetrachoric` *varlist* [*if*] [*in*] [*weight*] [, *options*]

| *options* | description |
|---|---|
| Main | |
| `available` | compute pairwise correlations using all available data |
| `format(`*%fmt*`)` | display format for correlations; default is `%8.4g` |
| `notable` | suppress display of correlations |
| `posdef` | modify correlation matrix to be positive definite or positive semidefinite |

`by` may be used with `tetrachoric`; see [D] **by**.

`fweights` are allowed; see [U] **11.1.6 weight**.

# Description

`tetrachoric` computes estimates of the tetrachoric correlation coefficients of the binary variables in *varlist*. All the variables should have values of 0, 1, or missing.

Tetrachoric correlations assume a latent bivariate-normal distribution ($X_1$,$X_2$) for each pair of variables ($v_1$,$v_2$), with a threshold model for the manifest variables, $v_i = 1$ if and only if $X_i > 0$. The means and variances of the latent variables are not identified, but the correlation, $r$, of $X_1$ and $X_2$ can be estimated from the joint distribution of $v_1$ and $v_2$. These are called the tetrachoric correlation coefficients.

`tetrachoric` computes pairwise estimates of the tetrachoric correlations following Edwards and Edwards (1984).

The pairwise correlation matrix is returned as `r(corr)` and can be used to perform a factor analysis or a principal-component analysis of binary variables using the `factormat` or `pcamat` commands; see [MV] **factor** and [MV] **pca**.

# Options

Main

`available` computes the tetrachoric correlation of two binary variables *var1* and *var2* using all observations that are not missing on *var1* and *var2*. The default is to use the observations that are not missing for all variables in *varlist*.

`format(`*%fmt*`)` specifies the display format for the correlations. The default is `format(%8.4g)`.

`notable` suppresses the display of the correlation matrix.

`posdef` modifies the correlation matrix so that it is positive semidefinite, i.e., a proper correlation matrix. The modified result is the correlation matrix associated with the least-squares approximation of the tetrachoric correlation matrix by a positive-semidefinite matrix.

# Remarks

Remarks are presented under the headings

*Association in 2-by-2 tables*
*Factor analysis of dichotomous variables*
*Tetrachoric correlations with simulated data*

## Association in 2-by-2 tables

Although a wide variety of measures of association in cross tabulations have been proposed, such measures are essentially equivalent (monotonically related) in the special case of $2 \times 2$ tables—there is only one degree of freedom for nonindependence. Still, some measures have more desirable properties than others. Here we compare two measures: the standard Pearson correlation coefficient and the tetrachoric correlation coefficient. Given asymmetric row or column margins, Pearson correlations are limited to a range smaller than $-1$ to 1, while tetrachoric correlations are still able to span the range from $-1$ to 1. To illustrate, consider the following set of tables for two binary variables, X and Z:

| | Z = 0 | Z = 1 | |
|---|---|---|---|
| X = 0 | $20 - a$ | $10 + a$ | 30 |
| X = 1 | $a$ | $10 - a$ | 10 |
| | 20 | 20 | 40 |

For $a$ equal to 0, 1, 2, 5, 8, 9, and 10, the Pearson and tetrachoric correlations for the above table are

| $a$ | 0 | 1 | 2 | 5 | 8 | 9 | 10 |
|---|---|---|---|---|---|---|---|
| Pearson | 0.577 | 0.462 | 0.346 | 0 | −0.346 | −0.462 | −0.577 |
| Tetrachoric | 1.000 | 0.792 | 0.607 | 0 | −0.607 | −0.792 | −1.000 |

The restricted range for the Pearson correlation is especially unfortunate when you try to analyze the association between binary variables using models developed for continuous data, such as factor analysis and principal-component analysis.

The tetrachoric correlation of two variables $(Y_1, Y_2)$ can be thought of as the Pearson correlation of two latent bivariate-normal distributed variables $(Y_1^*, Y_2^*)$ with threshold measurement models $Y_i = (Y_i^* > c_i)$ for unknown cutpoints $c_i$. Or equivalently, $Y_i = (Y_i^{**} > 0)$ where the latent bivariate-normal $(Y_1^{**}, Y_2^{**})$ are shifted versions of $(Y_1^*, Y_2^*)$ so that the cutpoints are zero. Obviously, you must judge whether assuming underlying latent variables is meaningful for the data. If this assumption is justified, tetrachoric correlations have two advantages. First you have an intuitive understanding of the size of correlations that are substantively interesting in your field of research, and this intuition is based on correlations that range from $-1$ to 1. Second since the tetrachoric correlation for binary variables estimates the Pearson correlation of the latent continuous variables (assumed multivariate-normal distributed), you can use the tetrachoric correlations to analyze multivariate relationships between the dichotomous variables. When doing so, remember that you must interpret the model in terms of the underlying continuous variables.

▷ Example 1

To illustrate tetrachoric correlations, we examine three binary variables from the `familyvalues` dataset (described in example 2).

```
. use http://www.stata-press.com/data/r9/familyvalues
(Attitudes on gender, relationships and family)
. tabulate RS075 RS076
```

| fam att: women in charge bad | fam att: trad division of labor 0 | 1 | Total |
|---|---|---|---|
| 0 | 1,564 | 979 | 2,543 |
| 1 | 119 | 632 | 751 |
| Total | 1,683 | 1,611 | 3,294 |

```
. correlate  RS074 RS075 RS076
(obs=3291)
```

| | RS074 | RS075 | RS076 |
|---|---|---|---|
| RS074 | 1.0000 | | |
| RS075 | 0.0396 | 1.0000 | |
| RS076 | 0.1595 | 0.3830 | 1.0000 |

```
. tetrachoric RS074 RS075 RS076
Tetrachoric correlations (N=3291)
```

| Variable | RS074 | RS075 | RS076 |
|---|---|---|---|
| RS074 | 1 | | |
| RS075 | .07413 | 1 | |
| RS076 | .2476 | .6854 | 1 |

As usual, the tetrachoric correlation coefficients are larger (in absolute value) and more dispersed than the Pearson correlations.

◁

## Factor analysis of dichotomous variables

▷ Example 2

Factor analysis is a popular model for measuring latent continuous traits. The standard estimators are appropriate only for continuous unimodal data. Because of the skewness implied by Bernoulli-distributed variables (especially when the probability is distributed unevenly), a factor analysis of a Pearson correlation matrix can be rather misleading when used in this context. A factor analysis of a matrix of tetrachoric correlations is more appropriate under these conditions (Uebersax 2000). We illustrate this with data on gender, relationship, and family attitudes of spouses using the Households in the Netherlands survey 1995 (Weesie et al., 1995). For attitude variables, it seems reasonable to assume that agreement or disagreement is just a coarse measurement of more nuanced underlying attitudes.

To demonstrate, we examine a few of the variables from the `familyvalues` dataset.

```
. use http://www.stata-press.com/data/r9/familyvalues
(Attitudes on gender, relationships and family)
. describe RS056-RS063
              storage  display     value
variable name   type   format      label      variable label
-----------------------------------------------------------------------------
RS056           byte   %9.0g                  fam att: should be together
RS057           byte   %9.0g                  fam att: should fight for relat
RS058           byte   %9.0g                  fam att: should avoid conflict
RS059           byte   %9.0g                  fam att: woman better nurturer
RS060           byte   %9.0g                  fam att: both spouses money goo
RS061           byte   %9.0g                  fam att: woman techn school goo
RS062           byte   %9.0g                  fam att: man natural breadwinne
RS063           byte   %9.0g                  fam att: common leisure good
. summarize RS056-RS063
    Variable |       Obs        Mean    Std. Dev.       Min        Max
-------------+--------------------------------------------------------
       RS056 |      3298    .5630685    .4960816          0          1
       RS057 |      3296    .5400485    .4984692          0          1
       RS058 |      3283    .6387451    .4804374          0          1
       RS059 |      3308     .654474    .4756114          0          1
       RS060 |      3302    .3906723     .487975          0          1
-------------+--------------------------------------------------------
       RS061 |      3293    .7102946    .4536945          0          1
       RS062 |      3307    .5857272    .4926705          0          1
       RS063 |      3298    .5379018     .498637          0          1
. correlate RS056-RS063
(obs=3221)
             |    RS056    RS057    RS058    RS059    RS060    RS061    RS062
-------------+---------------------------------------------------------------
       RS056 |   1.0000
       RS057 |   0.1350   1.0000
       RS058 |   0.2377   0.0258   1.0000
       RS059 |   0.1816   0.0097   0.2550   1.0000
       RS060 |  -0.1020  -0.0538  -0.0424   0.0126   1.0000
       RS061 |  -0.1137   0.0610  -0.1375  -0.2076   0.0706   1.0000
       RS062 |   0.2014   0.0285   0.2273   0.4098  -0.0793  -0.2873   1.0000
       RS063 |   0.2057   0.1460   0.1049   0.0911   0.0179  -0.0233   0.0975
             |    RS063
-------------+---------
       RS063 |   1.0000
```

Skewness in these data is relatively modest. For comparison, here are the tetrachoric correlations:

(*Continued on next page*)

```
. tetrachoric RS056-RS063
Tetrachoric correlations (N=3221)
```

| Variable | RS056 | RS057 | RS058 | RS059 | RS060 | RS061 |
|---|---|---|---|---|---|---|
| RS056 | 1 | | | | | |
| RS057 | .2118 | 1 | | | | |
| RS058 | .3754 | .04228 | 1 | | | |
| RS059 | .2936 | .01612 | .4076 | 1 | | |
| RS060 | -.1638 | -.08661 | -.07053 | .02144 | 1 | |
| RS061 | -.1999 | .1054 | -.2557 | -.3986 | .1272 | 1 |
| RS062 | .3144 | .0455 | .3604 | .6128 | -.1285 | -.5147 |
| RS063 | .3186 | .2278 | .1703 | .1496 | .02896 | -.04049 |

| Variable | RS062 | RS063 |
|---|---|---|
| RS062 | 1 | |
| RS063 | .1547 | 1 |

Again we see that the tetrachoric correlations are generally larger in absolute value than the Pearson correlations. The Edwards-and-Edwards estimator implemented in `tetrachoric` is computationally efficient, but the resulting correlation matrix need not be positive semidefinite—a mathematical property of any real correlation matrix. Positive definiteness is also required by commands for the analyses of correlation matrices, such as `factormat` and `pcamat`; see [MV] **factor** and [MV] **pca**. The `posdef` option of `tetrachoric` tests for positive definiteness and projects the estimated correlation matrix to a positive-semidefinite matrix if needed.

```
. tetrachoric RS056-RS063, notable posdef
(correlation matrix is positive definite)
. matrix C = r(corr)
```

This time, we suppressed the display of the correlations with the `notable` option and requested that the correlation matrix be positive semidefinite with the `posdef` option. The message "(correlation matrix is positive definite)" indicates that the correlation matrix was indeed positive definite. We placed the resulting tetrachoric correlation matrix into a matrix, `C`, so that we can perform a factor analysis upon it.

`tetrachoric` with the `posdef` option asserted that `C` was positive definite. We can verify this by using a familiar characterization of symmetric positive-definite matrices: all eigenvalues are real and positive.

```
. matrix symeigen eigenvectors eigenvalues = C
. matrix list eigenvalues
eigenvalues[1,8]
           e1          e2          e3          e4          e5          e6
r1  2.6395204   1.3604896   1.0539467   .77691907    .7265117   .56344917

           e7          e8
r1  .52993798   .34922542
```

We can proceed with a factor analysis on the matrix `C`. We use `factormat` and select iterated principal factors as the estimation method; see [MV] **factor**.

```
. factormat C, n(3221) ipf factor(2)
(obs=3221)
Factor analysis/correlation                        Number of obs    =     3221
    Method: iterated principal factors             Retained factors =        2
    Rotation: (unrotated)                          Number of params =       15
```

| Factor | Eigenvalue | Difference | Proportion | Cumulative |
|---|---|---|---|---|
| Factor1 | 2.11808 | 1.43959 | 0.7574 | 0.7574 |
| Factor2 | 0.67849 | 0.48108 | 0.2426 | 1.0000 |
| Factor3 | 0.19741 | 0.06653 | 0.0706 | 1.0706 |
| Factor4 | 0.13088 | 0.11126 | 0.0468 | 1.1174 |
| Factor5 | 0.01962 | 0.09914 | 0.0070 | 1.1244 |
| Factor6 | -0.07952 | 0.00874 | -0.0284 | 1.0960 |
| Factor7 | -0.08826 | 0.09189 | -0.0316 | 1.0644 |
| Factor8 | -0.18015 | . | -0.0644 | 1.0000 |

```
    LR test: independent vs. saturated:  chi2(28) = 4831.55 Prob>chi2 = 0.0000
Factor loadings (pattern matrix) and unique variances
```

| Variable | Factor1 | Factor2 | Uniqueness |
|---|---|---|---|
| RS056 | 0.5485 | 0.4173 | 0.5250 |
| RS057 | 0.1062 | 0.4250 | 0.8081 |
| RS058 | 0.5374 | 0.0808 | 0.7047 |
| RS059 | 0.7025 | -0.1578 | 0.4816 |
| RS060 | -0.1359 | -0.0582 | 0.9781 |
| RS061 | -0.5512 | 0.3012 | 0.6055 |
| RS062 | 0.7875 | -0.2047 | 0.3380 |
| RS063 | 0.2849 | 0.3954 | 0.7625 |

◁

## ▷ Example 3

We noted in example 2 that the matrix of Edwards-and-Edwards estimates of the tetrachoric correlation coefficients need not be positive definite. Here is an example.

```
. use http://www.stata-press.com/data/r9/familyvalues
(Attitudes on gender, relationships and family)
. tetrachoric RS056-RS063 in 1/20, format(%7.4f) posdef
Tetrachoric correlations (N=18)
```

| Variable | RS056 | RS057 | RS058 | RS059 | RS060 | RS061 | RS062 |
|---|---|---|---|---|---|---|---|
| RS056 | 1.0000 | | | | | | |
| RS057 | 0.5560 | 1.0000 | | | | | |
| RS058 | 0.3189 | 0.2819 | 1.0000 | | | | |
| RS059 | 0.3484 | 0.3097 | 0.0624 | 1.0000 | | | |
| RS060 | -0.5350 | -0.4534 | -0.7210 | 0.0623 | 1.0000 | | |
| RS061 | 0.3504 | 0.5034 | -0.1042 | -0.2078 | -0.1005 | 1.0000 | |
| RS062 | 0.1121 | -0.0406 | 0.0116 | 0.4126 | -0.2421 | -0.7485 | 1.0000 |
| RS063 | 0.3919 | 0.4979 | 0.4570 | 0.0088 | -0.9003 | -0.0393 | 0.5066 |

| Variable | RS063 |
|---|---|
| RS063 | 1.0000 |

```
(correlation matrix is positive semidefinite; 2 negative eigenvalues set to 0)
```

```
. matrix C2 = r(corr)

. matrix symeigen eigenvectors2 eigenvalues2 = C2

. matrix list eigenvalues2

eigenvalues2[1,8]
            e1          e2          e3          e4          e5          e6
r1   3.2277201   2.0731818   1.3419062   .73916854   .45994512    .1580782

            e7          e8
r1  -7.814e-17  -3.971e-16
```

The estimated tetrachoric correlation matrix is rank-2 deficient. With this C2 matrix, we can only use models of correlation that allow for singular cases.

◁

## Tetrachoric correlations with simulated data

### ▷ Example 4

We use drawnorm (see [D] **drawnorm**) to generate a sample of 1,000 observations from a bivariate normal distribution with means $-1$ and 1, unit variances, and correlation 0.4.

```
. clear

. set seed 1245

. matrix m = ( 1, -1 )

. matrix V  = ( 1, 0.4 \ 0.4, 1 )

. drawnorm c1 c2, n(1000) means(m) cov(V)
(obs 1000)
```

Now consider the measurement model assumed by the tetrachoric correlations. We only observe whether c1 and c2 are greater than zero,

```
. generate d1 = (c1 > 0)

. generate d2 = (c2 > 0)

. tabulate d1 d2
```

| d1 | d2 0 | 1 | Total |
|---|---|---|---|
| 0 | 138 | 11 | 149 |
| 1 | 709 | 142 | 851 |
| Total | 847 | 153 | 1,000 |

We want to estimate the correlation of c1 and c2 from the binary variables d1 and d2. Pearson's correlation of the binary variables d1 and d2 is 0.092—a seriously biased estimate of the underlying correlation $\rho = 0.4$.

```
. correlate d1 d2
(obs=1000)
```

| | d1 | d2 |
|---|---|---|
| d1 | 1.0000 | |
| d2 | 0.0920 | 1.0000 |

The tetrachoric correlation coefficient of `d1` and `d2` estimates the Pearson correlation of the latent continuous variables, `c1` and `c2`.

```
. tetrachoric d1 d2
Tetrachoric correlations (N=1000)

    Variable |       d1          d2
-------------+----------------------
          d1 |        1
          d2 |    .3468           1
```

The estimate of the tetrachoric correlation of `d1` and `d2`, .3468, is much closer to the underlying correlation, .4, between `c1` and `c2`. Moreover, the tetrachoric correlation estimator is relatively insensitive to the marginal distributions. We dichotomize the two continuous variables at different cutpoints and compute the Pearson and tetrachoric correlations again.

```
. replace d1 = (c1 >  1)
(366 real changes made)
. replace d2 = (c2 > -1)
(362 real changes made)
. tabulate d1 d2
           |          d2
        d1 |         0          1 |     Total
-----------+----------------------+----------
         0 |       306        209 |       515
         1 |       179        306 |       485
-----------+----------------------+----------
     Total |       485        515 |     1,000

. correlate d1 d2
(obs=1000)

             |       d1       d2
-------------+------------------
          d1 |   1.0000
          d2 |   0.2251   1.0000

. tetrachoric d1 d2
Tetrachoric correlations (N=1000)

    Variable |       d1          d2
-------------+----------------------
          d1 |        1
          d2 |    .3455           1
```

We conclude from this example—in fact, it holds rather robustly—that Pearson's correlation coefficient may be strongly affected by the margins, while the tetrachoric correlation coefficient remains relatively unchanged.

◁

(*Continued on next page*)

## Saved Results

`tetrachoric` saves in `r()`:

Scalars

| | |
|---|---|
| `r(rho)` | tetrachoric correlation coefficient between variables 1 and 2 |
| `r(N)` | number of observations |
| `r(nneg)` | number of negative eigenvalues (`posdef` only) |

Matrices

| | |
|---|---|
| `r(corr)` | tetrachoric correlation matrix |

## Methods and Formulas

`tetrachoric` is implemented as an ado-file.

The tetrachoric correlation $\rho$ of two binary variables, $X$ and $Y$, with frequencies $n_{ij}$, $i, j = 0, 1$, is estimated as (Edwards and Edwards 1984; see also Digby 1983)

$$\widehat{\rho} = \frac{\alpha^{\pi/4} - 1}{\alpha^{\pi/4} + 1}$$

where

$$\alpha = \frac{n_{00}n_{11}}{n_{01}n_{10}}$$

if all $n_{ij} > 0$. If $n_{00} = 0$ or $n_{11} = 0$, $\widehat{\rho} = -1$; if $n_{01} = 0$ or $n_{10} = 0$, $\widehat{\rho} = 1$.

A correlation matrix, $C$, computed by `tetrachoric` need not be positive definite (i.e., not be a proper correlation matrix) if different observations are used to estimate tetrachoric correlations for different pairs of variables. In this case, `tetrachoric` optionally returns the correlation matrix $\widetilde{C}$ of the positive-semidefinite matrix closest to $C$ in the Frobenius norm; see Golub and Van Loan (1996, 55) for details.

## References

Digby, P. G. N. 1983. Approximating the tetrachoric correlation coefficient. *Biometrics* 39: 753–757.

Golub, G. H. and C. F. Van Loan. 1996. *Matrix Computations.* 3rd ed. Baltimore: The Johns Hopkins University Press.

Edwards, J. H. and A. W. F. Edwards. 1984. Approximating the tetrachoric correlation coefficient. *Biometrics* 40: 563.

Uebersax, J. 2000. Estimating a Latent Trait Model by Factor Analysis of Tetrachoric Correlations. http://ourworld.compuserve.com/homepages/jsuebersax/irt.htm

Weesie, J., M. Kalmijn, W. Bernasco, and D. Giesen. 1995. *Households in the Netherlands 1995.* [datafile]. Utrecht, Netherlands: ISCORE.

## Also See

**Complementary:** [MV] **factor**, [MV] **pca**;
[R] **tabulate twoway**

**Related:** [R] **correlate**

# Title

**tobit** — Tobit regression

# Syntax

<u>tob</u>it *depvar* [*indepvars*] [*if*] [*in*] [*weight*], ll[(*#*)] ul[(*#*)] [*options*]

| *options* | description |
|---|---|
| Model | |
| *ll[(*#*)] | left-censoring limit |
| *ul[(*#*)] | right-censoring limit |
| <u>off</u>set(*varname*) | include *varname* in model with coefficient constrained to 1 |
| SE/Robust | |
| vce(*vcetype*) | *vcetype* may be <u>boot</u>strap or <u>jack</u>knife |
| Reporting | |
| <u>l</u>evel(*#*) | set confidence level; default is level(95) |
| Max options | |
| *maximize_options* | control the maximization process; seldom used |

*You must specify at least one of ll[(*#*)] or ul[(*#*)].

bootstrap, by, jackknife, rolling, statsby, stepwise, and xi are allowed; see [U] **11.1.10 Prefix commands**.

aweights and fweights are allowed; see [U] **11.1.6 weight**.

See [U] **20 Estimation and postestimation commands** for additional capabilities of estimation commands.

# Description

tobit fits a model of *depvar* on *indepvars* where the censoring values are fixed.

# Options

Model

ll[(*#*)] and ul[(*#*)] indicate the lower and upper limits for censoring, respectively. You may specify one or both. Observations with *depvar* ≤ ll() are left-censored; observations with *depvar* ≥ ul() are right-censored; and remaining observations are not censored. You do not have to specify the censoring values at all. It is enough to type ll, ul, or both. When you do not specify a censoring value, tobit assumes that the lower limit is the minimum observed in the data (if ll is specified) and the upper limit is the maximum (if ul is specified).

offset(*varname*); see [R] **estimation options**.

SE/Robust

vce(*vcetype*); see [R] ***vce_option***.

Reporting

level(*#*); see [R] **estimation options**.

Max options

*maximize_options*: iterate(#), [no]log, trace, tolerance(#), ltolerance(#); see [R] **maximize**. These options are seldom used.

Unlike most maximum likelihood commands, tobit defaults to nolog—it suppresses the iteration log.

# Remarks

Tobit estimation was originally developed by Tobin (1958). A consumer durable was purchased if a consumer's desire was high enough, where desire was measured by the dollar amount spent by the purchaser. If no purchase was made, the measure of desire was censored at zero.

## ▷ Example 1

We will demonstrate tobit using an artificial example, which, in the process, will allow us to emphasize the assumptions underlying the estimation. We have a dataset containing the mileage ratings and weights of 74 cars. There are no censored variables in this dataset, but we are going to create one. Before that, however, the relationship between mileage and weight in our complete data is

```
. use http://www.stata-press.com/data/r9/auto
(1978 Automobile Data)
. generate wgt = weight/1000
. regress mpg wgt
      Source |       SS       df       MS              Number of obs =      74
-------------+------------------------------           F(  1,    72) =  134.62
       Model |  1591.99024     1  1591.99024           Prob > F      =  0.0000
    Residual |  851.469221    72  11.8259614           R-squared     =  0.6515
-------------+------------------------------           Adj R-squared =  0.6467
       Total |  2443.45946    73  33.4720474           Root MSE      =  3.4389

------------------------------------------------------------------------------
         mpg |      Coef.   Std. Err.      t    P>|t|     [95% Conf. Interval]
-------------+----------------------------------------------------------------
         wgt |  -6.008687   .5178782   -11.60   0.000    -7.041058   -4.976316
       _cons |   39.44028   1.614003    24.44   0.000     36.22283    42.65774
------------------------------------------------------------------------------
```

(We divided weight by 1,000 simply to make discussing the resulting coefficients easier. We find that each additional 1,000 pounds of weight reduces mileage by 6 mpg.)

mpg in our data ranges from 12 to 41. Let us now pretend that our data were censored in the sense that we could not observe a mileage rating below 17 mpg. If the true mpg is 17 or less, all we know is that the mpg is less than or equal to 17:

```
. replace mpg=17 if mpg<=17
(14 real changes made)

. tobit mpg wgt, ll

Tobit regression                                  Number of obs   =         74
                                                  LR chi2(1)      =      72.85
                                                  Prob > chi2     =     0.0000
Log likelihood = -164.25438                       Pseudo R2       =     0.1815

------------------------------------------------------------------------------
         mpg |      Coef.   Std. Err.      t    P>|t|     [95% Conf. Interval]
-------------+----------------------------------------------------------------
         wgt |   -6.87305   .7002559    -9.82   0.000    -8.268658   -5.477442
       _cons |   41.49856    2.05838    20.16   0.000     37.39621     45.6009
-------------+----------------------------------------------------------------
      /sigma |   3.845701   .3663309                      3.115605    4.575797
------------------------------------------------------------------------------
  Obs. summary:         18  left-censored observations at mpg<=17
                        56     uncensored observations
                         0 right-censored observations
```

The `replace` before estimation was not really necessary—we remapped all the mileage ratings below 17 to 17 merely to reassure you that `tobit` was not somehow using uncensored data. We typed `ll` after `tobit` to inform `tobit` that the data were left-censored. `tobit` found the minimum of `mpg` in our data and assumed that was the censoring point. Alternatively, we could have dispensed with `replace` and typed `ll(17)`, informing `tobit` that all values of the dependent variable 17 and below are really censored at 17. In either case, at the bottom of the table, we are informed that there are, as a result, 18 left-censored observations.

On these data, our estimate is now a reduction of 6.9 mpg per 1,000 extra pounds of weight as opposed to 6.0. The parameter reported as `/sigma` is the estimated standard error of the regression; the resulting 3.8 is comparable to the estimated root mean squared error reported by `regress` of 3.4.

◁

## ❑ Technical Note

You would never want to throw away information by purposefully censoring variables. The `regress` estimates are in every way preferable to those of `tobit`. Our example is solely designed to illustrate the relationship between `tobit` and `regress`. If you have uncensored data, use `regress`. If your data are censored, you have no choice but to use `tobit`.

❑

## ▷ Example 2

`tobit` can also fit models that are censored from above. This time, let's assume that we do not observe the actual mileage rating of cars yielding 24 mpg or better—we know only that it is at least 24. (Also assume that we have undone the change to `mpg` we made in the previous example.)

```
. use http://www.stata-press.com/data/r9/auto, clear
(1978 Automobile Data)

. generate wgt = weight/100

. regress mpg wgt
  (output omitted)
```

```
. tobit mpg wgt, ul(24)
Tobit regression                                  Number of obs   =         74
                                                  LR chi2(1)      =      90.72
                                                  Prob > chi2     =     0.0000
Log likelihood =  -129.8279                       Pseudo R2       =     0.2589

------------------------------------------------------------------------------
         mpg |      Coef.   Std. Err.      t    P>|t|     [95% Conf. Interval]
-------------+----------------------------------------------------------------
         wgt |  -.5080645    .043493   -11.68   0.000    -.5947459   -.4213831
       _cons |   36.08037   1.432056    25.19   0.000     33.22628    38.93445
-------------+----------------------------------------------------------------
      /sigma |   2.385357   .2444604                      1.898148    2.872566
------------------------------------------------------------------------------
  Obs. summary:          0  left-censored observations
                        51     uncensored observations
                        23 right-censored observations at mpg>=24
```

◁

## ▷ Example 3

tobit can also fit models that are censored from both sides (the so-called two-limit tobit):

```
. tobit mpg wgt, ll(17) ul(24)
Tobit regression                                  Number of obs   =         74
                                                  LR chi2(1)      =      77.60
                                                  Prob > chi2     =     0.0000
Log likelihood = -104.25976                       Pseudo R2       =     0.2712

------------------------------------------------------------------------------
         mpg |      Coef.   Std. Err.      t    P>|t|     [95% Conf. Interval]
-------------+----------------------------------------------------------------
         wgt |  -.5764448   .0724542    -7.96   0.000    -.7208457   -.4320438
       _cons |   38.07468   2.255917    16.88   0.000     33.57865    42.57072
-------------+----------------------------------------------------------------
      /sigma |   2.886337   .3952143                      2.098676    3.673998
------------------------------------------------------------------------------
  Obs. summary:         18  left-censored observations at mpg<=17
                        33     uncensored observations
                        23 right-censored observations at mpg>=24
```

◁

James Tobin (1918–2002) was an American economist who after education and research at Harvard moved to Yale, where he was on the faculty from 1950 to 1988. He made many outstanding contributions to economics and was awarded the Nobel Prize in 1981 "for his analysis of financial markets and their relations to expenditure decisions, employment, production and prices". He appeared thinly disguised as a character in Herman Wouk's novel *The Caine Mutiny* (1951) who thwarts the ambition of Willie Keith to be first in his class at midshipman school: "A mandarin-like midshipman named Tobin, with a domed forehead, measured quiet speech, and a mind like a sponge, was ahead of the field by a spacious percentage".

## Saved Results

tobit saves in e():

Scalars

| | | | |
|---|---|---|---|
| e(N) | number of observations | e(df_r) | residual degrees of freedom |
| e(N_unc) | number of uncensored obs. | e(r2_p) | pseudo-$R$-squared |
| e(N_lc) | number of left-censored obs. | e(chi2) | $\chi^2$ |
| e(N_rc) | number of right-censored obs. | e(ll) | log likelihood |
| e(llopt) | contents of ll(), if specified | e(ll_0) | log likelihood, constant-only model |
| e(ulopt) | contents of ul(), if specified | e(converged) | 1 if converged, 0 otherwise |
| e(df_m) | model degrees of freedom | | |

Macros

| | | | |
|---|---|---|---|
| e(cmd) | tobit | e(chi2type) | LR; type of model $\chi^2$ test |
| e(depvar) | name of dependent variable | e(vce) | *vcetype* specified in vce() |
| e(censored) | variable specified in censored() | e(vcetype) | title used to label Std. Err. |
| e(wtype) | weight type | e(crittype) | optimization criterion |
| e(wexp) | weight expression | e(properties) | b V |
| e(title) | title in estimation output | e(predict) | program used to implement predict |
| e(offset) | offset | | |

Matrices

| | | | |
|---|---|---|---|
| e(b) | coefficient vector | e(V) | variance–covariance matrix of the estimators |

Functions

| | |
|---|---|
| e(sample) | marks estimation sample |

## Methods and Formulas

See the *Methods and Formulas* of [R] **intreg**.

See Tobin (1958) for the original derivation of the tobit model. An introductory description of the tobit model can be found in, for instance, Johnston and DiNardo (1997, 436–441), Kmenta (1997, 562–566), Long (1997, 196–210), and Maddala (1992, 338–342).

## References


Amemiya, T. 1973. Regression analysis when the dependent variable is truncated Normal. *Econometrica* 41: 997–1016.

——. 1984. Tobit models: A survey. *Journal of Econometrics* 24: 3–61.

Cong, R. 2000. sg144: Marginal effects of the tobit model. *Stata Technical Bulletin* 56: 27–34. Reprinted in *Stata Technical Bulletin Reprints*, vol. 10, pp. 189–197.

Drukker, D. M. 2002. Bootstrapping a conditional moments test for normality after tobit estimation. *Stata Journal* 2: 125–139.

Goldberger, A. S. 1983. Abnormal selection bias. In *Studies in Econometrics, Time Series, and Multivariate Statistics*, ed. S. Karlin, T. Amemiya, and L. A. Goodman, 67–84. New York: Academic Press.

Hurd, M. 1979. Estimation in truncated samples when there is heteroscedasticity. *Journal of Econometrics* 11: 247–258.

Johnston, J. and J. DiNardo. 1997. *Econometric Methods*. 4th ed. New York: McGraw–Hill.

Kendall, M. G. and A. Stuart. 1973. *The Advanced Theory of Statistics*, vol. 2. New York: Hafner.

Kmenta, J. 1997. *Elements of Econometrics*. 2nd ed. Ann Arbor: University of Michigan Press.

Long, J. S. 1997. *Regression Models for Categorical and Limited Dependent Variables*. Thousand Oaks, CA: Sage.

Maddala, G. S. 1992. *Introduction to Econometrics*. 2nd ed. New York: Macmillan.

McDonald, J. and R. Moffitt. 1980. The use of tobit analysis. *Review of Economics and Statistics* 62: 318–321.

Shiller, R. J. 1999. The ET Interview: Professor James Tobin. *Econometric Theory* 15: 867–900.

Stewart, M. B. 1983. On least-squares estimation when the dependent variable is grouped. *Review of Economic Studies* 50: 737–753.

Tobin, J. 1958. Estimation of relationships for limited dependent variables. *Econometrica* 26: 24–36.

## Also See

**Complementary:** [R] **tobit postestimation**

**Related:** [R] **intreg**; [R] **cnreg**, [R] **heckman**, [R] **ivtobit**, [R] **oprobit**, [R] **regress**, [SVY] **svy: intreg**, [XT] **xtintreg**, [XT] **xttobit**

**Background:** [U] **11.1.10 Prefix commands**,
[U] **20 Estimation and postestimation commands**,
[R] **estimation options**, [R] **maximize**, [R] ***vce_option***

# Title

**tobit postestimation** — Postestimation tools for tobit

# Description

The following postestimation commands are available for `tobit`:

| command | description |
|---|---|
| `adjust` | adjusted predictions of $\mathbf{x}\boldsymbol{\beta}$ |
| `estat` | AIC, BIC, VCE, and estimation sample summary |
| `estimates` | cataloging estimation results |
| `hausman` | Hausman's specification test |
| `lincom` | point estimates, standard errors, testing, and inference for linear combinations of coefficients |
| `linktest` | link test for model specification |
| `lrtest` | likelihood-ratio test |
| `mfx` | marginal effects or elasticities |
| `nlcom` | point estimates, standard errors, testing, and inference for nonlinear combinations of coefficients |
| `predict` | predictions, residuals, influence statistics, and other diagnostic measures |
| `predictnl` | point estimates, standard errors, testing, and inference for generalized predictions |
| `test` | Wald tests for simple and composite linear hypotheses |
| `testnl` | Wald tests of nonlinear hypotheses |

See the corresponding entries in the *Stata Base Reference Manual* for details.

# Syntax for predict

`predict` [*type*] *newvar* [*if*] [*in*] [, *statistic* `nooffset`]

| *statistic* | description |
|---|---|
| `xb` | $\mathbf{x}_j\mathbf{b}$, fitted values; the default |
| `stdp` | standard error of the prediction |
| `stdf` | standard error of the forecast |
| `pr(`*a*,*b*`)` | $\Pr(a < y_j < b)$ |
| `e(`*a*,*b*`)` | $E(y_j \| a < y_j < b)$ |
| `ystar(`*a*,*b*`)` | $E(y_j^*)$, $y_j^* = \max\{a, \min(y_j, b)\}$ |

These statistics are available both in and out of sample; type `predict ... if e(sample) ...` if wanted only for the estimation sample.

where *a* and *b* may be numbers or variables; *a* missing ($a \geq .$) means $-\infty$, and *b* missing ($b \geq .$) means $+\infty$; see [U] **12.2.1 Missing values**.

## Options for predict

`xb`, the default, calculates the linear prediction.

`stdp` calculates the standard error of the prediction, which can be thought of as the standard error of the predicted expected value or mean for the observation's covariate pattern. This is also referred to as the standard error of the fitted value.

`stdf` calculates the standard error of the forecast, which is the standard error of the point prediction for a single observation. It is commonly referred to as the standard error of the future or forecast value. By construction, the standard errors produced by `stdf` are always larger than those produced by `stdp`; see [R] **regress** *Methods and Formulas*.

`pr(`*a*`,`*b*`)` calculates $\Pr(a < \mathbf{x}_j\mathbf{b} + u_j < b)$, the probability that $y_j|\mathbf{x}_j$ would be observed in the interval $(a, b)$.

*a* and *b* may be specified as numbers or variable names; *lb* and *ub* are variable names;
`pr(20,30)` calculates $\Pr(20 < \mathbf{x}_j\mathbf{b} + u_j < 30)$;
`pr(`*lb*`,`*ub*`)` calculates $\Pr(lb < \mathbf{x}_j\mathbf{b} + u_j < ub)$; and
`pr(20,`*ub*`)` calculates $\Pr(20 < \mathbf{x}_j\mathbf{b} + u_j < ub)$.

*a* missing ($a \geq .$) means $-\infty$; `pr(.,30)` calculates $\Pr(-\infty < \mathbf{x}_j\mathbf{b} + u_j < 30)$;
`pr(`*lb*`,30)` calculates $\Pr(-\infty < \mathbf{x}_j\mathbf{b} + u_j < 30)$ in observations for which $lb \geq .$
and calculates $\Pr(lb < \mathbf{x}_j\mathbf{b} + u_j < 30)$ elsewhere.

*b* missing ($b \geq .$) means $+\infty$; `pr(20,.)` calculates $\Pr(+\infty > \mathbf{x}_j\mathbf{b} + u_j > 20)$;
`pr(20,`*ub*`)` calculates $\Pr(+\infty > \mathbf{x}_j\mathbf{b} + u_j > 20)$ in observations for which $ub \geq .$
and calculates $\Pr(20 < \mathbf{x}_j\mathbf{b} + u_j < ub)$ elsewhere.

`e(`*a*`,`*b*`)` calculates $E(\mathbf{x}_j\mathbf{b} + u_j \mid a < \mathbf{x}_j\mathbf{b} + u_j < b)$, the expected value of $y_j|\mathbf{x}_j$ conditional on $y_j|\mathbf{x}_j$ being in the interval $(a, b)$, meaning that $y_j|\mathbf{x}_j$ is censored.
*a* and *b* are specified as they are for `pr()`.

`ystar(`*a*`,`*b*`)` calculates $E(y_j^*)$, where $y_j^* = a$ if $\mathbf{x}_j\mathbf{b} + u_j \leq a$, $y_j^* = b$ if $\mathbf{x}_j\mathbf{b} + u_j \geq b$, and $y_j^* = \mathbf{x}_j\mathbf{b} + u_j$ otherwise, meaning that $y_j^*$ is truncated. *a* and *b* are specified as they are for `pr()`.

`nooffset` is relevant only if you specified `offset(`*varname*`)`. It modifies the calculations made by `predict` so that they ignore the offset variable; the linear prediction is treated as $\mathbf{x}_j\mathbf{b}$ rather than as $\mathbf{x}_j\mathbf{b} + \text{offset}_j$.

## Methods and Formulas

All postestimation commands listed above are implemented as ado-files.

## Also See

**Complementary:** [R] **tobit**; [R] **adjust**, [R] **estimates**, [R] **hausman**, [R] **lincom**, [R] **linktest**, [R] **lrtest**, [R] **mfx**, [R] **nlcom**, [R] **predictnl**, [R] **test**, [R] **testnl**

**Background:** [U] **13.5 Accessing coefficients and standard errors**, [U] **20 Estimation and postestimation commands**, [R] **estat**, [R] **predict**

## Title

**total** — Estimate totals

## Syntax

`total` *varlist* [*if*] [*in*] [*weight*] [, *options*]

| *options* | description |
|---|---|
| if/in/over | |
| `over(`*varlist*[, `nolabel`]`)` | group over subpopulations defined by *varlist*; optionally, suppress group labels |
| SE/Cluster | |
| `vce(`*vcetype*`)` | *vcetype* may be `bootstrap` or `jackknife` |
| `cluster(`*varname*`)` | adjust standard errors for intragroup correlation |
| Reporting | |
| `level(`*#*`)` | set confidence level; default is `level(95)` |
| `noheader` | suppress table header |
| `nolegend` | suppress table legend |

`svy` may be used with `total`; see [SVY] **svy: total**.

`fweight`s, `pweight`s, and `iweight`s are allowed; see [U] **11.1.6 weight**.

See [U] **20 Estimation and postestimation commands** for additional capabilities of estimation commands.

## Description

`total` produces estimates of totals, along with standard errors.

## Options

if/in/over

`over(`*varlist* [, `nolabel`]`)` specifies that estimates be computed for multiple subpopulations, which are identified by the different values of the variables in *varlist*.

When this option is supplied with a single variable name, such as `over(`*varname*`)`, the value labels of *varname* are used to identify the subpopulations. If *varname* does not have labeled values (or there are unlabeled values), the values themselves are used, provided that they are non-negative integers. Noninteger values, negative values, and labels that are not valid Stata names are substituted with a default identifier.

When `over()` is supplied with multiple variable names, each subpopulation is assigned a unique default identifier.

`nolabel` specifies that value labels attached to the variables identifying the subpopulations be ignored.

SE/Cluster

`vce(vcetype)`; see [R] ***vce_option***.

`cluster(varname)`; see [R] **estimation options**.

Reporting

`level(#)`; see [R] **estimation options**.

`noheader` prevents the table header from being displayed. This option implies `nolegend`.

`nolegend` prevents the table legend identifying the subpopulations from being displayed.

# Remarks

## ▷ Example 1

Suppose that we collected data on incidence of heart attacks. The variable `heartatk` indicates whether a person ever had a heart attack (1 means yes; 0 means no). We can then estimate the total number of persons who have had heart attacks for each `sex` in the population represented by the data we collected.

```
. use http://www.stata-press.com/data/r9/total
. total heartatk [pw=swgt], over(sex)
Total estimation                    Number of obs    =    4946

          Male: sex = Male
        Female: sex = Female

--------------------------------------------------------------
        Over |      Total   Std. Err.    [95% Conf. Interval]
-------------+------------------------------------------------
heartatk     |
        Male |     944559   104372.3       739943     1149175
      Female |     581590   82855.59     419156.3    744023.7
--------------------------------------------------------------
```

◁

# Saved Results

`total` saves in `e()`:

Scalars

| | | | |
|---|---|---|---|
| `e(N)` | number of observations | `e(df_r)` | sample degrees of freedom |
| `e(N_over)` | number of subpopulations | | |

Macros

| | | | |
|---|---|---|---|
| `e(cmd)` | `total` | `e(over)` | *varlist* from `over()` |
| `e(wtype)` | weight type | `e(over_labels)` | labels from `over()` variables |
| `e(wexp)` | weight expression | `e(over_namelist)` | names from `e(over_labels)` |
| `e(title)` | title in estimation output | `e(properties)` | `b V` |

Matrices

| | |
|---|---|
| `e(b)` | vector of total estimates |
| `e(V)` | (co)variance estimates |
| `e(_N)` | vector of numbers of nonmissing observations |

Functions

| | |
|---|---|
| `e(sample)` | marks estimation sample |

## Methods and Formulas

`total` is implemented as an ado-file.

Let $y$ denote the variable on which to calculate the total and $y_j, j = 1, \ldots, n$, denote an individual observation on $y$. Let $w_j$ be the weight, and if no weight is specified, define $w_j = 1$ for all $j$. The sum of the weights is an estimate of the population size:

$$\widehat{N} = \sum_{j=1}^{n} w_j$$

If the population values of $y$ are denoted by $Y_j, j = 1, \ldots, N$, the associated population total is

$$Y = \sum_{j=1}^{N} Y_j = N\overline{y}$$

where $\overline{y}$ is the population mean. The total is estimated as

$$\widehat{Y} = \widehat{N}\overline{y}$$

The variance estimator for the total is

$$\widehat{V}(\widehat{Y}) = \widehat{N}^2\widehat{V}(\overline{y})$$

where $\widehat{V}(\overline{y})$ is the variance estimator for the mean; see [R] **mean**. The standard error of the total is the square root of the variance.

If $x$, $x_j$, $\overline{x}$, and $\widehat{X}$ are similarly defined for another variable (observed jointly with $y$), the covariance estimator between $\widehat{X}$ and $\widehat{Y}$ is

$$\widehat{\mathrm{Cov}}(\widehat{X}, \widehat{Y}) = \widehat{N}^2\widehat{\mathrm{Cov}}(\overline{x}, \overline{y})$$

where $\widehat{\mathrm{Cov}}(\overline{x}, \overline{y})$ is the covariance estimator between two means; see [R] **mean**.

## References


Cochran, W. G. 1977. *Sampling Techniques.* 3rd ed. New York: Wiley.

Stuart, A. and J. K. Ord. 1994. *Kendall's Advanced Theory of Statistics, Vol. I.* 6th ed. London: Arnold.


## Also See

**Complementary:** [R] **total postestimation**

**Related:** [SVY] **svy: total**;
[R] **mean**, [R] **proportion**, [R] **ratio**

**Background:** [U] **20 Estimation and postestimation commands**,
[R] **estimation options**, [R] ***vce_option***

# Title

**total postestimation** — Postestimation tools for total

# Description

The following postestimation commands are available for `total`:

| command | description |
|---|---|
| `estat` | VCE |
| `estimates` | cataloging estimation results |
| `lincom` | point estimates, standard errors, testing, and inference for linear combinations of coefficients |
| `nlcom` | point estimates, standard errors, testing, and inference for nonlinear combinations of coefficients |
| `test` | Wald tests for simple and composite linear hypotheses |
| `testnl` | Wald tests of nonlinear hypotheses |

See the corresponding entries in the *Stata Base Reference Manual* for details.

# Methods and Formulas

All postestimation commands listed above are implemented as ado-files.

# Also See

**Complementary:** [R] **total**; [R] **estimates**, [R] **lincom**, [R] **nlcom**, [R] **test**, [R] **testnl**

**Related:** [SVY] **svy: total postestimation**

**Background:** [U] **13.5 Accessing coefficients and standard errors**,
[U] **13.6 Accessing results from Stata commands**,
[U] **20 Estimation and postestimation commands**,
[R] **estat**

## Title

**translate** — Print and translate logs

## Syntax

*Print log and SMCL files*

`print` *filename* [ `,` `like(`*ext*`)` `name(`*windowname*`)` *override_options* ]

*Translate log files to SMCL files and vice versa*

`translate` $filename_{in}$ $filename_{out}$ [ `,` `translator(`*tname*`)` `name(`*windowname*`)`

*override_options* `replace` ]

*View current translate parameter settings*

`translator query` [*tname*]

*Change current translate parameter settings*

`translator set` [*tname setopt setval*]

*Return parameter settings to default values*

`translator reset` *tname*

*List current mappings from one extension to another*

`transmap query` [ `.`*ext* ]

*Specify that files with one extension be treated the same as files with another extension*

`transmap define` `.`$ext_{new}$ `.`$ext_{old}$

*filename* in `print`, in addition to being a filename to be printed, may be specified as `@Results` to mean the Results window and `@Viewer` to mean the Viewer window.

$filename_{in}$ in `translate` may be specified just as *filename* in `print`.

*tname* in `translator` specifies the name of a translator; see the `translator()` option under *Options for translate*.

## Description

`print` prints log, SMCL, and text files. Although there is considerable flexibility in how `print` (and `translate`, which `print` uses) can be set to work, they have already been set up and should just work:

```
. print mylog.smcl
. print mylog.log
```

Unix users may discover that they need to do a bit of setup before `print` works; see *Printing files, Unix* below. European Unix users may also wish to modify the default paper size. All users can tailor `print` and `translate` to their needs.

`print` may also be used to print the current contents of the Results window or the Viewer. For instance, the current contents of the Results window could be printed by typing

```
. print @Results
```

`translate` translates log and SMCL files from one format to another, the other typically being suitable for printing. `translate` can also translate SMCL logs (logs created by typing, say, `log using mylog`) to ASCII text:

```
. translate mylog.smcl mylog.log
```

You can use `translate` to recover a log when you have forgotten to start one. You may type

```
. translate @Results mylog.txt
```

to capture as ASCII text what is currently shown in the Results window. Unix(console) users cannot do this.

This entry provides a general overview of `print` and `translate` and covers in detail the printing and translation of text (nongraphic) files.

# Options for print

`like(`*ext*`)` specifies how the file should be translated to a form suitable for printing. The default is to determine the translation method based on the extension of *filename*. Thus `mylog.smcl` is translated according to the rule for translating `smcl` files, `myfile.txt` is translated according to the rule for translating `txt` files, and so on. (These rules are, in fact, `translate`'s `smcl2prn` and `txt2prn` translators, but put that aside for the moment.)

Rules for the following extensions are predefined:

| | |
|---|---|
| `.txt` | Assume input file contains ASCII text |
| `.log` | Assume input file contains Stata log ASCII text |
| `.smcl` | Assume input file contains SMCL |

To print a file that has an extension different from those listed above, you can define a new extension, but you do not have to do that. Assume that you wish to print the file `read.me`, which you know to contain ASCII text. If you were just to type `print read.me`, you would be told that Stata cannot translate `.me` files. (You would literally be told that the translator for `me2prn` was not found.) You could type `print read.me, like(txt)` to tell `print` to print `read.me` like a `.txt` file.

On the other hand, you could type

```
. transmap define .me .txt
```

to tell Stata that `.me` files are always to be treated like `.txt` files. If you did that, Stata would remember the new rule, even in future sessions.

When you specify option `like()`, you override the recorded rules. So, if you were to type `print mylog.smcl, like(txt)`, the file would be printed as ASCII text (meaning all the SMCL commands would show).

`name(`*windowname*`)` specifies which window to print when printing a Viewer. The default is for Stata to print the topmost Viewer (Unix(GUI) users: see the second technical note in *Printing files, Unix*). The `name()` option is ignored when printing the Results window.

The window name is located inside parentheses in the window title. For example, if the title for a Viewer window is *Viewer (#1) [help print]*, the name for the window is `#1`.

*override_options* refers to `translate`'s override options. `print` uses `translate` to translate the file into a format suitable for sending to the printer. To find out what you can override for *X* files (files ending in .*X*), type `translator query` *X*`2prn`. For instance, to find out what you can override when printing SMCL files, type

```
. translator query smcl2prn
  (output omitted)
```

In the omitted output (which varies across operating systems), you might learn that there is an `rmargin` *#* tunable value, which specifies the right margin in inches. You could specify the *override_option* `rmargin(`*#*`)` to temporarily override the default value, or you could type `translator set smcl2prn rmargin` *#* beforehand to permanently reset the value.

Alternatively, on some computers with some translators, you might discover that nothing can be set.

## Options for translate

`translator(`*tname*`)` specifies the name of the translator to be used to translate the file. The available translators are

| *tname* | input | output |
|---|---|---|
| `smcl2ps` | SMCL | PostScript |
| `log2ps` | Stata ASCII text log | PostScript |
| `txt2ps` | generic ASCII text file | PostScript |
| `Viewer2ps` | Viewer window | PostScript |
| `Results2ps` | Results window | PostScript |
| `smcl2prn` | SMCL | default printer format |
| `log2prn` | Stata ASCII text log | default printer format |
| `txt2prn` | generic ASCII text log | default printer format |
| `Results2prn` | Results window | default printer format |
| `Viewer2prn` | Viewer window | default printer format |
| `smcl2txt` | SMCL | generic ASCII text file |
| `smcl2log` | SMCL | Stata ASCII text log |
| `Results2txt` | Results window | generic ASCII text file |
| `Viewer2txt` | Viewer window | generic ASCII text file |
| `smcl2pdf` | SMCL | PDF (Macintosh only) |
| `log2pdf` | Stata ASCII text log | PDF (Macintosh only) |
| `txt2pdf` | generic ASCII text log | PDF (Macintosh only) |
| `Results2pdf` | Results window | PDF (Macintosh only) |
| `Viewer2pdf` | Viewer window | PDF (Macintosh only) |

If `translator()` is not specified, `translate` determines which translator to use based on extensions of the filenames specified. Typing `translate myfile.smcl myfile.ps` would use the `smcl2ps` translator. Typing `translate myfile.smcl myfile.ps, translate(smcl2prn)` would override the default and use the `smcl2prn` translator.

Actually, when you type `translate` *a*.*b* *c*.*d*, `translate` looks up .*b* in the `transmap` extension-synonym table. If .*b* is not found, the translator *b*2*d* is used. If .*b* is found in the table, the mapped extension is used (call it $b'$), and then the translator $b'2d$ is used. For example,

| command | translator used |
|---|---|
| translate myfile.smcl myfile.ps | `smcl2ps` |
| translate myfile.odd myfile.ps | `odd2ps`, which does not exist, so error |
| transmap define .odd .txt<br>translate myfile.odd myfile.ps | `txt2ps` |

You can list the mappings that `translate` uses by typing `transmap query`.

`name(`*windowname*`)` specifies which window to translate when translating a Viewer. The default is for Stata to translate the topmost Viewer. The `name()` option is ignored when translating the Results window.

The window name is located inside parentheses in the window title. For example, if the title for a Viewer window is *Viewer (#1) [help print]*, the name for the window is `#1`.

*override_options* overrides any of the default options of the specified or implied translator. To find out what you can override for, say, `log2ps`, type

```
. translator query log2ps
  (output omitted)
```

In the omitted output, you might learn that there is an `rmargin #` tunable value, which, in the case of `log2ps`, specifies the right margin in inches. You could specify the *override_option* `rmargin(#)` to temporarily override the default value or type `translator set log2ps rmargin #` beforehand to permanently reset the value.

`replace` specifies that *filename*$_{\text{out}}$ be replaced if it already exists.

# Remarks

Remarks below are presented under the headings

*Printing files*
*Printing files, Windows and Macintosh*
*Printing files, Unix*
*Translating files from one format to another*

## Printing files

Printing should be easy; just type

```
. print mylog.smcl
. print mylog.log
```

You can use `print` to print SMCL files, ASCII text files, and even the contents of the Results and Viewer windows:

```
. print @Results
. print @Viewer
. print @Viewer, name(#2)
```

For information about printing and translating graph files, see [G] **graph print** and see [G] **graph export**.

## Printing files, Windows and Macintosh

When you type `print`, you are using the same facility that you would be using if you had selected **Print** from the **File** menu. If you try to print a file that Stata does not know about, Stata will complain:

```
. print read.me
translator me2prn not found
(perhaps you need to specify the like() option)
r(111);
```

In that case, you could type

```
. print read.me, like(txt)
```

to indicate that you wanted `read.me` sent to the printer in the same fashion as if the file were named `readme.txt`, or you could type

```
. transmap define .me .txt
. print read.me
```

In this case, you are telling Stata once and for all that you want files ending in `.me` to be treated in the same way as files ending in `.txt`. Stata will remember this mapping, even across sessions. To clear the `.me` mapping, type

```
. transmap define .me
```

To see all the mappings, type

```
. transmap query
```

To print to a file, use the `translate` command, not `print`:

```
. translate mylog.smcl mylog.prn
```

`translate` prints to a file using the Windows print driver when the new filename ends in `.prn`. Under Macintosh, the `prn` translators are the same as the `pdf` translators. We suggest that you simply use the `.pdf` file extension when printing to a file.

## Printing files, Unix

Stata assumes that you have a PostScript printer attached to your Unix computer and that the Unix command `lpr(1)` can be used to send PostScript files to it, but you can change this. On your Unix system, typing

```
mycomputer$ lpr < filename
```

may not be sufficient to print PostScript files. For instance, perhaps on your system you would need to type

```
mycomputer$ lpr -Plexmark < filename
```

or

```
mycomputer$ lpr -Plexmark filename
```

or something else. To set the print command to be `lpr -Plexmark` *filename* and to state that the printer expects to receive PostScript files, type

```
. printer define prn ps "lpr -Plexmark @"
```

To set the print command to `lpr -Plexmark` < *filename* and to state that the printer expects to receive ASCII text files, type

```
. printer define prn txt "lpr -Plexmark < @"
```

That is, just type the command necessary to send files to your printer and include an `@` sign where the filename should be substituted. Two file formats are available: `ps` and `txt`. The default setting, as shipped from the factory, is

```
. printer define prn ps "lpr < @"
```

We will return to the `printer` command in the technical note that follows because it has some other capabilities you should know about.

In any case, after you redefine the default printer, the following should just work:

```
. print mylog.smcl
. print mylog.log
```

If you try to print a file that Stata does not know about, it will complain:

```
. print read.me
translator me2prn not found
r(111);
```

In that case, you could type

```
. print read.me, like(txt)
```

to indicate that you wanted **read.me** sent to the printer in the same fashion as if the file were named **readme.txt**, or you could type

```
. transmap define .me .txt
. print read.me
```

In this case, you are telling Stata once and for all that you want files ending in `.me` to be treated in the same way as files ending in `.txt`. Stata will remember this setting for `.me`, even across sessions.

If you want to clear the `.me` setting, type

```
. transmap define .me
```

If you want to see all your settings, type

```
. transmap query
```

## ❑ Technical Note

The syntax of the `printer` command is

`printer` `define` *printername* [ { `ps` | `txt` } "*Unix command with @*" ]

`printer` `query` [*printername*]

You may define multiple printers. By default, `print` uses the printer named `prn`, but `print` has the syntax

`print` *filename* [ , `like(`*ext*`)` `printer(`*printername*`)` *override_options* ]

so, if you define multiple printers, you may route your output to them.

For instance, if you have a second printer on your system, you might type

```
. printer define lexmark ps "lpr -Plexmark < @"
```

After doing that, you could type

```
. print myfile.smcl, printer(lexmark)
```

Any printers that you set will be remembered even across sessions. You can delete printers:

```
. printer define lexmark
```

You can list all the defined printers by typing `printer query`, and you can list the definition of a particular printer, say, `prn`, by typing `printer query prn`.

The default printer `prn` we have predefined for you is

```
. printer define prn ps "lpr < @"
```

meaning that we assume it is a PostScript printer and that the Unix command `lpr(1)`, without options, is sufficient to cause files to print. Feel free to change the default definition. If you change it, the change will be remembered across sessions. ❑

## ❑ Technical Note

Unix(GUI) users should note that X-Windows does not have the concept of a window z-order, which prevents Stata from determining which window is the topmost window. Instead, Stata determines which window is topmost based on which window has the focus. However, some window managers will set the focus to a window without bringing the window to the top. What Stata considers the topmost window may not appear topmost visually. For this reason, you should always use the `name()` option to ensure the correct window is printed. ❑

## ❑ Technical Note

When you select the Results window to print from the **Print** menu or toolbar button, the result is exactly the same as if you were to issue the `print` command. When you select a Viewer window to print from the **Print** menu or toolbar button, the result is exactly the same as if you were to issue the `print` command with a `name()` option. ❑

The translation to PostScript format is done by `translate` and, in particular, is performed by the translators `smcl2ps`, `log2ps`, and `txt2ps`. There are lots of tunable parameters in each of these translators. You can display the current values of these tunable parameters for, say, `smcl2ps` by typing

```
. translator query smcl2ps
  (output omitted)
```

and you can set any of the tunable parameters (for instance, setting `smcl2ps`'s `rmargin` value to 1) by typing

```
. translator set smcl2ps rmargin 1
  (output omitted)
```

Any settings you make will be remembered across sessions. You can reset `smcl2ps` to be as it was when Stata was shipped by typing

```
. translator reset smcl2ps
```

## Translating files from one format to another

If you have a SMCL log, which you might have created by previously typing `log using mylog`, you can translate it to an ASCII text log by typing

```
. translate myfile.smcl myfile.log
```

and you can translate it to a PostScript file by typing

```
. translate myfile.smcl myfile.ps
```

`translate` translates files from one format to another, and, in fact, `print` uses `translate` to produce a file suitable for sending to the printer.

When you type

```
. translate a.b c.d
```

`translate` looks for the predefined translator *b*2*d* and uses that to perform the translation. If there is a `transmap` synonym for *b*, however, the mapped value $b'$ is used: $b'2d$.

Only certain translators exist, and they are listed under the description of the `translate()` option in *Options for translate* above, or you can type

```
. translator query
```

for a complete (and perhaps more up-to-date) list.

Anyway, `translate` forms the name *b*2*d* or $b'2d$, and if the translator does not exist, `translate` issues an error. With the `translator()` option, you can specify exactly which translator to use, and, in that case, it does not matter how your files are named.

The only other thing to know is that some translators have tunable parameters that affect how they perform their translation. You can type

```
. translator query translator_name
```

to find out what those parameters are. Some translators have no tunable parameters, and some have many:

```
. translator query smcl2ps
```

```
-------------------------------------------------------------------------
                header | on
            headertext |
                  logo | on
                  user |
           projecttext |
             cmdnumber | on
-------------------------------------------------------------------------
              fontsize | 9                            lmargin | 1.00
              pagesize | letter                       rmargin | 1.00
             pagewidth | 8.50                         tmargin | 1.00
            pageheight | 11.00                        bmargin | 1.00
-------------------------------------------------------------------------
                scheme | monochrome
-------------------------------------------------------------------------
     cust1_result_color |   0   0   0    cust2_result_color |   0   0   0
   cust1_standard_color |   0   0   0  cust2_standard_color |   0   0   0
      cust1_error_color |   0   0   0     cust2_error_color | 255   0   0
      cust1_input_color |   0   0   0     cust2_input_color |   0   0   0
       cust1_link_color |   0   0   0      cust2_link_color |   0   0 255
     cust1_hilite_color |   0   0   0    cust2_hilite_color |   0   0   0
      cust1_result_bold | on                cust2_result_bold | on
    cust1_standard_bold | off             cust2_standard_bold | off
```

| | | | |
|---|---|---|---|
| cust1_error_bold | on | cust2_error_bold | on |
| cust1_input_bold | off | cust2_input_bold | off |
| cust1_link_bold | off | cust2_link_bold | off |
| cust1_hilite_bold | on | cust2_hilite_bold | on |
| cust1_link_underline | on | cust2_link_underline | on |
| cust1_hilite_underline | off | cust2_hilite_underline | off |

You can temporarily override any setting by specifying option *setopt*(*setval*) on the `translate` (or `print`) command. For instance, you can type

```
. translate ..., ... cmdnumber(off)
```

or you can reset the value permanently by typing

```
. translator set smcl2ps setopt setval
```

For instance,

```
. translator set smcl2ps cmdnumber off
```

If you reset a value, Stata will remember the change, even in future sessions.

Windows and Macintosh users: The `smcl2ps` (and the other `*2ps` translators) are not used by `print`, even when you have a PostScript printer attached to your computer. Instead, the Windows or Macintosh print driver is used. Resetting `smcl2ps` values will not affect printing; instead, you change the defaults in the Printers Control Panel in Windows and by selecting **Page Setup...** from the **File** menu in Macintosh. You can, however, `translate` files yourself using the `smcl2ps` translator and the other `*2ps` translators.

## Saved Results

`transmap query` .*ext* saves in macro `r(suffix)` the mapped extension (without the leading period) or saves *ext* if the *ext* is not mapped.

`translator query` *translatorname* saves *setval* in macro `r(`*setopt*`)` for every *setopt*, *setval* pair.

`printer query` *printername* (Unix only) saves in macro `r(suffix)` the "filetype" of the input that the printer expects (currently "`ps`" or "`txt`") and, in macro `r(command)`, the command to send output to the printer.

## Also See

**Related:** [R] **log**, [G] **graph export**, [G] **graph print**, [P] **smcl**

**Complementary:** [R] **view**

**Background:** [U] **15 Printing and preserving output**

# Title

**treatreg** — Treatment-effects model

# Syntax

*Basic syntax*

treatreg *depvar* [*indepvars*], treat(*depvar_t* = *indepvars_t*) [twostep]

*Full syntax for maximum likelihood estimates only*

treatreg *depvar* [*indepvars*] [*if*] [*in*] [*weight*] ,
treat(*depvar_t* = *indepvars_t* [, noconstant]) [*treatreg_ml_options*]

*Full syntax for two-step consistent estimates only*

treatreg *depvar* [*indepvars*] [*if*] [*in*] ,
treat(*depvar_t* = *indepvars_t* [, noconstant]) twostep [*treatreg_ts_options*]

| *treatreg_ml_options* | description |
|---|---|
| Model | |
| noconstant | suppress (treatment) constant term |
| constraints(*constraints*) | apply specified linear constraints |
| SE/Robust | |
| vce(*vcetype*) | *vcetype* may be oim, robust, opg, bootstrap, or jackknife |
| robust | synonym for vce(robust) |
| cluster(*varname*) | adjust standard errors for intragroup correlation |
| Reporting | |
| level(#) | set confidence level; default is level(95) |
| first | report first-step probit estimates |
| noskip | perform likelihood-ratio test |
| hazard(*newvar*) | create *newvar* containing hazard from treatment equation |
| Max options | |
| *maximize_options* | control maximization process; seldom used |

| *treatreg_ts_options* | description |
|---|---|
| Model | |
| noconstant | suppress (treatment) constant term |
| SE/Robust | |
| vce(*vcetype*) | *vcetype* may be bootstrap or jackknife |
| Reporting | |
| level(*#*) | set confidence level; default is level(95) |
| first | report first-step probit estimates |
| hazard(*newvar*) | create *newvar* containing hazard from treatment equation |

*depvar*, *indepvars*, *depvar*$_t$, and *indepvars*$_t$ may contain time-series operators; see [U] **11.4.3 Time-series varlists**.

bootstrap, by, jackknife, rolling, statsby, and xi are allowed; see [U] **11.1.10 Prefix commands**.

pweights, aweights, fweights, and iweights are allowed with maximum likelihood estimation; see [U] **11.1.6 weight**. No weights are allowed if twostep is specified.

See [U] **20 Estimation and postestimation commands** for additional capabilities of estimation commands.

# Description

treatreg fits a treatment-effects model using either a two-step consistent estimator or full maximum likelihood. The treatment-effects model considers the effect of an endogenously chosen binary treatment on another endogenous continuous variable, conditional on two sets of independent variables.

# Options

Model

treat(*depvar*$_t$ = *indepvars*$_t$ [ , noconstant ]) specifies the variables and options for the treatment equation. It is an integral part of specifying a treatment-effects model and is required.

twostep specifies that two-step consistent estimates of the parameters, standard errors, and covariance matrix be produced, instead of the default maximum likelihood estimates.

noconstant, constraints(*constraints*); see [R] **estimation options**.

SE/Robust

vce(*vcetype*); see [R] ***vce_option***.

robust, cluster(*varname*); see [R] **estimation options**.

Reporting

level(*#*); see [R] **estimation options**.

first specifies that the first-step probit estimates of the treatment equation be displayed before estimation.

`noskip` specifies that a full maximum likelihood model with only a constant for the regression equation be fitted. This model is not displayed but is used as the base model to compute a likelihood-ratio test for the model test statistic displayed in the estimation header. By default, the overall model test statistic is an asymptotically equivalent Wald test that all the parameters in the regression equation are zero (except the constant). For many models, this option can substantially increase estimation time.

`hazard(`*newvar*`)` will create a new variable containing the hazard from the treatment equation. The hazard is computed from the estimated parameters of the treatment equation.

Max options

*maximize_options*: `difficult`, `technique(`*algorithm_spec*`)`, `iterate(`*#*`)`, `[no]log`, `trace`, `gradient`, `showstep`, `hessian`, `shownrtolerance`, `tolerance(`*#*`)`, `ltolerance(`*#*`)`, `gtolerance(`*#*`)`, `nrtolerance(`*#*`)`, `nonrtolerance`, `from(`*init_specs*`)`; see [R] **maximize**. These options are seldom used.

## Remarks

The treatment-effects model estimates the effect of an endogenous binary treatment, $z_j$, on a continuous, fully-observed variable $y_j$, conditional on the independent variables $x_j$ and $w_j$. The primary interest is in the regression function

$$y_j = \mathbf{x}_j\boldsymbol{\beta} + \delta z_j + \epsilon_j$$

where $z_j$ is an endogenous dummy variable indicating whether the treatment is assigned or not. The binary decision to obtain the treatment $z_j$ is modeled as the outcome of an unobserved latent variable, $z_j^*$. It is assumed that $z_j^*$ is a linear function of the exogenous covariates $\mathbf{w}_j$ and a random component $u_j$. Specifically,

$$z_j^* = \mathbf{w}_j\boldsymbol{\gamma} + u_j$$

and the observed decision is

$$z_j = \begin{cases} 1, & \text{if } z_j^* > 0 \\ 0, & \text{otherwise} \end{cases}$$

where $\epsilon$ and $u$ are bivariate normal with mean zero and covariance matrix

$$\begin{bmatrix} \sigma & \rho \\ \rho & 1 \end{bmatrix}$$

There are many variations of this model in the literature. Maddala (1983) derives the maximum likelihood and two-step estimators of the version implemented here and also gives a brief review of several empirical applications of this model. Barnow et al. (1981) provide another useful derivation of this model. Barnow et al. (1981) concentrate on deriving the conditions for which the self-selection bias of the simple OLS estimator of the treatment effect, $\delta$, is nonzero and of a specific sign.

### ▷ Example 1

We will illustrate `treatreg` using a subset of the Mroz dataset distributed with Berndt (1991). This dataset contains 753 observations on women's labor supply. Our subsample is of 250 observations, with 150 market laborers and 100 nonmarket laborers.

```
. use http://www.stata-press.com/data/r9/labor
. describe
Contains data from http://www.stata-press.com/data/r9/labor.dta
  obs:           250
 vars:            15                          18 Apr 2005 05:01
 size:        16,000 (98.3% of memory free)

              storage  display     value
variable name   type   format      label      variable label

lfp             float  %9.0g                  1 if woman worked in 1975
whrs            float  %9.0g                  wife's hours of work
kl6             float  %9.0g                  # of children younger than 6
k618            float  %9.0g                  # of children between 6 and 18
wa              float  %9.0g                  wife's age
we              float  %9.0g                  wife's education attainment
ww              float  %9.0g                  wife's wage
hhrs            float  %9.0g                  husband's hours worked in 1975
ha              float  %9.0g                  husband's age
he              float  %9.0g                  husband's educational attainment
hw              float  %9.0g                  husband's wage
faminc          float  %9.0g                  family income
wmed            float  %9.0g                  wife's mother's educational
                                                attainment
wfed            float  %9.0g                  wife's father's educational
                                                attainment
cit             float  %9.0g                  1 if live in large city

Sorted by:
. summarize
    Variable |       Obs        Mean    Std. Dev.       Min        Max
-------------+--------------------------------------------------------
         lfp |       250          .6    .4908807          0          1
        whrs |       250      799.84    915.6035          0       4950
         kl6 |       250        .236    .5112234          0          3
        k618 |       250       1.364    1.370774          0          8
          wa |       250       42.92    8.426483         30         60
-------------+--------------------------------------------------------
          we |       250      12.352    2.164912          5         17
          ww |       250     2.27523     2.59775          0     14.631
        hhrs |       250    2234.832    600.6702        768       5010
          ha |       250      45.024    8.171322         30         60
          he |       250      12.536    3.106009          3         17
-------------+--------------------------------------------------------
          hw |       250    7.494435    4.636192     1.0898     40.509
      faminc |       250    23062.54    12923.98       3305      91044
        wmed |       250       9.136    3.536031          0         17
        wfed |       250       8.608    3.751082          0         17
         cit |       250        .624    .4853517          0          1
```

We will assume that the wife went to college if her educational attainment is more than 12 years. Let wc be the dummy variable indicating whether the individual went to college. With this definition, our sample contains the following distribution of college education:

```
. generate wc = 0
. replace wc = 1 if we > 12
(69 real changes made)
```

```
. tab wc

         wc |      Freq.     Percent        Cum.
------------+-----------------------------------
          0 |        181       72.40       72.40
          1 |         69       27.60      100.00
------------+-----------------------------------
      Total |        250      100.00
```

We will model the wife's wage as a function of her age, whether the family was living in a big city, and whether she went to college. An ordinary least-squares estimation produces the following results:

```
. regress ww wa cit wc

      Source |       SS       df       MS              Number of obs =     250
-------------+------------------------------           F(  3,   246) =    4.82
       Model |  93.2398568     3  31.0799523           Prob > F      =  0.0028
    Residual |  1587.08776   246  6.45157627           R-squared     =  0.0555
-------------+------------------------------           Adj R-squared =  0.0440
       Total |  1680.32762   249  6.74830369           Root MSE      =    2.54

------------------------------------------------------------------------------
          ww |      Coef.   Std. Err.      t    P>|t|     [95% Conf. Interval]
-------------+----------------------------------------------------------------
          wa |  -.0104985   .0192667    -0.54   0.586    -.0484472    .0274502
         cit |   .1278922   .3389058     0.38   0.706    -.5396351    .7954195
          wc |   1.332192   .3644344     3.66   0.000     .6143819    2.050001
       _cons |   2.278337   .8432385     2.70   0.007     .6174488    3.939225
------------------------------------------------------------------------------
```

Is 1.332 a consistent estimate of the marginal effect of a college education on wages? If individuals choose whether or not to attend college and the error term of the model that gives rise to this choice is correlated with the error term in the wage equation, then the answer is no. (See Barnow et al. 1981 for a good discussion of the existence and sign of selectivity bias.) We might suspect that individuals with higher abilities, either innate or due to the circumstances of their birth, would be more likely to go to college and to earn higher wages. Such ability is, of course, unobserved. Furthermore, if the error term in our model for going to college is correlated with ability, and the error term in our wage equation is correlated with ability, the two terms should be positively correlated. These conditions make the problem of signing the selectivity bias equivalent to an omitted-variable problem. In the case at hand, because we would expect the correlation between the omitted variable and a college education to be positive, we suspect that OLS is biased upward.

To account for the bias, we fit the treatment-effects model. We model the wife's college decision as a function of her mother's and her father's educational attainment. Thus we are interested in fitting the model

$$\texttt{ww} = \beta_0 + \beta_1\texttt{wa} + \beta_2\texttt{cit} + \delta\texttt{wc} + \epsilon$$

$$wc^* = \gamma_0 + \gamma_1\texttt{wmed} + \gamma_2\texttt{wfed} + u$$

where

$$\texttt{wc} = \begin{cases} 1, & wc^* > 0, \text{ i.e., wife went to college} \\ 0, & \text{otherwise} \end{cases}$$

and where $\epsilon$ and $u$ have a bivariate normal distribution with zero mean and covariance matrix

$$\begin{bmatrix} \sigma & \rho \\ \rho & 1 \end{bmatrix}$$

The following output gives the maximum likelihood estimates of the parameters of this model:

```
. treatreg ww wa cit, treat(wc=wmed wfed)
Iteration 0:   log likelihood = -707.07237
Iteration 1:   log likelihood = -707.07215
Iteration 2:   log likelihood = -707.07215
Treatment-effects model -- MLE                  Number of obs      =       250
                                                Wald chi2(3)       =      4.11
Log likelihood = -707.07215                     Prob > chi2        =    0.2501

------------------------------------------------------------------------------
             |      Coef.   Std. Err.      z    P>|z|     [95% Conf. Interval]
-------------+----------------------------------------------------------------
ww           |
          wa |  -.0110424   .0199652    -0.55   0.580    -.0501735    .0280887
         cit |    .127636   .3361938     0.38   0.704    -.5312917    .7865638
          wc |   1.271327   .7412951     1.72   0.086    -.1815842    2.724239
       _cons |   2.318638   .9397573     2.47   0.014     .4767478    4.160529
-------------+----------------------------------------------------------------
wc           |
        wmed |   .1198055   .0320056     3.74   0.000     .0570757    .1825352
        wfed |   .0961886   .0290868     3.31   0.001     .0391795    .1531977
       _cons |  -2.631876   .3309128    -7.95   0.000    -3.280453   -1.983299
-------------+----------------------------------------------------------------
     /athrho |   .0178668   .1899898     0.09   0.925    -.3545063    .3902399
    /lnsigma |   .9241584   .0447455    20.65   0.000     .8364588    1.011858
-------------+----------------------------------------------------------------
         rho |   .0178649   .1899291                     -.3403659     .371567
       sigma |   2.519747   .1127473                      2.308179    2.750707
      lambda |   .0450149   .4786442                     -.8931105    .9831404
------------------------------------------------------------------------------
LR test of indep. eqns. (rho = 0):   chi2(1) =     0.01   Prob > chi2 = 0.9251
------------------------------------------------------------------------------
```

In the input, we specified that the continuous dependent variable, `ww` (wife's wage), is a linear function of `cit` and `wa`. Note the syntax for the treatment variable. The treatment `wc` is not included in the first variable list; it is specified in the `treat()` option. In this example, `wmed` and `wfed` are specified as the exogenous covariates in the treatment equation.

The output has the form of many two-equation estimators in Stata. We note that our conjecture that the OLS estimate was biased upward is verified. But it is perhaps more interesting that the size of the bias is negligible, and the likelihood-ratio test at the bottom of the output indicates that we cannot reject the null hypothesis that the two error terms are uncorrelated. This result might be due to several specification errors. We ignored the selectivity bias due to the endogeneity of entering the labor market. We have also written both the wage equation and the college-education equation in linear form, ignoring any higher power terms or interactions.

The results for the two ancillary parameters require explanation. For numerical stability during optimization, `treatreg` does not directly estimate $\rho$ or $\sigma$. Instead, `treatreg` estimates the inverse hyperbolic tangent of $\rho$,

$$\operatorname{atanh}\rho = \frac{1}{2}\ln\left(\frac{1+\rho}{1-\rho}\right)$$

and $\ln\sigma$. Also, `treatreg` reports $\lambda = \rho\sigma$, along with an estimate of the standard error of the estimate and confidence interval.

◁

## ▷ Example 2

Stata also produces a two-step estimator of the model with the `twostep` option. Maximum likelihood estimation of the parameters can be time consuming with large datasets, and the two-step estimates may provide a good alternative in such cases. Continuing with the women's wage model, we can obtain the two-step estimates with consistent covariance estimates by typing

```
. treatreg ww wa cit, treat(wc=wmed wfed) twostep
Treatment-effects model -- two-step estimates   Number of obs      =       250
                                                Wald chi2(3)       =      3.67
                                                Prob > chi2        =    0.2998
```

| | Coef. | Std. Err. | z | P>\|z\| | [95% Conf. | Interval] |
|---|---|---|---|---|---|---|
| ww | | | | | | |
| wa | -.0111623 | .020152 | -0.55 | 0.580 | -.0506594 | .0283348 |
| cit | .1276102 | .33619 | 0.38 | 0.704 | -.53131 | .7865305 |
| wc | 1.257995 | .8007428 | 1.57 | 0.116 | -.3114319 | 2.827422 |
| _cons | 2.327482 | .9610271 | 2.42 | 0.015 | .4439031 | 4.21106 |
| wc | | | | | | |
| wmed | .1198888 | .0319859 | 3.75 | 0.000 | .0571976 | .1825801 |
| wfed | .0960764 | .0290581 | 3.31 | 0.001 | .0391236 | .1530292 |
| _cons | -2.631496 | .3308344 | -7.95 | 0.000 | -3.279919 | -1.983072 |
| hazard | | | | | | |
| lambda | .0548738 | .5283928 | 0.10 | 0.917 | -.9807571 | 1.090505 |
| rho | 0.02178 | | | | | |
| sigma | 2.5198211 | | | | | |
| lambda | .05487379 | .5283928 | | | | |

The reported `lambda` ($\lambda$) is the parameter estimate on the hazard from the augmented regression, which is derived in Maddala (1983) and presented in *Methods and Formulas* below.

◁

## ❑ Technical Note

The difference in expected earnings between participants and nonparticipants is

$$E\left(y_j \mid z_j = 1\right) - E\left(y_j \mid z_j = 0\right) = \delta + \rho\sigma \left[ \frac{\phi(\mathbf{w}_j\gamma)}{\Phi(\mathbf{w}_j\gamma)\{1 - \Phi(\mathbf{w}_j\gamma)\}} \right]$$

where $\phi$ is the standard normal density and $\Phi$ is the standard normal cumulative distribution function. If the correlation between the error terms, $\rho$, is zero, the problem reduces to one estimable by OLS and the difference is simply $\delta$. Since $\rho$ is positive in the example, least squares overestimates the treatment effect, $\delta$.

❑

## Saved Results

treatreg saves in e():

Scalars

| | | | |
|---|---|---|---|
| e(N) | number of observations | e(chi2) | $\chi^2$ |
| e(k) | number of parameters | e(chi2_c) | $\chi^2$ for comparison test |
| e(k_eq) | number of equations | e(p_c) | $p$-value for comparison test |
| e(k_dv) | number of dependent variables | e(p) | significance |
| e(df_m) | model degrees of freedom | e(rho) | $\rho$ |
| e(ll) | log likelihood | e(rank) | rank of e(V) |
| e(N_clust) | number of clusters | e(ic) | number of iterations |
| e(lambda) | $\lambda$ | e(rc) | return code |
| e(selambda) | standard error of $\lambda$ | e(converged) | 1 if converged, 0 otherwise |
| e(sigma) | estimate of sigma | | |

Macros

| | | | |
|---|---|---|---|
| e(cmd) | treatreg | e(vce) | *vcetype* specified in vce() |
| e(depvar) | name of dependent variable | e(vcetype) | title used to label Std. Err. |
| e(hazard) | variable containing hazard | e(opt) | type of optimization |
| e(wtype) | weight type | e(method) | ml or twostep |
| e(wexp) | weight expression | e(ml_method) | type of ml method |
| e(title) | title in estimation output | e(user) | name of likelihood-evaluator program |
| e(clustvar) | name of cluster variable | e(technique) | maximization technique |
| e(chi2type) | Wald or LR; type of model $\chi^2$ test | e(crittype) | optimization criterion |
| e(chi2_ct) | Wald or LR; type of model $\chi^2$ test corresponding to e(chi2_c) | e(properties) | b V |
| | | e(predict) | program used to implement predict |

Matrices

| | | | |
|---|---|---|---|
| e(b) | coefficient vector | e(V) | variance–covariance matrix of the estimators |
| e(ilog) | iteration log (up to 20 iterations) | | |
| e(gradient) | gradient vector | | |

Functions

| | |
|---|---|
| e(sample) | marks estimation sample |

## Methods and Formulas

treatreg is implemented as an ado-file. Maddala (1983, 117–122) derives both the maximum likelihood and the two-step estimator implemented here. Greene (2003, 787–789) also provides an introduction to the treatment-effects model.

The primary regression equation of interest is

$$y_j = \mathbf{x}_j\boldsymbol{\beta} + \delta z_j + \epsilon_j$$

where $z_j$ is a binary decision variable that is assumed to stem from an unobservable latent variable:

$$z_j^* = \mathbf{w}_j\boldsymbol{\gamma} + u_j$$

The decision to obtain the treatment is made according to the rule

$$z_j = \begin{cases} 1, & \text{if } z_j^* > 0 \\ 0, & \text{otherwise} \end{cases}$$

where $\epsilon$ and $u$ are bivariate normal with mean zero and covariance matrix

$$\begin{bmatrix} \sigma & \rho \\ \rho & 1 \end{bmatrix}$$

The likelihood function for this model is given in Maddala (1983, 122). Greene (2000, 180) discusses the standard method of reducing a bivariate normal to a function of a univariate normal and the correlation $\rho$. The following is the log likelihood for observation $j$,

$$\ln L_j = \begin{cases} \ln \Phi \left\{ \dfrac{\mathbf{w}_j\boldsymbol{\gamma} + (y_j - \mathbf{x}_j\boldsymbol{\beta} - \delta)\rho/\sigma}{\sqrt{1-\rho^2}} \right\} - \dfrac{1}{2}\left( \dfrac{y_j - \mathbf{x}_j\boldsymbol{\beta} - \delta}{\sigma} \right)^2 - \ln(\sqrt{2\pi}\sigma), & z_j = 1 \\ \ln \Phi \left\{ \dfrac{-\mathbf{w}_j\boldsymbol{\gamma} - (y_j - \mathbf{x}_j\boldsymbol{\beta})\rho/\sigma}{\sqrt{1-\rho^2}} \right\} - \dfrac{1}{2}\left( \dfrac{y_j - \mathbf{x}_j\boldsymbol{\beta}}{\sigma} \right)^2 - \ln(\sqrt{2\pi}\sigma), & z_j = 0 \end{cases}$$

where $\Phi()$ is the cumulative distribution function of the standard normal distribution.

In the maximum likelihood estimation, $\sigma$ and $\rho$ are not directly estimated. Rather $\ln\sigma$ and $\operatorname{atanh}\rho$ are directly estimated, where

$$\operatorname{atanh}\rho = \frac{1}{2}\ln\left(\frac{1+\rho}{1-\rho}\right)$$

The standard error of $\lambda = \rho\sigma$ is approximated through the delta method, which is given by

$$\operatorname{Var}(\lambda) \approx \mathbf{D}\operatorname{Var}\big\{(\operatorname{atanh}\rho \quad \ln\sigma)\big\}\,\mathbf{D}'$$

where $\mathbf{D}$ is the Jacobian of $\lambda$ with respect to $\operatorname{atanh}\rho$ and $\ln\sigma$.

Maddala (1983, 120–122) also derives the two-step estimator. In the first stage, probit estimates are obtained of the treatment equation

$$\Pr(z_j = 1 \mid \mathbf{w}_j) = \Phi(\mathbf{w}_j\boldsymbol{\gamma})$$

From these estimates, the hazard, $h_j$, for each observation $j$ is computed as

$$h_j = \begin{cases} \phi(\mathbf{w}_j\widehat{\boldsymbol{\gamma}})/\Phi(\mathbf{w}_j\widehat{\boldsymbol{\gamma}}), & z_j = 1 \\ -\phi(\mathbf{w}_j\widehat{\boldsymbol{\gamma}})/\big\{1 - \Phi(\mathbf{w}_j\widehat{\boldsymbol{\gamma}})\big\}, & z_j = 0 \end{cases}$$

where $\phi$ is the standard normal density function. If

$$d_j = h_j(h_j + \widehat{\boldsymbol{\gamma}}\mathbf{w}_j)$$

then

$$E\,(y_j \mid z_j) = \mathbf{x}_j\boldsymbol{\beta} + \delta\mathbf{z}_j + \rho\sigma h_j$$
$$\operatorname{Var}\,(y_j \mid z_j) = \sigma^2\left(1 - \rho^2 d_j\right)$$

The two-step parameter estimates of $\boldsymbol{\beta}$ and $\delta$ are obtained by augmenting the regression equation with the hazard $h$. Thus the regressors become $[\,\mathbf{x}\ \mathbf{z}\ h\,]$, and the additional parameter estimate $\beta_h$ is obtained on the variable containing the hazard. A consistent estimate of the regression disturbance variance is obtained using the residuals from the augmented regression and the parameter estimate on the hazard

$$\widehat{\sigma}^{\,2} = \frac{\mathbf{e}'\mathbf{e} + \beta_h^2 \sum_{j=1}^N d_j}{N}$$

The two-step estimate of $\rho$ is then

$$\widehat{\rho} = \frac{\beta_h}{\widehat{\sigma}}$$

To understand how the consistent estimates of the coefficient covariance matrix based on the augmented regression are derived, let $\mathbf{A} = [\,\mathbf{x}\ \mathbf{z}\ h\,]$ and $\mathbf{D}$ be a square diagonal matrix of size $N$ with $(1 - \widehat{\rho}^{\,2} d_j)$ on the diagonal elements.

$$\mathbf{V}_{\text{twostep}} = \widehat{\sigma}^{\,2}(\mathbf{A}'\mathbf{A})^{-1}(\mathbf{A}'\mathbf{D}\mathbf{A} + \mathbf{Q})(\mathbf{A}'\mathbf{A})^{-1}$$

where

$$\mathbf{Q} = \widehat{\rho}^{\,2}(\mathbf{A}'\mathbf{D}\mathbf{A})\mathbf{V}_{\mathbf{p}}(\mathbf{A}'\mathbf{D}\mathbf{A})$$

and $\mathbf{V}_{\mathbf{p}}$ is the variance–covariance estimate from the probit estimation of the treatment equation.

## References


Barnow, B., G. Cain, and A. Goldberger. 1981. Issues in the analysis of selectivity bias. *Evaluation Studies Review Annual* 5. Beverly Hills: Sage.

Berndt, E. R. 1991. *The Practice of Econometrics.* New York: Addison–Wesley.

Cong, R. and D. M. Drukker. 2000. sg141: Treatment-effects model. *Stata Technical Bulletin* 55: 25–33. Reprinted in *Stata Technical Bulletin Reprints*, vol. 10, pp. 159–169.

Greene, W. H. 2000. *Econometric Analysis.* 4th ed. Upper Saddle River, NJ: Prentice Hall.

——. 2003. *Econometric Analysis.* 5th ed. Upper Saddle River, NJ: Prentice Hall.

Maddala, G. S. 1983. *Limited-dependent and Qualitative Variables in Econometrics.* Cambridge: Cambridge University Press.


## Also See


**Complementary:** [R] **treatreg postestimation**; [R] **constraint**

**Related:** [R] **heckman**, [R] **probit**

**Background:** [U] **11.1.10 Prefix commands**,
[U] **20 Estimation and postestimation commands**,
[R] **estimation options**, [R] **maximize**, [R] ***vce_option***

# Title

**treatreg postestimation** — Postestimation tools for treatreg

# Description

The following postestimation commands are available for `treatreg`:

| command | description |
|---|---|
| `adjust`[1] | adjusted predictions of $\mathbf{x}\beta$ |
| *`estat` | AIC, BIC, VCE, and estimation sample summary |
| `estimates` | cataloging estimation results |
| `lincom` | point estimates, standard errors, testing, and inference for linear combinations of coefficients |
| `lrtest` | likelihood-ratio test |
| `mfx` | marginal effects or elasticities |
| `nlcom` | point estimates, standard errors, testing, and inference for nonlinear combinations of coefficients |
| `predict` | predictions, residuals, influence statistics, and other diagnostic measures |
| `predictnl` | point estimates, standard errors, testing, and inference for generalized predictions |
| *`suest` | seemingly unrelated estimation |
| `test` | Wald tests for simple and composite linear hypotheses |
| `testnl` | Wald tests of nonlinear hypotheses |

[1] `adjust` does not work with time-series operators.

*`estat ic` and `suest` may not be used after `treatreg, twostep`.

See the corresponding entries in the *Stata Base Reference Manual* for details.

# Syntax for predict

*After ML or twostep*

`predict` [*type*] *newvar* [*if*] [*in*] [, *statistic*]

*After ML*

`predict` [*type*] { *stub*`*` | $newvar_{\text{reg}}$ $newvar_{\text{treat}}$ $newvar_{\text{athrho}}$ $newvar_{\text{lnsigma}}$ }

[*if*] [*in*] , `scores`

| *statistic* | description |
|---|---|
| xb | $\mathbf{x}_j\mathbf{b}$, linear prediction; the default |
| stdp | standard error of the prediction |
| stdf | standard error of the forecast |
| yctrt | $E(y_j \mid \text{treatment} = 1)$ |
| ycntrt | $E(y_j \mid \text{treatment} = 0)$ |
| ptrt | $\Pr(\text{treatment} = 1)$ |
| xbtrt | linear prediction for treatment equation |
| stdptrt | standard error of the linear prediction for treatment equation |

These statistics are available both in and out of sample; type `predict ... if e(sample) ...` if wanted only for the estimation sample.

## Options for predict

`xb`, the default, calculates the linear prediction, $\mathbf{x}_j\mathbf{b}$.

`stdp` calculates the standard error of the prediction, which can be thought of as the standard error of the predicted expected value or mean for the observation's covariate pattern. This is also referred to as the standard error of the fitted value.

`stdf` calculates the standard error of the forecast, which is the standard error of the point prediction for a single observation. It is commonly referred to as the standard error of the future or forecast value. By construction, the standard errors produced by `stdf` are always larger than those produced by `stdp`; see [R] **regress postestimation** *Methods and Formulas*.

`yctrt` calculates the expected value of the dependent variable conditional on the presence of the treatment: $E(y_j \mid \text{treatment} = 1)$.

`ycntrt` calculates the expected value of the dependent variable conditional on the absence of the treatment: $E(y_j \mid \text{treatment} = 0)$.

`ptrt` calculates the probability of the presence of the treatment:
$\Pr(\text{treatment} = 1) = \Pr(\mathbf{w}_j\boldsymbol{\gamma} + u_j > 0)$.

`xbtrt` calculates the linear prediction for the treatment equation.

`stdptrt` calculates the standard error of the linear prediction for the treatment equation.

`scores`, not available with `twostep`, calculates equation-level score variables.

The first new variable will contain $\partial \ln L/\partial(\boldsymbol{x}_j\boldsymbol{\beta})$.

The second new variable will contain $\partial \ln L/\partial(\boldsymbol{w}_j\boldsymbol{\gamma})$.

The third new variable will contain $\partial \ln L/\partial \text{ atanh } \rho$.

The fourth new variable will contain $\partial \ln L/\partial \ln \sigma$.

(*Continued on next page*)

# Remarks

▷ Example 1

The default statistic produced by `predict` after `treatreg` is the expected value of the dependent variable from the underlying distribution of the regression model. For example 1 in [R] **treatreg**, this model is

$$\texttt{ww} = \beta_0 + \beta_1\texttt{wa} + \beta_2\texttt{cit} + \delta\texttt{wc} + \epsilon$$

Several other interesting aspects of the treatment effects model can be explored with `predict`. Continuing with our wage model, the wife's expected wage, conditional on attending college, can be obtained with the `yctrt` option. The wife's expected wages, conditional on not attending college, can be obtained with the `ycntrt` option. Thus the difference in expected wages between participants and nonparticipants is the difference between `yctrt` and `ycntrt`. For the case at hand, we have the following calculation:

```
. predict wwctrt, yctrt
. predict wwcntrt, ycntrt
. generate diff = wwctrt - wwcntrt
. summarize diff
    Variable |       Obs        Mean    Std. Dev.       Min        Max
-------------+--------------------------------------------------------
        diff |       250    1.356912    .0134202    1.34558   1.420173
```

◁

# Methods and Formulas

All postestimation commands listed above are implemented as ado-files.

# Also See

**Complementary:** [R] **treatreg**; [R] **adjust**, [R] **estimates**, [R] **lincom**, [R] **lrtest**, [R] **mfx**, [R] **nlcom**, [R] **predictnl**, [R] **suest**, [R] **test**, [R] **testnl**

**Background:** [U] **13.5 Accessing coefficients and standard errors**, [U] **20 Estimation and postestimation commands**, [R] **estat**, [R] **predict**

# Title

**truncreg** — Truncated regression

# Syntax

`truncreg` *depvar* [*indepvars*] [*if*] [*in*] [*weight*] [, *options*]

| *options* | description |
|---|---|
| Model | |
| `noconstant` | suppress constant term |
| `ll(`*varname* \| *#*`)` | lower limit for left truncation |
| `ul(`*varname* \| *#*`)` | upper limit for right truncation |
| Model 2 | |
| `offset(`*varname*`)` | include *varname* in model with coefficient constrained to 1 |
| `marginal` | estimate the marginal effects |
| `at(`*matname*`)` | point at which to estimate marginal effect |
| `constraints(`*constraints*`)` | apply specified linear constraints |
| SE/Robust | |
| `vce(`*vcetype*`)` | *vcetype* may be `oim`, `robust`, `opg`, `bootstrap`, or `jackknife` |
| `robust` | synonym for `vce(robust)` |
| `cluster(`*varname*`)` | adjust standard errors for intragroup correlation |
| Reporting | |
| `level(`*#*`)` | set confidence level; default is `level(95)` |
| `noskip` | perform likelihood-ratio test |
| Max options | |
| *maximize_options* | control maximization process; seldom used |

*depvar* and *indepvars* may contain time-series operators; see [U] **11.4.3 Time-series varlists**.
`bootstrap`, `by`, `jackknife`, `rolling`, `statsby`, and `xi` are allowed; see [U] **11.1.10 Prefix commands**.
`aweight`s, `fweight`s, and `pweight`s are allowed; see [U] **11.1.6 weight**.
See [U] **20 Estimation and postestimation commands** for additional capabilities of estimation commands.

# Description

`truncreg` fits a regression model of *depvar* on *varlist* from a sample drawn from a restricted part of the population. Under the normality assumption for the whole population, the error terms in the truncated regression model have a truncated normal distribution, which is a normal distribution that has been scaled upward so that the distribution integrates to one over the restricted range.

(*Continued on next page*)

# Options

Model

`noconstant`; see [R] **estimation options**.

`ll(`*varname* | *#*`)` and `ul(`*varname* | *#*`)` indicate the lower and upper limits for truncation, respectively. You may specify one or both. Observations with *depvar* ≤ `ll()` are left-truncated, observations with *depvar* ≥ `ul()` are right-truncated, and the remaining observations are not truncated. See [R] **tobit** for a more detailed description.

Model 2

`offset(`*varname*`)`; see [R] **estimation options**.

`marginal` estimates the marginal effects in the subpopulation. You may specify `marginal` when you fit the model or redisplay results.

`at(`*matname*`)` specifies the point at which to estimate the marginal effects. The default is to estimate the effects at the means of the independent variables, where the means are computed over the subpopulation of observations that are not truncated. If there are $k$ independent variables, *matname* should be $1 \times k$. You can specify `at()` when you fit the model or redisplay the results.

`at()` implies `marginal`; specifying `marginal at()` is equivalent to typing `at()` by itself.

`constraints(`*constraints*`)`; see [R] **estimation options**.

SE/Robust

`vce(`*vcetype*`)`; see [R] ***vce_option***.

`robust`, `cluster(`*varname*`)`; see [R] **estimation options**.

Reporting

`level(`*#*`)`; see [R] **estimation options**.

`noskip` specifies that a full maximum-likelihood model with only a constant for the regression equation be fitted. This model is not displayed, but is used as the base model to compute a likelihood-ratio test for the model test statistic displayed in the estimation header. By default, the overall model test statistic is an asymptotically equivalent Wald test of all the parameters in the regression equation being zero (except the constant). For many models, this option can substantially increase estimation time.

Max options

*maximize_options*: `difficult`, `technique(`*algorithm_spec*`)`, `iterate(`*#*`)`, [`no`]`log`, `trace`, `gradient`, `showstep`, `hessian`, `shownrtolerance`, `tolerance(`*#*`)`, `ltolerance(`*#*`)`, `gtolerance(`*#*`)`, `nrtolerance(`*#*`)`, `nonrtolerance`, `from(`*init_specs*`)`; see [R] **maximize**. These options are seldom used, but you may use the `ltol(`*#*`)` option to relax the convergence criterion; the default is `1e-6` during specification searches.

# Remarks

Remarks are presented under the headings



## Introduction

Truncated regression fits a model of a dependent variable on independent variables from a restricted part of a population. Truncation is essentially a characteristic of the distribution from which the sample data are drawn. If $x$ has a normal distribution with mean $\mu$ and standard deviation $\sigma$, the density of the truncated normal distribution is

$$f\left(x \mid a < x < b\right) = \frac{f(x)}{\Phi\left(\frac{b-\mu}{\sigma}\right) - \Phi\left(\frac{a-\mu}{\sigma}\right)} = \frac{\frac{1}{\sigma}\phi\left(\frac{x-\mu}{\sigma}\right)}{\Phi\left(\frac{b-\mu}{\sigma}\right) - \Phi\left(\frac{a-\mu}{\sigma}\right)}$$

where $\phi$ and $\Phi$ are the density and distribution functions of the standard normal distribution.

Compared with the mean of the untruncated variable, the mean of the truncated variable is greater if the truncation is from below, and the mean of the truncated variable is smaller if the truncation is from above. Moreover, truncation reduces the variance compared with the variance in the untruncated distribution.

### ▷ Example 1

We will demonstrate `truncreg` using a subset of the Mroz dataset distributed with Berndt (1991). This dataset contains 753 observations on women's labor supply. Our subsample is of 250 observations, with 150 market laborers and 100 nonmarket laborers.

```
. use http://www.stata-press.com/data/r9/laborsub
. describe
Contains data from http://www.stata-press.com/data/r9/laborsub.dta
  obs:           250
 vars:             6                          25 Sep 2004 18:36
 size:         2,750 (99.6% of memory free)
-------------------------------------------------------------------------------
              storage  display     value
variable name   type   format      label      variable label
-------------------------------------------------------------------------------
lfp             byte   %9.0g                  1 if woman worked in 1975
whrs            int    %9.0g                  Wife's hours of work
kl6             byte   %9.0g                  # of children younger than 6
k618            byte   %9.0g                  # of children between 6 and 18
wa              byte   %9.0g                  Wife's age
we              byte   %9.0g                  Wife's educational attainment
-------------------------------------------------------------------------------
Sorted by:
```

```
. summarize, sep(0)
    Variable |       Obs        Mean    Std. Dev.       Min        Max
-------------+--------------------------------------------------------
         lfp |       250          .6    .4908807          0          1
        whrs |       250      799.84    915.6035          0       4950
         kl6 |       250        .236    .5112234          0          3
        k618 |       250       1.364    1.370774          0          8
          wa |       250       42.92    8.426483         30         60
          we |       250      12.352    2.164912          5         17
```

We first perform ordinary least-squares estimation on the market laborers.

```
. regress whrs kl6 k618 wa we if whrs >0
      Source |       SS       df       MS              Number of obs =     150
-------------+------------------------------           F(  4,   145) =    2.80
       Model |  7326995.15     4  1831748.79           Prob > F      =  0.0281
    Residual |  94793104.2   145  653745.546           R-squared     =  0.0717
-------------+------------------------------           Adj R-squared =  0.0461
       Total |   102120099   149  685369.794           Root MSE      =  808.55

------------------------------------------------------------------------------
        whrs |      Coef.   Std. Err.      t    P>|t|     [95% Conf. Interval]
-------------+----------------------------------------------------------------
         kl6 |  -421.4822   167.9734    -2.51   0.013    -753.4748   -89.48953
        k618 |  -104.4571   54.18616    -1.93   0.056    -211.5538    2.639666
          wa |  -4.784917   9.690502    -0.49   0.622     -23.9378    14.36797
          we |   9.353195   31.23793     0.30   0.765    -52.38731     71.0937
       _cons |   1629.817   615.1301     2.65   0.009     414.0371    2845.597
------------------------------------------------------------------------------
```

Now we use `truncreg` to perform truncated regression with truncation from below zero.

```
. truncreg whrs kl6 k618 wa we, ll(0)
(note: 100 obs. truncated)
Fitting full model:
Iteration 0:   log likelihood = -1205.6992
Iteration 1:   log likelihood = -1200.9873
Iteration 2:   log likelihood = -1200.9159
Iteration 3:   log likelihood = -1200.9157
Iteration 4:   log likelihood = -1200.9157
Truncated regression
Limit:   lower =          0                     Number of obs =    150
         upper =       +inf                     Wald chi2(4)  =  10.05
Log likelihood = -1200.9157                     Prob > chi2   = 0.0395

------------------------------------------------------------------------------
        whrs |      Coef.   Std. Err.      z    P>|z|     [95% Conf. Interval]
-------------+----------------------------------------------------------------
eq1          |
         kl6 |  -803.0042   321.3614    -2.50   0.012    -1432.861   -173.1474
        k618 |   -172.875   88.72898    -1.95   0.051    -346.7806    1.030579
          wa |  -8.821123   14.36848    -0.61   0.539    -36.98283    19.34059
          we |   16.52873   46.50375     0.36   0.722    -74.61695    107.6744
       _cons |    1586.26    912.355     1.74   0.082    -201.9233    3374.442
-------------+----------------------------------------------------------------
sigma        |
       _cons |   983.7262   94.44303    10.42   0.000     798.6213    1168.831
------------------------------------------------------------------------------
```

If we assume that our data were censored, the tobit model is

```
. tobit whrs kl6 k618 wa we, ll(0)
Tobit regression                                  Number of obs   =        250
                                                  LR chi2(4)      =      23.03
                                                  Prob > chi2     =     0.0001
Log likelihood = -1367.0903                       Pseudo R2       =     0.0084

------------------------------------------------------------------------------
        whrs |      Coef.   Std. Err.      t    P>|t|     [95% Conf. Interval]
-------------+----------------------------------------------------------------
         kl6 |  -827.7657   214.7407    -3.85   0.000     -1250.731   -404.8008
        k618 |  -140.0192   74.22303    -1.89   0.060     -286.2129    6.174547
          wa |  -24.97919   13.25639    -1.88   0.061     -51.08969    1.131317
          we |   103.6896   41.82393     2.48   0.014      21.31093    186.0683
       _cons |   589.0001   841.5467     0.70   0.485     -1068.556    2246.556
-------------+----------------------------------------------------------------
      /sigma |   1309.909   82.73335                      1146.953    1472.865
------------------------------------------------------------------------------
  Obs. summary:        100  left-censored observations at whrs<=0
                       150     uncensored observations
                         0 right-censored observations
```

◁

❑ Technical Note

Whether truncated regression is more appropriate than the ordinary least-squares estimation depends on the purpose of that estimation. If we are interested in the mean of wife's working hours conditional on the subsample of market laborers, least-squares estimation is appropriate. However if we are interested in the mean of wife's working hours regardless of market or nonmarket labor status, least-squares estimates could be seriously misleading.

Truncation and censoring are different concepts. A sample has been censored if no observations have been systematically excluded but some of the information contained in them has been suppressed. In a truncated distribution, only the part of the distribution above (or below, or between) the truncation points is relevant to our computations. We need to scale it up by the probability that an observation falls in the range that interests us to make the distribution integrate to one. The censored distribution used by tobit, however, is a mixture of discrete and continuous distributions. Instead of rescaling over the observable range, we simply assign the full probability from the censored regions to the censoring points. The truncated regression model is sometimes less well behaved than the tobit model. Davidson and MacKinnon (1993) provide an example where truncation results in more inconsistency than censoring.

❑

## Marginal effects

You can obtain the marginal effects in the truncation model in the subpopulation by specifying the `marginal` option. `at(`*matname*`)` specifies the points at which to estimate the marginal effects. The default is to estimate the effects around the means of the independent variables, where the means are based on the subpopulation of observations that are not truncated.

▷ Example 2

The marginal effects around the mean of the independent variables conditional on being in the subpopulation for our previous truncated regression are given by

```
. truncreg whrs kl6 k618 wa we, marginal ll(0) nolog
(note: 100 obs. truncated)

Marginal effects of truncated regression
Prediction: E(whrs | whrs > 0)
Limit:    lower =          0                     Number of obs =     150
          upper =       +inf                     Wald chi2(4)  =   10.05
Log likelihood = -1200.9157                      Prob > chi2   = 0.0395

------------------------------------------------------------------------------
    whrs |      dF/dx   Std. Err.      z    P>|z|      X_at   [    95% C.I.   ]
---------+--------------------------------------------------------------------
     kl6 |  -521.9979    199.4016   -2.62   0.009   .173333  -912.818 -131.178
    k618 |  -112.3785    56.48681   -1.99   0.047   1.31333  -223.091 -1.66638
      wa |  -5.734226    9.319849   -0.62   0.538   42.7867  -24.0008  12.5323
      we |    10.7446     30.2098    0.36   0.722     12.64  -48.4655  69.9547
------------------------------------------------------------------------------
```

We can obtain the marginal effects around the means of the independent variables of the entire population by using the `at()` option. Notice that we can specify the `marginal` and the `at()` options when results are redisplayed.

```
. mat accum junk = kl6 k618 wa we, means(B) nocons
(obs=250)

. truncreg, marginal at(B)

Marginal effects of truncated regression
Prediction: E(whrs | whrs > 0)
Limit:    lower =          0                     Number of obs =     150
          upper =       +inf                     Wald chi2(4)  =   10.05
Log likelihood = -1200.9157                      Prob > chi2   = 0.0395

------------------------------------------------------------------------------
    whrs |      dF/dx   Std. Err.      z    P>|z|      X_at   [    95% C.I.   ]
---------+--------------------------------------------------------------------
     kl6 |  -506.3126    188.0831   -2.69   0.007      .236  -874.949 -137.676
    k618 |  -109.0017    54.42745   -2.00   0.045     1.364  -215.677 -2.32583
      wa |   -5.56192    9.010092   -0.62   0.537     42.92  -23.2214  12.0975
      we |   10.42174    29.22635    0.36   0.721    12.352  -46.8608  67.7043
------------------------------------------------------------------------------
```

◁

## ❑ Technical Note

Whether the marginal effect or the coefficient itself is of interest depends on the purpose of the study. If the analysis is confined to the subpopulation, then marginal effects are of interest. Otherwise, it is the coefficients themselves that are of interest.

❑

## Saved Results

`truncreg` saves in `e()`:

Scalars

| | | | |
|---|---|---|---|
| `e(N)` | number of observations | `e(sigma)` | estimate of sigma |
| `e(N_bf)` | number of obs. before truncation | `e(p)` | significance |
| `e(chi2)` | model $\chi^2$ | `e(rank)` | rank of `e(V)` |
| `e(df_m)` | model degrees of freedom | `e(ic)` | number of iterations |
| `e(ll)` | log likelihood | `e(rc)` | return code |
| `e(ll_0)` | log likelihood, constant-only model | `e(converged)` | `1` if converged, `0` otherwise |
| `e(N_clust)` | number of clusters | | |

Macros

| | | | |
|---|---|---|---|
| `e(cmd)` | `truncreg` | `e(vcetype)` | title used to label Std. Err. |
| `e(llopt)` | contents of `ll()`, if specified | `e(opt)` | type of optimization |
| `e(ulopt)` | contents of `ul()`, if specified | `e(ml_method)` | type of `ml` method |
| `e(depvar)` | name of dependent variable | `e(chi2type)` | `Wald` or `LR`; type of model $\chi^2$ test |
| `e(wtype)` | weight type | `e(user)` | name of likelihood-evaluator program |
| `e(wexp)` | weight expression | `e(technique)` | maximization technique |
| `e(title)` | title in estimation output | `e(crittype)` | optimization criterion |
| `e(clustvar)` | name of cluster variable | `e(properties)` | `b V` |
| `e(offset)` | offset | `e(predict)` | program used to implement `predict` |
| `e(vce)` | *vcetype* specified in `vce()` | | |

Matrices

| | | | |
|---|---|---|---|
| `e(b)` | coefficient vector | `e(dfdx)` | marginal effects |
| `e(V)` | variance–covariance matrix of the estimators | `e(means)` | means of independent variables |
| | | `e(at)` | points for calculating marginal effects |
| `e(V_dfdx)` | variance–covariance matrix of the marginal effects | `e(ilog)` | iteration log (up to 20 iterations) |
| | | `e(dummy)` | indicator for dummy variables |

Functions

| | |
|---|---|
| `e(sample)` | marks estimation sample |

## Methods and Formulas

`truncreg` is implemented as an ado-file. Greene (2003, 756–761) and Davidson and MacKinnon (1993, 534–537) provide introductions to the truncated regression model.

Let $\mathbf{y} = \mathbf{X}\boldsymbol{\beta} + \epsilon$ be the model. $\mathbf{y}$ represents continuous outcomes either observed or not observed. Our model assumes that $\epsilon \sim N(\mathbf{0}, \sigma^2\mathbf{I})$.

Let $a$ be the lower limit and $b$ be the upper limit. The log likelihood is

$$\ln L = -\frac{n}{2}\log(2\pi\sigma^2) - \frac{1}{2\sigma^2}\sum_{j=1}^{n}(y_j - \mathbf{x}_j\boldsymbol{\beta})^2 - \sum_{j=1}^{n}\log\left\{\Phi\left(\frac{b-\mathbf{x}_j\boldsymbol{\beta}}{\sigma}\right) - \Phi\left(\frac{a-\mathbf{x}_j\boldsymbol{\beta}}{\sigma}\right)\right\}$$

The marginal effects at $\mathbf{x}_j$ with truncation from below are

$$\frac{\partial E\left(y_j \mid y_j > a\right)}{\partial \mathbf{x}_j} = \boldsymbol{\beta}\left(1 - \lambda_j^2 + \alpha_j\lambda_j\right)$$

where

$$\alpha_j = (a - \mathbf{x}_j\boldsymbol{\beta})/\sigma$$
$$\lambda_j = \phi(\alpha_j)/\{1 - \Phi(\alpha_j)\}$$

The marginal effects at $\mathbf{x}_j$ with truncation from above are

$$\frac{\partial E(y_j \mid y_j < b)}{\partial \mathbf{x}_j} = \boldsymbol{\beta}\left(1 - \lambda_j^2 + \alpha_j\lambda_j\right)$$

where

$$\alpha_j = (b - \mathbf{x}_j\boldsymbol{\beta})/\sigma$$
$$\lambda_j = -\phi(\alpha_j)/\Phi(\alpha_j)$$

The marginal effects at $\mathbf{x}_j$ with truncation from both above and below are

$$\frac{\partial E[y_j \mid a < y_j < b]}{\partial \mathbf{x}_j} = \boldsymbol{\beta}\left[1 - \lambda_j^2 - \alpha_{j2}\lambda_j - \frac{(b-a)\phi(\alpha_{j1})}{\sigma\{\Phi(\alpha_{j2}) - \Phi(\alpha_{j1})\}}\right]$$

where

$$\alpha_{j1} = (a - \mathbf{x}_j\boldsymbol{\beta})/\sigma$$
$$\alpha_{j2} = (b - \mathbf{x}_j\boldsymbol{\beta})/\sigma$$
$$\lambda_j = \frac{\phi(\alpha_{j2}) - \phi(\alpha_{j1})}{\Phi(\alpha_{j2}) - \Phi(\alpha_{j1})}$$

## References


Berndt, E. R. 1991. *The Practice of Econometrics.* New York: Addison–Wesley.

Cong, R. 1999. sg122: Truncated regression. *Stata Technical Bulletin* 52: 47–52. Reprinted in *Stata Technical Bulletin Reprints*, vol. 9, pp. 248–255.

Davidson, R. and J. G. MacKinnon. 1993. *Estimation and Inference in Econometrics.* New York: Oxford University Press.

Greene, W. H. 2003. *Econometric Analysis.* 5th ed. Upper Saddle River, NJ: Prentice Hall.


## Also See

**Complementary:** [R] **truncreg postestimation**; [R] **constraint**

**Related:** [R] **tobit**

**Background:** [U] **11.1.10 Prefix commands**,
[U] **20 Estimation and postestimation commands**,
[R] **estimation options**, [R] **maximize**, [R] ***vce_option***

# Title

**truncreg postestimation** — Postestimation tools for truncreg

# Description

The following postestimation commands are available for `truncreg`:

| command | description |
|---|---|
| `adjust`[1] | adjusted predictions of $\mathbf{x}\beta$ |
| `estat` | AIC, BIC, VCE, and estimation sample summary |
| `estimates` | cataloging estimation results |
| `lincom` | point estimates, standard errors, testing, and inference for linear combinations of coefficients |
| `lrtest` | likelihood-ratio test |
| `mfx` | marginal effects or elasticities |
| `nlcom` | point estimates, standard errors, testing, and inference for nonlinear combinations of coefficients |
| `predict` | predictions, residuals, influence statistics, and other diagnostic measures |
| `predictnl` | point estimates, standard errors, testing, and inference for generalized predictions |
| `suest` | seemingly unrelated estimation |
| `test` | Wald tests for simple and composite linear hypotheses |
| `testnl` | Wald tests of nonlinear hypotheses |

[1] `adjust` does not work with time-series operators.

See the corresponding entries in the *Stata Base Reference Manual* for details.

# Syntax for predict

`predict` [*type*] *newvar* [*if*] [*in*] [, *statistic* `nooffset`]

`predict` [*type*] { *stub** | $newvar_{\text{reg}}$ $newvar_{\text{lnsigma}}$ } [*if*] [*in*] , `scores`

| *options* | description |
|---|---|
| `xb` | $\mathbf{x}_j\mathbf{b}$, linear prediction; the default |
| `stdp` | standard error of the prediction |
| `stdf` | standard error of the forecast |
| `pr(`*a*`,`*b*`)` | $\Pr(a < y_j < b)$ |
| `e(`*a*`,`*b*`)` | $E(y_j \mid a < y_j < b)$ |
| `ystar(`*a*`,`*b*`)` | $E(y_j^*)$, $y_j^* = \max\{a, \min(y_j, b)\}$ |

These statistics are available both in and out of sample; type `predict ... if e(sample) ...` if wanted only for the estimation sample.

where *a* and *b* may be numbers or variables; *a* missing ($a \geq .$) means $-\infty$, and *b* missing ($b \geq .$) means $+\infty$; see [U] **12.2.1 Missing values**.

# Options for predict

xb, the default, calculates the linear prediction.

stdp calculates the standard error of the prediction, which can be thought of as the standard error of the predicted expected value or mean for the observation's covariate pattern. This is also referred to as the standard error of the fitted value.

stdf calculates the standard error of the forecast, which is the standard error of the point prediction for a single observation. It is commonly referred to as the standard error of the future or forecast value. By construction, the standard errors produced by stdf are always larger than those produced by stdp; see [R] **regress** *Methods and Formulas*.

pr(*a*,*b*) calculates $\Pr(a < \mathbf{x}_j\mathbf{b} + u_j < b)$, the probability that $y_j|\mathbf{x}_j$ would be observed in the interval $(a, b)$.

*a* and *b* may be specified as numbers or variable names; *lb* and *ub* are variable names;
pr(20,30) calculates $\Pr(20 < \mathbf{x}_j\mathbf{b} + u_j < 30)$;
pr(*lb*,*ub*) calculates $\Pr(lb < \mathbf{x}_j\mathbf{b} + u_j < ub)$; and
pr(20,*ub*) calculates $\Pr(20 < \mathbf{x}_j\mathbf{b} + u_j < ub)$.

*a* missing ($a \geq .$) means $-\infty$; pr(.,30) calculates $\Pr(-\infty < \mathbf{x}_j\mathbf{b} + u_j < 30)$;
pr(*lb*,30) calculates $\Pr(-\infty < \mathbf{x}_j\mathbf{b} + u_j < 30)$ in observations for which $lb \geq .$
and calculates $\Pr(lb < \mathbf{x}_j\mathbf{b} + u_j < 30)$ elsewhere.

*b* missing ($b \geq .$) means $+\infty$; pr(20,.) calculates $\Pr(+\infty > \mathbf{x}_j\mathbf{b} + u_j > 20)$;
pr(20,*ub*) calculates $\Pr(+\infty > \mathbf{x}_j\mathbf{b} + u_j > 20)$ in observations for which $ub \geq .$
and calculates $\Pr(20 < \mathbf{x}_j\mathbf{b} + u_j < ub)$ elsewhere.

e(*a*,*b*) calculates $E(\mathbf{x}_j\mathbf{b} + u_j \mid a < \mathbf{x}_j\mathbf{b} + u_j < b)$, the expected value of $y_j|\mathbf{x}_j$ conditional on $y_j|\mathbf{x}_j$ being in the interval $(a, b)$, meaning that $y_j|\mathbf{x}_j$ is censored.
*a* and *b* are specified as they are for pr().

ystar(*a*,*b*) calculates $E(y_j^*)$, where $y_j^* = a$ if $\mathbf{x}_j\mathbf{b} + u_j \leq a$, $y_j^* = b$ if $\mathbf{x}_j\mathbf{b} + u_j \geq b$, and $y_j^* = \mathbf{x}_j\mathbf{b} + u_j$ otherwise, meaning that $y_j^*$ is truncated. *a* and *b* are specified as they are for pr().

nooffset is relevant only if you specified offset(*varname*). It modifies the calculations made by predict so that they ignore the offset variable; the linear prediction is treated as $\mathbf{x}_j\mathbf{b}$ rather than as $\mathbf{x}_j\mathbf{b} + \text{offset}_j$.

scores calculates equation-level score variables.

The first new variable will contain $\partial \ln L/\partial(\mathbf{x}_j\boldsymbol{\beta})$.

The second new variable will contain $\partial \ln L/\partial \ln \sigma$.

# Methods and Formulas

All postestimation commands listed above are implemented as ado-files.

# Also See

**Complementary:** [R] **truncreg**; [R] **adjust**, [R] **estimates**, [R] **lincom**, [R] **lrtest**, [R] **mfx**, [R] **nlcom**, [R] **predictnl**, [R] **suest**, [R] **test**, [R] **testnl**

**Background:** [U] **13.5 Accessing coefficients and standard errors**, [U] **20 Estimation and postestimation commands**, [R] **estat**, [R] **predict**

## Title

**ttest** — Mean comparison tests

## Syntax

*One-sample mean comparison test*

ttest *varname* == # [*if*] [*in*] [, level(#)]

*Two-sample mean comparison test*

ttest $varname_1$ == $varname_2$ [*if*] [*in*] [, $options_1$]

*Group mean comparison test*

ttest *varname* [*if*] [*in*], by(*groupvar*) [$options_2$]

*Immediate form of one-sample mean comparison test*

ttesti $\#_{obs}$ $\#_{mean}$ $\#_{sd}$ $\#_{val}$ [, level(#)]

*Immediate form of two-sample mean comparison test*

ttesti $\#_{obs1}$ $\#_{mean1}$ $\#_{sd1}$ $\#_{obs2}$ $\#_{mean2}$ $\#_{sd2}$ [, $options_3$]

| $options_1$ | description |
|---|---|
| Main | |
| unpaired | treat data as unpaired |
| unequal | unpaired data have unequal variances |
| welch | use Welch's approximation |
| level(#) | set confidence level; default is level(95) |

| $options_2$ | description |
|---|---|
| Main | |
| * by(*groupvar*) | variable defining the groups |
| unequal | unpaired data have unequal variances |
| welch | use Welch's approximation |
| level(#) | set confidence level; default is level(95) |

* by(*groupvar*) is required.

| *options*$_3$ | description |
|---|---|
| Main | |
| unequal | unpaired data have unequal variances |
| welch | use Welch's approximation |
| level(#) | set confidence level; default is level(95) |

## Description

ttest performs $t$ tests on the equality of means. In the first form, ttest tests that *varname* has a mean of #. In the second form, ttest tests that *varname*$_1$ and *varname*$_2$ have the same mean. Data are assumed to be paired, but specifying unpaired changes this assumption. In the third form, ttest tests that *varname* has the same mean within the two groups defined by *groupvar*.

ttesti is the immediate form of ttest; see [U] **19 Immediate commands**.

For the equivalent of a two-sample $t$ test with sampling weights (pweights), use the svy: mean command with the by() option, and then use lincom; see [SVY] **svy: mean** and [SVY] **svy postestimation**.

## Options

Main

by(*groupvar*) specifies the *groupvar* that defines the two groups that ttest will use to test the hypothesis that their means are equal. Specifying by(*groupvar*) implies an unpaired (two-sample) $t$ test. Do not confuse the by() option with the by prefix; you can specify both.

unpaired specifies that the data be treated as unpaired. The unpaired option is used when the two sets of values to be compared are in different variables.

unequal specifies that the unpaired data not be assumed to have equal variances.

welch specifies that the approximate degrees of freedom for the test be obtained from Welch's formula rather than from Satterthwaite's approximation formula (1946), which is the default when unequal is specified. Specifying welch implies specifying unequal.

level(#) specifies the confidence level, as a percentage, for confidence intervals. The default is level(95) or as set by set level; see [U] **20.6 Specifying the width of confidence intervals**.

## Remarks

### ▷ Example 1: One-sample mean comparison test

In the first form, ttest tests whether the mean of the sample is equal to a known constant under the assumption of unknown variance. Assume that we have a sample of 74 automobiles. We know each automobile's average mileage rating and wish to test whether the overall average for the sample is 20 miles per gallon.

```
. use http://www.stata-press.com/data/r9/auto
(1978 Automobile Data)
```

```
. ttest mpg==20
One-sample t test

Variable |     Obs        Mean    Std. Err.   Std. Dev.   [95% Conf. Interval]
---------+--------------------------------------------------------------------
     mpg |      74     21.2973    .6725511    5.785503     19.9569    22.63769
------------------------------------------------------------------------------
    mean = mean(mpg)                                              t =   1.9289
Ho: mean = 20                                    degrees of freedom =       73

    Ha: mean < 20               Ha: mean != 20                Ha: mean > 20
 Pr(T < t) = 0.9712        Pr(|T| > |t|) = 0.0576         Pr(T > t) = 0.0288
```

The test indicates that the underlying mean is not 20 with a significance level of 5.8%.

◁

### ▷ Example 2: Two-sample mean comparison test

We are testing the effectiveness of a new fuel additive. We run an experiment with 12 cars. We run the cars without and with the fuel treatment. The results of the experiment are as follows:

| Without Treatment | With Treatment | Without Treatment | With Treatment |
|---|---|---|---|
| 20 | 24 | 18 | 17 |
| 23 | 25 | 24 | 28 |
| 21 | 21 | 20 | 24 |
| 25 | 22 | 24 | 27 |
| 18 | 23 | 23 | 21 |
| 17 | 18 | 19 | 23 |

By creating two variables called `mpg1` and `mpg2` representing mileage without and with the treatment, respectively, we can test the equality of means by typing

```
. use http://www.stata-press.com/data/r9/fuel
. ttest mpg1==mpg2
Paired t test

Variable |     Obs        Mean    Std. Err.   Std. Dev.   [95% Conf. Interval]
---------+--------------------------------------------------------------------
    mpg1 |      12          21    .7881701    2.730301    19.26525    22.73475
    mpg2 |      12       22.75    .9384465    3.250874    20.68449    24.81551
---------+--------------------------------------------------------------------
    diff |      12       -1.75    .7797144     2.70101    -3.46614   -.0338602
------------------------------------------------------------------------------
     mean(diff) = mean(mpg1 - mpg2)                               t =  -2.2444
 Ho: mean(diff) = 0                              degrees of freedom =       11

 Ha: mean(diff) < 0           Ha: mean(diff) != 0           Ha: mean(diff) > 0
 Pr(T < t) = 0.0232         Pr(|T| > |t|) = 0.0463          Pr(T > t) = 0.9768
```

We find that the means are statistically different from each other at any level greater than 4.6%.

◁

### ▷ Example 3: Group mean comparison test

Let's pretend that the preceding data were collected by running 24 cars: 12 cars with the additive and 12 without. Although we might be tempted to enter the data in the same way, we should resist (see the technical note below). Instead, we enter the data as 24 observations on `mpg` with an additional variable, `treated`, taking on 1 if the car received the fuel treatment and 0 otherwise:

```
. use http://www.stata-press.com/data/r9/fuel3

. ttest mpg, by(treated)

Two-sample t test with equal variances
------------------------------------------------------------------------------
   Group |     Obs        Mean    Std. Err.   Std. Dev.   [95% Conf. Interval]
---------+--------------------------------------------------------------------
       0 |      12          21    .7881701    2.730301    19.26525    22.73475
       1 |      12       22.75    .9384465    3.250874    20.68449    24.81551
---------+--------------------------------------------------------------------
combined |      24      21.875    .6264476    3.068954    20.57909    23.17091
---------+--------------------------------------------------------------------
    diff |               -1.75    1.225518               -4.291568    .7915684
------------------------------------------------------------------------------
    diff = mean(0) - mean(1)                                      t =  -1.4280
Ho: diff = 0                                     degrees of freedom =       22

    Ha: diff < 0                 Ha: diff != 0                 Ha: diff > 0
 Pr(T < t) = 0.0837         Pr(|T| > |t|) = 0.1673          Pr(T > t) = 0.9163
```

This time we do not find a statistically significant difference.

If we were not willing to assume that the variances were equal and wanted to use Welch's formula, we could type

```
. ttest mpg, by(treated) welch

Two-sample t test with unequal variances
------------------------------------------------------------------------------
   Group |     Obs        Mean    Std. Err.   Std. Dev.   [95% Conf. Interval]
---------+--------------------------------------------------------------------
       0 |      12          21    .7881701    2.730301    19.26525    22.73475
       1 |      12       22.75    .9384465    3.250874    20.68449    24.81551
---------+--------------------------------------------------------------------
combined |      24      21.875    .6264476    3.068954    20.57909    23.17091
---------+--------------------------------------------------------------------
    diff |               -1.75    1.225518                -4.28369    .7836902
------------------------------------------------------------------------------
    diff = mean(0) - mean(1)                                      t =  -1.4280
Ho: diff = 0                             Welch's degrees of freedom =  23.2465

    Ha: diff < 0                 Ha: diff != 0                 Ha: diff > 0
 Pr(T < t) = 0.0833         Pr(|T| > |t|) = 0.1666          Pr(T > t) = 0.9167
```

◁

### ❑ Technical Note

In two-group randomized designs, subjects will sometimes refuse the assigned treatment but still be measured for an outcome. In this case, take care to specify the group properly. You might be tempted to let *varname* contain missing where the subject refused, and thus let `ttest` drop such observations from the analysis. Zelen (1979) argues that it would be better to specify that the subject belongs to the group in which he or she was randomized, even though such inclusion will dilute the measured effect.

❑

## ❑ Technical Note

There is a second, inferior way to organize the data in the preceding example. Remember, we ran a test on 24 cars, 12 without the additive and 12 with. Nevertheless, we could have entered the data in the same way as we did when we had 12 cars, each run without and with the additive, by creating two variables—mpg1 and mpg2.

This is inferior because it suggests a connection that is not there. In the case of the 12-car experiment, there was most certainly a connection—it was the same car. In the 24-car experiment, however, it is arbitrary which mpg results appear next to which. Nevertheless, if our data are organized like this, ttest can accommodate us.

```
. use http://www.stata-press.com/data/r9/fuel
. ttest mpg1==mpg2, unpaired
Two-sample t test with equal variances
------------------------------------------------------------------------------
Variable |     Obs        Mean    Std. Err.   Std. Dev.   [95% Conf. Interval]
---------+--------------------------------------------------------------------
    mpg1 |      12          21    .7881701    2.730301    19.26525    22.73475
    mpg2 |      12       22.75    .9384465    3.250874    20.68449    24.81551
---------+--------------------------------------------------------------------
combined |      24      21.875    .6264476    3.068954    20.57909    23.17091
---------+--------------------------------------------------------------------
    diff |               -1.75    1.225518               -4.291568    .7915684
------------------------------------------------------------------------------
    diff = mean(mpg1) - mean(mpg2)                                t =  -1.4280
Ho: diff = 0                                     degrees of freedom =       22

    Ha: diff < 0                 Ha: diff != 0                 Ha: diff > 0
 Pr(T < t) = 0.0837         Pr(|T| > |t|) = 0.1673          Pr(T > t) = 0.9163
```

❑

## ▷ Example 4

ttest can be used to test the equality of a pair of means; see [R] **oneway** for testing the equality of more than two means.

Suppose that we have data on the 50 states. The dataset contains the median age of the population (medage) and the region of the country (region) for each state. Region 1 refers to the Northeast, region 2 to the North Central, region 3 to the South, and region 4 to the West. Using oneway, we can test the equality of all four means.

```
. use http://www.stata-press.com/data/r9/census
(1980 Census data by state)
. oneway medage region
                        Analysis of Variance
    Source              SS         df      MS            F     Prob > F
------------------------------------------------------------------------
Between groups      46.3961903      3   15.4653968      7.56     0.0003
 Within groups      94.1237947     46   2.04616945
------------------------------------------------------------------------
    Total           140.519985     49    2.8677548
Bartlett's test for equal variances:  chi2(3) =  10.5757  Prob>chi2 = 0.014
```

We find that the means are different, but we are only interested in testing whether the means for the East (region==1) and West (region==4) are different. We could use oneway,

```
. oneway medage region if region==1 | region==4
                        Analysis of Variance
    Source              SS         df      MS            F     Prob > F
------------------------------------------------------------------------
Between groups      46.241247      1   46.241247      20.02     0.0002
 Within groups     46.1969169     20   2.30984584
------------------------------------------------------------------------
    Total          92.4381638     21   4.40181733
Bartlett's test for equal variances:  chi2(1) =   2.4679  Prob>chi2 = 0.116
```

or we could use `ttest`:

```
. ttest medage if region==1 | region==4, by(region)
Two-sample t test with equal variances
------------------------------------------------------------------------------
   Group |     Obs        Mean    Std. Err.   Std. Dev.   [95% Conf. Interval]
---------+--------------------------------------------------------------------
      NE |       9    31.23333    .3411581    1.023474    30.44662    32.02005
    West |      13    28.28462    .4923577    1.775221    27.21186    29.35737
---------+--------------------------------------------------------------------
combined |      22    29.49091    .4473059    2.098051    28.56069    30.42113
---------+--------------------------------------------------------------------
    diff |            2.948718    .6590372                 1.57399    4.323445
------------------------------------------------------------------------------
    diff = mean(NE) - mean(West)                                  t =   4.4743
Ho: diff = 0                                     degrees of freedom =       20

    Ha: diff < 0                 Ha: diff != 0                 Ha: diff > 0
 Pr(T < t) = 0.9999         Pr(|T| > |t|) = 0.0002          Pr(T > t) = 0.0001
```

Note that the significance levels of both tests are the same.

◁

## Immediate form

### ▷ Example 5

`ttesti` is like `ttest`, except that we specify summary statistics rather than variables as arguments. For instance, we are reading an article which reports the mean number of sunspots per month as 62.6 with a standard deviation of 15.8. There are 24 months of data. We wish to test whether the mean is 75:

```
. ttesti 24 62.6 15.8 75
One-sample t test
------------------------------------------------------------------------------
         |     Obs        Mean    Std. Err.   Std. Dev.   [95% Conf. Interval]
---------+--------------------------------------------------------------------
       x |      24        62.6    3.225161        15.8    55.92825    69.27175
------------------------------------------------------------------------------
    mean = mean(x)                                                t =  -3.8448
Ho: mean = 75                                    degrees of freedom =       23

    Ha: mean < 75               Ha: mean != 75                 Ha: mean > 75
 Pr(T < t) = 0.0004         Pr(|T| > |t|) = 0.0008          Pr(T > t) = 0.9996
```

◁

▷ Example 6

There is no immediate form of `ttest` with paired data since the test is also a function of the covariance, a number unlikely to be reported in any published source. For nonpaired data, however, we might type

```
. ttesti 20 20 5  32 15 4
Two-sample t test with equal variances

------------------------------------------------------------------------------
         |     Obs        Mean    Std. Err.   Std. Dev.   [95% Conf. Interval]
---------+--------------------------------------------------------------------
       x |      20          20    1.118034           5    17.65993    22.34007
       y |      32          15    .7071068           4    13.55785    16.44215
---------+--------------------------------------------------------------------
combined |      52    16.92308    .6943785    5.007235    15.52905     18.3171
---------+--------------------------------------------------------------------
    diff |                   5    1.256135                2.476979    7.523021
------------------------------------------------------------------------------
    diff = mean(x) - mean(y)                                      t =   3.9805
Ho: diff = 0                                     degrees of freedom =       50

    Ha: diff < 0                 Ha: diff != 0                 Ha: diff > 0
 Pr(T < t) = 0.9999         Pr(|T| > |t|) = 0.0002          Pr(T > t) = 0.0001
```

If we had typed `ttesti 20 20 5 32 15 4, unequal`, the test would have assumed unequal variances. ◁

## Saved Results

`ttest` and `ttesti` save in `r()`:

Scalars

| | | | |
|---|---|---|---|
| `r(N_1)` | sample size $n_1$ | `r(t)` | $t$ statistic |
| `r(N_2)` | sample size $n_2$ | `r(sd_1)` | standard deviation for first variable |
| `r(p_l)` | lower one-sided $p$-value | `r(sd_2)` | standard deviation for second variable |
| `r(p_u)` | upper one-sided $p$-value | `r(mu_1)` | $\bar{x}_1$ mean for population 1 |
| `r(p)` | two-sided $p$-value | `r(mu_2)` | $\bar{x}_2$ mean for population 2 |
| `r(se)` | estimate of standard error | `r(df_t)` | degrees of freedom |

## Methods and Formulas

`ttest` and `ttesti` are implemented as ado-files.

See, for instance, Hoel (1984, 140–161) or Dixon and Massey (1983, 121–130) for an introduction and explanation of the calculation of these tests.

The test for $\mu = \mu_0$ for unknown $\sigma$ is given by

$$t = \frac{(\overline{x} - \mu_0)\sqrt{n}}{s}$$

The statistic is distributed as Student's $t$ with $n - 1$ degrees of freedom (Gosset 1908).

The test for $\mu_x = \mu_y$ when $\sigma_x$ and $\sigma_y$ are unknown but $\sigma_x = \sigma_y$ is given by

$$t = \frac{\overline{x} - \overline{y}}{\left\{ \frac{(n_x-1)s_x^2+(n_y-1)s_y^2}{n_x+n_y-2} \right\}^{1/2} \left( \frac{1}{n_x} + \frac{1}{n_y} \right)^{1/2}}$$

The result is distributed as Student's $t$ with $n_x + n_y - 2$ degrees of freedom.

You could perform `ttest` (without the `unequal` option) in a regression setting given that regression assumes a homoskedastic error model. To compare with the `ttest` command, denote the underlying observations on $x$ and $y$ by $x_j$, $j = 1, \dots, n_x$ and $y_j$, $j = 1, \dots, n_y$. In a regression framework, typing `ttest` without the `unequal` option is equivalent to

creating a new variable $z_j$ that represents the stacked observations on $x$ and $y$ (so that $z_j = x_j$ for $j = 1, \dots, n_x$ and $z_{n_x+j} = y_j$ for $j = 1, \dots, n_y$)

and then estimating the equation $z_j = \beta_0 + \beta_1 d_j + \epsilon_j$, where $d_j = 0$ for $j = 1, \dots, n_x$ and $d_j = 1$ for $j = n_x + 1, \dots, n_x + n_y$ (i.e., $d_j = 0$ when the $z$ observations represent $x$, and $d_j = 1$ when the $z$ observations represent $y$).

The estimated value of $\beta_1$, $b_1$, will equal $\overline{y} - \overline{x}$, and the reported $t$ statistic will be the same $t$ statistic as given by the formula above.

The test for $\mu_x = \mu_y$ when $\sigma_x$ and $\sigma_y$ are unknown and $\sigma_x \neq \sigma_y$ is given by

$$t = \frac{\overline{x} - \overline{y}}{\left( s_x^2/n_x + s_y^2/n_y \right)^{1/2}}$$

The result is distributed as Student's $t$ with $\nu$ degrees of freedom, where $\nu$ is given by (using Satterthwaite's formula)

$$\frac{\left( s_x^2/n_x + s_y^2/n_y \right)^2}{\frac{\left( s_x^2/n_x \right)^2}{n_x-1} + \frac{\left( s_y^2/n_y \right)^2}{n_y-1}}$$

Using Welch's formula (1947), the number of degrees of freedom is given by

$$-2 + \frac{\left( s_x^2/n_x + s_y^2/n_y \right)^2}{\frac{\left( s_x^2/n_x \right)^2}{n_x+1} + \frac{\left( s_y^2/n_y \right)^2}{n_y+1}}$$

The test for $\mu_x = \mu_y$ for matched observations (also known as paired observations, correlated pairs, or permanent components) is given by

$$t = \frac{\overline{d}\sqrt{n}}{s_d}$$

where $\overline{d}$ represents the mean of $x_i - y_i$ and $s_d$ represents the standard deviation. The test statistic $t$ is distributed as Student's $t$ with $n - 1$ degrees of freedom.

Note that you can also use `ttest` without the `unpaired` option in a regression setting since a paired comparison includes the assumption of constant variance. The `ttest` with an unequal variance assumption does not lend itself to an easy representation in regression settings and is not discussed here. $(x_j - y_j) = \beta_0 + \epsilon_j$.

> William Sealy Gosset (1876–1937) was born in Canterbury, England. He studied chemistry and mathematics at Oxford and obtained employment as a chemist with the brewers Guinness in Dublin. Gosset became very interested in statistical problems, which he discussed with Karl Pearson and later with Fisher and Neyman, and published several important papers under the pseudonym "Student", including that on the test that usually bears his name.

## References

Boland, P. J. 2000. William Sealy Gosset—alias 'Student' 1876–1937. In *Creators of Mathematics: The Irish Connection*, ed. K. Houston, 105–112. Dublin: University College Dublin Press.

Dixon, W. J. and F. J. Massey, Jr. 1983. *Introduction to Statistical Analysis*. 4th ed. New York: McGraw–Hill.

Gleason, J. R. 1999. sg101: Pairwise comparisons of means, including the Tukey wsd method. *Stata Technical Bulletin* 47: 31–37. Reprinted in *Stata Technical Bulletin Reprints*, vol. 8, pp. 225–233.

Gosset, W. S. [Student, pseud.] 1908. The probable error of a mean. *Biometrika* 6: 1–25.

——. 1943. *"Student's" Collected Papersr*, ed. E. S. Pearson and J. Wishart. London: Biometrika Office, University College.

Hoel, P. G. 1984. *Introduction to Mathematical Statistics*. 5th ed. New York: Wiley.

Pearson, E. S. and R. L. Plackett. 1990. *'Student': A Statistical Biography of William Sealy Gosset*. Oxford: Oxford University Press.

Preece, D. A. 1982. t is for trouble (and textbooks): A critique of some examples of the paired-samples t-test. *Statistician* 31: 169–195.

Satterthwaite, F. E. 1946. An approximate distribution of estimates of variance components. *Biometrics Bulletin* 2: 110–114.

Senn, S. J. and W. Richardson. 1994. The first $t$-test. *Statistics in Medicine* 13: 785–803.

Welch, B. L. 1947. The generalization of Student's problem when several different population variances are involved. *Biometrika* 34: 28–35.

Zelen, M. 1979. A new design for randomized clinical trials. *New England Journal of Medicine* 300: 1242–1245.

## Also See

**Related:** [R] **bitest**, [R] **ci**, [R] **oneway**, [R] **prtest**, [R] **sdtest**, [R] **signrank**, [MV] **hotelling**, [SVY] **svy: mean**, [SVY] **svy postestimation**

**Background:** [U] **19 Immediate commands**

## Title

**update** — Update Stata

## Syntax

*Report on update level of currently installed Stata*

`update`

*Set update source*

`update from` *location*

*Compare update level of currently installed Stata with that of source*

`update query` [ `, from(`*location*`)` ]

*Compare update level of ado-files of currently installed Stata with that of source*

`update ado` [ `, from(`*location*`) into(`*dirname*`)` ]

*Compare update level of executable of currently installed Stata with that of source*

`update executable` [ `, from(`*location*`) into(`*dirname*`) force` ]

*Perform update ado and update executable*

`update all` [ `, from(`*location*`)` ]

*Swap newly downloaded executable with currently running executable*

`update swap` [ `, clear` ]

*Set automatic updates (Windows and Macintosh only)*

`set update_query` { `on` | `off` }

`set update_interval` *#*

`set update_prompt` { `on` | `off` }

## Description

The `update` command reports on the current update level and installs official updates to Stata. Official updates are updates to Stata as it was originally shipped from StataCorp, not the additions to Stata published in, for instance, the *Stata Journal* (SJ) or *Stata Technical Bulletin* (STB). Those additions are installed using the `net` command; see [R] **net**.

`update` without arguments reports on the update level of the currently installed Stata.

`update from` sets an update source, where *location* is a directory name or URL. If you are on the Internet, type `update from http://www.stata.com`.

`update query` compares the update level of the currently installed Stata with that available from the update source and displays a report.

`update ado` compares the update level of the official ado-files of the currently installed Stata with those available from the update source. If the currently installed ado-files need updating, `update ado` copies and installs those files from the update source that are necessary to bring the ado-files up to date.

`update executable` compares the update level of the currently installed Stata executable with that available from the update source. If the currently installed Stata needs updating, `update executable` copies the new executable from the update source, but the last step of the installation—erasing the old executable and renaming the new executable—is left for you to perform. `update executable` displays instructions on how to do this.

`update all` does the same as `update ado` followed by `update executable`.

`update swap` automatically swaps the newly downloaded executable with the currently running executable. During this process, Stata will shut down to perform the swap before automatically launching the new executable. This command is only useful if it is preceded by either the `update executable` or the `update all` commands.

`set update_query` determines if `update query` is to be automatically performed when Stata is launched.

`set update_interval` sets the number of days to elapse before performing the next automatic `update query`. The interval starts from the last time an `update query` was performed (automatically or manually).

`set update_prompt` determines whether a dialog is to be displayed before performing an automatic `update query`. The dialog allows you to perform an `update query` now, perform one the next time Stata is launched, perform one after the next interval has passed, or disable automatic `update query`.

## Options

`from(`*location*`)` specifies the location of the update source. You can specify the `from()` option on the individual `update` commands or use the `update from` command. Which you do makes no difference.

`into(`*dirname*`)` specifies the name of the directory into which the updates are to be copied. *dirname* can be a directory name or a `sysdir` codeword, such as `UPDATES` or `STATA`; see [P] **sysdir**.

In the case of `update ado`, the default is `into(UPDATES)`, the official update directory. Network computer managers might want to specify `into()` if they want to download the updates but leave the last step—copying the files into the official directory—to do themselves.

In the case of `update executable`, the default is `into(STATA)`, the official Stata directory. Network computer managers might want to specify `into()` so that they can copy the update into a more accessible directory. In that case, the last step of copying the new executable over the existing executable would be left for them to perform.

`force` is used with `update executable` to force downloading a new executable even if, based on the date comparison, Stata does not think it is necessary. There is seldom a reason to specify this option. There is no such option for `update ado` because, if you wanted to force the reinstallation of all ado-file updates, you would need only to erase the `UPDATES` directory. You can type `sysdir list` to see where the `UPDATES` directory is on your computer; see [P] **sysdir**.

`clear` specifies that it is okay to shut down this Stata and start the new one, even though the data in memory have not been saved to disk.

## Remarks

`update` updates the two official components of Stata—its binary executable and its ado-files—from the official source: *http://www.stata.com*. Jumping ahead of the story, the easiest thing to do if you are connected to the Internet is to type

```
. update all
```

and follow the instructions. If Stata is up to date, `update all` will do nothing. Otherwise, it will download whatever is necessary and display detailed instructions on what, if anything, needs to be done next. If you first want to know what `update all` will do, type

```
. update query
```

`update query` will present a report comparing what you have installed with what is available and will recommend that you do nothing or that you type `update ado`, or `update executable`, or `update all`. You should never update your ado-files without also updating your executable (and vice versa) if both ado-files and executable updates are available.

If you want only a report on what you have installed without comparing with what is available, type

```
. update
```

`update` will show you what you have installed and where it is and will recommend that you type `update query` to compare that with what is available.

Before doing any of this, you can type

```
. update from http://www.stata.com
```

but that is not really necessary because *http://www.stata.com* is the default location.

In addition to using the `update` command, you may select **Official Updates** from the **Help** menu. The menu item does the same thing as the command, but it does not provide the file redirection option `into()`, which managers of networked computers may wish to use so that they can download the updates and then copy the files to the official locations for themselves.

For examples of using `update`, see

| | |
|---|---|
| Windows | [GSW] **19 Using the Internet** and |
| | [U] **28 Using the Internet to keep up to date** |
| Macintosh | [GSM] **19 Using the Internet** and |
| | [U] **28 Using the Internet to keep up to date** |
| Unix | [GSU] **19 Using the Internet** and |
| | [U] **28 Using the Internet to keep up to date** |

## Notes for multiuser system administrators

There are two types of updates: ado-file updates and the binary executable update. Stata's periodic updates are usually ado-file updates, but sometimes they are both, and even more occasionally, they are only a binary update.

By default, `update` installs the ado-file updates, which may include lots of small files. `update` is very careful about how it does this. First it downloads all the files you need to a temporary place. Next it closes the connection to *http://www.stata.com* and checks the files to make sure that they are complete. Only after all that does `update` copy the files to the official `UPDATES` directory; see [P] **sysdir**. This is all designed so that, should anything go wrong at any step along the way, no damage is done.

Updated binary executables, on the other hand, are just downloaded, and then you can either manually or automatically swap the executables. To manually swap the executables, (1) exit Stata, (2) rename the current executable, (3) rename the updated executable, (4) try Stata, and (5) erase the old executable. To automatically swap the executables, type `update swap`, and Stata will briefly quit to perform the copy before relaunching the newly downloaded executable.

`update` must have write access to both the `STATA` and `UPDATES` directories. You can obtain the names of these directories by typing `sysdir`. As system administrator, you must decide whether you are going to start Stata with such permissions (Unix users could first become superuser to ensure write access to the `STATA` and `UPDATES` directories) and trust Stata to do the right thing. That is what we recommend you do, but we provide the `into()` option for those who do not trust our recommendation.

To perform the final copying by hand, obtain the new executable by typing

```
. update executable, into(.)
```

That will place the new executable in the current directory. You need no special permissions to do this. Later you can copy the file into the appropriate place and give it the appropriate name. Type `update` without arguments; the default output will make it clear where the file goes and what its name must be. When you copy the file, be sure to make it executable by everybody.

To obtain the ado-file updates, make a new, empty directory, and then place the updates into it. For example, typing

```
. mkdir mydir
. update ado, into(mydir)
```

places the new, empty directory in the current directory under the name `mydir`. You need no special permissions to perform this step. Later, you can copy all the files in `mydir` to the official place. Type `update` without arguments; the default output will make it clear where the files go. When you copy the files, be sure to copy all of them and to make all of them readable by everybody.

(*Continued on next page*)

## Saved Results

update without a subcommand, update from, and update query save the following in r():

Scalars

| | | |
|---|---|---|
| r(inst_exe) | date of executable installed | (*) |
| r(avbl_exe) | date of executable available over web | (*) (**) |
| r(inst_ado) | date of ado-files installed | (*) |
| r(avbl_ado) | date of ado-files available over web | (*) (**) |

Macros

| | |
|---|---|
| r(dir_exe) | directory in which executable is stored |
| r(dir_ado) | directory in which ado-files are stored |
| r(name_exe) | name of the Stata executable |

Notes:

* Dates are stored as integers counting the number of days since January 1, 1960; see [U] **24 Dealing with dates**.

** These dates are not saved by update without a subcommand since update by itself reports information solely about the local computer and does not check what is available on the web.

## Also See

**Related:** [R] **net**, [P] **sysdir**

**Background:** [U] **28 Using the Internet to keep up to date**, [GSM] **19 Using the Internet**, [GSU] **19 Using the Internet**, [GSW] **19 Using the Internet**

## Title

*vce_option* — Alternative variance estimators

## Syntax

vce(*vcetype*) specifies how to perform variance–covariance estimation. Allowed are

vce(oim)

vce(robust)

vce(opg)

vce(bootstrap [ , *bootstrap_options* ])

vce(jackknife [ , *jackknife_options* ])

## Description

This entry describes the vce() option, which is common to most estimation commands. vce() specifies how the variance–covariance matrix of the estimators is to be estimated.

## Options

SE/Robust

vce(oim) is usually the default if vcetype() is not specified. oim stands for observed information matrix, which is the negative inverse of the matrix of second derivatives $-\mathbf{D}^{-1}$; see [R] **ml**.

vce(robust) uses the robust or sandwich estimator of variance, $\mathbf{D}^{-1}(\sum_j \mathbf{g}_j\mathbf{g}_j')\mathbf{D}^{-1}$. Estimation commands that allow option vce(robust) also allow option robust as a synonym; it does not matter which you specify. Many estimation commands that allow vce(robust) also allow a cluster(*varname*) option; see [R] **estimation options**.

vce(opg) uses the sum of the outer product of the gradient vectors $\sum_j \mathbf{g}_j\mathbf{g}_j'$; see [R] **ml**.

vce(bootstrap [ , *bootstrap_options* ]) uses a nonparametric bootstrap; see [R] **bootstrap**. After estimation with vce(bootstrap), see [R] **bootstrap postestimation** to obtain percentile-based or bias-corrected confidence intervals.

vce(jackknife [ , *jackknife_options* ]) uses the delete-one jackknife; see [R] **jackknife**.

## Remarks

Remarks are presented under the headings

*Prefix commands*
*Passing options in* vce()

## Prefix commands

Specifying vce(bootstrap) or vce(jackknife) is usually equivalent to using the corresponding prefix command. Here is an example using jackknife with regress.

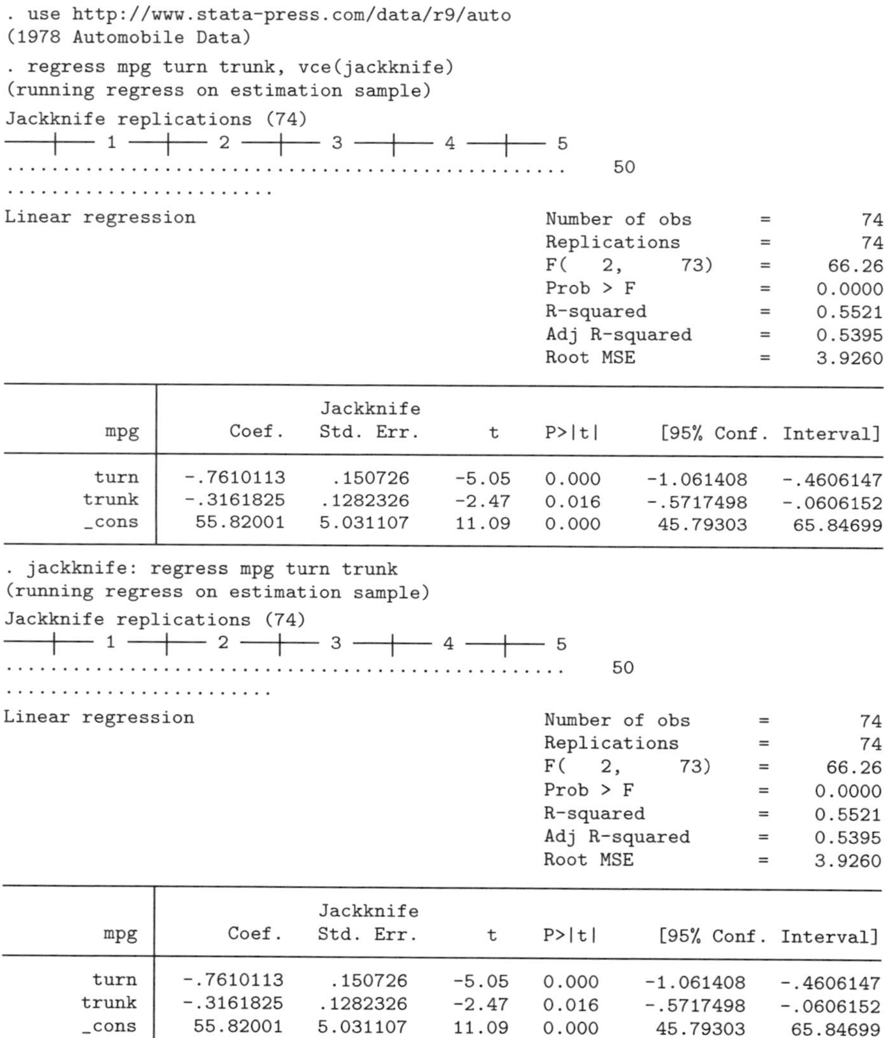

```
. use http://www.stata-press.com/data/r9/auto
(1978 Automobile Data)

. regress mpg turn trunk, vce(jackknife)
(running regress on estimation sample)

Jackknife replications (74)
----+--- 1 ---+--- 2 ---+--- 3 ---+--- 4 ---+--- 5
..................................................    50
........................

Linear regression                               Number of obs      =        74
                                                Replications       =        74
                                                F(   2,      73)   =     66.26
                                                Prob > F           =    0.0000
                                                R-squared          =    0.5521
                                                Adj R-squared      =    0.5395
                                                Root MSE           =    3.9260

------------------------------------------------------------------------------
             |              Jackknife
         mpg |      Coef.   Std. Err.      t    P>|t|     [95% Conf. Interval]
-------------+----------------------------------------------------------------
        turn |  -.7610113    .150726    -5.05   0.000    -1.061408   -.4606147
       trunk |  -.3161825   .1282326    -2.47   0.016    -.5717498   -.0606152
       _cons |   55.82001   5.031107    11.09   0.000     45.79303    65.84699
------------------------------------------------------------------------------

. jackknife: regress mpg turn trunk
(running regress on estimation sample)

Jackknife replications (74)
----+--- 1 ---+--- 2 ---+--- 3 ---+--- 4 ---+--- 5
..................................................    50
........................

Linear regression                               Number of obs      =        74
                                                Replications       =        74
                                                F(   2,      73)   =     66.26
                                                Prob > F           =    0.0000
                                                R-squared          =    0.5521
                                                Adj R-squared      =    0.5395
                                                Root MSE           =    3.9260

------------------------------------------------------------------------------
             |              Jackknife
         mpg |      Coef.   Std. Err.      t    P>|t|     [95% Conf. Interval]
-------------+----------------------------------------------------------------
        turn |  -.7610113    .150726    -5.05   0.000    -1.061408   -.4606147
       trunk |  -.3161825   .1282326    -2.47   0.016    -.5717498   -.0606152
       _cons |   55.82001   5.031107    11.09   0.000     45.79303    65.84699
------------------------------------------------------------------------------
```

In this case, it does not matter whether we specify option vce(jackknife) or instead use the jackknife prefix.

However, vce(jackknife) should be used in place of the jackknife prefix whenever available because they are not always equivalent. For example, to use the jackknife prefix with clogit properly, you must inform jackknife to omit observations from each group. Specifying vce(jackknife) does this automatically.

```
. use http://www.stata-press.com/data/r9/clogitid

. jackknife, cluster(id): clogit y x1 x2, group(id)
  (output omitted)
```

This extra information is automatically communicated to `jackknife` by `clogit` when the `vce()` option is specified.

```
. clogit y x1 x2, group(id) vce(jackknife)
(running clogit on estimation sample)
Jackknife replications (66)
----+--- 1 ---+--- 2 ---+--- 3 ---+--- 4 ---+--- 5
..................................................    50
................
Conditional (fixed-effects) logistic regression   Number of obs   =       369
                                                  Replications    =        66
                                                  F(   2,     65) =      4.58
                                                  Prob > F        =    0.0137
Log likelihood = -123.41386                       Pseudo R2       =    0.0355

------------------------------------------------------------------------------
             |              Jackknife
           y |      Coef.   Std. Err.      t    P>|t|     [95% Conf. Interval]
-------------+----------------------------------------------------------------
          x1 |    .653363   .3010608     2.17   0.034      .052103    1.254623
          x2 |   .0659169   .0487858     1.35   0.181    -.0315151    .1633489
------------------------------------------------------------------------------
```

## Passing options in vce()

Specifying the `vce(bootstrap)` or `vce(jackknife)` option also calls the corresponding prefix command. If you wish to specify additional options to the prefix command, you can include them within the `vce()` option. Below we request 300 bootstrap replications and save the replications in `bsreg.dta`:

```
. use http://www.stata-press.com/data/r9/auto, clear
(1978 Automobile Data)
. regress mpg turn trunk, vce(bootstrap, nodots seed(123) rep(300) saving(bsreg))
Linear regression                                Number of obs      =        74
                                                 Replications       =       300
                                                 Wald chi2(2)       =    127.28
                                                 Prob > chi2        =    0.0000
                                                 R-squared          =    0.5521
                                                 Adj R-squared      =    0.5395
                                                 Root MSE           =    3.9260

------------------------------------------------------------------------------
             |   Observed   Bootstrap                         Normal-based
         mpg |      Coef.   Std. Err.      z    P>|z|     [95% Conf. Interval]
-------------+----------------------------------------------------------------
        turn |  -.7610113   .1361786    -5.59   0.000    -1.027916   -.4941062
       trunk |  -.3161825   .1145728    -2.76   0.006     -.540741   -.0916239
       _cons |   55.82001    4.69971    11.88   0.000     46.60875    65.03127
------------------------------------------------------------------------------
```

*(Continued on next page)*

```
. bstat using bsreg
Bootstrap results                               Number of obs      =        74
                                                Replications       =       300

      command:  regress mpg turn trunk
```

| | Observed Coef. | Bootstrap Std. Err. | z | P>\|z\| | Normal-based [95% Conf. | Interval] |
|---|---|---|---|---|---|---|
| turn | -.7610113 | .1361786 | -5.59 | 0.000 | -1.027916 | -.4941062 |
| trunk | -.3161825 | .1145728 | -2.76 | 0.006 | -.540741 | -.0916239 |
| _cons | 55.82001 | 4.69971 | 11.88 | 0.000 | 46.60875 | 65.03127 |

## Methods and Formulas

By default, Stata's maximum likelihood estimators display standard errors based on variance estimates given by the inverse of the negative Hessian (second derivative) matrix. If `robust`, `cluster()`, or `pweights` is specified, standard errors are based on the robust variance estimator (see [U] **20.14 Obtaining robust variance estimates**); in this case, likelihood-ratio tests are not appropriate (see [SVY] **survey**), and the model $\chi^2$ is from a Wald test. If `vce(opg)` is specified, then the standard errors are based on the outer product of the gradients; this option has no effect on likelihood-ratio tests, though it does affect Wald tests.

If `vce(bootstrap)` or `vce(jackknife)` is specified, then the standard errors are based on the chosen replication method; in this case, the model $\chi^2$ or $F$ statistic is from a Wald test using the respective replication-based covariance matrix. The $t$ distribution is used in the coefficient table when the `vce(jackknife)` option is specified. `vce(bootstrap)` and `vce(jackknife)` are sometimes available with commands that are not maximum likelihood estimators.

## Also See

**Related:** [R] **bootstrap**, [R] **jackknife**, [R] **ml**

**Background:** [U] **20 Estimation and postestimation commands**,
[R] **estimation options**

# Title

**view** — View files and logs

# Syntax

*Display file in Viewer*

view [file] ["]*filename*["] [, asis adopath]

*Bring up browser pointed at specified URL*

view browse ["]*url*["]

*Display help results in Viewer*

view help [*topic_or_command_name*]

*Display search results in Viewer*

view search *keywords*

*Display news results in Viewer*

view news

*Display net results in Viewer*

view net [*netcmd*]

*Display ado results in Viewer*

view ado [*adocmd*]

*Display update results in Viewer*

view update [*updatecmd*]

*Programmer's analog to view file and view browse*

view view_d

*Programmer's analog to view help*

view help_d

*Programmer's analog to view search*

view search_d

*Programmer's analog to view net*

```
view net_d
```

*Programmer's analog to view ado*

```
view ado_d
```

*Programmer's analog to view update*

```
view update_d
```

## Description

`view` displays file contents in the Viewer.

`view file` displays the specified file. `file` is optional, so if you had a SMCL session log created by typing `log using mylog`, you could view it by typing `view mylog.smcl`. `view file` can properly display `.smcl` files (logs and the like), `.hlp` files, and ASCII text files. `view file`'s `asis` option specifies that the file be displayed as straight ASCII text, regardless of the *filename*'s extension.

`view browse` opens your browser pointed at *url*. Typing `view browse http://www.stata.com` would bring up your browser pointed to the *http://www.stata.com* web site.

`view help` does the same as the `help` command—see [R] **help**—but displays the result in the Viewer. For example, to review the help for Stata's `print` command, you could type `view help print`.

`view search` does the same as the `search` command—see [R] **search**—but displays the result in the Viewer. For instance, to search the online help for information on robust regression, you could type `view search robust regression`.

`view news` does the same as the `news` command—see [R] **news**—but displays the results in the Viewer. (`news` displays the latest news from *http://www.stata.com*.)

`view net` does the same as the `net` command—see [R] **net**—but displays the result in the Viewer. For instance, typing `view net search hausman test` would search the Internet for additions to Stata related to the Hausman test. Typing `view net from http://www.stata.com` would go to the Stata download site at *http://www.stata.com*.

`view ado` does the same as the `ado` command—see [R] **net**—but displays the result in the Viewer. For instance, typing `view ado dir` would show a list of files you have installed.

`view update` does the same as the `update` command—see [R] **update**—but displays the result in the Viewer. Typing `view update` would show the dates of what you have installed, and from there you could click to compare those dates with the latest updates available. Typing `view update query` would skip the first step and show the comparison.

The `view *_d` commands are more useful in programming contexts than they are interactively.

`view view_d` displays a dialog box from which you may type the name of a file or a URL to be displayed in the Viewer.

`view help_d` displays a help dialog box from which you may obtain interactive help on any Stata command.

`view search_d` displays a search dialog box from which you may obtain interactive help based on keywords.

`view net_d` displays a search dialog box from which you may search the Internet for additions to Stata (which you could then install).

`view ado_d` displays a dialog box from which you may search the user-written routines you have previously installed.

`view update_d` displays an update dialog box in which you may type the source from which updates are to be obtained.

## Options

`asis`, allowed with `view file`, specifies that the file be displayed as ASCII text, regardless of the *filename*'s extension. `net view`'s default action is to display files ending in `.smcl` and `.hlp` as SMCL.

`adopath`, allowed with `view file`, specifies that Stata search the `S_ADO` path for *filename* and display it, if found.

## Remarks

Most users access the Viewer by selecting **File > View...** and proceeding from there. The `view` command allows you to skip that step. Some common interactive uses of `view` are

```
. view mysession.smcl
. view mysession.log

. view help print
. view help regress

. view news
. view browse http://www.stata.com

. view net search hausman test

. view net
. view ado

. view update query
```

In addition, programmers find `view` useful for creating special effects.

## Also See

**Complementary:** [R] **help**, [R] **net**, [R] **news**, [R] **search**, [R] **update**, [D] **type**

**Related:** [P] **viewsource**

**Background:** [GSM] **5 Using the Viewer**, [GSU ] **5 Using the Viewer**, [GSW] **5 Using the Viewer**

# Title

**vwls** — Variance-weighted least squares

# Syntax

`vwls` *depvar indepvars* [*if*] [*in*] [*weight*] [, *options*]

| *options* | description |
|---|---|
| Model | |
| `noconstant` | suppress constant term |
| `sd(`*varname*`)` | variable containing estimate of conditional standard deviation |
| Reporting | |
| `level(`*#*`)` | set confidence level; default is `level(95)` |

`bootstrap`, `by`, `jackknife`, `rolling`, `statsby`, and `xi` are allowed; see [U] **11.1.10 Prefix commands**.
`fweights` are allowed; see [U] **11.1.6 weight**.
See [U] **20 Estimation and postestimation commands** for additional capabilities of estimation commands.

# Description

`vwls` estimates a linear regression using variance-weighted least squares. It differs from ordinary least-squares (OLS) regression in that it does not assume homogeneity of variance, but requires that the conditional variance of *depvar* be estimated prior to the regression. The estimated variance need not be constant across observations. `vwls` treats the estimated variance as if it were the true variance when it computes the standard errors of the coefficients.

You must supply an estimate of the conditional standard deviation of *depvar* to `vwls` using the `sd(`*varname*`)` option, or you must have grouped data with the groups defined by the *indepvars* variables. In the latter case, `vwls` treats all *indepvars* as categorical variables, computes the mean and standard deviation of *depvar* separately for each subgroup, and computes the regression of the subgroup means on *indepvars*.

`regress` with analytic weights can be used to produce another kind of "variance-weighted least squares"; see the following remarks for an explanation of the difference.

# Options

Model

`noconstant`; see [R] **estimation options**.

`sd(`*varname*`)` is an estimate of the conditional standard deviation of *depvar* (that is, it can vary observation by observation). All values of *varname* must be $>0$. If you specify `sd()`, you cannot use `fweights`.

If `sd()` is not given, the data will be grouped by *indepvars*. In this case, *indepvars* are treated as categorical variables, and the means and standard deviations of *depvar* for each subgroup are calculated and used for the regression. Any subgroup for which the standard deviation is zero is dropped.

Reporting

`level(#)`; see [R] **estimation options**.

## Remarks

The `vwls` command is intended for use with two special—and very different—types of data. The first contains data that consist of measurements from physical science experiments in which (1) all error is due solely to measurement errors and (2) the sizes of the measurement errors are known.

You can also use variance-weighted least-squares linear regression for certain problems in categorical data analysis, such as when all the independent variables are categorical and the outcome variable is either continuous or a quantity that can sensibly be averaged. If each of the subgroups defined by the categorical variables contains a reasonable number of subjects, then the variance of the outcome variable can be estimated independently within each subgroup. For the purposes of estimation, `vwls` treats each subgroup as a single observation, with the dependent variable being the subgroup mean of the outcome variable.

The `vwls` command fits the model

$$y_i = \mathbf{x}_i\boldsymbol{\beta} + \varepsilon_i$$

where the errors $\varepsilon_i$ are independent normal random variables with the distribution $\varepsilon_i \sim N(0, \nu_i)$. The independent variables $\mathbf{x}_i$ are assumed to be known without error.

As described above, `vwls` assumes that you already have estimates $s_i^2$ for the variances $\nu_i$. The error variance is not estimated in the regression. The estimates $s_i^2$ are used to compute the standard errors of the coefficients; see *Methods and Formulas* below.

In contrast, weighted ordinary least-squares regression assumes that the errors have the distribution $\varepsilon_i \sim N(0, \sigma^2/w_i)$, where the $w_i$ are known weights and $\sigma^2$ is an unknown parameter that is estimated in the regression. This is the difference from variance-weighted least squares: in weighted OLS, the magnitude of the error variance is estimated in the regression using all the data.

### ▷ Example 1

An artificial, but informative, example illustrates the difference between variance-weighted least squares and weighted OLS.

We measure the quantities $x_i$ and $y_i$ and estimate that the standard deviation of $y_i$ is $s_i$. We enter the data into Stata:

```
. input x y s
                x          y          s
  1.            1        1.2        0.5
  2.            2        1.9        0.5
  3.            3        3.2        1
  4.            4        4.3        1
  5.            5        4.9        1
  6.            6        6.0        2
  7.            7        7.2        2
  8.            8        7.9        2
  9. end
```

Since we want observations with smaller variance to carry larger weight in the regression, we compute an OLS regression with analytic weights proportional to the inverse of the squared standard deviations:

```
. regress y x [aweight=s^(-2)]
(sum of wgt is   1.1750e+01)

      Source |       SS       df       MS              Number of obs =       8
-------------+------------------------------           F(  1,     6) =  702.26
       Model |  22.6310183     1  22.6310183           Prob > F      =  0.0000
    Residual |  .193355117     6  .032225853           R-squared     =  0.9915
-------------+------------------------------           Adj R-squared =  0.9901
       Total |  22.8243734     7  3.26062477           Root MSE      =  .17952

------------------------------------------------------------------------------
           y |      Coef.   Std. Err.      t    P>|t|     [95% Conf. Interval]
-------------+----------------------------------------------------------------
           x |   .9824683   .0370739    26.50   0.000     .8917517    1.073185
       _cons |   .1138554   .1120078     1.02   0.349    -.1602179    .3879288
------------------------------------------------------------------------------
```

If we compute a variance-weighted least-squares regression using `vwls`, we get the same results for the coefficient estimates but very different standard errors:

```
. vwls y x, sd(s)
Variance-weighted least-squares regression           Number of obs   =       8
Goodness-of-fit chi2(6)    =     0.28                Model chi2(1)   =   33.24
Prob > chi2                =   0.9996                Prob > chi2     =  0.0000
------------------------------------------------------------------------------
           y |      Coef.   Std. Err.      z    P>|z|     [95% Conf. Interval]
-------------+----------------------------------------------------------------
           x |   .9824683    .170409     5.77   0.000     .6484728    1.316464
       _cons |   .1138554     .51484     0.22   0.825    -.8952124    1.122923
------------------------------------------------------------------------------
```

Although the values of $y_i$ were nicely linear with $x_i$, the `vwls` regression used the large estimates for the standard deviations to compute large standard errors for the coefficients. For weighted OLS regression, however, the scale of the analytic weights has no effect on the standard errors of the coefficients—only the relative proportions of the analytic weights affect the regression.

If we are sure of the sizes of our error estimates for $y_i$, using `vwls` is valid. However, if we can only estimate the relative proportions of error among the $y_i$, `vwls` is not appropriate.

◁

## ▷ Example 2

Let us now consider an example of the use of `vwls` with categorical data. Suppose that we have blood pressure data for $n = 400$ subjects, categorized by gender and race (black or white). Here is a description of the data:

```
. use http://www.stata-press.com/data/r9/bp, clear
. table gender race, c(mean bp sd bp freq) row col format(%8.1f)

---------------------------------
          |         Race
   Gender |  White  Black  Total
----------+----------------------
   Female |  117.1  118.5  117.8
          |   10.3   11.6   10.9
          |    100    100    200
          |
     Male |  122.1  125.8  124.0
          |   10.6   15.5   13.3
          |    100    100    200
          |
    Total |  119.6  122.2  120.9
          |   10.7   14.1   12.6
          |    200    200    400
---------------------------------
```

Performing a variance-weighted regression using `vwls` gives

```
. vwls bp gender race
Variance-weighted least-squares regression          Number of obs   =       400
Goodness-of-fit chi2(1)     =    0.88               Model chi2(2)   =     27.11
Prob > chi2                 =  0.3486               Prob > chi2     =    0.0000

------------------------------------------------------------------------------
          bp |      Coef.   Std. Err.      z    P>|z|     [95% Conf. Interval]
-------------+----------------------------------------------------------------
      gender |   5.876522   1.170241     5.02   0.000     3.582892    8.170151
        race |   2.372818   1.191683     1.99   0.046     .0371631    4.708473
       _cons |   116.6486   .9296297   125.48   0.000     114.8266    118.4707
------------------------------------------------------------------------------
```

By comparison, an OLS regression gives the following result:

```
. regress bp gender race
      Source |       SS       df       MS              Number of obs =       400
-------------+------------------------------           F(  2,   397) =     15.24
       Model |  4485.66639     2  2242.83319           Prob > F      =    0.0000
    Residual |  58442.7305   397  147.210908           R-squared     =    0.0713
-------------+------------------------------           Adj R-squared =    0.0666
       Total |  62928.3969   399   157.71528           Root MSE      =    12.133

------------------------------------------------------------------------------
          bp |      Coef.   Std. Err.      t    P>|t|     [95% Conf. Interval]
-------------+----------------------------------------------------------------
      gender |     6.1775   1.213305     5.09   0.000     3.792194    8.562806
        race |     2.5875   1.213305     2.13   0.034     .2021938    4.972806
       _cons |   116.4862   1.050753   110.86   0.000     114.4205     118.552
------------------------------------------------------------------------------
```

Note the larger value for the `race` coefficient (and smaller $p$-value) in the OLS regression. The assumption of homogeneity of variance in OLS means that the mean for black men pulls the regression line higher than in the `vwls` regression, which takes into account the larger variance for black men and reduces its effect on the regression.

◁

## Saved Results

`vwls` saves in `e()`:

| | |
|---|---|
| Scalars | |
| `e(N)` | number of observations |
| `e(df_m)` | model degrees of freedom |
| `e(chi2)` | model $\chi^2$ |
| `e(df_gf)` | goodness-of-fit degrees of freedom |
| `e(chi2_gf)` | goodness-of-fit $\chi^2$ |
| Macros | |
| `e(cmd)` | `vwls` |
| `e(depvar)` | name of dependent variable |
| `e(properties)` | `b V` |
| Matrices | |
| `e(b)` | coefficient vector |
| `e(V)` | variance–covariance matrix of the estimators |
| Functions | |
| `e(sample)` | marks estimation sample |

## Methods and Formulas

`vwls` is implemented as an ado-file.

Let $\mathbf{y} = (y_1, y_2, \ldots, y_n)'$ be the vector of observations of the dependent variable, where $n$ is the number of observations. When `sd()` is specified, let $s_1$, $s_2$, …, $s_n$ be the standard deviations supplied by `sd()`. For categorical data, when `sd()` is not given, the means and standard deviations of $y$ for each subgroup are computed, and $n$ becomes the number of subgroups, $\mathbf{y}$ is the vector of subgroup means, and $s_i$ are the standard deviations for the subgroups.

Let $\mathbf{V} = \text{diag}(s_1^2, s_2^2, \ldots, s_n^2)$ denote the estimate of the variance of $\mathbf{y}$. Then the estimated regression coefficients are

$$\mathbf{b} = (\mathbf{X}'\mathbf{V}^{-1}\mathbf{X})^{-1}\mathbf{X}'\mathbf{V}^{-1}\mathbf{y}$$

and their estimated covariance matrix is

$$\widehat{\text{Cov}}(\mathbf{b}) = (\mathbf{X}'\mathbf{V}^{-1}\mathbf{X})^{-1}$$

A statistic for the goodness of fit of the model is

$$Q = (\mathbf{y} - \mathbf{X}\mathbf{b})'\,\mathbf{V}^{-1}(\mathbf{y} - \mathbf{X}\mathbf{b})$$

where $Q$ has a $\chi^2$ distribution with $n - k$ degrees of freedom ($k$ is the number of independent variables plus the constant, if any).

## References


Grizzle, J. E., C. F. Starmer, and G. G. Koch. 1969. Analysis of categorical data by linear models. *Biometrics* 25: 489–504.

Press, W. H., S. A. Teukolsky, W. T. Vetterling, and B. P. Flannery. 1992. *Numerical Recipes in C: The Art of Scientific Computing*. 2nd ed. Cambridge: Cambridge University Press.

## Also See

**Complementary:** [R] **vwls postestimation**

**Related:** [R] **regress**

**Background:** [U] **11.1.10 Prefix commands**,
[U] **20 Estimation and postestimation commands**,
[R] **estimation options**

# Title

**vwls postestimation** — Postestimation tools for vwls

# Description

The following postestimation commands are available for `vwls`:

| command | description |
|---|---|
| `adjust` | adjusted predictions of $\mathbf{x}\beta$ |
| `estat` | VCE and estimation sample summary |
| `estimates` | cataloging estimation results |
| `lincom` | point estimates, standard errors, testing, and inference for linear combinations of coefficients |
| `linktest` | link test for model specification |
| `mfx` | marginal effects or elasticities |
| `nlcom` | point estimates, standard errors, testing, and inference for nonlinear combinations of coefficients |
| `predict` | predictions, residuals, influence statistics, and other diagnostic measures |
| `predictnl` | point estimates, standard errors, testing, and inference for generalized predictions |
| `test` | Wald tests for simple and composite linear hypotheses |
| `testnl` | Wald tests of nonlinear hypotheses |

See the corresponding entries in the *Stata Base Reference Manual* for details.

# Syntax for predict

`predict` [*type*] *newvar* [*if*] [*in*] [, `xb stdp`]

These statistics are available both in and out of sample; type `predict ... if e(sample) ...` if wanted only for the estimation sample.

# Options for predict

`xb`, the default, calculates the linear prediction.

`stdp` calculates the standard error of the linear prediction.

# Methods and Formulas

All postestimation commands listed above are implemented as ado-files.

# Also See

**Complementary:** [R] **vwls**; [R] **adjust**, [R] **estimates**, [R] **lincom**, [R] **linktest**, [R] **mfx**, [R] **nlcom**, [R] **predictnl**, [R] **test**, [R] **testnl**

**Background:** [U] **13.5 Accessing coefficients and standard errors**, [U] **20 Estimation and postestimation commands**, [R] **estat**, [R] **predict**

## Title

**which** — Display location and version for an ado-file

## Syntax

`which` *fname*[.*ftype*] [, `all`]

## Description

`which` looks for *fname*.*ftype* along the `S_ADO` path. If Stata finds the file, `which` displays the full path and filename, along with, if the file is ASCII, all lines in the file that begin with "`*!`" in the first column. If Stata cannot find the file, `which` issues the message "file not found along adopath" and sets the return code to 111. *ftype* must be a file type for which Stata usually looks along the ado-path to find. Allowable *ftype*s are

`.ado`, `.class`, `.dlg`, `.ihlp`, `.hlp`, `.idlg`, `.key`, `.mata`, `.mlib`, `.mo`, `.mnu`,

`.plugin`, `.scheme`, and `.style`

If *ftype* is omitted, `which` assumes `.ado`. When searching for `.ado` files, if Stata cannot find the file, Stata then checks to see if *fname* is a built-in Stata command, allowing for valid abbreviations. If it is, the message "built-in command" is displayed; if not, the message "command not found as either built-in or ado-file" is displayed, and the return code is set to 111.

For information about internal version control, see [P] **version**.

## Option

`all` forces `which` to report the location of all files matching the *fname*.*ftype* found along the search path. The default is to report just the first one found.

## Remarks

If you write programs, you know that you make changes to the programs over time. If you are like us, you also end up with multiple versions of the program stored on your disk, perhaps in different directories. You may even have given copies of your programs to other Stata users, and you may not remember which version of a program you or your friends are using. The `which` command helps you solve this problem.

### ▷ Example 1

The `which` command displays the path for *filename*`.ado` and any lines in the code that begin with "`*!`". For example, we might want information about the `notes` command, described in [D] **notes**, which is an ado-file written by StataCorp. Here is what happens when we type `which notes`:

```
. which notes
c:\stata\ado\base\n\notes.ado
*! version 1.0.3  04jan2001
```

`which` displays the path for the `notes.ado` file and also a line beginning with "`*!`" that indicates the version of the file. This is how we, at StataCorp, do version control—see [U] **18.11.1 Version** for an explanation of our version control numbers.

We do not need to be so formal. `which` will display anything typed after lines that begin with '`*!`'. For instance, we might write `myprog.ado`:

```
. which myprog
.\myprog.ado
*! first written 1/03/2005
*! bug fix on 1/05/2005 (no variance case)
*! updated 1/24/2005 to include noconstant option
*! still suspicious if variable takes on only two values
```

It does not matter where in the program the lines beginning with `*!` are—`which` will list them (in particular, our "still suspicious" comment was buried about fifty lines down in the code). All that is important is that the `*!` marker appear in the first two columns of a line.

◁

## ▷ Example 2

If we type `which` *command*, where *command* is a built-in command rather than an ado-file, Stata responds with

```
. which summarize
built-in command:  summarize
```

If *command* was not either a built-in command or an ado-file, Stata would respond with

```
. which junk
command junk not found as either built-in or ado-file
r(111);
```

◁

# Also See

**Related:** [P] **findfile**, [P] **unab**, [P] **version**

**Background:** [U] **17 Ado-files**,
[U] **18.11.1 Version**

# Title

**xi** — Interaction expansion

# Syntax

`xi` [ , `prefix(`*string*`)` `noomit` ] *term(s)*

`xi` [ , `prefix(`*string*`)` `noomit` ] : *any_stata_command varlist_with_terms* ...

where a *term* has the form

| | or | |
|---|---|---|
| i.*varname* | | I.*varname* |
| i.$varname_1$*i.$varname_2$ | | I.$varname_1$*I.$varname_2$ |
| i.$varname_1$*$varname_3$ | | I.$varname_1$*$varname_3$ |
| i.$varname_1$ \| $varname_3$ | | I.$varname_1$ \| $varname_3$ |

*varname*, $varname_1$, and $varname_2$ denote numeric or string categorical variables. $varname_3$ denotes a continuous, numeric variable.

# Description

`xi` expands terms containing categorical variables into indicator (also called dummy) variable sets by creating new variables, and, in the second syntax (`xi:` *any_stata_command*), executes the specified command with the expanded terms. The dummy variables created are

| | |
|---|---|
| i.*varname* | Creates dummies for categorical variable *varname* |
| i.$varname_1$*i.$varname_2$ | Creates dummies for categorical variables $varname_1$ and $varname_2$: and all interactions and main effects |
| i.$varname_1$*$varname_3$ | Creates dummies for categorical variable $varname_1$ and continuous variable $varname_3$: and all interactions and main effects |
| i.$varname_1$ \| $varname_3$ | Creates dummies for categorical variable $varname_1$ and continuous variable $varname_3$: all interactions and main effect of $varname_3$, but no main effect of $varname_1$ |

# Options

`prefix(`*string*`)` allows you to choose a prefix other than `_I` for the newly created interaction variables. The prefix cannot be longer than 4 characters. By default, `xi` will create interaction variables starting with `_I`. When you use `xi`, it drops all previously created interaction variables starting with the prefix specified in the `prefix(`*string*`)` option or with `_I` by default. Therefore, if you want to keep the variables with a certain prefix, specify a different prefix in the `prefix(`*string*`)` option.

`noomit` prevents `xi` from omitting groups. This option provides a way to generate an indicator variable for every category having one or more variables, which is useful when combined with the `noconstant` option of an estimation command.

## Remarks

Remarks are presented under the headings

*Background*
*Indicator variables for simple effects*
*Controlling the omitted dummy*
*Categorical variable interactions*
*Interactions with continuous variables*
*Using xi: Interpreting output*
*How xi names variables*
*xi as a command rather than a command prefix*
*Warnings*

`xi` provides a convenient way to include dummy or indicator variables when fitting a model (say with `regress`, `logistic`, etc.). For instance, assume that the categorical variable `agegrp` contains 1 for ages 20–24, 2 for ages 25–39, 3 for ages 40–44, etc. Typing

```
. xi: logistic outcome weight i.agegrp bp
```

estimates a logistic regression of `outcome` on `weight`, dummies for each `agegrp` category, and `bp`. That is, `xi` searches out and expands terms starting with "`i.`" or "`I.`" but ignores the other variables. `xi` will expand both numeric and string categorical variables, so if you had a string variable `race` containing "white", "black", and "other", typing

```
. xi: logistic outcome weight bp i.agegrp i.race
```

would include indicator variables for the race group as well.

The `i.` indicator variables `xi` expands may appear anywhere in the *varlist*, so

```
. xi: logistic outcome i.agegrp weight i.race bp
```

would fit the same model.

You can also create interactions of categorical variables; typing

```
xi: logistic outcome weight bp i.agegrp*i.race
```

fits a model with indicator variables for all `agegrp` and `race` combinations, including the `agegrp` and `race` main-effect terms (i.e., the terms that are created when you just type `i.agegrp i.race`).

You can interact dummy variables with continuous variables; typing

```
xi: logistic outcome bp i.agegrp*weight i.race
```

fits a model with indicator variables for all `agegrp` categories interacted with `weight`, plus the main-effect terms `weight` and `i.agegrp`.

You can get the interaction terms without the `agegrp` main effect (but with the `weight` main effect) by typing

```
xi: logistic outcome bp i.agegrp|weight i.race
```

You can also include multiple interactions:

```
xi: logistic outcome bp i.agegrp*weight i.agegrp*i.race
```

We will now back up and describe the construction of dummy variables in more detail.

## Background

The terms *continuous*, *categorical*, and *indicator* or *dummy* variables are used below. Continuous variables measure something—such as height or weight—and at least conceptually can take on any real number over some range. Categorical variables, on the other hand, take on a finite number of values, each denoting membership in a subclass—for example, excellent, good, and poor, which might be coded 0, 1, 2 or 1, 2, 3 or even "Excellent", "Good", and "Poor". An indicator or dummy variable—the terms are used interchangeably—is a special type of two-valued categorical variable that contains values 0, denoting false, and 1, denoting true. The information contained in any $k$-valued categorical variable can be equally well represented by $k$ indicator variables. Instead of a single variable recording values representing excellent, good, and poor, you can have three indicator variables, indicating the truth or falseness of "result is excellent", "result is good", and "result is poor".

`xi` provides a convenient way to convert categorical variables to dummy or indicator variables when you fit a model (say with `regress`, `logistic`, etc.).

### ▷ Example 1

For instance, assume that the categorical variable `agegrp` contains 1 for ages 20–24, 2 for ages 25–39, and 3 for ages 40–44. (There is no one over 44 in our data.) As it stands, `agegrp` would be a poor candidate for inclusion in a model even if we thought age affected the outcome. The reason is that the coding would restrict the effect of being in the second age group to be twice the effect of being in the first, and, similarly, the effect of being in the third to be three times the first. That is, if we fitted the model,

$$y = \beta_0 + \beta_1 \,\texttt{agegrp} + X\beta_2$$

the effect of being in the first age group is $\beta_1$, the second $2\beta_1$, and the third $3\beta_1$. If the coding 1, 2, and 3 is arbitrary, we could just as well have coded the age groups 1, 4, and 9, making the effects $\beta_1$, $4\beta_1$, and $9\beta_1$.

The solution is to convert the categorical variable `agegrp` to a set of indicator variables $a_1$, $a_2$, and $a_3$, where $a_i$ is 1 if the individual is a member of the $i$th age group and 0 otherwise. We can then fit the model:

$$y = \beta_0 + \beta_{11}a_1 + \beta_{12}a_2 + \beta_{13}a_3 + X\beta_2$$

The effect of being in age group 1 is now $\beta_{11}$; 2, $\beta_{12}$; and 3, $\beta_{13}$; and these results are independent of our (arbitrary) coding. The only difficulty at this point is that the model is unidentified in the sense that there are an infinite number of $(\beta_0, \beta_{11}, \beta_{12}, \beta_{13})$ that fit the data equally well.

To see this, pretend that $(\beta_0, \beta_{11}, \beta_{12}, \beta_{13}) = (1, 1, 3, 4)$. The predicted values of $y$ for the various age groups are

$$y = \begin{cases} 1 + 1 + X\beta_2 = 2 + X\beta_2 & \text{(age group 1)} \\ 1 + 3 + X\beta_2 = 4 + X\beta_2 & \text{(age group 2)} \\ 1 + 4 + X\beta_2 = 5 + X\beta_2 & \text{(age group 3)} \end{cases}$$

Now pretend that $(\beta_0, \beta_{11}, \beta_{12}, \beta_{13}) = (2, 0, 2, 3)$. The predicted values of $y$ are

$$y = \begin{cases} 2 + 0 + X\beta_2 = 2 + X\beta_2 & \text{(age group 1)} \\ 2 + 2 + X\beta_2 = 4 + X\beta_2 & \text{(age group 2)} \\ 2 + 3 + X\beta_2 = 5 + X\beta_2 & \text{(age group 3)} \end{cases}$$

These two sets of predictions are indistinguishable: for age group 1, $y = 2 + X\beta_2$ regardless of the coefficient vector used, and similarly for age groups 2 and 3. This arises because we have 3 equations and 4 unknowns. Any solution is as good as any other, and, for our purposes, we merely need to choose one of them. The popular selection method is to set the coefficient on the first indicator variable to 0 (as we have done in our second coefficient vector). This is equivalent to fitting the model

$$y = \beta_0 + \beta_{12}a_2 + \beta_{13}a_3 + X\beta_2$$

How we select a particular coefficient vector (identifies the model) does not matter. It does, however, affect the *interpretation* of the coefficients.

For instance, we could just as well choose to omit the second group. In our artificial example, this would yield $(\beta_0, \beta_{11}, \beta_{12}, \beta_{13}) = (4, -2, 0, 1)$ instead of $(2, 0, 2, 3)$. These coefficient vectors are the same in the sense that

$$y = \begin{cases} 2 + 0 + X\beta_2 = 2 + X\beta_2 = 4 - 2 + X\beta_2 & \text{(age group 1)} \\ 2 + 2 + X\beta_2 = 4 + X\beta_2 = 4 + 0 + X\beta_2 & \text{(age group 2)} \\ 2 + 3 + X\beta_2 = 5 + X\beta_2 = 4 + 1 + X\beta_2 & \text{(age group 3)} \end{cases}$$

But what does it mean that $\beta_{13}$ can just as well be 3 or 1? We obtain $\beta_{13} = 3$ when we set $\beta_{11} = 0$, so $\beta_{13} = \beta_{13} - \beta_{11}$ and $\beta_{13}$ measures the difference between age groups 3 and 1.

In the second case, we obtain $\beta_{13} = 1$ when we set $\beta_{12} = 0$, so $\beta_{13} - \beta_{12} = 1$ and $\beta_{13}$ measures the difference between age groups 3 and 2. There is no inconsistency. According to our $\beta_{12} = 0$ model, the difference between age groups 3 and 1 is $\beta_{13} - \beta_{11} = 1 - (-2) = 3$, exactly the same result we got in the $\beta_{11} = 0$ model.

◁

## ▷ Example 2

The issue of interpretation is important because it can affect the way we discuss results. Imagine that we are studying recovery after a coronary bypass operation. Assume that the age groups are (1) children under 13 (we have two of them), (2) young adults under 25 (we have a handful of them), (3) adults under 46 (of which we have even more), (4) mature adults under 56, (5) older adults under 65, and (6) elderly adults. We follow the prescription of omitting the first group, so all our results are reported relative to children under 13. While there is nothing statistically wrong with this, readers will be suspicious when we make statements like "compared with young children, older and elder adults ...". Moreover, it is likely that we will have to end each statement with "although results are not statistically significant" because we have only two children in our comparison group. Of course, even with results reported in this way, we can do reasonable comparisons (say with mature adults), but we will have to do extra work to perform the appropriate linear hypothesis test using Stata's `test` command.

In this case, it would be better to force the omitted group to be more reasonable, such as mature adults. There is, however, a generic rule for automatic comparison group selection that, while less popular, tends to work better than the omit-the-first-group rule. That rule is to omit the most prevalent group. The most prevalent is usually a reasonable baseline.

◁

In any case, the prescription for categorical variables is

1. Convert each $k$-valued categorical variable to $k$ indicator variables.

2. Drop one of the $k$ indicator variables; any one will do, but dropping the first is popular, dropping the most prevalent is probably better in terms of having the computer guess at a reasonable interpretation, and dropping a specified one often eases interpretation the most.

3. Fit the model on the remaining $k - 1$ indicator variables.

xi automates this procedure.

We will now consider each of xi's features in detail.

## Indicator variables for simple effects

When you type i.*varname*, xi internally tabulates *varname* (which may be a string or a numeric variable) and creates indicator (dummy) variables for each observed value, omitting the indicator for the smallest value. For instance, say that agegrp takes on the values 1, 2, 3, and 4. Typing

```
xi: logistic outcome i.agegrp
```

creates indicator variables named _Iagegrp_2, _Iagegrp_3, and _Iagegrp_4. (xi chooses the names and tries to make them readable; xi guarantees that the names are unique.) The expanded logistic model is

```
. logistic outcome _Iagegrp_2 _Iagegrp_3 _Iagegrp_4
```

Afterwards, you can drop the new variables xi leaves behind by typing 'drop _I*' (note the capitalization).

xi provides the following features when you type i.*varname*:

1. *varname* may be string or numeric.
2. Dummy variables are created automatically.
3. By default, the dummy-variable set is identified by dropping the dummy corresponding to the smallest value of the variable (how to specify otherwise is discussed below).
4. The new dummy variables are left in your dataset. By default, the names of the new dummy variables start with _I; therefore, you can drop them by typing 'drop _I*'. You do not have to do this; each time you use xi, any automatically generated dummies with the same prefix as the one specified in the prefix(*string*) option, or _I by default, are dropped and new ones created.
5. The new dummy variables have variable labels so that you can determine to what they correspond by typing 'describe'.
6. xi may be used with any Stata command (not just logistic).

## Controlling the omitted dummy

By default, i.*varname* omits the dummy corresponding to the smallest value of *varname*; in the case of a string variable, this is interpreted as dropping the first in an alphabetical, case-sensitive sort. xi provides two alternatives to dropping the first: xi will drop the dummy corresponding to the most prevalent value of *varname*, or xi will let you choose the particular dummy to be dropped.

To change xi's behavior to dropping the most prevalent dummy, type

```
. char _dta[omit] prevalent
```

although whether you type "prevalent" or "yes" or anything else does not matter. Setting this characteristic affects the expansion of all categorical variables in the dataset. If you resave your dataset, the prevalent preference will be remembered. If you want to change the behavior back to the default drop-the-first rule, type

```
. char _dta[omit]
```

to clear the characteristic.

Once you set `_dta[omit]`, `i.`*varname* omits the dummy corresponding to the most prevalent value of *varname*. Thus the coefficients on the dummies have the interpretation of change from the most prevalent group. For example,

```
. char _dta[omit] prevalent
. xi: regress y i.agegrp
```

might create `_Iagegrp_1` through `_Iagegrp_4`, resulting in `_Iagegrp_2` being omitted if `agegrp` = 2 is most common (as opposed to the default dropping of `_Iagegrp_1`). The model is then

$$\texttt{y} = b_0 + b_1\,\texttt{\_Iagegrp\_1} + b_3\,\texttt{\_Iagegrp\_3} + b_4\,\texttt{\_Iagegrp\_4} + u$$

Then

| | |
|---|---|
| Predicted `y` for `agegrp` $1 = b_0 + b_1$ | Predicted `y` for `agegrp` $3 = b_0 + b_3$ |
| Predicted `y` for `agegrp` $2 = b_0$ | Predicted `y` for `agegrp` $4 = b_0 + b_4$ |

Thus the model's reported $t$ or $Z$ statistics are for a test of whether each group is different from the most prevalent group.

Perhaps you wish to omit the dummy for `agegrp` 3 instead. You do this by setting the variable's `omit` characteristic:

```
. char agegrp[omit] 3
```

This overrides `_dta[omit]` if you have set it. Now when you type

```
. xi: regress y i.agegrp
```

`_Iagegrp_3` will be omitted, and you will fit the model:

$$\texttt{y} = b'_0 + b'_1\,\texttt{\_Iagegrp\_1} + b'_2\,\texttt{\_Iagegrp\_2} + b'_4\,\texttt{\_Iagegrp\_4} + u$$

Later if you want to return to the default omission, type

```
. char agegrp[omit]
```

to clear the characteristic.

In summary, `i.`*varname* omits the first group by default, but if you define

```
. char _dta[omit] prevalent
```

the default behavior changes to dropping the most prevalent group. Either way, if you define a characteristic of the form

. `char` *varname*`[omit]` *#*

or, if *varname* is a string,

. `char` *varname*`[omit]` *string-literal*

the specified value will be omitted.

| Examples: | `. char agegrp[omit] 1` | |
|---|---|---|
| | `. char race[omit] White` | (for `race` a string variable) |
| | `. char agegrp[omit]` | (to restore default for `agegrp`) |

## Categorical variable interactions

`i.`*varname*$_1$`*i.`*varname*$_2$ creates the dummy variables associated with the interaction of the categorical variables *varname*$_1$ and *varname*$_2$. The identification rules—which categories are omitted—are the same as those for `i.`*varname*. For instance, assume that `agegrp` takes on four values and `race` takes on three values. Typing

```
. xi: regress y i.agegrp*i.race
```

results in

| model: | dummies for: |
|---|---|
| $y = a + b_2$ `_Iagegrp_2` $+ b_3$ `_Iagegrp_3` $+ b_4$ `_Iagegrp_4` | (`agegrp`) |
| $+c_2$ `_Irace_2` $+ c_3$ `_Irace_3` | (`race`) |
| $+d_{22}$ `_IageXrac_2_2` $+ d_{23}$ `_IageXrac_2_3` | |
| $+d_{32}$ `_IageXrac_3_2` $+ d_{33}$ `_IageXrac_3_3` | (`agegrp*race`) |
| $+d_{42}$ `_IageXrac_4_2` $+ d_{43}$ `_IageXrac_4_3` | |
| $+u$ | |

That is, typing

```
. xi: regress y i.agegrp*i.race
```

is the same as typing

```
. xi: regress y i.agegrp i.race i.agegrp*i.race
```

While there are many other ways the interaction could have been parameterized, this method has the advantage that you can test the joint significance of the interactions by typing

```
. testparm _IageXrac*
```

When you perform the estimation step, whether you specify `i.agegrp*i.race` or `i.race*i.agegrp` makes no difference (other than in the names given to the interaction terms; in the first case, the names will begin with `_IageXrac`; in the second, `_IracXage`). Thus

```
. xi: regress y i.race*i.agegrp
```

fits the same model.

You may also include multiple interactions simultaneously:

```
. xi: regress y i.agegrp*i.race i.agegrp*i.sex
```

The model fitted is

| model: | dummies for: |
|---|---|
| $y = a + b_2$ `_Iagegrp_2` $+ b_3$ `_Iagegrp_3` $+ b_4$ `_Iagegrp_4` | (`agegrp`) |
| $+c_2$ `_Irace_2` $+ c_3$ `_Irace_3` | (`race`) |
| $+d_{22}$ `_IageXrac_2_2` $+ d_{23}$ `_IageXrac_2_3` | |
| $+d_{32}$ `_IageXrac_3_2` $+ d_{33}$ `_IageXrac_3_3` | (`agegrp*race`) |
| $+d_{42}$ `_IageXrac_4_2` $+ d_{43}$ `_IageXrac_4_3` | |
| $+e_2$ `_Isex_2` | (`sex`) |
| $+f_{22}$ `_IageXsex_2_2` $+ f_{23}$ `_IageXsex_2_3` $+ f_{24}$ `_IageXsex_2_4` | (`agegrp*sex`) |
| $+u$ | |

Note that the `agegrp` dummies are (correctly) included only once.

## Interactions with continuous variables

i.*varname*$_1$**varname*$_2$ (as distinguished from i.*varname*$_1$*i.*varname*$_2$—note the second i.) specifies an interaction of a categorical variable with a continuous variable. For instance,

```
. xi: regress y i.agegr*wgt
```

results in the model:

$$y = a + b_2\,\texttt{\_Iagegrp\_2} + b_3\,\texttt{\_Iagegrp\_3} + b_4\,\texttt{\_Iagegrp\_4} \quad \text{(agegrp dummies)}$$
$$+c\,\texttt{wgt} \quad \text{(continuous wgt effect)}$$
$$+d_2\,\texttt{\_IageXwgt\_2} + d_3\,\texttt{\_IageXwgt\_3} + d_4\,\texttt{\_IageXwgt\_4} \quad \text{(agegrp*wgt interactions)}$$
$$+u$$

A variation on this notation, using | rather than *, omits the agegrp dummies. Typing

```
. xi: regress y i.agegrp|wgt
```

fits the model

$$y = a' \;\; +c'\,\texttt{wgt} \quad \text{(continuous wgt effect)}$$
$$+d'_2\,\texttt{\_IageXwgt\_2} + d'_3\,\texttt{\_IageXwgt\_3} + d'_4\,\texttt{\_IageXwgt\_4} \quad \text{(agegrp*wgt interactions)}$$
$$+u'$$

The predicted values of y are

| agegrp*wgt model | agegrp\|wgt model | |
|---|---|---|
| $y = a + c\,\texttt{wgt}$ | $a' + c'\,\texttt{wgt}$ | if agegrp $= 1$ |
| $a + c\,\texttt{wgt} + b_2 + d_2\,\texttt{wgt}$ | $a' + c'\texttt{wgt} + d'_2\,\texttt{wgt}$ | if agegrp $= 2$ |
| $a + c\,\texttt{wgt} + b_3 + d_3\,\texttt{wgt}$ | $a' + c'\texttt{wgt} + d'_3\,\texttt{wgt}$ | if agegrp $= 3$ |
| $a + c\,\texttt{wgt} + b_4 + d_4\,\texttt{wgt}$ | $a' + c'\texttt{wgt} + d'_4\,\texttt{wgt}$ | if agegrp $= 4$ |

That is, typing

```
. xi: regress y i.agegrp*wgt
```

is equivalent to typing

```
. xi: regress y i.agegrp i.agegrp|wgt
```

Also note that, in either case, you do not need to specify separately the continuous variable wgt; it is included automatically.

## Using xi: Interpreting output

```
. xi: regress mpg i.rep78
i.rep78              _Irep78_1-5    (naturally coded; _Irep78_1 omitted)
  (output from regress appears)
```

Interpretation: i.rep78 expanded to the dummies _Irep78_1, _Irep78_2, ..., _Irep78_5. The numbers on the end are "natural" in the sense that _Irep78_1 corresponds to rep78 = 1, _Irep78_2 to rep78 = 2, and so on. Finally, the dummy for rep78 = 1 was omitted.

```
. xi: regress mpg i.make
i.make               _Imake_1-74    (_Imake_1 for make==AMC Concord omitted)
  (output from regress appears)
```

Interpretation: `i.make` expanded to `_Imake_1`, `_Imake_2`, ..., `_Imake_74`. The coding is not natural because `make` is a string variable. `_Imake_1` corresponds to one make, `_Imake_2` another, and so on. You can find out the coding by typing `describe`. `_Imake_1` for the AMC Concord was omitted.

## How xi names variables

By default, `xi` assigns to the dummy variables it creates names having the form

`_I`*stub*`_`*groupid*

You may subsequently refer to the entire set of variables by typing '`I`*stub*`*`'. For example,

| name | = _I + *stub* | | + _ + *groupid* | | Entire set |
|---|---|---|---|---|---|
| `_Iagegrp_1` | `_I` | `agegrp` | `_` | `1` | `_Iagegrp*` |
| `_Iagegrp_2` | `_I` | `agegrp` | `_` | `2` | `_Iagegrp*` |
| `_IageXwgt_1` | `_I` | `ageXwgt` | `_` | `1` | `_IageXwgt*` |
| `_IageXrac_1_2` | `_I` | `ageXrac` | `_` | `1_2` | `_IageXrac*` |
| `_IageXrac_2_1` | `_I` | `ageXrac` | `_` | `2_1` | `_IageXrac*` |

If you specify a prefix in the `prefix(`*string*`)` option, say, `_S`, `xi` will name the variables starting with the prefix

`_S`*stub*`_`*groupid*

## xi as a command rather than a command prefix

`xi` can be used as a command prefix or as a command by itself. In the latter form, `xi` merely creates the indicator and interaction variables. Typing

```
. xi: regress y i.agegrp*wgt
i.agegrp          _Iagegrp_1-4    (naturally coded; _Iagegrp_1 omitted)
i.agegrp*wgt      _IageXwgt_1-4    (coded as above)
 (output from regress appears)
```

is equivalent to typing

```
. xi i.agegrp*wgt
i.agegrp          _Iagegrp_1-4    (naturally coded; _Iagegrp_1 omitted)
i.agegrp*wgt      _IageXwgt_1-4    (coded as above)
. regress y _Iagegrp* _IageXwgt*
 (output from regress appears)
```

## Warnings

1. `xi` creates new variables in your dataset; most are `bytes`, but interactions with continuous variables will have the storage type of the underlying continuous variable. You may get the message "insufficient memory". If so, you will need to increase the amount of memory allocated to Stata's data areas; see [U] **6 Setting the size of memory**.

2. When using `xi` with an estimation command, you may get the message "matsize too small". If so, see [R] **matsize**.

## Saved Results

`xi` saves the following characteristics:

| | |
|---|---|
| `_dta[__xi__Vars__Prefix__]` | prefix names |
| `_dta[__xi__Vars__To__Drop__]` | variables created |

## Methods and Formulas

`xi` is implemented as an ado-file.

## References

Hendrickx, J. 1999. dm73: Using categorical variables in Stata. *Stata Technical Bulletin* 52: 2–8. Reprinted in *Stata Technical Bulletin Reprints*, vol. 9, pp. 51–59.

——. 2000. dm73.1: Contrasts for categorical variables: Update. *Stata Technical Bulletin* 54: 7. Reprinted in *Stata Technical Bulletin Reprints*, vol. 9, pp. 60–61.

——. 2001a. dm73.2: Contrasts for categorical variables: Update. *Stata Technical Bulletin* 59: 2–5. Reprinted in *Stata Technical Bulletin Reprints*, vol. 10, pp. 9–14.

——. 2001b. dm73.3: Contrasts for categorical variables: Update. *Stata Technical Bulletin* 61: 5. Reprinted in *Stata Technical Bulletin Reprints*, vol. 10, pp. 14–15.

## Also See

**Background:** [U] **11.1.10 Prefix commands**,
[U] **20 Estimation and postestimation commands**

# Title

**zinb** — Zero-inflated negative binomial regression

# Syntax

zinb *depvar* [*indepvars*] [*if*] [*in*] [*weight*] ,

inflate(*varlist*[ , offset(*varname*) ] | _cons) [*options* ]

| *options* | description |
|---|---|
| Model | |
| * inflate() | equation that determines whether the count is zero |
| noconstant | suppress constant term |
| exposure(*varname*$_e$) | include ln(*varname*$_e$) in model with coefficient constrained to 1 |
| offset(*varname*$_o$) | include *varname*$_o$ in model with coefficient constrained to 1 |
| constraints(*constraints*) | apply specified linear constraints |
| probit | use probit model to characterize excess zeros; default is logit |
| SE/Robust | |
| vce(*vcetype*) | *vcetype* may be oim, robust, opg, bootstrap, or jackknife |
| robust | synonym for vce(robust) |
| cluster(*varname*) | adjust standard errors for intragroup correlation |
| Reporting | |
| level(#) | set confidence level; default is level(95) |
| irr | report incidence-rate ratios |
| vuong | perform Vuong test |
| zip | perform ZIP likelihood-ratio test |
| Max options | |
| *maximize_options* | control maximization process; seldom used |

* inflate(*varlist*[ , offset(*varname*) ] | _cons) is required.

bootstrap, by, jackknife, rolling, statsby, and xi are allowed; see [U] **11.1.10 Prefix commands**.

fweights, iweights, and pweights are allowed; see [U] **11.1.6 weight**.

See [U] **20 Estimation and postestimation commands** for additional capabilities of estimation commands.

# Description

zinb estimates a zero-inflated negative binomial regression of *depvar* on *indepvars*, where *depvar* is a non-negative count variable.

(*Continued on next page*)

# Options

Model

`inflate(`*varlist*[`, offset(`*varname*`)`] | `_cons)` specifies the equation that determines whether the observed count is zero. Conceptually, omitting `inflate()` would be equivalent to fitting the model with `nbreg`.

`inflate(`*varlist*[`, offset(`*varname*`)`]`)` specifies the variables in the equation. You may optionally include an offset for this *varlist*.

`inflate(_cons)` specifies that the equation determining whether the count is zero contains only an intercept. To run a zero-inflated model of *depvar* with only an intercept in both equations, type `zinb` *depvar*`, inflate(_cons)`.

`noconstant`, `exposure(`$varname_e$`)`, `offset(`$varname_o$`)`, `constraints(`*constraints*`)`; see [R] **estimation options**.

`probit` requests that a probit, instead of logit, model be used to characterize the excess zeros in the data.

SE/Robust

`vce(`*vcetype*`)`; see [R] ***vce_option***.

`robust`, `cluster(`*varname*`)`; see [R] **estimation options**.

Reporting

`level(`*#*`)`; see [R] **estimation options**.

`irr` reports estimated coefficients transformed to incidence-rate ratios, that is, $e^{\beta_i}$ rather than $\beta_i$. Standard errors and confidence intervals are similarly transformed. This option affects how results are displayed, not how they are estimated or stored. `irr` may be specified at estimation or when replaying previously estimated results.

`vuong` specifies that the Vuong (1989) test of ZINB versus negative binomial be reported. This test statistic has a standard normal distribution with large positive values favoring the ZINB model and large negative values favoring the negative binomial model.

`zip` requests that a likelihood-ratio test comparing the zero-inflated negative binomial model with the zero-inflated Poisson model be included in the output.

Max options

*maximize_options*: `difficult`, `technique(`*algorithm_spec*`)`, `iterate(`*#*`)`, [`no`]`log`, `trace`, `gradient`, `showstep`, `hessian`, `shownrtolerance`, `tolerance(`*#*`)`, `ltolerance(`*#*`)`, `gtolerance(`*#*`)`, `nrtolerance(`*#*`)`, `nonrtolerance`, `from(`*init_specs*`)`; see [R] **maximize**. These options are seldom used.

# Remarks

See Long (1997, 242–247) and Greene (2003, 749–752) for a discussion of zero-modified count models. For information about the test developed by Vuong (1989), see Greene (2003) and Long (1997). Greene (1994) applied the test to zero-inflated Poisson and negative binomial models, and there is a description of that work in Greene (2003).

Negative binomial regression fits models of the number of occurrences (counts) of an event. You could use `nbreg` for this (see [R] **nbreg**), but in some count-data models, you might want to account for the prevalence of zero counts in the data.

For instance, you could count how many fish each visitor to a park catches. A large number of visitors may catch zero, as they do not fish (as opposed to being unsuccessful). You may be able to model whether a person fishes depending on a number of covariates related to fishing activity and model how many fish a person catches depending on a number of covariates having to do with the success of catching fish (type of lure/bait, time of day, temperature, season, etc.). This is the type of data for which the `zinb` command is useful.

The zero-inflated (or zero-altered) negative binomial model allows overdispersion through the splitting process that models the outcomes as zero or nonzero.

## ▷ Example 1

We have data on the number of fish caught by visitors to a national park. Some of the visitors do not fish, but we do not have the data on whether a person fished or not; we merely have data on how many fish were caught, together with several covariates. As our data have a preponderance of zeros (142 out of 250), we use the `zinb` command to model the outcome.

```
. use http://www.stata-press.com/data/r9/fish
. zinb count persons livebait, inf(child camper) vuong nolog
Zero-inflated negative binomial regression        Number of obs   =        250
                                                  Nonzero obs     =        108
                                                  Zero obs        =        142

Inflation model = logit                           LR chi2(2)      =      82.23
Log likelihood  = -401.5478                       Prob > chi2     =     0.0000
```

| count | Coef. | Std. Err. | z | P>\|z\| | [95% Conf. | Interval] |
|---|---|---|---|---|---|---|
| count | | | | | | |
| persons | .9742984 | .1034938 | 9.41 | 0.000 | .7714543 | 1.177142 |
| livebait | 1.557523 | .4124424 | 3.78 | 0.000 | .7491503 | 2.365895 |
| _cons | -2.730064 | .476953 | -5.72 | 0.000 | -3.664874 | -1.795253 |
| inflate | | | | | | |
| child | 3.185999 | .7468551 | 4.27 | 0.000 | 1.72219 | 4.649808 |
| camper | -2.020951 | .872054 | -2.32 | 0.020 | -3.730146 | -.3117567 |
| _cons | -2.695385 | .8929071 | -3.02 | 0.003 | -4.44545 | -.9453189 |
| /lnalpha | .5110429 | .1816816 | 2.81 | 0.005 | .1549535 | .8671323 |
| alpha | 1.667029 | .3028685 | | | 1.167604 | 2.380076 |

```
Vuong test of zinb vs. standard negative binomial: z =     5.59  Pr>z = 0.0000
```

In general, Vuong test statistics that are significantly positive favor the zero-inflated models, while those that are significantly negative favor the nonzero-inflated models. Thus in the above model, the zero-inflation is significant.

◁

# Saved Results

zinb saves in e():

Scalars

| | |
|---|---|
| e(N) | number of observations |
| e(k) | number of parameters |
| e(k_eq) | number of equations |
| e(k_dv) | number of dependent variables |
| e(N_zero) | number of zero observations |
| e(df_m) | model degrees of freedom |
| e(ll) | log likelihood |
| e(ll_0) | log likelihood, constant-only model |
| e(df_c) | degrees of freedom for comparison test |
| e(N_clust) | number of clusters |
| e(chi2) | $\chi^2$ |
| e(chi2_c) | $\chi^2$ for comparison test |
| e(p) | significance of model test |
| e(chi2_p) | $\chi^2$ test against Poisson model |
| e(chi2_cp) | $\chi^2$ for test of $\alpha = 0$ |
| e(vuong) | Vuong test statistic |
| e(rank) | rank of e(V) |
| e(ic) | number of iterations |
| e(rc) | return code |
| e(converged) | 1 if converged, 0 otherwise |

Macros

| | |
|---|---|
| e(cmd) | zinb |
| e(depvar) | name of dependent variable |
| e(inflate) | logit or probit |
| e(wtype) | weight type |
| e(wexp) | weight expression |
| e(title) | title in estimation output |
| e(clustvar) | name of cluster variable |
| e(offset1) | offset |
| e(offset2) | offset for inflate() |
| e(chi2_ct) | Wald or LR; type of model $\chi^2$ test corresponding to e(chi2_c) |
| e(chi2_cpt) | Wald or LR; type of model $\chi^2$ test corresponding to e(chi2_cp) |
| e(chi2_pt) | Wald or LR; type of model $\chi^2$ test corresponding to e(chi2_p) |
| e(vce) | *vcetype* specified in vce() |
| e(vcetype) | title used to label Std. Err. |
| e(opt) | type of optimization |
| e(ml_method) | type of ml method |
| e(user) | name of likelihood-evaluator program |
| e(technique) | maximization technique |
| e(crittype) | optimization criterion |
| e(properties) | b V |
| e(predict) | program used to implement predict |

Matrices

| | |
|---|---|
| `e(b)` | coefficient vector |
| `e(V)` | variance–covariance matrix of the estimators |
| `e(gradient)` | gradient vector |
| `e(ilog)` | iteration log (up to 20 iterations) |

Functions

| | |
|---|---|
| `e(sample)` | marks estimation sample |

# Methods and Formulas

Several models in the literature are (correctly) described as zero-inflated. The `zinb` command maximizes the log likelihood $\ln L$, defined by

$$
\begin{aligned}
m &= 1/\alpha \\
p_j &= 1/(1+\alpha\mu_j) \\
\xi_j^\beta &= \mathbf{x}_j\boldsymbol{\beta} + \text{offset}_j^\beta \\
\xi_j^\gamma &= \mathbf{z}_j\boldsymbol{\gamma} + \text{offset}_j^\gamma \\
\mu_j &= \exp(\xi_j^\beta) \\
\ln L &= \sum_{j\in S} w_j \ln\left[F(\xi_j^\gamma) + \left\{1 - F(\xi_j^\gamma)\right\} p_j^m\right] \\
&\quad + \sum_{j\notin S} w_j \Big[\ln\left\{1 - F(\xi_j^\gamma)\right\} + \ln\Gamma(m+y_j) - \ln\Gamma(y_j+1) \\
&\qquad - \ln\Gamma(m) + m\ln p_j + y_j\ln(1-p_j)\Big]
\end{aligned}
$$

where $w_j$ are the weights, $F$ is the logit link (or probit link if `probit` was specified), and $S$ is the set of observations for which the outcome $y_j = 0$.

# References


Greene, W. H. 1994. Accounting for excess zeros and sample selection in Poisson and negative binomial regression models. Working Paper no. EC-94-10, Department of Economics, Stern School of Business, New York University.

——. 2003. *Econometric Analysis*. 5th ed. Upper Saddle River, NJ: Prentice Hall.

Long, J. S. 1997. *Regression Models for Categorical and Limited Dependent Variables*. Thousand Oaks, CA: Sage.

Long, J. S. and J. Freese. 2001. Predicted probabilities for count models. *Stata Journal* 1: 51–57.

——. 2003. *Regression Models for Categorical Dependent Variables Using Stata*. rev. ed. College Station, TX: Stata Press.

Mullahy, J. 1986. Specification and testing of some modified count-data models. *Journal of Econometrics* 33: 341–365.

Vuong, Q. 1989. Likelihood ratio tests for model selection and non-nested hypotheses. *Econometrica*: 57: 307–334.

# Also See

**Complementary:** [R] **zinb postestimation**; [R] **constraint**

**Related:** [R] **zip**, [R] **ztnb**; [R] **glm**, [R] **nbreg**, [R] **poisson**, [XT] **xtnbreg**

**Background:** [U] **11.1.10 Prefix commands**, [U] **20 Estimation and postestimation commands**, [R] **estimation options**, [R] **maximize**, [R] *vce_option*

## Title

**zinb postestimation** — Postestimation tools for zinb

## Description

The following postestimation commands are available for `zinb`:

| command | description |
|---|---|
| `adjust` | adjusted predictions of $\mathbf{x}\boldsymbol{\beta}$ or $\exp(\mathbf{x}\boldsymbol{\beta})$ |
| `estat` | AIC, BIC, VCE, and estimation sample summary |
| `estimates` | cataloging estimation results |
| `lincom` | point estimates, standard errors, testing, and inference for linear combinations of coefficients |
| `lrtest` | likelihood-ratio test |
| `mfx` | marginal effects or elasticities |
| `nlcom` | point estimates, standard errors, testing, and inference for nonlinear combinations of coefficients |
| `predict` | predictions, residuals, influence statistics, and other diagnostic measures |
| `predictnl` | point estimates, standard errors, testing, and inference for generalized predictions |
| `suest` | seemingly unrelated estimation |
| `test` | Wald tests for simple and composite linear hypotheses |
| `testnl` | Wald tests of nonlinear hypotheses |

See the corresponding entries in the *Stata Base Reference Manual* for details.

## Syntax for predict

`predict` [*type*] *newvar* [*if*] [*in*] [, *statistic* `nooffset`]

`predict` [*type*] { *stub** | $newvar_{\text{reg}}$ $newvar_{\text{inflate}}$ $newvar_{\text{lnalpha}}$ } [*if*] [*in*] , `scores`

| *statistic* | description |
|---|---|
| `n` | number of events; the default |
| `ir` | incidence rate |
| `pr` | probability of a zero outcome |
| `xb` | linear prediction |
| `stdp` | standard error of the linear prediction |

## Options for predict

`n`, the default, calculates the predicted number of events, which is $\exp(\mathbf{x}_j\boldsymbol{\beta})$ if neither `offset()` nor `exposure()` was specified when the model was fitted; $\exp(\mathbf{x}_j\boldsymbol{\beta} + \text{offset}_j)$ if `offset()` was specified; or $\exp(\mathbf{x}_j\boldsymbol{\beta}) \times \text{exposure}_j$ if `exposure()` was specified.

`ir` calculates the incidence rate $\exp(\mathbf{x}_j\boldsymbol{\beta})$, which is the predicted number of events when exposure is 1. This is equivalent to specifying both `n` and `nooffset` options.

`pr` calculates the probability of a zero outcome $F(\mathbf{z}_j\boldsymbol{\gamma})$, where $F()$ is the logit or probit link function. If `offset()` was specified within the `inflate()` option, then $F(\mathbf{z}_j\boldsymbol{\gamma} + \text{offset}_j^{\gamma})$ is calculated.

`xb` calculates the linear prediction, which is $\mathbf{x}_j\boldsymbol{\beta}$ if neither `offset()` nor `exposure()` was specified; $\mathbf{x}_j\boldsymbol{\beta} + \text{offset}_j$ if `offset()` was specified; or $\mathbf{x}_j\boldsymbol{\beta} + \ln(\text{exposure}_j)$ if `exposure()` was specified; see `nooffset` below.

`stdp` calculates the standard error of the linear prediction.

`nooffset` is relevant only if you specified `offset()` or `exposure()` when you fitted the model. It modifies the calculations made by `predict` so that they ignore the offset or exposure variable; the linear prediction is treated as $\mathbf{x}_j\boldsymbol{\beta}$ rather than as $\mathbf{x}_j\boldsymbol{\beta} + \text{offset}_j$ or $\mathbf{x}_j\boldsymbol{\beta} + \ln(\text{exposure}_j)$. Specifying `predict ..., nooffset` is equivalent to specifying `predict ..., ir`.

`scores` calculates equation-level score variables.

The first new variable will contain $\partial \ln L/\partial(\mathbf{x}_j\boldsymbol{\beta})$.

The second new variable will contain $\partial \ln L/\partial(\mathbf{z}_j\boldsymbol{\gamma})$.

The third new variable will contain $\partial \ln L/\partial \ln \alpha$.

## Methods and Formulas

All postestimation commands listed above are implemented as ado-files.

## Also See

**Complementary:** [R] **zinb**; [R] **adjust**, [R] **estimates**, [R] **lincom**, [R] **lrtest**, [R] **mfx**, [R] **nlcom**, [R] **predictnl**, [R] **suest**, [R] **test**, [R] **testnl**

**Background:** [U] **13.5 Accessing coefficients and standard errors**, [U] **20 Estimation and postestimation commands**, [R] **estat**, [R] **predict**

# Title

**zip** — Zero-inflated Poisson regression

# Syntax

`zip` *depvar* [*indepvars*] [*if*] [*in*] [*weight*] ,

`inflate(`*varlist*[ , `offset(`*varname*`)` ] | `_cons)` [*options*]

| *options* | description |
|---|---|
| Model | |
| * `inflate()` | equation that determines whether the count is zero |
| `noconstant` | suppress constant term |
| `exposure(`*varname*$_e$`)` | include ln(*varname*$_e$) in model with coefficient constrained to 1 |
| `offset(`*varname*$_o$`)` | include *varname*$_o$ in model with coefficient constrained to 1 |
| `constraints(`*constraints*`)` | apply specified linear constraints |
| `probit` | use probit model to characterize excess zeros; default is logit |
| SE/Robust | |
| `vce(`*vcetype*`)` | *vcetype* may be `oim`, `robust`, `opg`, `bootstrap`, or `jackknife` |
| `robust` | synonym for `vce(robust)` |
| `cluster(`*varname*`)` | adjust standard errors for intragroup correlation |
| Reporting | |
| `level(`*#*`)` | set confidence level; default is `level(95)` |
| `irr` | report incidence-rate ratios |
| `vuong` | perform Vuong test |
| Max options | |
| *maximize_options* | control maximization process; seldom used |

* `inflate(`*varlist*[ , `offset(`*varname*`)` ] | `_cons)` is required.

`bootstrap`, `by`, `jackknife`, `rolling`, `statsby`, and `xi` are allowed; see [U] **11.1.10 Prefix commands**.

`fweight`s, `iweight`s, and `pweight`s are allowed; see [U] **11.1.6 weight**.

See [U] **20 Estimation and postestimation commands** for additional capabilities of estimation commands.

# Description

`zip` estimates a zero-inflated Poisson regression of *depvar* on *indepvars*, where *depvar* is a non-negative count variable.

(*Continued on next page*)

## Options

Model

inflate(*varlist*[, offset(*varname*)] | _cons) specifies the equation that determines whether the observed count is zero. Conceptually, omitting inflate() would be equivalent to fitting the model with poisson.

inflate(*varlist*[, offset(*varname*)]) specifies the variables in the equation. You may optionally include an offset for this *varlist*.

inflate(_cons) specifies that the equation determining whether the count is zero contains only an intercept. To run a zero-inflated model of *depvar* with only an intercept in both equations, type zip *depvar*, inflate(_cons).

noconstant, exposure(*varname*$_e$), offset(*varname*$_o$), constraints(*constraints*); see [R] **estimation options**.

probit requests that a probit, instead of logit, model be used to characterize the excess zeros in the data.

SE/Robust

vce(*vcetype*); see [R] ***vce_option***.

robust, cluster(*varname*); see [R] **estimation options**.

Reporting

level(*#*); see [R] **estimation options**.

irr reports estimated coefficients transformed to incidence-rate ratios, that is, $e^b$ rather than $b$. Standard errors and confidence intervals are similarly transformed. This option affects how results are displayed, not how they are estimated or stored. irr may be specified at estimation or when replaying previously estimated results.

vuong specifies that the Vuong (1989) test of ZIP versus Poisson (or ZIP versus negative binomial) be reported. This test statistic has a standard normal distribution with large positive values favoring the ZIP (ZINB) model and large negative values favoring the Poisson (negative binomial) model.

Max options

*maximize_options*: difficult, technique(*algorithm_spec*), iterate(*#*), [no]log, trace, gradient, showstep, hessian, shownrtolerance, tolerance(*#*), ltolerance(*#*), gtolerance(*#*), nrtolerance(*#*), nonrtolerance, from(*init_specs*); see [R] **maximize**. These options are seldom used.

## Remarks

See Long (1997, 242–247) and Greene (2003, 749–752) for a discussion of zero-modified count models. For information about the test developed by Vuong (1989), see Greene (2003) and Long (1997). Greene (1994) applied the test to zero-inflated Poisson and negative binomial models, as described in Greene (2003).

Poisson regression fits models of the number of occurrences (counts) of an event. You could use poisson for this (see [R] **poisson**), but in some count-data models, you might want to account for the prevalence of zero counts in the data.

For instance, you might count how many fish each visitor to a park catches. A large number of visitors may catch zero, as they do not fish (as opposed to being unsuccessful). You may be able to model whether a person fishes depending on a number of covariates related to fishing activity and model how many fish a person catches depending on a number of covariates having to do with the success of catching fish (type of lure/bait, time of day, temperature, season, etc.). This is the type of data for which the `zip` command is useful.

The zero-inflated (or zero-altered) Poisson model allows overdispersion through the splitting process that models the outcomes as zero or nonzero.

### ▷ Example 1

We have data on the number of fish caught by visitors to a national park. Some of the visitors do not fish, but we do not have the data on whether a person fished or not; we merely have data on how many fish were caught together with several covariates. As our data have a preponderance of zeros (142 out of 250), we use the `zip` command to model the outcome.

```
. use http://www.stata-press.com/data/r9/fish
. zip count persons livebait, inf(child camper) vuong nolog
Zero-inflated Poisson regression                  Number of obs   =        250
                                                  Nonzero obs     =        108
                                                  Zero obs        =        142
Inflation model = logit                           LR chi2(2)      =     506.48
Log likelihood  = -850.7014                       Prob > chi2     =     0.0000

------------------------------------------------------------------------------
             |      Coef.   Std. Err.      z    P>|z|     [95% Conf. Interval]
-------------+----------------------------------------------------------------
count        |
     persons |   .8068853   .0453288    17.80   0.000     .7180424    .8957281
    livebait |   1.757289   .2446082     7.18   0.000     1.277866    2.236713
       _cons |  -2.178472   .2860289    -7.62   0.000    -2.739078   -1.617865
-------------+----------------------------------------------------------------
inflate      |
       child |   1.602571   .2797719     5.73   0.000     1.054228    2.150913
      camper |  -1.015698    .365259    -2.78   0.005    -1.731593   -.2998038
       _cons |  -.4922872   .3114562    -1.58   0.114     -1.10273    .1181558
------------------------------------------------------------------------------
Vuong test of zip vs. standard Poisson:            z =     3.95  Pr>z = 0.0000
```

In general, Vuong test statistics that are significantly positive favor the zero-inflated models, while those that are significantly negative favor the nonzero-inflated models. Thus in the above model, the zero-inflation is significant.

◁

*(Continued on next page)*

# Saved Results

zip saves in e():

Scalars

| | |
|---|---|
| e(N) | number of observations |
| e(k) | number of parameters |
| e(k_eq) | number of equations |
| e(k_dv) | number of dependent variables |
| e(N_zero) | number of zero observations |
| e(df_m) | model degrees of freedom |
| e(ll) | log likelihood |
| e(ll_0) | log likelihood, constant-only model |
| e(ll_c) | log likelihood, comparison model |
| e(df_c) | degrees of freedom for comparison test |
| e(N_clust) | number of clusters |
| e(chi2) | $\chi^2$ |
| e(chi2_c) | $\chi^2$ for comparison test |
| e(p) | significance of model test |
| e(vuong) | Vuong test statistic |
| e(rank) | rank of e(V) |
| e(ic) | number of iterations |
| e(rc) | return code |
| e(converged) | 1 if converged, 0 otherwise |

Macros

| | |
|---|---|
| e(cmd) | zip |
| e(depvar) | name of dependent variable |
| e(inflate) | logit or probit |
| e(wtype) | weight type |
| e(wexp) | weight expression |
| e(title) | title in estimation output |
| e(clustvar) | name of cluster variable |
| e(offset1) | offset |
| e(offset2) | offset for inflate() |
| e(chi2type) | Wald or LR; type of model $\chi^2$ test |
| e(chi2_ct) | Wald or LR; type of model $\chi^2$ test corresponding to e(chi2_c) |
| e(vce) | *vcetype* specified in vce() |
| e(vcetype) | title used to label Std. Err. |
| e(opt) | type of optimization |
| e(ml_method) | type of ml method |
| e(user) | name of likelihood-evaluator program |
| e(technique) | maximization technique |
| e(crittype) | optimization criterion |
| e(properties) | b V |
| e(predict) | program used to implement predict |

Matrices

| | |
|---|---|
| e(b) | coefficient vector |
| e(V) | variance–covariance matrix of the estimators |
| e(gradient) | gradient vector |
| e(ilog) | iteration log (up to 20 iterations) |

Functions

| | |
|---|---|
| e(sample) | marks estimation sample |

## Methods and Formulas

Several models in the literature are (correctly) described as zero-inflated. The `zip` command maximizes the log-likelihood $\ln L$, defined by

$$\begin{aligned}
\xi_j^\beta &= \mathbf{x}_j\boldsymbol{\beta} + \text{offset}_j^\beta \\
\xi_j^\gamma &= \mathbf{z}_j\boldsymbol{\gamma} + \text{offset}_j^\gamma \\
\ln L &= \sum_{i\in S} w_j \ln\left[F(\xi_j^\gamma) + \left\{1 - F(\xi_j^\gamma)\right\}\exp(-\lambda_j)\right] + \\
&\qquad \sum_{i\notin S} w_j \Big[\ln\left\{1 - F(\xi_j^\gamma)\right\} - \lambda_j + \xi_j^\beta y_j - \ln(y_j!)\Big]
\end{aligned}$$

where $w_j$ are the weights, $F$ is the logit link (or probit link if `probit` was specified), and $S$ is the set of observations for which the outcome $y_j = 0$.

## References


Greene, W. H. 1994. Accounting for excess zeros and sample selection in Poisson and negative binomial regression models. Working Paper no. EC-94-10, Department of Economics, Stern School of Business, New York University.

——. 2003. *Econometric Analysis*. 5th ed. Upper Saddle River, NJ: Prentice Hall.

Lambert, D. 1992. Zero-inflated Poisson regression, with an application to defects in manufacturing. *Technometrics* 34: 1–14.

Long, J. S. 1997. *Regression Models for Categorical and Limited Dependent Variables*. Thousand Oaks, CA: Sage.

Long, J. S. and J. Freese. 2001. Predicted probabilities for count models. *Stata Journal* 1: 51–57.

——. 2003. *Regression Models for Categorical Dependent Variables Using Stata*. rev. ed. College Station, TX: Stata Press.

Mullahy, J. 1986. Specification and testing of some modified count-data models. *Journal of Econometrics* 33: 341–365.

Vuong, Q. 1989. Likelihood ratio tests for model selection and non-nested hypotheses. *Econometrica*: 57: 307–334.


## Also See


**Complementary:** [R] **zip postestimation**; [R] **constraint**

**Related:** [R] **zinb**, [R] **ztp**; [R] **glm**, [R] **nbreg**, [R] **poisson**,
[XT] **xtnbreg**

**Background:** [U] **11.1.10 Prefix commands**,
[U] **20 Estimation and postestimation commands**,
[R] **estimation options**, [R] **maximize**, [R] ***vce_option***

## Title

**zip postestimation** — Postestimation tools for zip

## Description

The following postestimation commands are available for `zip`:

| command | description |
|---|---|
| `adjust` | adjusted predictions of $\mathbf{x}\beta$ or $\exp(\mathbf{x}\beta)$ |
| `estat` | AIC, BIC, VCE, and estimation sample summary |
| `estimates` | cataloging estimation results |
| `lincom` | point estimates, standard errors, testing, and inference for linear combinations of coefficients |
| `lrtest` | likelihood-ratio test |
| `mfx` | marginal effects or elasticities |
| `nlcom` | point estimates, standard errors, testing, and inference for nonlinear combinations of coefficients |
| `predict` | predictions, residuals, influence statistics, and other diagnostic measures |
| `predictnl` | point estimates, standard errors, testing, and inference for generalized predictions |
| `suest` | seemingly unrelated estimation |
| `test` | Wald tests for simple and composite linear hypotheses |
| `testnl` | Wald tests of nonlinear hypotheses |

See the corresponding entries in the *Stata Base Reference Manual* for details.

## Syntax for predict

`predict` [*type*] *newvar* [*if*] [*in*] [, *statistic* `nooffset`]

`predict` [*type*] {*stub** | $newvar_{\text{reg}}$ $newvar_{\text{inflate}}$} [*if*] [*in*] , `scores`

| *statistic* | description |
|---|---|
| `n` | number of events; the default |
| `ir` | incidence rate |
| `pr` | probability of a zero outcome |
| `xb` | linear prediction |
| `stdp` | standard error of the linear prediction |

## Options for predict

`n`, the default, calculates the predicted number of events, which is $\exp(\mathbf{x}_j\boldsymbol{\beta})$ if neither `offset()` nor `exposure()` was specified when the model was fitted; $\exp(\mathbf{x}_j\boldsymbol{\beta} + \text{offset}_j)$ if `offset()` was specified; or $\exp(\mathbf{x}_j\boldsymbol{\beta}) \times \text{exposure}_j$ if `exposure()` was specified.

`ir` calculates the incidence rate $\exp(\mathbf{x}_j\boldsymbol{\beta})$, which is the predicted number of events when exposure is 1. This is equivalent to specifying both `n` and `nooffset` options.

`pr` calculates the probability of a zero outcome $F(\mathbf{z}_j\boldsymbol{\gamma})$, where $F()$ is the logit or probit link function. If `offset()` was specified within the `inflate()` option, then $F(\mathbf{z}_j\boldsymbol{\gamma} + \text{offset}_j^{\gamma})$ is calculated.

`xb` calculates the linear prediction, which is $\mathbf{x}_j\boldsymbol{\beta}$ if neither `offset()` nor `exposure()` was specified; $\mathbf{x}_j\boldsymbol{\beta} + \text{offset}_j$ if `offset()` was specified; or $\mathbf{x}_j\boldsymbol{\beta} + \ln(\text{exposure}_j)$ if `exposure()` was specified; see `nooffset` below.

`stdp` calculates the standard error of the linear prediction.

`nooffset` is relevant only if you specified `offset()` or `exposure()` when you fitted the model. It modifies the calculations made by `predict` so that they ignore the offset or exposure variable; the linear prediction is treated as $\mathbf{x}_j\boldsymbol{\beta}$ rather than as $\mathbf{x}_j\boldsymbol{\beta} + \text{offset}_j$ or $\mathbf{x}_j\boldsymbol{\beta} + \ln(\text{exposure}_j)$. Specifying `predict ..., nooffset` is equivalent to specifying `predict ..., ir`.

`scores` calculates equation-level score variables.

The first new variable will contain $\partial \ln L / \partial(\mathbf{x}_j\boldsymbol{\beta})$.

The second new variable will contain $\partial \ln L / \partial(\mathbf{z}_j\boldsymbol{\gamma})$.

## Methods and Formulas

All postestimation commands listed above are implemented as ado-files.

## Also See

**Complementary:** [R] **zip**; [R] **adjust**, [R] **estimates**, [R] **lincom**, [R] **lrtest**, [R] **mfx**, [R] **nlcom**, [R] **predictnl**, [R] **suest**, [R] **test**, [R] **testnl**

**Background:** [U] **13.5 Accessing coefficients and standard errors**, [U] **20 Estimation and postestimation commands**, [R] **estat**, [R] **predict**

# Title

**ztnb** — Zero-truncated negative binomial regression

# Syntax

`ztnb` *depvar* [*indepvars*] [*if*] [*in*] [*weight*] [, *options*]

| *options* | description |
|---|---|
| Model | |
| `noconstant` | suppress constant term |
| `dispersion(mean)` | parameterization of dispersion; `dispersion(mean)` is the default |
| `dispersion(constant)` | constant dispersion for all observations |
| `exposure(`$varname_e$`)` | include ln($varname_e$) in model with coefficient constrained to 1 |
| `offset(`$varname_o$`)` | include $varname_o$ in model with coefficient constrained to 1 |
| `constraints(`*constraints*`)` | apply specified linear constraints |
| SE/Robust | |
| `vce(`*vcetype*`)` | *vcetype* may be `oim`, `robust`, `opg`, `bootstrap`, or `jackknife` |
| `robust` | synonym for `vce(robust)` |
| `cluster(`*varname*`)` | adjust standard errors for intragroup correlation |
| Reporting | |
| `level(`*#*`)` | set confidence level; default is `level(95)` |
| `nolrtest` | suppress likelihood-ratio test |
| `irr` | report incidence-rate ratios |
| Max options | |
| *maximize_options* | control the maximization process; seldom used |

`bootstrap`, `by`, `jackknife`, `rolling`, `statsby`, and `xi` are allowed; see [U] **11.1.10 Prefix commands**.
`fweights`, `iweights`, and `pweights` are allowed; see [U] **11.1.6 weight**.
See [U] **20 Estimation and postestimation commands** for additional capabilities of estimation commands.

# Description

`ztnb` fits a zero-truncated negative binomial regression model of *depvar* on *indepvars*, where *depvar* is a positive count variable.

# Options

Model

`noconstant`; see [R] **estimation options**.

`dispersion(mean | constant)` specifies the parameterization of the model. `dispersion(mean)`, the default, yields a model with dispersion equal to $1+\alpha\exp(\mathbf{x}_j\boldsymbol{\beta}+\text{offset}_j)$; that is, the dispersion is a function of the expected mean: $\exp(\mathbf{x}_j\boldsymbol{\beta}+\text{offset}_j)$. `dispersion(constant)` has dispersion equal to $1+\delta$; that is, it is a constant for all observations.

exposure(*varname*$_e$), offset(*varname*$_o$), constraints(*constraints*); see [R] **estimation options**.

SE/Robust

vce(*vcetype*); see [R] ***vce_option***.

robust, cluster(*varname*); see [R] **estimation options**.

Reporting

level(*#*); see [R] **estimation options**.

nolrtest suppresses fitting the Poisson model. Without this option, a comparison Poisson model is fitted, and the likelihood is used in a likelihood-ratio test of the null hypothesis that the dispersion parameter is zero.

irr reports estimated coefficients transformed to incidence-rate ratios, i.e., $e^b$ rather than $b$. Standard errors and confidence intervals are similarly transformed. This option affects how results are displayed, not how they are estimated or stored. You can specify irr at estimation or when you replay previously estimated results.

Max options

*maximize_options*: difficult, technique(*algorithm_spec*), iterate(*#*), [no]log, trace, gradient, showstep, hessian, shownrtolerance, tolerance(*#*), ltolerance(*#*), gtolerance(*#*), nrtolerance(*#*), nonrtolerance, from(*init_specs*); see [R] **maximize**. These options are seldom used.

## Remarks

Grogger and Carson (1991) showed that overdispersion causes inconsistent estimation of the mean in the truncated Poisson model, so they proposed the zero-truncated negative binomial model as an alternative. If the data exhibit zero-truncation but not overdispersion, the zero-truncated Poisson model is appropriate; see [R] **ztp**. See Long (1997, chapter 8) for an introduction to negative binomial and zero-truncated negative binomial regression.

ztnb fits two different parameterizations of zero-truncated negative binomial models, namely the mean-dispersion and constant-dispersion models. They are equivalent to those modeled by nbreg; see [R] **nbreg**.

### ▷ Example 1

We illustrate the zero-truncated negative binomial model using the HCFA's 1997 MedPar dataset (Hilbe 1999). The data represent 475 patients in the state of Arizona who were assigned a diagnostic-related group (DRG) of patients having a ventilator. Length of stay, the dependent variable, is a positive integer; it cannot have zero values. The variables used in the study include

| | |
|---|---|
| provnum | Provider number (Hospital identifier) |
| died | Patient died while in hospital |
| hmo | Patient is a member of an HMO |
| los | Length of stay |
| type1 | Emergency admission |
| type2 | Urgent admission (first available bed) |
| type3 | Elective admission |

```
. use http://www.stata-press.com/data/r9/medpar
. ztnb los died hmo type2-type3, nolog cluster(provnum)
Zero-truncated negative binomial regression       Number of obs   =       1495
Dispersion     = mean                             Wald chi2(4)    =      36.01
Log likelihood = -4737.535                        Prob > chi2     =     0.0000
                               (Std. Err. adjusted for 54 clusters in provnum)
```

| los | Coef. | Robust Std. Err. | z | P>\|z\| | [95% Conf. | Interval] |
|---|---|---|---|---|---|---|
| died | -.2521884 | .061533 | -4.10 | 0.000 | -.3727908 | -.1315859 |
| hmo | -.0754173 | .0533132 | -1.41 | 0.157 | -.1799091 | .0290746 |
| type2 | .2685095 | .0666474 | 4.03 | 0.000 | .137883 | .3991359 |
| type3 | .7668101 | .2183505 | 3.51 | 0.000 | .338851 | 1.194769 |
| _cons | 2.224028 | .034727 | 64.04 | 0.000 | 2.155964 | 2.292091 |
| /lnalpha | -.630108 | .0764019 | | | -.779853 | -.480363 |
| alpha | .5325343 | .0406866 | | | .4584734 | .6185588 |

Because observations within the same hospital (`provnum`) are very likely to be correlated, we specified the `cluster(provnum)` option. The results show that whether the patient died in the hospital and the type of admission have significant effects on the patient's length of stay.

◁

## ▷ Example 2

In this example, we compare the zero-truncated Poisson model with the zero-truncated negative binomial model. The following data appeared in Rodríguez (1993).

```
. use http://www.stata-press.com/data/r9/rod93
. quietly tab cohort, gen(coh)
. ztp deaths coh2 coh3, exp(exposure) nolog
Zero-truncated Poisson regression                 Number of obs   =         21
                                                  LR chi2(2)      =      49.16
                                                  Prob > chi2     =     0.0000
Log likelihood = -2159.5098                       Pseudo R2       =     0.0113
```

| deaths | Coef. | Std. Err. | z | P>\|z\| | [95% Conf. | Interval] |
|---|---|---|---|---|---|---|
| coh2 | -.3020296 | .0573367 | -5.27 | 0.000 | -.4144076 | -.1896517 |
| coh3 | .074248 | .0589754 | 1.26 | 0.208 | -.0413417 | .1898377 |
| _cons | -3.899522 | .0411385 | -94.79 | 0.000 | -3.980152 | -3.818892 |
| exposure | (exposure) | | | | | |

```
. ztnb deaths coh2 coh3, exp(exposure) nolog
Zero-truncated negative binomial regression       Number of obs   =         21
                                                  LR chi2(2)      =       0.32
Dispersion     = mean                             Prob > chi2     =     0.8535
Log likelihood = -130.05845                       Pseudo R2       =     0.0012

------------------------------------------------------------------------------
      deaths |      Coef.   Std. Err.      z    P>|z|     [95% Conf. Interval]
-------------+----------------------------------------------------------------
        coh2 |  -.2628645    .817489    -0.32   0.748    -1.865114    1.339384
        coh3 |  -.4587405   .8167563    -0.56   0.574    -2.059553    1.142072
       _cons |  -2.181593   .5831368    -3.74   0.000     -3.32452   -1.038666
    exposure | (exposure)
-------------+----------------------------------------------------------------
    /lnalpha |   .8916014   .4062167                      .0954313    1.687772
-------------+----------------------------------------------------------------
       alpha |   2.439032   .9907757                      1.100133    5.407417
------------------------------------------------------------------------------
Likelihood-ratio test of alpha=0:  chibar2(01) = 4058.90 Prob>=chibar2 = 0.000
```

Comparing the zero-truncated Poisson and zero-truncated negative binomial models, we see that **ztnb** has a much larger log likelihood. To test $\alpha = 0$, **ztnb** performs a likelihood-ratio test. The large $\chi^2$ value of 4058.90 strongly suggests that the zero-truncated negative binomial model is more appropriate.

◁

## ❑ Technical Note

The usual Gaussian test of $\alpha = 0$ is omitted since this test occurs on the boundary, invalidating the usual theory associated with such tests. However, the likelihood-ratio test of $\alpha = 0$ has been modified to be valid on the boundary. In particular, the null distribution of the likelihood-ratio test statistic is not the usual $\chi_1^2$, but rather a 50 : 50 mixture of a $\chi_0^2$ (point mass at zero) and a $\chi_1^2$, denoted as $\overline{\chi}_{01}^2$. See Gutierrez, Carter, and Drukker (2001) for more details.

❑

(*Continued on next page*)

# Saved Results

ztnb saves in e():

Scalars

| | | | |
|---|---|---|---|
| e(N) | number of observations | e(N_clust) | number of clusters |
| e(k) | number of parameters | e(chi2) | $\chi^2$ |
| e(k_eq) | number of equations | e(chi2_c) | $\chi^2$ for comparison test |
| e(k_dv) | number of dependent variables | e(p) | significance |
| e(df_m) | model degrees of freedom | e(rank) | rank of e(V) |
| e(r2_p) | pseudo-$R$-squared | e(rank0) | rank of e(V) for constant-only model |
| e(ll) | log likelihood | | |
| e(ll_0) | log likelihood, constant-only model | e(ic) | number of iterations |
| e(ll_c) | log likelihood, comparison model | e(rc) | return code |
| e(alpha) | the value of alpha | e(converged) | 1 if converged, 0 otherwise |

Macros

| | | | |
|---|---|---|---|
| e(cmd) | ztnb | e(chi2type) | Wald or LR; type of model $\chi^2$ test |
| e(depvar) | name of dependent variable | e(chi2_ct) | Wald or LR; type of model $\chi^2$ test corresponding to e(chi2_c) |
| e(wtype) | weight type | | |
| e(wexp) | weight expression | e(opt) | type of optimization |
| e(title) | title in estimation output | e(ml_method) | type of ml method |
| e(clustvar) | name of cluster variable | e(user) | name of likelihood-evaluator program |
| e(offset) | offset | e(technique) | maximization technique |
| e(dispers) | mean or constant | e(crittype) | optimization criterion |
| e(vce) | *vcetype* specified in vce() | e(properties) | b V |
| e(vcetype) | title used to label Std. Err. | e(predict) | program used to implement predict |

Matrices

| | | | |
|---|---|---|---|
| e(b) | coefficient vector | e(V) | variance–covariance matrix of the estimators |
| e(ilog) | iteration log (up to 20 iterations) | | |
| e(gradient) | gradient vector | | |

Functions

| | |
|---|---|
| e(sample) | marks estimation sample |

# Methods and Formulas

ztnb is implemented as an ado-file.

## Mean-dispersion model

A negative binomial distribution can be regarded as a gamma mixture of Poisson random variables. The number of times something occurs, $y_j$, is distributed as Poisson$(\nu_j \mu_j)$. That is, its conditional likelihood is

$$f(y_j \mid \nu_j) = \frac{(\nu_j \mu_j)^{y_j} e^{-\nu_j \mu_j}}{\Gamma(y_j + 1)}$$

where $\mu_j = \exp(\mathbf{x}_j \boldsymbol{\beta} + \text{offset}_j)$ and $\nu_j$ is an unobserved parameter with a Gamma$(1/\alpha, \alpha)$ density:

$$g(\nu) = \frac{\nu^{(1-\alpha)/\alpha} e^{-\nu/\alpha}}{\alpha^{1/\alpha}\Gamma(1/\alpha)}$$

This gamma distribution has mean 1 and variance $\alpha$, where $\alpha$ is the ancillary parameter.

The unconditional likelihood for the $j$th observation is therefore

$$f(y_j) = \int_0^\infty f(y_j \mid \nu) g(\nu)\, d\nu = \frac{\Gamma(m+y_j)}{\Gamma(y_j+1)\Gamma(m)} p_j^m (1-p_j)^{y_j}$$

where $p_j = 1/(1+\alpha\mu_j)$ and $m = 1/\alpha$. Solutions for $\alpha$ are handled by searching for $\ln\alpha$ since $\alpha$ must be greater than zero. The conditional probability of observing $y_j$ events given that $y_j > 0$ is

$$\Pr(Y = y_j \mid y_j > 0, \mathbf{x}_j) = \frac{f(y_j)}{1-(1+\alpha\mu_j)^{-\alpha^{-1}}}$$

The log likelihood (with weights $w_j$ and offsets) is given by

$$m = 1/\alpha \qquad p_j = 1/(1+\alpha\mu_j) \qquad \mu_j = \exp(\mathbf{x}_j\boldsymbol{\beta} + \text{offset}_j)$$

$$\ln L = \sum_{j=1}^n w_j \Big[ \ln\{\Gamma(m+y_j)\} - \ln\{\Gamma(y_j+1)\} - \ln\{\Gamma(m)\} + m\ln(p_j) + y_j\ln(1-p_j) - \ln(1-p_j^m) \Big]$$

### Constant-dispersion model

The constant-dispersion model assumes that $y_j$ is conditionally distributed as Poisson$(\mu_j^*)$, where $\mu_j^* \sim \text{Gamma}(\mu_j/\delta, \delta)$ for some dispersion parameter $\delta$ (by contrast, the mean-dispersion model assumes $\mu_j^* \sim \text{Gamma}(1/\alpha, \alpha\mu_j)$). The log likelihood is given by

$$m_j = \mu_j/\delta \qquad p = 1/(1+\delta)$$

$$\ln L = \sum_{j=1}^n w_j \Big[ \ln\{\Gamma(m_j+y_j)\} - \ln\{\Gamma(y_j+1)\} - \ln\{\Gamma(m_j)\} + m_j\ln(p) + y_j\ln(1-p) - \ln(1-p^{m_j}) \Big]$$

with everything else defined as before in the calculations for the mean-dispersion model.

## References


Cameron, A. C. and P. K. Trivedi. 1998. *Regression Analysis of Count Data.* Cambridge: Cambridge University Press.

Grogger, J. T. and R. T. Carson. 1991. Models for truncated counts. In *Journal of Applied Econometrics* 6: 225–238.

Gutierrez, R. G., S. L. Carter, and D. M. Drukker. 2001. On boundary-value likelihood-ratio tests. *Stata Technical Bulletin* 60: 15–18. Reprinted in *Stata Technical Bulletin Reprints*, vol. 10, pp. 269–273.

Hilbe, J. 1998. sg91: Robust variance estimators for MLE Poisson and negative binomial regression. *Stata Technical Bulletin* 45: 26–28. Reprinted in *Stata Technical Bulletin Reprints*, vol. 8, pp. 177–180.

——. 1999. sg102: Zero-truncated Poisson and negative binomial regression. *Stata Technical Bulletin* 47: 37–40. Reprinted in *Stata Technical Bulletin Reprints*, vol. 8, pp. 233–236.

Long, J. S. 1997. *Regression Models for Categorical and Limited Dependent Variables.* Thousand Oaks, CA: Sage.

Rodríguez, G. 1993. sbe10: An improvement to poisson. *Stata Technical Bulletin* 11: 11–14. Reprinted in *Stata Technical Bulletin Reprints*, vol. 2, pp. 94–98.

Simonoff, J. S. 2003 *Analyzing Categorical Data.* New York: Springer.

# Also See

**Complementary:** [R] **ztnb postestimation**; [R] **constraint**

**Related:** [R] **glm**, [R] **nbreg**, [R] **poisson**, [R] **zinb**, [R] **zip**, [R] **ztp**, [SVY] **svy: nbreg**, [XT] **xtnbreg**

**Background:** [U] **11.1.10 Prefix commands**, [U] **20 Estimation and postestimation commands**, [R] **estimation options**, [R] **maximize**, [R] ***vce_option***

## Title

**ztnb postestimation** — Postestimation tools for ztnb

## Description

The following postestimation commands are available for `ztnb`:

| command | description |
|---|---|
| `adjust` | adjusted predictions of $\mathbf{x}\boldsymbol{\beta}$ or $\exp(\mathbf{x}\boldsymbol{\beta})$ |
| `estat` | AIC, BIC, VCE, and estimation sample summary |
| `estimates` | cataloging estimation results |
| `lincom` | point estimates, standard errors, testing, and inference for linear combinations of coefficients |
| `lrtest` | likelihood-ratio test |
| `mfx` | marginal effects or elasticities |
| `nlcom` | point estimates, standard errors, testing, and inference for nonlinear combinations of coefficients |
| `predict` | predictions, residuals, influence statistics, and other diagnostic measures |
| `predictnl` | point estimates, standard errors, testing, and inference for generalized predictions |
| `suest` | seemingly unrelated estimation |
| `test` | Wald tests for simple and composite linear hypotheses |
| `testnl` | Wald tests of nonlinear hypotheses |

See the corresponding entries in the *Stata Base Reference Manual* for details.

## Syntax for predict

`predict` [*type*] *newvar* [*if*] [*in*] [, *statistic* `nooffset`]

`predict` [*type*] *stub** | $newvar_{\text{reg}}$ $newvar_{\text{disp}}$ [*if*] [*in*] , `scores`

| *statistic* | description |
|---|---|
| `n` | number of events; the default |
| `ir` | incidence rate |
| `cm` | estimate of the conditional mean, $E(n \mid n > 0)$ |
| `xb` | linear prediction |
| `stdp` | standard error of the linear prediction |

## Options for predict

`n`, the default, calculates the predicted number of events, which is $\exp(\mathbf{x}_j\boldsymbol{\beta})$ if neither `offset()` nor `exposure()` was specified when the model was fitted; $\exp(\mathbf{x}_j\boldsymbol{\beta} + \text{offset}_j)$ if `offset()` was specified; or $\exp(\mathbf{x}_j\boldsymbol{\beta}) \times \text{exposure}_j$ if `exposure()` was specified.

`ir` calculates the incidence rate $\exp(\mathbf{x}_j\boldsymbol{\beta})$, which is the predicted number of events when exposure is 1. This is equivalent to specifying both `n` and `nooffset` options.

`cm` calculates the estimate of the conditional mean.

$$E(y_j \mid y_j > 0) = \frac{E(y_j)}{\Pr(y_j > 0)}$$

`xb` calculates the linear prediction, which is $\mathbf{x}_j\boldsymbol{\beta}$ if neither `offset()` nor `exposure()` was specified; $\mathbf{x}_j\boldsymbol{\beta} + \text{offset}_j$ if `offset()` was specified; or $\mathbf{x}_j\boldsymbol{\beta} + \ln(\text{exposure}_j)$ if `exposure()` was specified; see `nooffset` below.

`stdp` calculates the standard error of the linear prediction.

`nooffset` is relevant only if you specified `offset()` or `exposure()` when you fitted the model. It modifies the calculations made by `predict` so that they ignore the offset or exposure variable; the linear prediction is treated as $\mathbf{x}_j\boldsymbol{\beta}$ rather than as $\mathbf{x}_j\boldsymbol{\beta} + \text{offset}_j$ or $\mathbf{x}_j\boldsymbol{\beta} + \ln(\text{exposure}_j)$. Specifying `predict ..., nooffset` is equivalent to specifying `predict ..., ir`.

`scores` calculates equation-level score variables.

The first new variable will contain $\partial \ln L/\partial(\mathbf{x}_j\boldsymbol{\beta})$.

The second new variable will contain $\partial \ln L/\partial(\ln\alpha)$ for `dispersion(mean)`.

The second new variable will contain $\partial \ln L/\partial(\ln\delta)$ for `dispersion(constant)`.

# Methods and Formulas

All postestimation commands listed above are implemented as ado-files.

In the following, we use the same notation as in [R] **ztnb**.

## Mean-dispersion model

The equation-level scores are given by

$$\begin{aligned}
\text{score}(\mathbf{x}\boldsymbol{\beta})_j &= p_j(y_j - \mu_j) - \frac{p_j^{(m+1)}\mu_j}{1 - p_j^m} \\
\text{score}(\tau)_j &= -\, m\left\{\frac{\alpha(\mu_j - y_j)}{1 + \alpha\mu_j} - \ln(1 + \alpha\mu_j) + \psi(y_j + m) - \psi(m)\right\} \\
&\quad - \frac{p_j^m}{1 - p_j^m}\{m\ln(p_j) + \mu_j p_j\}
\end{aligned}$$

where $\tau_j = \ln\alpha_j$ and $\psi(z)$ is the digamma function.

## Constant-dispersion model

The equation-level scores are given by

$$\begin{aligned}
\text{score}(\mathbf{x}\boldsymbol{\beta})_j &= m_j\left\{\psi(y_j + m_j) - \psi(m_j) + \ln(p) + \frac{p^{m_j}\ln(p)}{1 - p^{m_j}}\right\} \\
\text{score}(\tau)_j &= y_j - (y_j + m_j)(1 - p) - \text{score}(\mathbf{x}\boldsymbol{\beta})_j - \frac{\mu_j p}{1 - p^{m_j}}
\end{aligned}$$

where $\tau_j = \ln\delta_j$.

# Also See

**Complementary:** [R] **ztnb**; [R] **adjust**, [R] **estimates**, [R] **lincom**, [R] **lrtest**, [R] **mfx**, [R] **nlcom**, [R] **predictnl**, [R] **suest**, [R] **test**, [R] **testnl**

**Background:** [U] **13.5 Accessing coefficients and standard errors**,
[U] **20 Estimation and postestimation commands**,
[R] **estat**, [R] **predict**

# Title

**ztp** — Zero-truncated Poisson regression

# Syntax

`ztp` *depvar* [*indepvars*] [*if*] [*in*] [*weight*] [, *options*]

| *options* | description |
|---|---|
| Model | |
| `noconstant` | suppress constant term |
| `exposure(`*varname*$_e$`)` | include ln(*varname*$_e$) in model with coefficient constrained to 1 |
| `offset(`*varname*$_o$`)` | include *varname*$_o$ in model with coefficient constrained to 1 |
| `constraints(`*constraints*`)` | apply specified linear constraints |
| SE/Robust | |
| `vce(`*vcetype*`)` | *vcetype* may be `oim`, `robust`, `opg`, `bootstrap`, or `jackknife` |
| `robust` | synonym for `vce(robust)` |
| `cluster(`*varname*`)` | adjust standard errors for intragroup correlation |
| Reporting | |
| `level(`*#*`)` | set confidence level; default is `level(95)` |
| `irr` | report incidence-rate ratios |
| Max options | |
| *maximize_options* | control the maximization process; seldom used |

*depvar* and *indepvars* may contain time-series operators; see [U] **11.4.3 Time-series varlists**.

`bootstrap`, `by`, `jackknife`, `rolling`, `statsby`, and `xi` are allowed; see [U] **11.1.10 Prefix commands**.

`fweights`, `iweights`, and `pweights` are allowed; see [U] **11.1.6 weight**.

See [U] **20 Estimation and postestimation commands** for additional capabilities of estimation commands.

# Description

`ztp` fits a zero-truncated Poisson regression of *depvar* on *indepvars*, where *depvar* is a positive count variable.

# Options for ztp

Model

`noconstant`, `exposure(`*varname*$_e$`)`, `offset(`*varname*$_o$`)`, `constraints(`*constraints*`)`; see [R] **estimation options**.

SE/Robust

`vce(`*vcetype*`)`; see [R] ***vce_option***.

`robust`, `cluster(`*varname*`)`; see [R] **estimation options**.

Reporting

`level(#)`; see [R] **estimation options**.

`irr` reports estimated coefficients transformed to incidence-rate ratios, that is, $e^{\beta_i}$ rather than $\beta_i$. Standard errors and confidence intervals are similarly transformed. This option affects how results are displayed, not how they are estimated or stored. `irr` may be specified at estimation or when replaying previously estimated results.

Max options

*maximize_options*: `difficult`, `technique(`*algorithm_spec*`)`, `iterate(#)`, `[no]log`, `trace`, `gradient`, `showstep`, `hessian`, `shownrtolerance`, `tolerance(#)`, `ltolerance(#)`, `gtolerance(#)`, `nrtolerance(#)`, `nonrtolerance`, `from(`*init_specs*`)`; see [R] **maximize**. These options are seldom used.

# Remarks

Zero-truncated Poisson regression is used to model the number of occurrences of an event when that number is restricted to be positive. If zero is an admissible value for the dependent variable, then standard Poisson regression may be more appropriate; see [R] **poisson**. Zero-truncated poisson regression was first proposed by Grogger and Carson (1991). See Long (1997, chapter 8) for an introduction to Poisson and zero-truncated Poisson regression.

Suppose that you counted the number of times a patient has medical treatments, and those patients who have at least one treatment are randomly selected into the sample. You might examine the length of hospital stay of patients, which has to be positive. In those cases, the `ztp` command should be used instead of `poisson`.

## ▷ Example 1

We will illustrate the zero-truncated Poisson model using the dataset of running shoes for a sample of runners who registered an online running log, which is used by Simonoff (2003). A running-shoe marketing executive is interested in knowing how the number of running shoes purchased related to other factors such as gender, marital status, age, education, income, typical number of runs per week, average miles run per week, and the preferred type of running. These data are naturally truncated. We fit a zero-truncated Poisson regression model based on distance running, runs per week, miles run per week, gender, and age.

```
. use http://www.stata-press.com/data/r9/runshoes
. ztp shoes distance rpweek mpweek male age, nolog
Zero-truncated Poisson regression                 Number of obs   =         60
                                                  LR chi2(5)      =      27.10
                                                  Prob > chi2     =     0.0001
Log likelihood = -84.079517                       Pseudo R2       =     0.1388

------------------------------------------------------------------------------
       shoes |      Coef.   Std. Err.      z    P>|z|     [95% Conf. Interval]
-------------+----------------------------------------------------------------
    distance |   .5101724   .2336767     2.18   0.029     .0521746    .9681703
      rpweek |   .1822472   .1146021     1.59   0.112    -.0423688    .4068631
      mpweek |   .0143156   .0099102     1.44   0.149    -.0051081    .0337393
        male |   .1314745   .2480547     0.53   0.596    -.3547039    .6176528
         age |    .011772    .010266     1.15   0.252    -.0083489    .0318929
       _cons |  -1.174196   .6524461    -1.80   0.072    -2.452967    .1045745
------------------------------------------------------------------------------
```

◁

## Saved Results

`ztp` saves in `e()`:

Scalars

| | | | |
|---|---|---|---|
| `e(N)` | number of observations | `e(ll_0)` | log likelihood, constant-only model |
| `e(k)` | number of parameters | `e(N_clust)` | number of clusters |
| `e(k_eq)` | number of equations | `e(rc)` | return code |
| `e(k_dv)` | number of dependent variables | `e(chi2)` | $\chi^2$ |
| `e(df_m)` | model degrees of freedom | `e(p)` | significance |
| `e(r2_p)` | pseudo-$R$-squared | `e(ic)` | number of iterations |
| `e(ll)` | log likelihood | `e(rank)` | rank of `e(V)` |

Macros

| | | | |
|---|---|---|---|
| `e(cmd)` | `ztp` | `e(vcetype)` | title used to label Std. Err. |
| `e(depvar)` | name of dependent variable | `e(opt)` | type of optimization |
| `e(wtype)` | weight type | `e(ml_method)` | type of `ml` method |
| `e(wexp)` | weight expression | `e(user)` | name of likelihood-evaluator program |
| `e(title)` | title in estimation output | `e(technique)` | maximization technique |
| `e(clustvar)` | name of cluster variable | `e(crittype)` | optimization criterion |
| `e(offset)` | offset | `e(properties)` | `b V` |
| `e(chi2type)` | `Wald` or `LR`; type of model $\chi^2$ test | `e(predict)` | program used to implement `predict` |
| `e(vce)` | *vcetype* specified in `vce()` | | |

Matrices

| | | | |
|---|---|---|---|
| `e(b)` | coefficient vector | `e(V)` | variance–covariance matrix of the estimators |
| `e(ilog)` | iteration log (up to 20 iterations) | | |
| `e(gradient)` | gradient vector | | |

Functions

| | |
|---|---|
| `e(sample)` | marks estimation sample |

## Methods and Formulas

`ztp` is implemented as an ado-file.

The conditional probability of observing $y_j$ events given that $y_j > 0$ is

$$\Pr(Y = y_j \mid y_j > 0, \mathbf{x}_j) = \frac{e^{-\lambda}\lambda^{y_j}}{y_j!(1-e^{-\lambda})}$$

The log likelihood (with weights $w_j$ and offsets) is given by

$$\xi_j = \mathbf{x}_j\boldsymbol{\beta} + \text{offset}_j$$

$$f(y_j) = \frac{e^{-\exp(\xi_j)}e^{\xi_j y_j}}{y_j!\{1-e^{-\exp(\xi_j)}\}}$$

$$\ln L = \sum_{j=1}^{n} w_j \left[-e^{\xi_j} + \xi_j y_j - \ln(y_j!) - \ln\{1-e^{-\exp(\xi_j)}\}\right]$$

## References

Cameron, A. C. and P. K. Trivedi. 1998. *Regression Analysis of Count Data*. Cambridge: Cambridge University Press.

Grogger, J. T. and R. T. Carson. 1991. Models for truncated counts. In *Journal of Applied Econometrics* 6: 225–238.

Hilbe, J. 1998. sg91: Robust variance estimators for MLE Poisson and negative binomial regression. *Stata Technical Bulletin* 45: 26–28. Reprinted in *Stata Technical Bulletin Reprints*, vol. 8, pp. 177–180.

——. 1999. sg102: Zero-truncated Poisson and negative binomial regression. *Stata Technical Bulletin* 47: 37–40. Reprinted in *Stata Technical Bulletin Reprints*, vol. 8, pp. 233–236.

Hilbe, J. and D. H. Judson. 1998. sg94: Right, left, and uncensored Poisson regression. *Stata Technical Bulletin* 46: 18–20. Reprinted in *Stata Technical Bulletin Reprints*, vol. 8, pp. 186–189.

Long, J. S. 1997. *Regression Models for Categorical and Limited Dependent Variables*. Thousand Oaks, CA: Sage.

Simonoff, J. S. 2003 *Analyzing Categorical Data*. New York: Springer.

## Also See

**Complementary:** [R] **ztp postestimation**; [R] **constraint**

**Related:** [R] **glm**, [R] **nbreg**, [R] **poisson**, [R] **ztnb**; [R] **constraint**, [ST] **epitab**, [SVY] **svy: poisson**, [XT] **xtpoisson**

**Background:** [U] **11.1.10 Prefix commands**, [U] **20 Estimation and postestimation commands**, [R] **estimation options**, [R] **maximize**, [R] ***vce_option***

# Title

**ztp postestimation** — Postestimation tools for ztp

# Description

The following postestimation commands are available for `ztp`:

| command | description |
|---|---|
| `adjust` | adjusted predictions of $\mathbf{x}\beta$ or $\exp(\mathbf{x}\beta)$ |
| `estat` | AIC, BIC, VCE, and estimation sample summary |
| `estimates` | cataloging estimation results |
| `lincom` | point estimates, standard errors, testing, and inference for linear combinations of coefficients |
| `lrtest` | likelihood-ratio test |
| `mfx` | marginal effects or elasticities |
| `nlcom` | point estimates, standard errors, testing, and inference for nonlinear combinations of coefficients |
| `predict` | predictions, residuals, influence statistics, and other diagnostic measures |
| `predictnl` | point estimates, standard errors, testing, and inference for generalized predictions |
| `suest` | seemingly unrelated estimation |
| `test` | Wald tests for simple and composite linear hypotheses |
| `testnl` | Wald tests of nonlinear hypotheses |

See the corresponding entries in the *Stata Base Reference Manual* for details.

# Syntax for predict

`predict` [*type*] *newvar* [*if*] [*in*] [, *statistic* `nooffset`]

| *statistic* | description |
|---|---|
| `n` | number of events; the default |
| `ir` | incidence rate |
| `cm` | estimate of the conditional mean, $E(n \mid n > 0)$ |
| `xb` | linear prediction |
| `stdp` | standard error of the linear prediction |
| `score` | first derivative of the log likelihood with respect to $\mathbf{x}_j\beta$ |

These statistics are available both in and out of sample; type `predict ... if e(sample) ...` if wanted only for the estimation sample.

## Options for predict

`n`, the default, calculates the predicted number of events, which is $\exp(\mathbf{x}_j\boldsymbol{\beta})$ if neither `offset()` nor `exposure()` was specified when the model was fitted; $\exp(\mathbf{x}_j\boldsymbol{\beta} + \text{offset}_j)$ if `offset()` was specified; or $\exp(\mathbf{x}_j\boldsymbol{\beta}) \times \text{exposure}_j$ if `exposure()` was specified.

`ir` calculates the incidence rate $\exp(\mathbf{x}_j\boldsymbol{\beta})$, which is the predicted number of events when exposure is 1. This is equivalent to specifying both `n` and `nooffset` options.

`cm` calculates the estimate of the conditional mean.

$$E(y_j \mid y_j > 0) = \frac{E(y_j)}{\Pr(y_j > 0)}$$

`xb` calculates the linear prediction, which is $\mathbf{x}_j\boldsymbol{\beta}$ if neither `offset()` nor `exposure()` was specified; $\mathbf{x}_j\boldsymbol{\beta} + \text{offset}_j$ if `offset()` was specified; or $\mathbf{x}_j\boldsymbol{\beta} + \ln(\text{exposure}_j)$ if `exposure()` was specified; see `nooffset` below.

`stdp` calculates the standard error of the linear prediction.

`score` calculates the equation-level score, $\partial \ln L / \partial(\mathbf{x}_j\boldsymbol{\beta})$.

`nooffset` is relevant only if you specified `offset()` or `exposure()` when you fitted the model. It modifies the calculations made by `predict` so that they ignore the offset or exposure variable; the linear prediction is treated as $\mathbf{x}_j\boldsymbol{\beta}$ rather than as $\mathbf{x}_j\boldsymbol{\beta} + \text{offset}_j$ or $\mathbf{x}_j\boldsymbol{\beta} + \ln(\text{exposure}_j)$. Specifying `predict ..., nooffset` is equivalent to specifying `predict ..., ir`.

## Methods and Formulas

All postestimation commands listed above are implemented as ado-files.

In the following, we use the same notation as in [R] **ztp**.

The equation-level scores are given by

$$\text{score}(\mathbf{x}\boldsymbol{\beta})_j = y_j - e^{\xi_j} - \frac{e^{-\exp(\xi_j)} e^{\xi_j}}{1 - e^{-\exp(\xi_j)}}$$

## Also See

**Complementary:** [R] **ztp**; [R] **adjust**, [R] **estimates**, [R] **lincom**, [R] **lrtest**, [R] **mfx**, [R] **nlcom**, [R] **predictnl**, [R] **suest**, [R] **test**, [R] **testnl**

**Background:** [U] **13.5 Accessing coefficients and standard errors**, [U] **20 Estimation and postestimation commands**, [R] **estat**, [R] **predict**

# Author Index

This is the author index for the *Stata Base Reference Manual*.

## A



## B

# C



# D

# E



# F



# G

# H

## I



## J



## K



## L

# M



# N

# O



# P



# R

# S

# T



# U



# V



# W

## Z

# Subject Index

This is the subject index for the 3-volume *Stata Base Reference Manual*. Readers may also want to see the combined subject index in the *Stata Quick Reference and Index*, which indexes the *Getting Started with Stata for Macintosh Manual*, the *Getting Started with Stata for Unix Manual*, the *Getting Started with Stata for Windows Manual*, the *Stata User's Guide*, the *Stata Data Management Reference Manual*, the *Stata Graphics Reference Manual*, the *Stata Programming Reference Manual*, the *Stata Longitudinal/Panel Data Reference Manual*, the *Stata Multivariate Statistics Reference Manual*, the *Stata Survey Data Reference Manual*, the *Stata Survival and Epidemiological Tables Reference Manual*, the *Stata Time-Series Reference Manual*, and this manual.

Readers interested in Mata topics should see the index at the end of the *Mata Reference Manual*.

Semicolons set off the most important entries from the rest. Sometimes no entry will be set off with semicolons; this means all entries are equally important.

## A



## B

# C

# D

# E



# F

# G



# H

# I



# J

# M

# N



# O

# P

# Q



# R

# S

# U



# V



# W

# X



# Z